Lecture Notes in Electrical Engineering 1035

The book series *Lecture Notes in Electrical Engineering* (LNEE) publishes the latest developments in Electrical Engineering—quickly, informally and in high quality. While original research reported in proceedings and monographs has traditionally formed the core of LNEE, we also encourage authors to submit books devoted to supporting student education and professional training in the various fields and applications areas of electrical engineering. The series cover classical and emerging topics concerning:

- Communication Engineering, Information Theory and Networks
- Electronics Engineering and Microelectronics
- Signal, Image and Speech Processing
- Wireless and Mobile Communication
- Circuits and Systems
- Energy Systems, Power Electronics and Electrical Machines
- Electro-optical Engineering
- Instrumentation Engineering
- Avionics Engineering
- Control Systems
- Internet-of-Things and Cybersecurity
- Biomedical Devices, MEMS and NEMS

For general information about this book series, comments or suggestions, please contact leontina.dicecco@springer.com.

To submit a proposal or request further information, please contact the Publishing Editor in your country:

China

Jasmine Dou, Editor (jasmine.dou@springer.com)

India, Japan, Rest of Asia

Swati Meherishi, Editorial Director (Swati.Meherishi@springer.com)

Southeast Asia, Australia, New Zealand

Ramesh Nath Premnath, Editor (ramesh.premnath@springernature.com)

USA, Canada

Michael Luby, Senior Editor (michael.luby@springer.com)

All other Countries

Leontina Di Cecco, Senior Editor (leontina.dicecco@springer.com)

**** This series is indexed by EI Compendex and Scopus databases. ****

Nagender Kumar Suryadevara · Boby George ·
Krishanthi P. Jayasundera ·
Subhas Chandra Mukhopadhyay
Editors

Sensing Technology

Proceedings of ICST'15

Editors
Nagender Kumar Suryadevara
School of Computer
and Information Sciences
University of Hyderabad
Hyderabad, Telangana, India

Boby George
Department of Electrical Engineering
Indian Institute of Technology Madras
Chennai, Tamil Nadu, India

Krishanthi P. Jayasundera
Macquarie University
Sydney, NSW, Australia

Subhas Chandra Mukhopadhyay
School of Engineering
Macquarie University
Sydney, NSW, Australia

ISSN 1876-1100 ISSN 1876-1119 (electronic)
Lecture Notes in Electrical Engineering
ISBN 978-3-031-29873-8 ISBN 978-3-031-29871-4 (eBook)
https://doi.org/10.1007/978-3-031-29871-4

This Springer imprint is published by the registered company Springer Nature Switzerland AG
The registered company address is: Gewerbestrasse 11, 6330 Cham, Switzerland

Preface

We are pleased to present the proceedings of the 15th International Conference on Sensing Technologies (ICST) 2022. As expected, we received submissions focusing on advanced research studies related to sensors, sensing techniques, associated systems, and new sensor applications. It is pleasing to note that, as in previous years of the ICST, researchers from different continents presented their outcomes during the conference. We gratefully acknowledge the authors' significant contributions.

We thank Macquarie University, Sydney, Australia, for the kind support provided to organize and execute the conference. The technical program committee and the reviewers did an excellent job through their quick and quality review. The complete process of paper submission, review process, and final submission was managed by the academic conference management system, EDAS.

Springer book series Lecture Notes in Electrical Engineering (LNEE) which publishes the latest original research reported in proceedings, Scopus, and Compendex indexing has accepted to publish the proceedings in the LNEE. We thank Springer for their kind support.

Yours sincerely

Nagender Kumar Suryadevara
Boby George
Krishanthi P. Jayasundera
Subhas Chandra Mukhopadhyay
The Conference Organizers

ICST'15 – Organizing Committee

Honorary Chairs

Darren Bagnall	Macquarie University, Australia
Iain Collings	Macquarie University, Australia
K. T. V. Grattan	City University, UK
Emil Petriu	University of Ottawa, Canada
D. P. Tsai	AS, Taiwan
P. Sallis	AUT, NZ

General Chairs

Subhas Mukhopadhyay	MQ, Australia
K. P. Jayasundera	UTS, Australia

Technical Program Chairs

B. George	IITM, India
G. Brooker	USYD, Australia
O. Postolache	ETA, Portugal
N. K. Suryadevara	University of Hyderabad, India

Regional Program Chairs

America

H. Leung	University of Calgary, Canada

Europe

I. Matias	PUN, Spain

Middle-East

C. Gooneratne KAUST

Australia

D. Preethichandra CQU, Australia

India

B. George IIT Madras, India

Taiwan

Joe-Air Jiang NTU, Taiwan

Japan

S. Yamada Komatsu University, Japan

Focused Session Chairs

T. Islam JMI, India
Sayan Kanungo BITS Pilani, Hyderabad, India

Publicity Chairs

Shantanu Pal Queensland University of Technology, Australia

International Advisory Committee

Tong Sun City University London, UK
Elena Gaura Coventry University, UK
Wan-Young Chung PNKU, Korea
Zhi Liu Shandong University, China
Peter Xu Auckland University, New Zealand
Goutam Chakraborty MITS, India
Joyanta Roy NIT, Kolkata, India

Technical Committee Members

Mohd Syaifudin Abdul Rahman	Malaysian Agricultural Research & Development Institute (MARDI), Malaysia
Nasrin Afsarimanesh	Curtin University, Australia
Md Eshrat Alahi	NTNU, Norway
Khalid Arif	Massey University, New Zealand
Francisco Arregui	Universidad Publica de Navarra, Spain
Norhana Arsad	Universiti Kebangsaan Malaysia, Malaysia
Ranjit Barai	Nanyang Technological University, Singapore
S. Bhadra	McGill University, Canada
Aniruddha Bhattacharjya	Guru Nanak Institute of Technology (GNIT), India
David Chavez	Pontificia Universidad Catolica del Peru, Peru
Bryan Chin	Auburn University, USA
Komkrit Chomsuwan	King Mongkut's University of Technology Thonburi, Thailand
Cheng-Hsin Chuang	Southern Taiwan University of Science and Technology, Taiwan
Eduardo Cordova-Lopez	Liverpool John Moores University, UK
Jesus Corres	Public University of Navarra, Spain
Tiziana D'Orazio	National Research Council, Italy
Matthew D'Souza	The University of Queensland, Australia
Saakshi Dhanekar	Indian Institute of Technology (IIT) Delhi, India
Robin Dykstra	Victoria University of Wellington, New Zealand
Bernd Eichberger	Graz University of Technology, Austria
Mala Ekanayake	Central Queensland University, Australia
Maria Fazio	University of Messina, Italy
Cristian Fosalau	Technical University of Iasi, Romania
Boby George	Indian Institute of Technology Madras, India
Avik Ghose	Tata Consultancy Services, India
Boris Ginzburg	Soreq NRC, Israel
Roman Gruden	DHBW Stuttgart, Germany
Maki Habib	The American University in Cairo, Egypt
Qingbo He	University of Science and Technology of China, P.R. China
Chi-Hung Hwang	Instrument Technology Research Center, Taiwan
Ikuo Ihara	Nagaoka University of Technology, Japan
Satoshi Ikezawa	Tokyo University of Agriculture, Japan
Tarikul Islam	Jamia Millia Islamia University, India
Joe-Air Jiang	National Taiwan University, Taiwan
Olfa Kanoun	Chemnitz University of Technology, Germany

Mateusz Smietana	Warsaw University of Technology, Poland
Janusz Smulko	Gdansk University of Technology, Poland
Aiguo Song	Southeast University, P.R. China
Rakesh Srivastava	Indian Institute of Technology Banaras Hindu University, India
Dan Mihai Stefanescu	Romanian Measurement Society, Romania
Qingquan Sun	The University of Alabama, USA
Nagender Suryadevara	University of Hyderabad, India
Akshya Swain	University of Auckland, New Zealand
K. Tashiro	Shinshu University, Japan
Om Thakur	NSIT, Delhi University, India
Guiyun Tian	Newcastle University, UK
Wei-Chen Tu	Chung Yuan Christian University, Taiwan
Ioan Tuleasca	The Open Polytechnic in New Zealand, New Zealand
Ramanarayanan Viswanathan	University of Mississippi, USA
Huaqing Wang	Beijing University of Chemical Technology, P.R. China
Peng Wang	Case Western Reserve University, USA
Shibin Wang	The State Key Laboratory for Manufacturing Systems Engineering, Xi'an Jiaotong University, P.R. China
Daniel Watzenig	Graz University of Technology, Austria
Wei Wei	Xi'an University of Technology, P.R. China
Ruqiang Yan	Xi'an Jiaotong University, P.R. China
Bo-Ru Yang	Sun Yat-sen University, P.R. China
Min Yao	Tsinghua University, P.R. China
Mehmet Rasit Yuce	Monash University, Australia
Hong Zeng	Southeast University, P.R. China
Liye Zhao	Southeast University, P.R. China
Zhongkui Zhu	Soochow University, P.R. China
Arcady Zhukov	Basque Country University, UPV/EHU, Spain
Syed Muzahir Abbas	Macquarie University, Australia
Hyoun Woo Kim	Hanyang University, Korea
Sajad Abolpour Moshizi	Macquarie University, Australia
Takehito Azuma	Utsuminiya University, Japan
Sandrine Bernardini	Aix-Marseille University—CNRS, IM2NP, France
Nilanjan Biswas	Birla Institute of Technology, Mesra, India
Nan-Kuang Chen	Liaocheng University, China
Yongqiang Cheng	University of Hull, UK
Cheng-Hsin Chuang	National Sun Yat-sen University, Taiwan

Contents

About the Editors

Nagender K. Suryadevara Dr. Nagender Kumar Suryadevara received his Ph.D. degree from the School of Engineering and Advanced Technology, Massey University, New Zealand, in 2014. He is Associate Professor at School of Computer and Information Sciences, University of Hyderabad, India. He has authored/co-authored three books, edited three books, and published over 60 papers in various International journals, conferences, and book chapters. He has supervised over 100 graduate and postgraduate students. He has examined over 30 postgraduate theses. He has delivered 42 presentations including keynote, tutorial, and special lectures. His research interests include Internet of Things, time series data mining, and ambient-assisted living environment. He is Senior Member of IEEE.

Subhas Chandra Mukhopadhyay Subhas (M'97, SM'02, F'11) holds a B.E.E. (gold medalist), M.E.E., Ph.D. (India), and Doctor of Engineering (Japan). He has over 31 years of teaching, industrial, and research experience. Currently, he is working as Professor of Mechanical/Electronics Engineering, Macquarie University, Australia, and is Discipline Leader of the Mechatronics Engineering Degree Program. He is also Director of International Engagement for the School of Engineering at Macquarie University. His fields of interest include smart sensors and sensing technology, wireless sensors and network (WSN), Internet of Things (IoT), wearable sensors, medical devices, health care, environmental monitoring, and mechatronics and robotics.

He has supervised 55 postgraduate students and over 150 Honors students. He has examined 76 postgraduate theses. He has published over 450 papers in different international journals and conference proceedings, written ten books and fifty-two book chapters, and edited twenty conference proceedings. He has also edited thirty-five books with Springer-Verlag and thirty-two journal special issues. He has received various awards, most notably: the Australian Research Field Leader in Engineering and Computer Science 2020; Distinguished Lecturer, IEEE Sensors Council 2020–2022; Outstanding Volunteer by IEEE

R10, 2019; World Famous Professor by Government of Indonesia, 2018; Certificate of Distinction from IEEE Sensors Council, 2017; IETE R.S. Khandpur Award, India, 2016; and Best Performing Topical Editor of IEEE Sensors Journal from 2013 to 2018, six years consecutively. He has organized over 20 international conferences as either General Chairs/Co-chairs or technical program chair.

He has delivered 426 presentations including keynote, invited, tutorial, and special lectures. He is Fellow of IEEE (USA), Fellow of IET (UK), Fellow of IETE (India), Topical Editor of IEEE Sensors journal, and Associate Editor of IEEE Transactions on Instrumentation and Measurements and IEEE Review of Biomedical Engineering. He is Editor-in-Chief of the International Journal on Smart Sensing and Intelligent Systems. He was Distinguished Lecturer of the IEEE Sensors Council from 2017 to 2022. He is Founding Chair of the IEEE Sensors Council New South Wales Chapter.

More details can be available at:

https://researchers.mq.edu.au/en/persons/subhas-mukhopadhyay.

https://scholar.google.com/citations?hl=en&user=bpwXxYEAAAAJ&view_op=list_works.

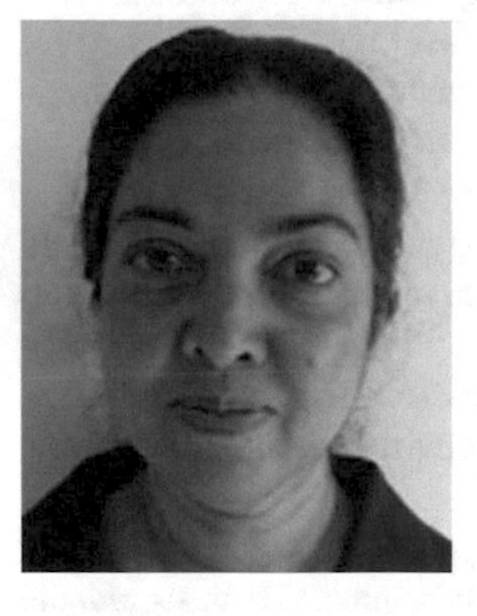

Krishanthi P. Jayasundera Dr. Krishanthi is Synthetic Chemist graduated from the University of Peradeniya, Sri Lanka, and a Ph.D. in Chemistry from Kanazawa University, Japan. She worked in Massey University, New Zealand, and University of Technology in Sydney, Australia, as Research Fellow on projects focused on chemical synthesis/development of bioassays for architecturally interesting molecules that have biological or medicinal significance. She also worked as Technical Program Co-chair to organize national and international conferences on sensing technology. At present, she is Senior Researcher at Macquarie University, Australia. She has published over 30 journal and conference proceedings papers and edited over 10 conference proceedings.

Boby George Boby George received the M.Tech. and Ph.D. degrees in Electrical Engineering from the Indian Institute of Technology (IIT) Madras, Chennai, India, in 2003 and 2007, respectively. He was Postdoctoral Fellow with the Institute of Electrical Measurement and Measurement Signal Processing, Technical University of Graz, Graz, Austria, from 2007 to 2010.

He joined the faculty of the Department of Electrical Engineering at IIT Madras in 2010. Currently, he is working as Professor there. His areas of interest include magnetic and electric field-based sensing approaches, sensor interface circuits/signal conditioning circuits, sensors, and instrumentation for automotive and industrial applications. He has co-authored more than 75 IEEE transactions/journals and more than 100 in tier-1 conference proceedings. He is Associate Editor for IEEE Sensors Journal, IEEE Transactions on Industrial Electronics, and IEEE Transactions on Instrumentation and Measurement.

A Novel Carbon Paste Electrode for Convenient and Efficient Auditory Brainstem Response Acquisition: A Pilot Study

Xin Wang[1,2,3], Qiong Tian[1,3], Yangjie Xu[4], Mingxing Zhu[1,5], Yingying Wang[1,2,3], Yuchao He[1,3], Shixiong Chen[1,2,3](✉), Zhiyuan Liu[1,2,3](✉), and Guanglin Li[1,3]

[1] CAS Key Laboratory of Human-Machine Intelligence-Synergy Systems, Shenzhen Institutes of Advanced Technology, Chinese Academy of Sciences, Shenzhen 518055, China
{sx.chen,zy.liu1}@siat.ac.cn

[2] Shenzhen College of Advanced Technology, University of Chinese Academy of Sciences, Shenzhen 518055, China

[3] Guangdong-Hong Kong-Macao Joint Laboratory of Human-Machine Intelligence-Synergy Systems, Shenzhen 518055, China

[4] Interdisciplinary Centre for Security, Reliability and Trust, University of Luxembourg, 1855 Luxembourg, Luxembourg

[5] School of Electronics and Information Engineering, Harbin Institute of Technology, Shenzhen 518055, China

Abstract. Auditory Brainstem Response (ABR) is the commonly used tool in Otology for clinical hearing loss and disease diagnoses. Wet electrodes, served as the gold standard, are widely used in clinical settings to acquire ABR signals for superior performance. However, it is a complex and time-consuming process to fix the electrodes by applying bandages or adhesive tapes. Moreover, wet electrodes might cause allergy and uncomfortable for patients, especially for children. Therefore, a novel anhydrous carbon paste electrode (CPE) with adhesion and biocompatibility was proposed to solve the aforementioned problems. The impedance of unfixed CPE and Conductive Paste-based electrode was recorded, and the ABR waveforms obtained by both two methods were systematically compared to further investigate and verify the feasibility of the proposed unfixed CPE. The results showed that the proposed CPE could achieve high-quality ABR recording under unfixed condition, which was comparable with the gold standard method. Besides, the analyses of the latency and amplitude of the ABRs showed that unfixed CPE could acquire higher amplitudes which would help with wave recognition, while the latency was similar. This proved that the unfixed CPE could obtain ABR with the same information compared with that of Paste-based ABR, since the physicians always made the diagnosis by analyzing the latency. Furtherly, we compared the correlation coefficient of the ABRs under different averaging times, and the results showed that unfixed CPE was comparable, even better than the conventional method.

Keywords: Auditory Brainstem Response · Carbon paste electrode · Adhesion

Xin Wang and Qiong Tian contribute equally to the paper.

N. K. Suryadevara et al. (Eds.): ICST 2022, LNEE 1035, pp. 1–10, 2023.
https://doi.org/10.1007/978-3-031-29871-4_1

1 Introduction

Hearing loss refers to a decrease in hearing sensitivity or hearing dysfunction due to damage to the physiological structure of the auditory system [1]. In the clinic, there are several methods for hearing loss detection. Among them, Auditory Brainstem Response (ABR) is one of the most used methods based on its non-invasive and objective. ABR is a sound stimulus-evoked response recorded on the subject's scalp [2], typically consisting of five to seven waves named I to VII, with wave I, III, and V mostly used in clinical diagnosis [3]. Wet electrode is widely used in physiological signal acquisition, including ABR. Definitely, wet electrode has its advantage of the signal's quality because of the usage of the conductive paste. However, this gold standard method does have some limitations in ABR recording. Firstly, the placement of the electrode is complex and time-consuming [2]. Adhesive tape or bandages are needed to fix the electrode on the subjects, which might cause uncomfortable or even hurt for children [4]. Besides, for the new-born, it's hard for them to stay quiet for a long time so a sedative might be needed [5]. Secondly, it cannot be denied that the usage of the conductive paste will greatly decrease the impedance between the skin and the electrode which will improve the signal's quality. But it is also because the use of the conductive paste, it may cause allergy and uncomfortable monitoring in ABR recording [6]. Besides, it was reported that with time goes by, the signal quality will decrease because the conductive paste dehydrates [7]. Therefore, a time-saving, efficient, and biocompatible solution should be proposed to solve the above-mentioned problems.

Many researchers try to improve the efficiency of ABR recording by designing new stimuli [8, 9] and algorithms [10]. Only several studies focus on the improvement of the electrodes. Yanz and Dodds proposed an ear-canal electrode for the measurements of ABR which could yield a large wave I but without loss in wave V [11]. Besio et al. introduced a tripolar concentric ring electrode that could achieve a recognizable ABR waveform with only 100 averaging times [12]. But it should be pointed out that this result was achieved with the usage of conductive paste. None of them can fulfill the requirements of convenient and efficient ABR acquisition.

In this paper, we proposed an anhydrous carbon paste electrode (CPE) that was biocompatible and could be used without adhesive tape or bandage for fixation, and its ability on ABR recording was systematically investigated. First, the impedances of the CPE and gold standard (Paste) group were studied. Then the characteristics of ABR including quality, amplitude and latency were studied. Finally, the time-saving of CPE-based ABR was focused on by evaluating the averaging time and the time for electrode placement.

2 Methods

2.1 Subjects

Three subjects (Two males and one female, with a mean age of 22.67 ± 1.15) without cognitive impairments and abnormal muscle function were recruited. The consent forms were sent out and clearly explained to all subjects. These signed consent forms were afterward collected back before conducting the experiments involved. After that, the subjects were asked to sit at a comfortable desk with their eyes closed. The skin on which the electrodes would be placed was first swiped by an alcohol pad to lower the impedance of the skin-electrode. The ABR acquisition was conducted in an electromagnetically shielding room for eliminating the interference of the power-line, while the recording of impedance had no requirements on the environment. All the experimental protocols were approved by the Institutional Review Board (IRB) of Shenzhen Institutes of Advanced Technology, Chinese Academy of Sciences (*SIAT-IRB-190615-H0352*).

2.2 The Experimental Scheme

The experimental scheme was shown in Fig. 1. Firstly, we prepared the electrodes for the experiments. Basically, two groups of electrodes were prepared, namely carbon paste-based electrode (called CPE later) and Conductive Paste electrode (called Paste later). What should be pointed out was that all the data obtained by the CPE in this paper was under unfixed condition while the Paste was fixed by a bandage or adhesive tape. The processing to construct CPE was discussed in [13]. In order to control the variable, we used a disk electrode and a customized shell to control the volume of the materials used for the electrodes. Three electrodes were used to acquire ABR signals from the subjects, with their placements on the forehead, left mastoid and right mastoid respectively. It should be pointed out that the stimulus was given on the left ear of the subjects, therefore, the electrode placed on the right mastoid was served as reference electrode while the left two electrodes were differential electrodes. The ABR signal was obtained by a customized system which was first reported in [14]. Based on the embedded PCM5102 (TI Inc., USA), the system could achieve the production of stimuli by programming. Besides, the data transmission between the customized ABR acquisition system and PC was achieved by WIFI. The data was offline analyzed on PC and the stimulus was given 1000 times at the rate of 30 Hz and stimulus level of 65 dB nHL.

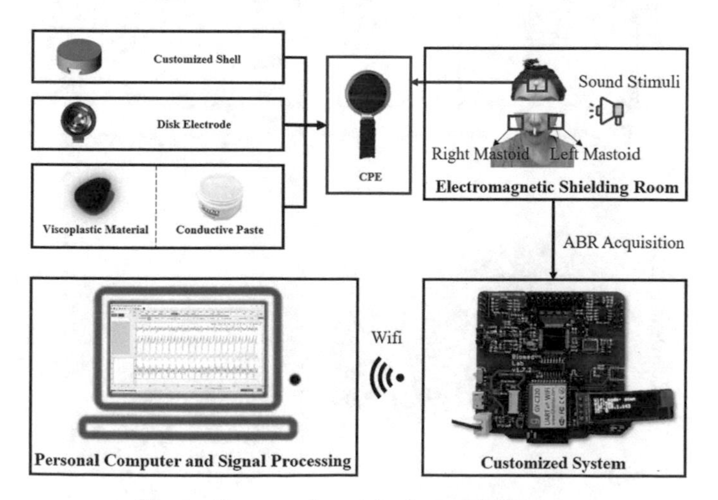

Fig. 1. The experimental scheme of this paper

2.3 Signal Processing and Analysis

The acquired ABR signals were acquired by using a customized GUI software (Mathworks Inc., USA) and analyzed by Matlab. The raw data was filtered by a 3 order butterworth filtering with cut-off frequencies of 100 Hz and 1500 Hz, and was segmented into epochs according to the stimulus onset. These epochs were then averaged to finally obtain the targeted ABR signals. The ABR signals averaged by 10 different averaging times from 100 to 1000, with an increment of 100, were obtained to further analyze the time-consuming issue. It should be noted that the ABR averaged by 1000 times was treated as the final ABR signal. Besides, the clinical indexes of ABR, like latency and amplitude, were also calculated and considered to investigate the performance of characteristics wave morphology.

3 Results

3.1 The Comparison of the Impedance

In this part, the impedance of different electrode groups was studied for all subjects. The total impedance of an RC parallel circuit model was suggested by Wang et al. [15], and the results were shown in Fig. 2. Based on the fact of the frequency band of ABR, the impedance was recorded from 100 Hz to 3000 Hz. For the individual, the impedance of CPE under unfixed condition was greater than that of Paste under fixed condition, which could be seen in the upper panel of Fig. 2. The average impedance of all subjects was also plotted in the lower panel of Fig. 2, from which we could notice that the similar trends in individuals. Overall, the impedance of CPE was a little bit higher than that of Paste.

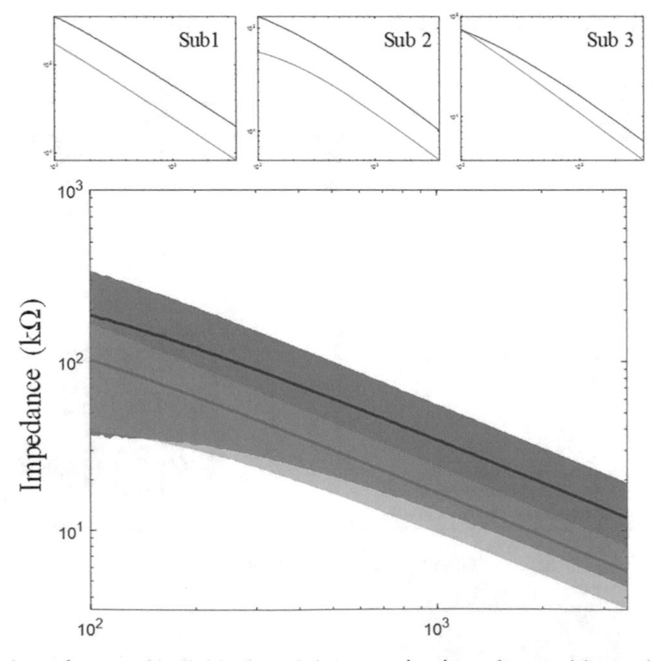

Fig. 2. The impedances of individuals and the averaging impedance with standard deviation

3.2 The Comparison of the Quality of ABR

Figure 3 presented the ABR waveforms obtained from three subjects by two groups of electrodes, with the red line representing CPE-based ABR and the blue line representing Paste-based ABR. The waveform of the ABR shared similar quality, with all waveforms having five recognizable characteristics waves (I–V) which were marked out in Fig. 3. Furthermore, to investigate the improvement in the clinical implication of the acquired ABR signals, the amplitude and latency were also taken into consideration. For the latency of ABR, the results showed that ABR obtained by different electrodes had close latencies, as shown in Table 1. For wave I, the latency was 1.442 ± 0.250 ms for CPE and 1.421 ± 0.201 ms for Paste. Similarly, for the latency of CPE-based wave II to V, they were 2.630 ± 0.125 ms, 3.630 ± 0.125 ms, 4.776 ± 0.157 ms, and 5.526 ± 0.130 ms, respectively, while that of Paste-based were 2.588 ± 0.036 ms, 3.609 ± 0.036 ms, 4.921 ± 0.130 ms, and 5.588 ± 0.095 ms, respectively. Considering the fact that the inter-wave latency made sense in the diagnosis of auditory diseases, inter-wave latencies of I–III, I–V, and III–V were also taken into consideration, as shown in Table 2. The results showed that the inter-wave latency shared the same trend as the individual wave latency. For CPE-based ABR, the inter-wave latencies of wave I–III, I–V, and III–V were 2.188 ± 0.217 ms, 4.084 ± 0.252 ms, and 1.896 ± 0.036 ms respectively, while for Paste-based ABR, they were 2.188 ± 0.226 ms, 4.167 ± 0.295 ms, 1.979 ± 0.072 ms respectively. Since the amplitude of ABR was also an index in the clinical diagnosis, we analyzed the amplitude of each characteristic wave of ABR obtained by different electrodes.

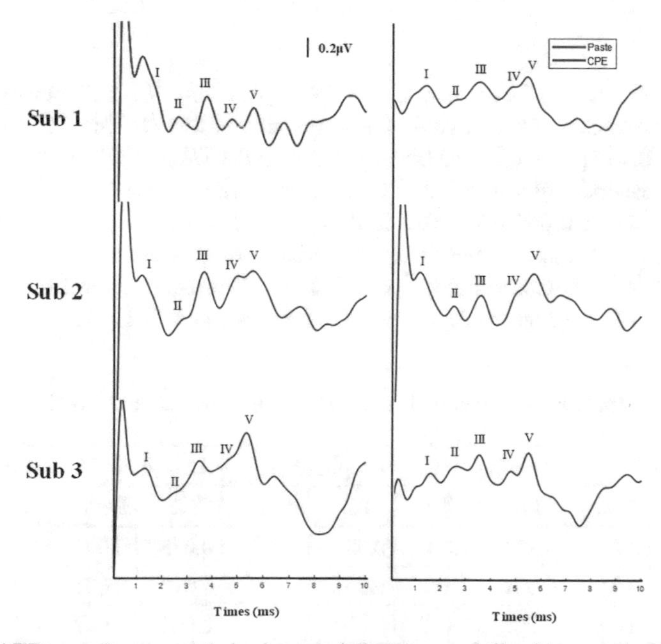

Fig. 3. The ABR waveforms and their characteristics waves of all subjects under the stimulation rate of 30 Hz and the stimulus level of 65 dB nHL

Table 1. The latency of ABR obtained by CPE and Paste.

	$Latency_I$		$Latency_{II}$		$Latency_{III}$		$Latency_{IV}$		$Latency_V$	
	CPE	Paste	CPE	Paste	CPE	Paste	CPE	Paste	CPE	Paste
Sub1	1.692	1.505	2.630	2.630	3.755	3.630	4.755	4.880	5.630	5.567
Sub2	1.192	1.192	2.755	2.567	3.630	3.630	4.942	5.067	5.567	5.692
Sub3	1.442	1.567	2.505	2.567	3.505	3.567	4.630	4.817	5.380	5.505
Mean	1.442	1.421	2.630	2.588	3.630	3.609	4.776	4.921	5.526	5.588
SD	0.250	0.201	0.125	0.036	0.125	0.036	0.157	0.130	0.130	0.095

Table 2. The inter-wave latency of ABR obtained by CPE and Paste.

	Inter-wave $_{I-III}$		Inter-wave $_{I-V}$		Inter-wave III-V	
	CPE	Paste	CPE	Paste	CPE	Paste
Sub1	2.063	2.125	3.938	4.062	1.875	1.937
Sub2	2.438	2.438	4.375	4.500	1.937	2.063
Sub3	2.063	2.000	3.938	3.938	1.875	1.938
Mean	2.188	2.188	4.084	4.167	1.896	1.979
SD	0.217	0.226	0.252	0.295	0.036	0.072

As shown in Table 3, the amplitude of ABR acquired by CPE had greater amplitudes than that of Paste, except for wave II. The amplitude of five waves of CPE-based ABR were 0.450 ± 0.147 µV, 0.041 ± 0.069 µV, 0.265 ± 0.137 µV, 0.032 ± 0.044 µV, 0.450 ± 0.072 µV, respectively, while that of Paste-based ABR were 0.242 ± 0.173 µV, 0.059 ± 0.074 µV, 0.24 ± 0.066 µV, 0.013 ± 0.021 µV, 0.324 ± 0.073 µV, respectively. It should be noticed that sometimes the characteristic waves were not recognizable based on the fact of bad ABR morphology or individual difference. Therefore, in the Table, the amplitudes of unrecognizable waves were recorded as 0.001 µV.

Table 3. The amplitude of ABR obtained by CPE and Paste.

	Amplitude I		Amplitude II		Amplitude III		Amplitude IV		Amplitude V	
S	CPE	Paste	CPE	Paste	CPE	Paste	CPE	Paste	CPE	Paste
1	0.454	0.205	0.121	0.001	0.33	0.167	0.083	0.001	0.373	0.285
2	0.595	0.431	0.001	0.143	0.357	0.296	0.013	0.001	0.461	0.278
3	0.302	0.091	0.001	0.033	0.108	0.257	0.001	0.037	0.515	0.408
M	0.450	0.242	0.041	0.059	0.265	0.240	0.032	0.013	0.450	0.324
SD	0.147	0.173	0.069	0.074	0.137	0.066	0.044	0.021	0.072	0.073

3.3 The Correlation Coefficient of ABR

In this part, taking the ABR averaged with 1000 times as final ABR signals, the correlation coefficients (CCs) was calculated for different averaging times. As shown in Fig. 4, the CCs of both electrode groups were plotted with nine different averaging times, namely 100 to 900 times, with an increment of 100. From the curves, we could notice that both curves shared similar trends, with the CC of CPE (blue line) higher than that of Paste (Orange line). The CCs of CPE under different averaging times were 0.63 ± 0.05, 0.74 ± 0.06, 0.84 ± 0.03, 0.91 ± 0.00, 0.95 ± 0.02, 0.96 ± 0.01, 0.97 ± 0.01, 0.98 ± 0.01, 0.99 ± 0.00, respectively, while that of Paste were 0.60 ± 0.04, 0.76 ± 0.01, 0.80 ± 0.04, 0.85 ± 0.05, 0.89 ± 0.03, 0.93 ± 0.03, 0.95 ± 0.02, 0.98 ± 0.01, 0.99 ± 0.01, respectively.

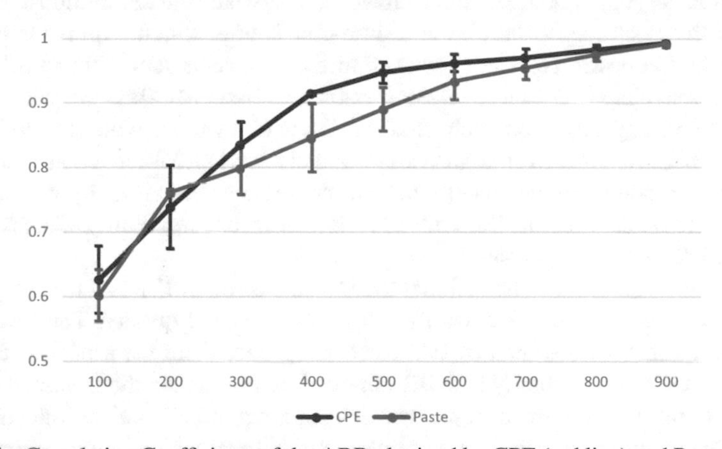

Fig. 4. The Correlation Coefficients of the ABR obtained by CPE (red line) and Paste (blue line)

3.4 The Time Cost in Electrode Placement

In this part, we tested the time cost in the electrode placement. For the Paste, it typically needed adhesive tape or more to fix it on the subjects. In this experiment, we only used one adhesive tape to fix the electrode without using adhesive tape to fix the electrode wire. The padding of the electrodes was not taken into consideration. As shown in Table 4, the mean time for the electrode placement of CPE was 5.667 ± 0.577 s, while that of Paste was 12.000 ± 1.000 s.

Table 4. The time costed in electrode placement.

The Time Costed in Electrode Placement (s)					
	Sub1	Sub2	Sub3	Mean	SD
CPE	6	5	6	5.667	0.577
Paste	11	12	13	12.000	1.000

4 Discussion

In the hospital, the acquisition of ABR is usually achieved by using wet electrodes which is regarded as the gold standard method. However, it still has some limitations. Firstly, the fixation of the electrode is complex and time-consuming, which requires experienced physicians and adhesive tape. For those children or patients with skin sensibility, the usage of adhesive tape may cause uncomfortable and harm [4]. Besides, wet electrodes might cause allergy based on their characteristics of hydrous which provides it the capacity for high-quality physiological signal acquisition [2]. Moreover, as reported, the signal quality recorded by wet electrodes will decrease as time goes by because of the dehydration [7]. Therefore, a CPE with adhesion and biocompatibility was proposed to solve the aforementioned problems.

In this study, the results showed that the impedance of CPE was a little higher than Paste, but this fact was not necessarily affecting the signal quality. Then we studied on the quality of ABR obtained by two electrode groups from the aspect of the waveform. We found that the quality of ABR obtained by the two methods shared a similar quality. Therefore, it was worth to further investigating the clinical significance of the ABR by analyzing the details like amplitude and latency which helped physicians to diagnose. The research found that the CPE showed high-quality ABR signals recording as the gold standard method did. Besides, CPE showed a little bit better performance on the amplitude of ABR waves, which would help with the hearing threshold evaluation that was diagnosed by the appearance of wave V. Moreover, when considering the latency, we found that CPE had little influence on it, especially on the inter-wave latency, which meant that CPE-based ABR could provide high amplitude but without changing its latency. This fact suggested that CPE-based ABR provided the physicians with a useful suggestion as the gold standard method did, but with high amplitude and without changes in latency. Further, the CCs were also agreed with the above-mentioned results, with a little better performance than Paste-based ABR, which indicated that the CPE-based ABR could achieve higher related ABR with twice less averaging times. The time cost in electrode placement also proved the CPE-based ABR could achieve ABR acquisition with around 53% less time. Yanz et al. proposed an ear-canal electrode for the measurements of ABR which could yield a large wave I but without loss in wave V [11]. In our study, CPE-based ABR also provided larger wave I, and moreover, larger wave V was achieved. Besio et al. introduced a tripolar concentric ring electrode that could achieve ABR acquisition with only 100 averaging times [16]. It should be noticed that this result was achieved with the usage of conductive paste. The CPE could also achieve ABR acquisition with less averaging time without conductive paste. Compared with the gold standard and the existing electrodes for ABR, the proposed CPE could achieve high-quality ABR with convenience, efficiency, and biocompatibility. Moreover, based on its anhydrous, it had been proved in [17] that the CPE could acquire signals after 48 days, and it was believed that the life span of the CPE could be much longer. Also in [17], the CPE was used in the sweating condition, and the CPE could still acquire the Alpha wave of EEG. Since this is a pilot study, future work will recruit more subjects, improve the CPE, and arrange more experiments and conditions.

Acknowledgment. This work was supported in part by the Key-Area Research and Development Program of Guangdong Province (#2020B0909020004), National Natural Science Foundation of China (#81927804, #62101538), Shenzhen Governmental Basic Research Grant (#JCYJ20180507182241622), Science and Technology Planning Project of Shenzhen (#JSGG20210713091808027, #JSGG20211029095801002), China Postdoctoral Science Foundation (2022M710968), SIAT Innovation Program for Excellent Young Researchers (E1G027), CAS President's International Fellowship Initiative Project (2022VEA0012).

References

1. W. H. Organization: Deafness and hearing loss, Fact sheet N∘ 300, WHO Media Centre, Geneva, Switzerland (2015)
2. Li, X.: Fundamental and Application of Auditory Evoked Response. Peoples Military Medical Press, Beijing, China (2007)
3. Ji, F., Chen, A., Zhao, Y., Liu, X., Zhou, Q.: "Application of electrocochleography and ABR in the diagnosis of auditory neuropathy," Lin Chuang er bi yan hou tou Jing wai ke za zhi=. J. Clin. Otorhinolaryngol. Head Neck Surg. **24**, 447–449 (2010)
4. Kelly, K.J., Pearson, M.L., Kurup, V.P., Havens, P.L., Byrd, R.S., Setlock, M.A., et al.: A cluster of anaphylactic reactions in children with spina bifida during general anesthesia: epidemiologic features, risk factors, and latex hypersensitivity. J. Allergy Clin. Immunol. **94**, 53–61 (1994)
5. Stach, B.A., Ramachandran, V.: Clinical Audiology: An Introduction: Plural Publishing (2021)
6. Luo, J., Xing, Y., Sun, C., Fan, L., Shi, H., Zhang, Q., et al.: A bio-adhesive ion-conducting organohydrogel as a high-performance non-invasive interface for bioelectronics. Chem. Eng. J. **427**, 130886 (2022)
7. Wang, X., Liu, S., Zhu, M., Wang, X., Liu, Z., Jiang, Y., et al.: Performance of flexible non-contact electrodes in bioelectrical signal measurements. In: 2019 IEEE International Conference on Real-time Computing and Robotics (RCAR), pp. 175–179 (2019)
8. Chen, S., Deng, J., Bian, L., Li, G.: Stimulus frequency otoacoustic emissions evoked by swept tones. Hearing Res. **306**, 104–114 (2013)
9. Elberling, C., Don, M., Kristensen, S.G.B.: Auditory brainstem responses to chirps delivered by an insert earphone with equalized frequency response. The J. Acoust. Soc. Am. **132**(2), EL149–EL154 (2012)
10. McKearney, R.M., Bell, S.L., Chesnaye, M.A., Simpson, D.M.: Auditory brainstem response detection using machine learning: a comparison with statistical detection methods. Ear Hear. **43**(3), 949–960 (2021)
11. Yanz, J.L., Dodds, H.J.: An ear-canal electrode for the measurement of the human auditory brain stem response. Ear Hear. **6**(2), 98–104 (1985)

12. Steele, P., Mercier, J., Nasrollaholhosseini, S., Bartels, R., Besio, W.: Modeling tripolar concentric ring electrode (TCRE) sensor and acquisition of auditory brainstem response. In: 2017 IEEE SENSORS, pp. 1–3 (2017)
13. Wang, X., Tian, Q., Pi, Y., Xu, Y., Zhu, M., Wang, X., et al.: A pilot study on long-term physiological signal monitoring using anhydrous viscoplastic electrodes. In: 2021 43rd Annual International Conference of the IEEE Engineering in Medicine & Biology Society (EMBC), pp. 6767–6770 (2021)
14. Wang, X., Zhu, M., Samuel, O.W., Wang, X., Zhang, H., Yao, J., et al.: The effects of random stimulation rate on measurements of auditory brainstem response. Front. Hum. Neurosci. **14**, 78 (2020)
15. Wang, Y., Jiang, L., Ren, L., Pingao, H., Yu, M., Tian, L., et al.: Towards improving the quality of electrophysiological signal recordings by using microneedle electrode arrays. IEEE Trans. Biomed. Eng. **68**, 3327–3335 (2021)
16. Nasrollaholhosseini, S.H., Mercier, J., Fischer, G., Besio, W.G.: Electrode–electrolyte interface modeling and impedance characterizing of tripolar concentric ring electrode. IEEE Trans. Biomed. Eng. **66**, 2897–2905 (2019)
17. Wang, X., et al.: Towards optimizing the quality of long-term physiological signals monitoring by using anhydrous carbon paste electrode. IEEE Trans. Biomed. Eng. **70**(2), 423–435 (2023)

Shiro-Neko, A Stationmaster Robot that Operates an Unmanned Station

Momi Kinoshita, Kantaro Baba, Zahido Subaru, and Mikiko Sode Tanaka(✉)

International College of Technology, Hisayasu 2-270, Kanazawa, Japan
{b2870126,b2870045,b2870100}@planet.kanazawa-it.ac.jp,
sode@neptune.kanazawa-it.ac.jp

Abstract. In recent years, Japan has seen a declining birthrate and an aging population. As a result, it is becoming increasingly difficult to maintain public transportation in rural areas. In order to reduce maintenance costs, railroad stations are becoming increasingly unmanned. Unmanned stations have a variety of security and safety issues. These are problems that must be solved. Therefore, we are considering the installing of stationmaster robots. In Japan, stationmaster robots are being installed at large stations in central Tokyo to assist people. However, installation at unmanned stations has not progressed. The reason is that many people in rural areas are elderly and unfamiliar with IT. Therefore, we have devised a way to lower the threshold and are studying the introduction of stationmaster robots. For example, we have devised ways to make the robot more approachable, such as using colors to give the weather forecast for the next few hours, or giving casual greetings. The functions to be provided are: tourist information, weather forecasts, event announcements, emergency notifications, explanations on how to buy boarding tickets, boarding ticket purchases, boarding ticket gate operations, and notification of train departure and arrival times.

Keywords: Natural language processing · semantic understanding · explainable AI · speech recognition · speech synthesis

1 Introduction

In 2021, total population of Japan is 125,682,000, it has decrease by 62,000 people in one year. In detail, the number of people less than 15 years old is 14,489,000 in total, and it is about 11% of the total population. The number of people 15–65 years old is 74,658,000 and it is about 60% of the total population. Finally, the number of people greater than 65 years old is 36,175,000 and it is about 30%. In addition, the largest demographic japan is people 60–65 years old. We will look at this data by comparing it to data from the USA. The total population of the USA is 331,449,281. The group 0–15 years old contains 61,039,314 people, and it is about 18% of the total population, the group 15–65 years old contains 214,685,514 people and it is about 65% of the total population. Finally, the number of greater than 65 years old is 53,340,089 and it is about 16%. The largest demographic group in USA is 20–25 years old. It means Japan has a

N. K. Suryadevara et al. (Eds.): ICST 2022, LNEE 1035, pp. 11–16, 2023.
https://doi.org/10.1007/978-3-031-29871-4_2

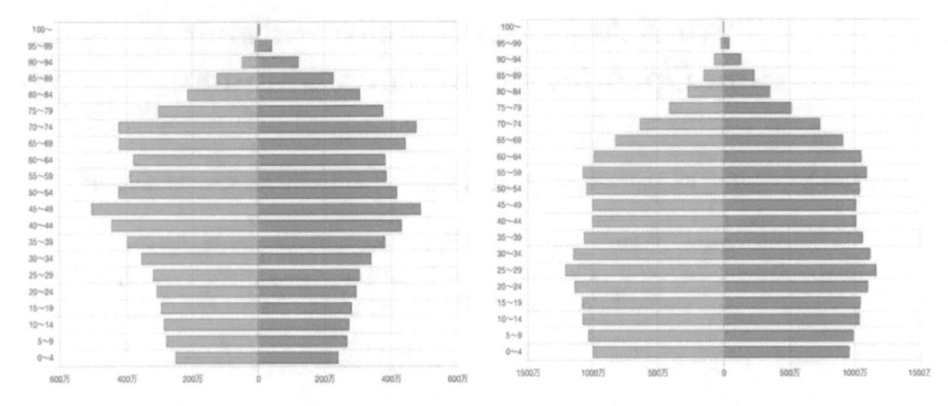

(a) Population distribution of Japan[1] **(b) Population distribution of the USA[2]**

Fig. 1. Comparison of population distribution between the USA and Japan

very small young generation so there are many old people. In addition, in Japan there are 17.67 billion trains' passengers in one years (Fig. 1).

With a declining population, it is becoming increasingly difficult to maintain public transportation in rural areas. In order to reduce maintenance costs, railroad stations are becoming increasingly unmanned. Unmanned stations pose a variety of security and safety issues. These are problems that must be solved. Therefore, we are considering the installation of stationmaster robots. In Japan, stationmaster robots are being installed at large stations in central Tokyo to assist people [3, 4]. However, installation at unmanned stations has not progressed. The reason is that many people in rural areas are elderly and unfamiliar with IT. Therefore, we have devised a way to lower the threshold and are studying the introduction of stationmaster robots. For example, we have devised ways to make the robot more approachable, such as using colors to give the weather forecast for the next few hours, or giving casual greetings. The functions to be provided are: directions, weather forecasts, event announcements, emergency notifications, and explanations on how to buy tickets, ticket purchase, ticket gate operations, and notification of train departure and arrival times.

This paper introduces the function and mechanism of the stationmaster robot Shironeko. The results of experiments are also explained.

2 Stationmaster Robot Shirokane

2.1 Stationmaster Robot Shironeko Concept

The concept of Stationmaster Robot Shironeko is friendliness. Since it is used in local cities where there are many elderly people, we thought that familiar things such as cats would be good. Figure 2 shows an example illustration. She has a voice conversation function. The display is connected to assist in voice conversation.

Service is based on voice dialogue. She starts the conversation when she is spoken to. Figure 3 is an example of utilization. When asked about the weather, she will answer

Fig. 2. Image of Shironeko, the Stationmaster Robot

(a) Example of teaching the weather (b) Emergency call case

Fig. 3. Stationmaster robot Shironeko utilization example

with the current weather, the weather three hours later, and the on the next day. When a user encounters trouble, she will make an emergency call. In the middle of the main body of the stationmaster robot Shironeko, there is a tablet for displaying notifications, train information, and other information. This tablet is for people who cannot hear well, as many of them are elderly. Also, the conventional stationmaster robots are in the same place and only give information to the person who asks for the information. On the other hand, the stationmaster robot Shironeko does more work such as station safety management by walking around. This is a big difference from other stationmaster robots.

2.2 Functional Overview of Shiro-Neko Stationmaster Robot

The stationmaster robot Shironeko installed at the unmanned station starts the service 15 min before the first train and stops the service 15 min after the last train. This robot stationmaster detects people with a heat sensor. When it detects a person, it approaches the person and asks "Are you in trouble?". If a user is looking at the stationmaster robot Shironeko, the white cat's display shows four choices:

1 How to buy a ticket
2 Arrival time and departure time
3 Weather forecast
4 Others

Users can talk by voice, but they can also make inquiries using the display. When all operations are completed, it automatically returns to the initial position. Figure 4 shows the hardware configuration diagram and Fig. 5 shows the operation flow.

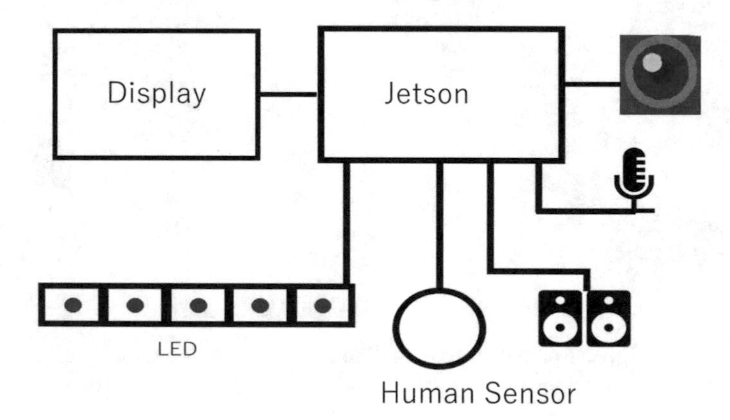

Fig. 4. Hardware configuration diagram of Shironeko stationmaster robot

She has a tourist information function, if there are request for things such as sightseeing routes, she will a guide users to the sightseeing route tailored to that person. For example, she will ask a user what they like with guest, such as "What are you interested in?" Or "Is it okay for you to walk a little?" And she will guide user to the route that suits them.

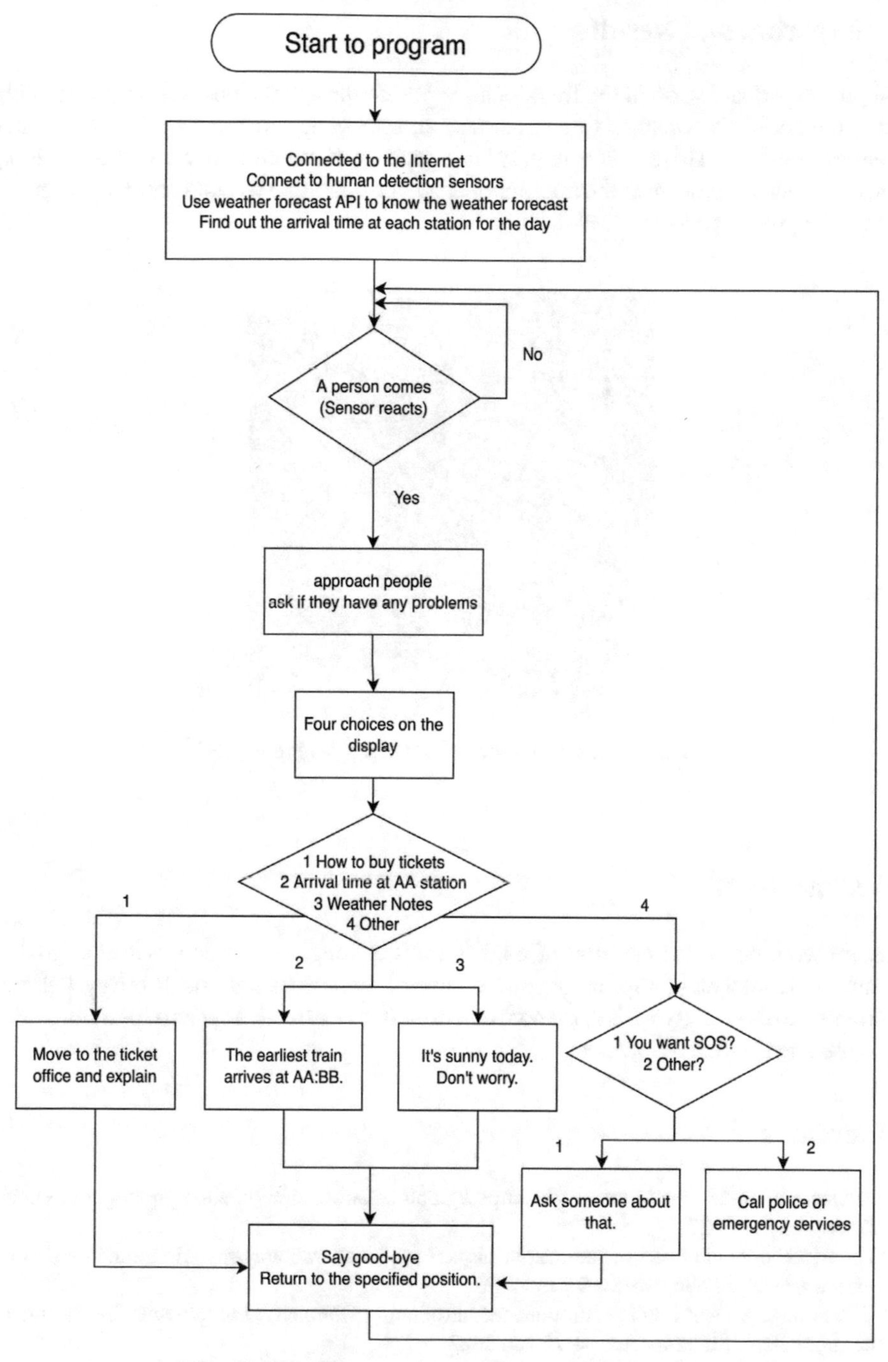

Fig. 5. Stationmaster robot Shironeko's operation flow

3 Experimental Results

We prototyped and verified the functionality. We confirmed the operation by voice conversation about the weather forecast function, emergency transmission function, and greeting function. There was a microphone pick-up problem that we could not input voice without speaking near the microphone, but the operation worked correctly. Figure 6 shows the prototype we created.

Fig. 6. Stationmaster Robot Shironeko Prototype

4 Conclusion

We are working on the creation of a stationmaster robot, Shironeko, with the goal of ensuring the safety and improving convenience of unmanned stations. It is a system that is friendly to the elderly with voice conversation. It is an effective system for maintaining an unmanned station at low cost.

References

1. Japanese pyramid seen in the graph: https://graphtochart.com/population/japan-pyramid.php. Access 15 Jun 2022
2. American pyramid seen in the graph: https://graphtochart.com/population/united-states-of-america-pyramid.php, Access 15 Jun 2022
3. The birth of a robot tourist stationmaster Introducing Ohno city: https://www.fukuishimbun.co.jp/articles/-/1195846. Access 15 Jun 2022
4. AI female station staff make an outstanding performance. The impact of "Takanawa Gateway": https://toyokeizai.net/articles/-/336547?page=4. Access 15 Jun 2022

Analysis of the SO Adsorption on Boron, Nitrogen, and Phosphorous Doped Monolayer Silicene-A First Principle Study

Aditya Tiwari[1], Naresh Bahadursha[1], Joshna Palepu[1], and Sayan Kanungo[1,2](✉)

[1] Department of Electrical and Electronics Engineering, Birla Institute of Technology and Sciences-Pilani, Hyderabad Campus, Hyderabad, India
sayan.kanungo@hyderabad.bits-pilani.ac.in

[2] Materials Center for Sustainable Energy and Environment, Birla Institute of Technology and Sciences-Pilani, Hyderabad Campus, Hyderabad, India

Abstract. In this work, for the first time, the effects of nonmetallic dopants (Boron, Nitrogen, Phosphorus) in monolayer two-dimensional (2D) Silicene (Si) have been extensively investigated for SO adsorption using Density Functional Theory (DFT) based on first principle approach. It has been observed that the SO is preferentially adsorbed on the Silicon atoms, where the presence of neighboring dopant atoms alters the local charge distributions at the adsorption site and tends to enhance the molecular adsorption. Next, the influence of SO adsorption on the structural and electronic property of doped Si is systematically analyzed. The results exhibit that SO adsorption changes the spatial distribution of electronic states near the Fermi level resulting in significant modulation in the bandgap and effective masses of the doped lattice. Among different doping species, Nitrogen demonstrates the strongest molecular adsorption and charge transfer, and after molecular adsorption induces the largest density of states peak near the Fermi level.

Keywords: DFT · Silicene · Doping · Molecular Adsorption · Sulfur Monoxide

1 Introduction

The development of compact, extremely sensitive, and selective gas sensors with short reaction times is aiding environmental monitoring, pollution control, food safety, and industrial automation applications [1]. Due to their exceptionally high surface-to-volume ratio and a large number of reactive surface sites, crystalline two-dimensional (2D) materials have shown tremendous potential for electrochemical gas sensor design, where a small number of gas-molecule adsorption can result in a large change in electronic conductivity [2]. Since its discovery in 2004, the first 2D material graphene (Gr) has been thoroughly investigated for its application as a gas sensor [3]. However, the pristine graphene shows little sensitivity toward virtually every ambient gas [4]. In this context, the introduction of suitable substitutional doping shows significant prospects for simultaneously improving the sensitivity and selectivity toward specific gas [5].

N. K. Suryadevara et al. (Eds.): ICST 2022, LNEE 1035, pp. 17–23, 2023.
https://doi.org/10.1007/978-3-031-29871-4_3

Similarly, other Gr-like 2D materials with a large active area are also widely in demand for detecting gas molecules for various applications. In this context, Silicene (Si), a monolayer of hexagonally arranged silicon atoms synthesized in 2010 [6], has already been theoretically predicted [7] and synthesized [8]. In contrast with purely sp^2 hybridization of carbon atoms in Gr, the mixed sp^2-sp^3 hybridization of silicon atoms can be observed in Si. Moreover, the π-bond in Si is essentially due to the overlapping of the $3p_z$ orbital, while the π-bond of Gr is due to the overlapping of the $2p_z$ orbitals. Consequently, relatively weaker silicon-silicon bonds can be observed in Si, compared to carbon-carbon bonds in Gr, which leads to a highly chemically reactive nature in Si towards molecular adsorptions compared to Gr [9].

On the other hand, both metallic and non-metallic doping is extensively investigated for Gr-like 2D materials. However, the introduction of metallic dopants into a semi-metal like Gr and Si tends to induce metallic nature in the crystal lattice [10]. This has encouraged the researcher to look into a variety of non-metallic dopants in Gr to increase gas-molecule adsorption while preserving the semiconducting nature. Consequently, Boron (B), nitrogen (N), and phosphorus (P) doping in Gr are found to be highly advantageous for electrochemical gas sensing applications [11].

To this effect, to date, only a few attempts have been made to investigate the effects of B, N, and P doping in Si [12]. However, to the best of the authors' knowledge, no attempts have been made to systematically explore different non-metallic doping species in Si for environmental pollutant gas sensing applications. In this context, for the first time, the present work attempt to systematically analyze the comparative effects of B, N, and P doping in Si for Sulfur Di-Oxide (SO) sensing.

2 Computational Methodology

Pristine Si monolayer exhibits a hexagonal honeycomb lattice structure, where each silicon atom forms bonds with three neighboring atoms, and two silicon atoms in the Si unit cell represent two different sub-lattices. The electrical characteristics of Si are very much similar to Gr in which both the conduction band edge (π- bands) and valence band edge (π*-bands) cross linearly at the fermi-level of the Brillouin zone.

In this work, the Si is considered in a $4 \times 4 \times 1$ supercell configuration for studying gas-molecule adsorption by repeating the unit cell in the in-plane direction, with a 50 Å vacuum in the out-of-plane direction to eliminate the artificial interaction of host lattices with its periodic images. The $4 \times 4 \times 1$ supercell offers a good trade-off between computational cost and accuracy of the result [11]. The doped lattices have been created by replacing one silicon atom with the impurity atom (B/N/P) in the monolayer Si thus creating a 3.125% of doping concentration. Furthermore, single gas-molecule adsorption on the dopant atom site is investigated throughout this work, with just the geometrical configuration associated with the lowest binding energy for a given gas-molecule and lattice interaction being taken into account.

In this study, Synopsys Quantum-Atomistix Wise's Tool Kit (ATK) and Virtual Nano Lab (VNL) simulation software packages were utilized to perform density functional theory (DFT) based first-principle calculations. A linear combination of atomic orbital (LCAO) double-zeta polarized basis set with a density mesh cut-off energy of 125 Hartree

[11, 13] and a Monkhorst-Pack grid of $3 \times 3 \times 1$ is used to sample the Brillouin zone. To begin, the geometries are relaxed using the LBFGS (Limited Memory Broyden Fletcher Goldfarb Shanno) algorithm, which has pressure and forces tolerances of 0.001 eV/3 and 0.01 eV/, respectively [14]. Following the geometrical optimization of the system with LDA, GGA correlation interaction and Revised Purdew Burke and Ernzerhof (RPBE) functional were used to calculate various properties [14]. The monolayer of pure Si has a lattice vector of 3.802 and a Si-Si bond length of 2.26 after optimization. The computed values are quite close to the experimentally determined lattice vector of 3.8 Å and Si-Si bond length of 2.27 Å [7, 8]. The inherent incompleteness of the Linear Combination of Atomic Orbitals (LCAO) basis generates Basis Set Superposition Error when two subsystems, such as the host lattice and the gas-molecule, are involved (BSSE) [13]. The energy of the lattice/gas-molecule system is overstated as a result of the artificial contact between the subsystems. The counterpoise (CP) adjustment is used during the binding energy calculation to circumvent the BSSE limitation. Finally, to account for the Van der Waals (VdW) forces between the graphene lattice and the gas molecule, the Grimme DFT-D3 empirical correction is applied [13].

The binding energy ($E_{Binding}$) of specific gas-molecule adsorption on the doped Si is calculated as follows,

$$E_{Binding} = E_{Gas\_and\_Host} - (E_{Gas} + E_{Host}) \tag{1}$$

where $E_{Gas\_and\_Host}$, E_{Gas}, and E_{Host} are the ground-state energies of gas-molecule/doped Si binding structure after incorporating CP correction, isolated gas-molecule, and doped Si, respectively. For any stable gas-molecule adsorption, $E_{Binding} < 0$. The subsequent charge transfer ($Q_{Transfer}$) between the gas-molecule and doped Si lattice is given by,

$$Q_{Transfer} = Q_{Gas\_before\_Adsorption} - Q_{Gas\_after\_Adsorption} \tag{2}$$

where $Q_{Gas\_before\_Adsorption}$, and $Q_{Gas\_after\_Adsorption}$ are the Mulliken charges in gas molecules before and after adsorption in doped Si lattice. If $Q_{Transfer} > 0$, the electron is transferred from gas-molecule to lattice. Whereas, $Q_{Transfer} < 0$, suggests the gas molecule acts as an acceptor type of impurity.

3 Results and Discussion

Three distinct dopant species are used as substitutional impurities in the supercell of the pristine Si lattice in this study. As a result, three different configurations with boron (B), nitrogen (N), and phosphorous (P) doped lattice are identified and depicted in Fig. 1(a). The size compatibility of the dopant atoms for substitutional doping is the greatest way to retain the structural integrity of any 2D material system. The typical atomic radius of Si, B, N, and P are 1.10 Å, 0.85 Å, 0.65 Å, and 1.00 Å, respectively. Subsequently, after different doping with different impurity (B/N/P) atoms, the different interatomic bond lengths are observed to be in the range of 1.83–2.28 Å, compared to the Si-Si bond length of 2.26 Å in pristine Si. Moreover, the lattice vector of B, N, and P doped Si systems are 3.781 Å, 3.739 Å, and 3.784 Å respectively which are slightly different from the pure Si lattice vector of 3.802 Å. It is to be noted that the unit cell lattice vector experiences the highest change (of −1.66%) in the case of N doped system. These findings suggest that the N-doping affects the structural stability of Si lattice slightly more than that of B and P impurity atoms. This greater modulation in bond length caused in the N-doped system which also has the highest instability compared to B and P-doped systems is because the N atom has the highest mismatch in atomic radius compared to the Si atom whereas the P-doped system shows better structural stability due P and Si atoms having an almost equal atomic radius.

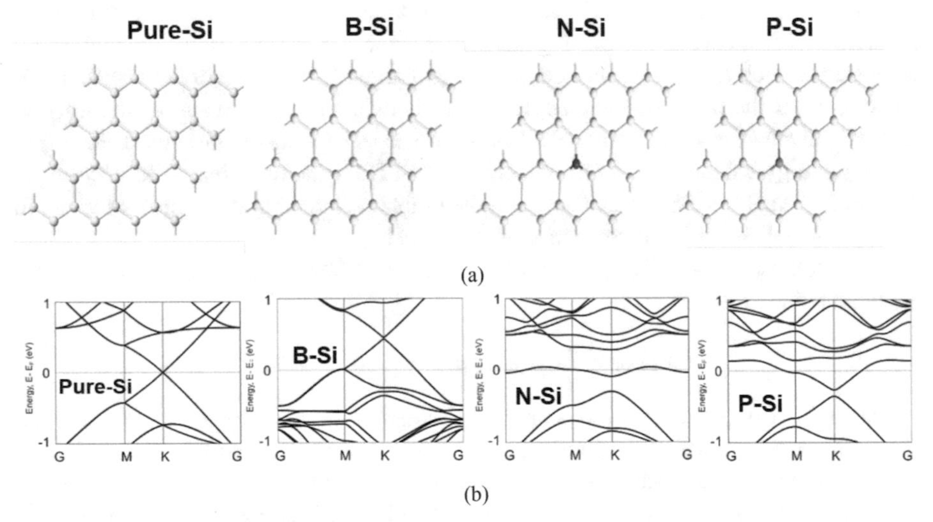

Fig. 1. Plots of: (a) schematic representation of 4 × 4 × 1 supercell, and (b) energy band structure of Pure Si, B-Si, N-Si, P-Si.

The energy band diagram of Si and its homogeneously doped counterparts are shown in Fig. 1 (b). We can see that the single doping of boron (trivalent) and nitrogen/phosphorous (pentavalent) shifts the Fermi level in the valance band and conduction band respectively, along with some degenerated band-gap opening in the range of 0.02 eV (B-doping) to 0.210 eV (N-doping). Furthermore, the doping is also introducing non-linearity in the E-K structure at the point where CBM and VBM meet at the Fermi level. Therefore, the work successfully is able to modulate the electrical and structural properties of pure Si lattice.

Next, the SO adsorption is investigated in multiple positions and orientations on the lattice, and the most stable configurations (highest SO adsorption energy) of gas-host systems are considered for further analysis, which is depicted in Fig. 2 (a). The results suggest SO gas adsorption affects the overall periodicity of the host crystal by introducing local distortions in the lattice atoms at the adsorption site to displace in the out-of-plane direction w.r.t the host lattice. It is interesting to note that in the doped Si, the SO molecule is preferentially adsorbed on the Silicon atoms situating near the dopant atoms (B, N, P), which suggests the dopants are enhancing the binding energy of SO adsorption, as shown in Table 1. This trend can be appreciated from the fact that a higher electron deficiency in the adsorption site (Silicon atom) is beneficial for the adsorption of electron-rich molecules like SO. Since the electronegativity of boron and silicon are very similar thus the binding energy of B-doped Si crystal is comparable to the pristine Si host lattice. In contrast, the relatively highest electronegativity of the N atom is facilitating the electron depletion on the neighboring Silicon atoms, resulting in significantly larger adsorption energy compared to pristine as well as B or P doped Si. In this context, it is worth mentioning that the binding energies greater than −1.0 eV usually denote chemisorption and a negative sign in charge transfer signifies acceptor-type molecular doping due to SO adsorption.

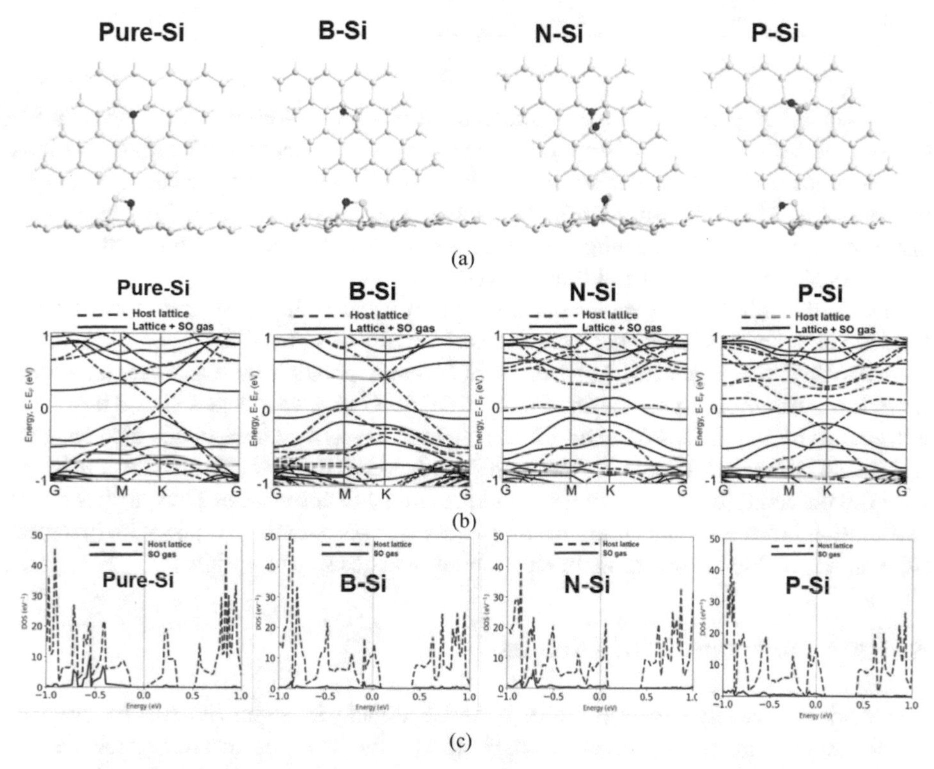

Fig. 2. Plots of: (a) schematic representation of 4 × 4 × 1 supercell after SO adsorption, (b) energy band structure before and after SO adsorption, and (c) contributions of lattice and adsorbed SO in DOS after SO adsorption of Pure Si, B-Si, N-Si, P-Si.

On the other hand, Table 1 indicates that the charge transfers between the silicon atom and SO molecule is highest in the case of Pure Si lattice, and is minimum in the case of B doped host lattice. In contrast, a comparable charge transfer with respect to pristine Si can be observed for the N and P doped lattices. This trend suggests the relative position of the conduction band and valance band edges with respect to the Fermi-level plays a significant role in charge transfer between lattice electronic states and molecular orbitals. Next, the effect of SO adsorption on the electronic properties is assessed from the energy band (E-k) structures before and after SO adsorption and is depicted in Fig. 2 (b), and the subsequent values are also tabulated in Table 1.

Table 1. Properties of Doped Si Lattice in Presence of Gas Molecule.

	Binding Energy (eV)	Gas/Lattice Dis-tance (Å)	Electron (Mulliken Charge) Transfer from SO	Band-gap (eV) and Its Modu-lation after SO Adsorption
Pristine Si	−2.46	1.693	−0.506	0.35 ($\Delta = 0.33$)
B-Si	−2.56	1.558	−0.334	0.27 ($\Delta = 0.06$)
N-Si	−3.71	1.592	−0.469	0.34 ($\Delta = 0.13$)
P-Si	−3.40	1.532	−0.451	0.29 ($\Delta = 0.20$)

The qualitative nature of the E-k structure of pristine as well as doped Si is significantly modulating after SO adsorption, which suggests a strong atomic orbital overlaps between the host lattices and adsorbed molecule. At the same time, the energy bandgap is increasing with SO adsorption in each case, where the highest modulation can be observed in pristine Si. In contrast, the highest and smallest bandgap modulations can be observed in the B-doped and P-doped systems.

Next, the influence of molecular adsorption on the electronic transport in pristine/doped Si has been analyzed from the total density of states (TDOS) profiles, and is shown in Fig. 2(c). In general, the adsorbed SO tends to populate the states near the Fermi level. Interestingly, a larger contribution from SO is observed for doped Si, and the effect is found to be the most prominent for the N-doped system. These trends also suggest significant atomic orbital interactions with the adsorbed SO molecule and are consistent with the observed modulation in E-k structure after SO adsorption. Thus, a significant modulation in the electronic transport properties is expected after SO adoption in doped Si, which is highly encouraging for electrochemical gas sensor design.

4 Summary and Future Scopes

The work extensively explored the B, N and P doping in Si for SO adsorption. The results demonstrate that the substitutional doping with these non-metallic species introduces a finite bandgap in Si without significantly altering the energy band profiles. The SO adsorption acts as acceptor type molecular doping in pristine as well as doped Si. The relatively higher electronegativity in the investigated non-metallic doping species creates electron deficiency in the neighboring Silicon atoms near the adsorption site, subsequently enhancing the overall molecular adsorption. The N doping is appearing to be the most beneficial for enhancing the SO adsorption in Si. In essence, the research provides a detailed theoretical understanding of the SO adsorption and the general transduction mechanism for molecular adsorption in presence of non-metallic substitutional doping in Si, which could be useful for synthesizing and fabricating high-performance electrochemical gas sensors.

Acknowledgments. The work is supported by the Start-up Research Grant (SRG) by DST-SERB (Grant No. SRG/2020/000547) awarded to Sayan Kanungo.

Conflict of Interests. There are no conflicts to declare.

References

1. Bag, A., Lee, N.E.: Gas sensing with heterostructures based on two-dimensional nanostructured materials: a review. J. Mater. Chem. C **7**, 13367–13383 (2019)
2. Yang, S., Jiang, C., Wei, S.: Gas sensing in 2D materials. Appl. Phys. Rev. **4**, 021304 (2017)
3. Basu, S., Chatterjee, S., Saha, M., Bandyopadhay, S., Mistry, K.K., Sengupta, K.: Study of electrical characteristics of porous alumina sensors for detection of low moisture in gases. Sens. Actuators, B Chem. **79**, 182–186 (2001)
4. Gao, H., Liu, Z.: DFT study of NO adsorption on pristine graphene. RSC Adv. **7**, 13082 (2017)
5. Dai, J., Yuan, J., Giannozzi, P.: Gas adsorption on Graphene doped with B, N, Al, and S: A theoretical study. Appl. Phys. Lett. **95**, 232105 (2009)
6. Zhou, M., Lu, Y.H., Cai, Y.Q., Zhang, C., Feng, Y.P.: Adsorption of gas molecules on transition metal embedded graphene: a search for high-performance graphene-based catalysts and gas sensors. Nanotechnology **22**, 385502 (2011)
7. Cahangirov, S., Topsakal, M., Akturk, E., Sahin, H., Ciraci, S.: Two- and one-dimensional honeycomb structures of silicon and germanium. Phys. Rev. Lett. **102**, 236804 (2009)
8. Gao, J., Zhao, J.: Initial geometries, interaction mechanism and high stability of silicene on Ag(111) surface. Sci. Rep. **2**, 1–8 (2012)
9. Xia, F., Wang, H., Xiao, D., Dubey, M., Ramasubramaniam, A.: Two-dimensional material nanophotonics. Nat. Photonics **8**, 899–907 (2014)
10. Esrafili, M.D.: Boron and nitrogen co-doped graphene nanosheets for NO and NO2 gas sensing. Phys. Lett. A **383**, 1607–1614 (2019)
11. Tiwari, A., Palepu, J., Choudhury, A., Bhattacharya, S., Kanungo, S.: Theoretical analysis of the NH3, NO, and NO2 adsorption on boron-nitrogen and boron-phosphorous co-doped monolayer graphene – A comparative study. FlatChem **34**, 100392 (2022)
12. Hernández Cocoletzi, H., Castellanos Águila, J.E.: DFT studies on the Al, B, and P doping of silicene. Superlattices Microstruct. **114**, 242–250 (2018)
13. Tiwari, A., Bahadursha, N., Palepu, J., Chakraborty, S., Kanungo, S.: Comparative analysis of Boron, nitrogen, and phosphorous doping in monolayer of semi-metallic Xenes (Graphene, Silicene, and Germanene)-A first principle calculation based approach. Mater. Sci. Semicond. Process. **153**, 107121 (2023)
14. QuantumATK version Q-2020.12: Synopsys QuantumATK (2020). https://quantumwise.com

Micro Sized Interdigital Capacitor for Gases Detection Based on Graphene Oxide Coating

Ignacio Vitoria[1,2(✉)], Dayron Armas[1], Carlos Coronel[1], Manuel Algarra[1,3], Carlos Ruiz Zamarreño[1,2], Ignacio R. Matias[1,2], and Subhas C. Mukhopadhyay[4]

[1] Public University of Navarre, 31006 Pamplona, NA, Spain
ignacio.vitoria@unavarra.es

[2] Institute of Smart Cities, Jeronimo de Ayanz Building, 31006 Pamplona, NA, Spain

[3] INAMAT2 – Institute for Advanced Materials and Mathematics, 31006 Pamplona, Spain

[4] School of Engineering, Macquarie University, Sydney, NSW 2109, Australia

Abstract. A micro sized interdigital capacitor sensible to CO_2 and NO is studied in this work. The photolithography technique enables to obtain fingers with dimensions of 10 × 500 μm and separated 7 μm between them. The deposition of a film composed of graphene oxide particles as the dielectrics of the capacitor allows to measure the gas concentration of CO_2 and NO mixed with N_2. The sensors were characterized in a gas chamber with a constant flow, obtaining promising results in changes of capacitance at 100 Hz. The sensors have a good linearity and sensitivity with a $R^2 = 0.996$ and $5.026 \cdot 10^{-1}$ pF/% v/v for CO_2 and $R^2 = 0.972$ and $1.433 \cdot 10^{-1}$ pF/ppb for NO.

Keywords: Interdigital · Gas Sensor · Graphene Oxide · Photolithography

1 Introduction

The increase in popularity of trends such as internet of things and smart cities, has open up a path of new types of sensors [1, 2]. Among them, the interdigital capacitor sensors are one of the most employed due to its advantageous characteristics such as small size, robustness and easy readout circuit [3]. Furthermore, the photolithography technique enables the miniaturization of the interdigital sensors, improving the performance of them [4] with smaller size and higher capacitance. Another advantage of this type of sensors is the incorporation of a great variety of dielectrics to the capacitor. These dielectrics are chosen for its sensitive properties to different factors such as temperature, humidity, presence of contaminants or gas detection, enabling the fabrication of sensors that cover a wide variety of fields [5]. Particularly, gas sensors play an important role in our daily live, employed in applications such as indoor air quality monitorization, city contamination, diagnosing diseases or controlling production processes among others [6].

Graphene oxide, (GO) has been proposed as the dielectric of the interdigital sensor. Nanomaterials based in graphenoids structures, with a single atomic layer of sp^2 hybridized carbon atoms offer a unique 2-D structure, thermal, mechanical, optical, and

N. K. Suryadevara et al. (Eds.): ICST 2022, LNEE 1035, pp. 24–30, 2023.
https://doi.org/10.1007/978-3-031-29871-4_4

excellent electronic properties [7]. GO creates an intrinsic dipole moment, is displaced by lattice vibrations and creates very high permittivity [8]. It is well known that the homogeneous dispersion of GO can satisfactorily increase the dielectric permittivity [9].

In this work, the fabrication of interdigital sensors by the photolithography technique is explained. The incorporation of a film with a thin film of GO particles allows using the interdigital capacitor as a gas sensor, producing changes in the capacitance at different gas concentrations. The performance of the sensor to CO_2 and NO gases is studied since previous GO sensors have obtained a good response [10, 11]. The results show a novel micro sized interdigital sensor with an extremely high sensitivity to NO at room temperature.

2 Materials and Methods

2.1 Interdigital Capacitor Fabrication

The photolithography technique is employed in the fabrication of the interdigital capacitors. This technique enables the deposition of a metallic thin film in top of a substrate with a desired pattern design. In this case, a polypropylene substrate with a thickness of 250 μm has chosen due to its flexible property, but other substrates such as glass, silicon or polymers could be used.

The fabrication of the capacitor is composed of various steps that are summarized in Fig. 1. First, the surface's substrate is cleaned with propanol and Kimtech™ wipes to remove dust and particles from it (Fig. 1.1). Cleanness is an important factor during all the fabrication process, since undesired particles can affect negatively in the results, causing short-circuits or open-circuits in the future electrodes. Therefore, the fabrication of the devices was conducted in a clean room.

On top of the cleaned substrate, an adhesive is deposited to obtain a good adherence for the next film. The Ti-Prime adhesive from the company MicroChemical GmbH was employed using the spin coating technique. In this technique, the centrifugal forces produced during the spinning process spread uniformly the solution forming a thin film. Few drops of the adhesive were placed to cover the substrate with the help of a pipette. The substrate was spined at 2,000 rpm during 30 s in the WS-650SZ-6NPP/Lite equipment from Laurell. A curing process was needed to fix the film into the surface, so the device was placed on a hot plate at 60 °C for 60 s.

In the next step, a film is deposited with the photoresist resin using the spin coating technique. The employed resin was AZ® MIR 701 14 CP from Merck at 100% concentration. The spinning process was first 10 s at 300 rpm and later 40 s at 2,500 rpm producing a yellowish color film (see Fig. 1.2). Also, the same curing process is repeated as it was done in the previous film. The resin is sensitive to UV light, so during the fabrication process the room was illuminated with a red light.

Once the resin is deposited, the device is ready for the engraving of the desired design using the PicoMaster 100-4PICO Litho BV equipment (Fig. 1.3). The configuration parameters for the process are a 900 nm resolution, an exposure energy of 250 mJ/cm^2 and a red power laser of 220 μW. After the engraving, the device is subjected again to the curing process to fix and stabilize the film. In this part of the process the resin parts that were attacked by the laser lose the yellowish color and become transparent, exposing the engraving design to the naked eye.

In the next step, the transparent parts of the resin are removed from the device (Fig. 1.4). A solution is prepared combining the AZ 400K developer with deionize water in a 1:5 ratio and mixing in a magnetic stirrer for 1 h. The device is immersed in the solution for 150 s, cleaned with deionize water and dried with compressed air. The immersed time of the device in the solution is a critical parameter since the attacked resin will not totally be removed with less time and with more time, the no-attacked resin will start to disintegrate. Again, the curing procedure is repeated (60° for 60 s).

Another film is deposited to build the electrodes, see Fig. 1.5. In this work silver was chosen as the material due its electrical characteristics and good adhesion to different substrates, although other metals could have been employed. The sputtering DC equipment K675XD from Quorum technologies was used for the deposition of the film with a thickness between 200 and 400 nm. The parameters of the procedure were 150 s at 35 mA and $7{\cdot}10^{-3}$ bar.

After the metallic deposition, the resin layer is removed with the stripping process, see Fig. 1.6. The device is immersed in the solutions AZ100 from the company Merck for 10 min. Later the device is cleaned in soapy water for 2 min, and again in ultrapure water for 2 min. During the three immersions, the fluids and device are agitated in an ultrasonic bath. Later the device is cleaned with propanol and dried with a compressed air gun.

In the next step, two wires are soldered on each electrode of the device (Fig. 1.7). The conductive epoxy adhesive CW2400 from Chemtronics is employed. A curing process of 50 °C for 10 min is needed to harden the adhesive. The capacitor is checked with a multimeter searching for short circuits. The presence of traces of silver that have remained between the electrodes can be removed connecting the capacitor to a high-power supply. The high currents passing through the defects will evaporate them. Also, the device was observed trough an optical microscope DM 2500 M from Leica to detect visual mistakes in the fabrication process.

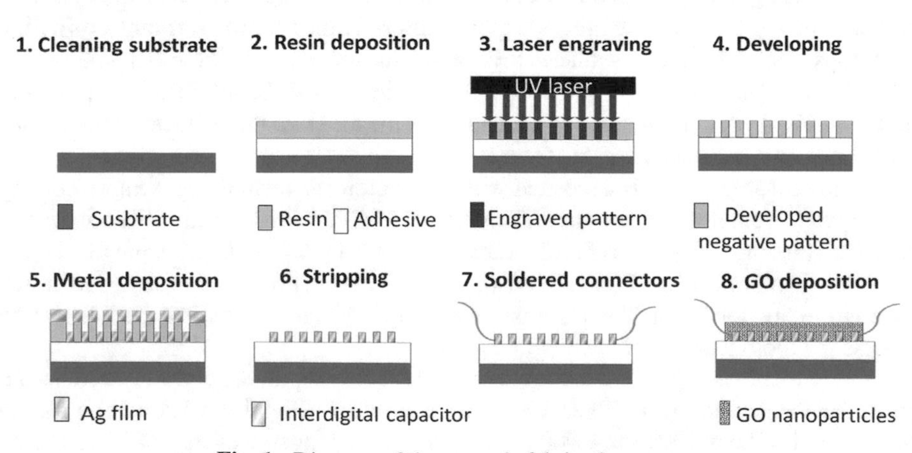

Fig. 1. Diagram of the sensor's fabrication process

Finally, the GO layer is added. Graphene oxide (GO, 4–10% edge-oxidized) was purchased from Sigma-Aldrich (Barcelona, Spain) and used without further purification. GO (3.8 mg) was suspended in deionized H_2O (1 ml), filtered through a 200 μm filter to homogenize the solution. 500 μL of this solution was deposited on the interdigital capacitor (Fig. 1.8) to form a film of GO nanoparticles.

2.2 Experimental Setup

Figure 2 shows the setup employed for the characterization of the sensor response to gases. The sensor is placed inside an ad-hoc gas chamber made in stainless steel and with an inner dimension of 14 × 2 × 3 cm. It has two tubes for the gas input and output that enable to maintain a constant flow through the chamber. A sealing gasket closes the chamber and prevent gases from scaping.

A gas panel is used to obtain the desired composition of gases. Various mass flow controllers of different gases and gas mixers lead to a single output connected to the gas chamber. The different flows of each controller enable to obtain different concentrations of gases.

The capacitance of the sensor is measured with the impedance analyzer E4990A from KeySight. The studied frequency range is between 10 and 100,000 Hz. The sensor is connected to the impedance analyzer using pass-trough cables into the chamber.

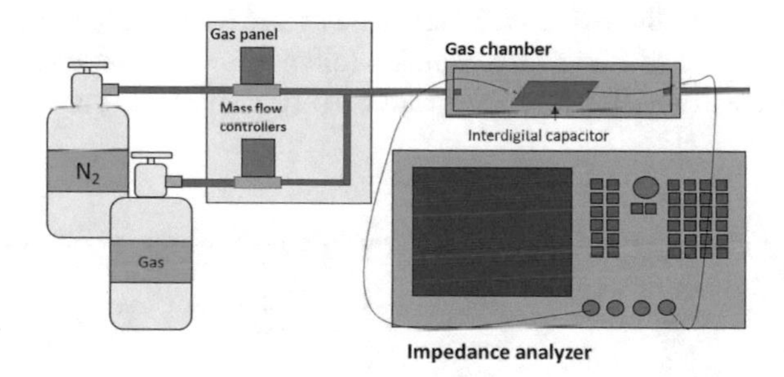

Fig. 2. Experimental setup for the characterization of the sensors to different gases

3 Results

The interdigital capacitor has been designed with a total dimension of 27 × 15 mm, see Fig. 3. It is composed by 560 interdigital blocks connected in parallel to increase the capacity. A gap of 100 μm was incorporated between the blocks to facilitate the removal of the deposited metal in the fabrication process. Each interdigital block has 16 pairs of fingers (making a total of 8,960 pairs) whose dimensions are 10x500 μm and are separated 7 μm between each other. Figure 3c shows the clear gap between the metallic fingers that prevents any short-circuits in the capacitor.

Once the interdigital capacitor sensor is coated with the sensitive GO it is characterized in the gas chamber trough different gas concentrations. Two gases were measured

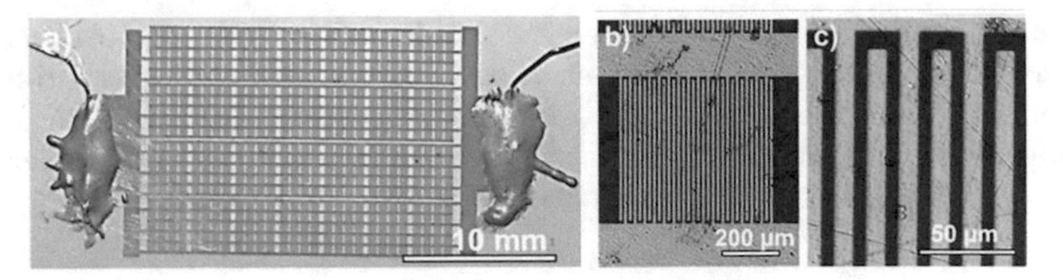

Fig. 3. a) Image of the sensor, b) microscope image of an interdigital block, c) microscope image of the fingers [12].

for this work, carbon dioxide, CO_2, and nitric oxide, NO. During the measurements, the flow of the mix of gases was keep constant at 0.2 L/min with nitrogen, N_2, as the gas carrier.

The different concentrations of CO_2 were done with different percentage between 0–100% of CO_2 in a mix of N_2 and CO_2. Figure 4a shows the results of the capacitance at different concentration and frequencies. It can be seen that the capacitance of the sensor is linear to the gas concentration with R^2 higher than 0.99. Also, it can be noted that at lower frequencies the sensitivities are higher with a $5.026 \cdot 10^{-1}$ pF/%CO_2 at 100 Hz. This behavior can be appreciated in Fig. 4b where the capacitance response at two different concentrations (5% and 100% of CO_2) is displayed. The shapes of the curves are similar to an exponential curve with higher capacitances at lower frequencies. Also, the difference between the two concentrations is more accentuated at low frequencies with small differences at higher frequencies (although is not represented in Fig. 4a, the sensitivities are $1.94 \cdot 10^{-2}$, $8.55 \cdot 10^{-3}$ and $6.16 \cdot 10^{-3}$ pF/%CO_2 at 1,000, 10,000 and 100,000 Hz respectively).

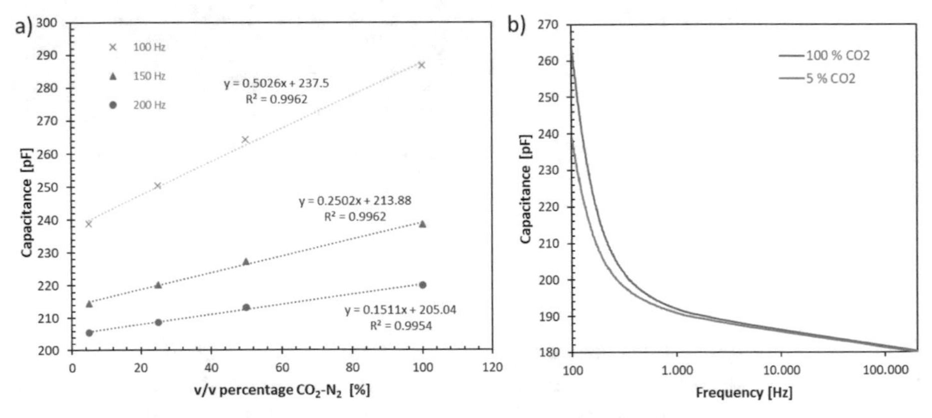

Fig. 4. a) Capacitance of the sensor at different frequencies and gas concentrations of CO_2 with N_2 as carrier gas, b) Capacitance of the sensor at different frequencies at two concentrations of CO_2.

The response to NO gases has also been characterized. In this case, due to the high toxicity of the gas, it is used at low concentrations (parts per billion in weight, ppb w/w) in a mix with N_2. The concentrations were modified changing the flow of N_2 and the mix of NO and N_2. In this case, the response to the NO was similar to the previous gas, linear to the concentrations (see Fig. 5a) and with an exponential shape of the curves (see Fig. 5b). The main difference between them was the higher response of the sensor ($1.433 \cdot 10^{-1}$ pF/ppb NO at 100 Hz) since the concentration of NO were seven orders of magnitude smaller than CO_2 and have changes in the capacitance of the same order. Although is not displayed in Fig. 5a, the sensitivities of the sensor at high frequencies were $1.25 \cdot 10^{-2}$, $8.09 \cdot 10^{-3}$ and $6.18 \cdot 10^{-3}$ pF/ppb of NO at 1, 10 and 100 kHz respectively.

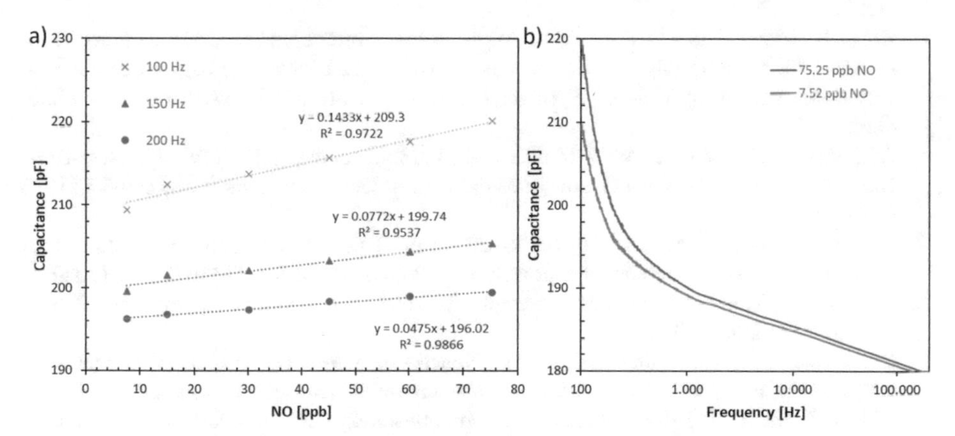

Fig. 5. a) Capacitance of the sensor at different frequencies and gas concentrations of NO with N_2 as carrier gas, b) Capacitance of the sensor at different frequencies at two concentrations of NO.

4 Conclusions

A new micro sized interdigital capacitor has been successfully fabricated. Fingers with dimensions of 10×500 μm and a separation between them of 7 μm has been obtained by the photolithography technique. The reduction of the dimension enables to obtain a great capacitance (approx. 200 pF at 100 Hz) in a small area. The dielectric composed of GO particles allows to use the device as a gas sensor for CO_2 and NO. A good linearity and sensitivity have been found in all cases. The sensitivities at 100 Hz were $5.026 \cdot 10^{-1}$ pF/%CO_2 and $1.433 \cdot 10^{-1}$ pF/ppb of NO.

Improvements in the interdigital design could be implemented in future works to obtain greater capacities. Also, other gases could be studied to obtain the cross sensitivity of the sensor as well as greater sensitivities.

This study opens the door to new micro sized gas capacitor sensors that can be employed in real applications in fields like the internet of things and smart cities.

Acknowledgement. This research was funded Agencia Estatal de Investigación (PID2019-106231RB-I00 and PID2021-122613OB-100) the Ministry of Science, Innovation and Universities of Spain (PRE2020-091797 and PEJ2018-002958-P) and Institute of Smart Cities PhD Student grants.

References

1. Ghobakhloo, M.: Industry 4.0, digitization, and opportunities for sustainability. J. Clean. Prod. **252**, 119869 (2020)
2. Bacco, M., Delmastro, F., Ferro, E., Gotta, A.: Environmental monitoring for smart cities. IEEE Sens. J. **17**, 7767–7774 (2017)
3. Smith, R. L.: "Interdigital Sensors and Transducers," Eng. Handbook, vol. 92, (Second edn.), pp. 151-1–151-11 (2004)
4. Liu, N., Gao, Y.: Recent progress in micro-supercapacitors with in-plane interdigital electrode architecture. Small **13**, 1–10 (2017)
5. Roy, J.K., Mukhopadhyay, S.C.: Some Applications of Interdigital Sensor for Future Technologies. In: Mukhopadhyay, S.C., George, B., Roy, J.K., Islam, T. (eds.) Interdigital Sensors. SSMI, vol. 36, pp. 383–407. Springer, Cham (2021). https://doi.org/10.1007/978-3-030-62684-6_16
6. Afsarimanesh, N., Nag, A., Alahi, M.E.E., Han, T., Mukhopadhyay, S.C.: Interdigital sensors: Biomedical, environmental and industrial applications. Sens. Actuators A Phys. **305**, 111923 (2020)
7. Algarra, M., et al.: Insights into the formation of N doped 3D-graphene quantum dots. Spectroscopic and computational approach. J. Colloid Interface Sci. **561**, 678–686 (2020)
8. Ramanathan, T., et al.: Functionalized graphene sheets for polymer nanocomposites. Nature Nanotech. **3**(6), 327–331 (2008)
9. Romasanta, L.J., Hernández, M., López-Manchado, M.A., Verdejo, R.: Functionalised graphene sheets as effective high dielectric constant fillers. Nanoscale Res. Lett. **6**, 1–6 (2011)
10. Akhter, F., Hasan, S., Eshrat, M., Alahi, E., Mukhopadhyay, S.C.: Temperature and humidity compensated graphene oxide (GO) coated interdigital sensor for carbon dioxide (CO2) gas sensing. In: Mukhopadhyay, S.C., George, B., Kumar Roy, J., Islam, T. (eds.) Interdigital Sensors: Progress over the Last Two Decades, pp. 329–349. Springer International Publishing, Cham (2021). https://doi.org/10.1007/978-3-030-62684-6_13
11. Fowler, J.D., Allen, M.J., Tung, V.C., Yang, Y., Kaner, R.B., Weiller, B.H.: Practical chemical sensors from chemically derived graphene. ACS Nano **3**, 301–306 (2009)
12. Vitoria, I., Armas, D., Coronel, C., Ozcariz, A., Zamarreño, C.R, Matias I.R.: Micro sized interdigital capacitor for humidity detection based on agarose coating. In: Smart Cities: Future Trends and Challenges (2021)

AE-Sleep: An Adaptive Enhancement Sleep Quality System Utilizing Data Mining and Adaptive Model

Nguyen Thi Phuoc Van(✉), Dao Minh Son, and Koji Zettsu

Big Data Integration Research Center, NICT, Tokyo, Japan
{phuocvan,dao,zettsu}@nict.go.jp, van.nguyen@usq.edu.au

Abstract. A data-driven approach is a new trend in healthcare. The development of wearable and non-wearable sensors makes it easy to collect and study physical data from people to improve their health. Sleep quality is one of the most critical parameters that significantly impact a person's health. Many researchers focused on building models to predict sleep attributes or classify sleep stages. They just used data from small groups to solve prediction problems. A few research concentrates on investigating adaptive model which can reuse the knowledge from public data to enhance the sleep quality prediction of an individual. Moreover, very few studies find the correlation metrics between features (activities, environmental information, behaviours and so on) with sleep attributes to extract hints to help people to improve their sleep quality. Motivated by these remarks, in this paper, we proposed an adaptive enhancement sleep quality system which includes (1) visualised tools to extract hints to improve sleep quality; (2) an adaptive sleep efficiency prediction model which can learn from both personal and public data. In addition, in this study, the biological data, such as activities, emotions, and environmental parameters, are considered and analyzed. The data is examined from a personal level to a group level. Based on this analysis, different approaches are considered to build a predictive model for sleep efficiency. Our proposed framework shows promising results when examined on public and private data sets.

Keyword: Sleeping efficiency; personalized prediction model, stacking generation, wearable sensors, adaptive modelling

1 Introduction

There are a lot of problems including the high level of sugar blood, concentrating/ working performance reduction [1–3] which caused by poor sleep. Daily activities have a positive effect on the sleep quality of humans [4,5]. Reference [4] showed that the good sleepers group gets involved more in physical activities

Centre for Health Research, University of Southern Queensland, Toowoomba, Australia

N. K. Suryadevara et al. (Eds.): ICST 2022, LNEE 1035, pp. 31–47, 2023.
https://doi.org/10.1007/978-3-031-29871-4_5

(PA) and enhances their social engagements. The positive correlations between activities and sleep quality are discussed in many works [6–10]. Feifei Wang and Szilvia Boros [9] provided a systematic review of the effect of PA on sleep quality for the duration of 8 years (from 2010 to 2018). They looked at 1408 articles and reported that moderate activities had a high impact on sleep quality; physical exercise is suggested as an efficient approach to solve sleeping issues. The increasing number of steps in a week might improve the sleep quality of menopausal women and secondary school students [8,11]. The work in [10] reviewed 40 studies on the effect of activities on the sleep quality of older adults in the 60+ age group. This paper reveals that exercise categories positively influence different sleeping attributes. The sleeping outcomes showed a significant enhancement when older people participated in moderate-intensity exercise thrice per week from 12 weeks to 6 months.

Besides the effect of activities on sleep quality, the environment also has a significant impact on the sleep quality of people. Zhongping *etal.* developed the thermal comfort model sleeping environment. The surrounding environment in the rural area of China has an association with the sleep quality of residents[12]. This research points out that the high level of residential greenness was associated with better sleep quality; this association could be modified if people were exposed to air pollution. Air pollution is a severe issue in many countries and negatively affects human health. The article [13] studies the association between sleep duration of 16, 889 students and air pollution in the duration of five years-from 2013 to 2018. The analysis data result of this study report that the reduction of sleep duration of students in Beijing, China, had been associated with air pollution.

The association between sleeping quality activities and the environment is quite obvious. Reference [14] presented the systematic review of models to predict sleeping quality from biological data and environmental information. This literature gave different prediction models for sleeping quality, from the time-series approach to machine learning. This article also suggested that there is still a research gap in finding the optimised model to predict sleeping quality/efficiency.

In consideration of current trends of sleep quality prediction research, we still see the following research gaps in sleep quality enhancement studies:

- There are a lot of apps [15–17] to give sleep tracking cycle, sleep states, social sharing, snoring issues, and waking up suggestions to users. but very less research pays attention to building a favourable and friendly tool that can show the important factors that affected sleep quality or give instructions/ suggestions to users.
- Even though, many studies work on sleep quality prediction. However, to the knowledge of authors, no one had come up with the adaptive personal model which is created from personal and public data to utilized public data to boost the personal model.

- Although several studies had demonstrated the association between environment parameters i.e air qualities or green level of surroundings [12,13], very less works introduce friendly visualized associations between different factors to sleep attributes. This tool is very important to convince users to follow the suggestion of the system to enhance their sleep.

To address the above challenges, we proposed a novel adaptive enhancement sleep quality system named AE-Sleep in this work. This system brings a promising solution to enhance people's sleep quality and can be extended to other healthcare enhancement tools. To demonstrate the performance of the proposed framework, we will examine two data sets to find the correlation between activities, environmental parameters, and sleeping attributes. From analysing results, different approaches to form prediction models for sleep efficiency will be compared with our proposed model to illustrate the prominence of our model. The rest of the paper is organised as follows: Sect. 4.1 presents the overview block diagram of the proposed system, and the following Section discusses Related works. In this section, the sleep attributes and sleep prediction models are considered carefully. Section 4 describes two data sets used in this work. Moreover, the data analysis results for both data sets are also reported in this Section. The following section concerns the sleep efficiency prediction result of our proposed framework on two data sets. The last section is the conclusion and future work.

2 Related Work

2.1 Sleep Monitoring

The Pittsburgh Sleep Quality Index (PSQI) is the popular standard to measure sleep quality. In the $PSQI$ assessment, a person is requested to answer 19 self-rated questions. Results of answers are used to form 7 scores, including sleep quality, sleep latency, sleep duration, sleep efficiency, sleep disturbances, use of sleep medication, and daytime dysfunction. Seven cores are added to create the global $PSQI$ score. This type of data is easy to collect large data but is quite subjective.

Polysomnography (PSG) is the global standard in sleep medicine. In this method, brainwaves, heartbeats, respiratory, oxygen levels, and eye movement are measured by different wearable sensors. Based on those bio-signals, specialists will label stages of sleep at every minute or 30 second as "Deep", "REM", "Light", "Wake", and diagnose obstructive sleep apnea syndrome ($OSAS$) and other sleep disorders. Since the PSG method needs a lot of special sensors/equipment, it is only used in sleep laboratories and requires a specialist to label data. For some people, the measured sleep quality in the lab does not reflect their actual sleep pattern since they are pretty sensitive to the new sleeping environment, and their sleep is disturbed by attached sensors to their bodies.

Actigraph/ commercial smartwatch is a low-cost, objective method of monitoring sleep behaviour using the integrated $three-axis$ accelerometer sensor, a processor, and a memory. Actigraph wearable device is easy to use in daily life to collect vital signs data such as heartbeat, breathing rate, sleep stages, and sleep scores continuously [18]. Yusuf *etal.* [19] mentioned in their work that there is only a 2% difference in sleep efficiency between actigraphy with proportional integration mode algorithm and PSG exams. Reference [20] presented a combination of acceleration and cardiological measurements to classify sleep/wake stages. The result was then validated by PSG and showed a perfect match.

A contactless sensor is an alternative way of sleep monitoring. Sensing systems may be put on the mattress to measure sleep stage, respiratory signals, and heartbeat [21]. Another way to remotely monitor the sleeping stage and vital signs is to use radio-based sensors[22,23]. The sensor sends radio signals toward the human position and receives the reflected radio signals from the human. The reflected signals are modulated by the movements of a human. Analysing the sent and received signals allows one to get helpful information like breathing rate and body movements. These data will enable us to predict the sleep stages of an individual.

2.2 Sleep Attributes

The different sleep quality assessments lead to various sleep attributes. In this study, we discuss some primary objective sleep attributes which were mentioned in the sleep quality assessment literature [21,24]. Most major attributes are summarized as follows.

- Sleep duration or total sleeping time (TST) is defined as the total time that a person sleep per day.
- Fragmentation is the number of awakenings after initial sleeping time and before the final awakening. This attribute is counted by body movements and out-of-bed time.
- Sleep efficiency is the proportion between TNT and on-bed time.
- Sleep stages include Deep, light, rapid eyes movement (REM), and wake. These stages are use to examine sleep duration, fragmentation, sleep efficiency and sleeping disorders.

2.3 Prediction Models for Sleep Study

Several studies used machine learning to solve the classification of sleep stages and sleep quality. In 2016, Aarti Sathyanarayana *etal.* [25] used an actigraphy wearable sensor to collect data from 92 people in the duration of one week. The physical activity during the awake time was used to predict sleep quality which was evaluated by sleep efficiency (good/not good). They showed that advanced machine learning techniques like convolution neural networks (CNN) achieved the highest accuracy in predicting good or poor sleep. Reference [26] work on the classification problem for two stages of sleep/awake from actigraphy data. They used the public data (MESA), synchronised PSG and actigraphy data.

This study implemented various machine learning/ deep learning algorithms to classify awake and sleep. They discovered that CNN and $LSTM$ produced higher accuracy than other algorithms. Recently, Phat *etal.* [27] applied deep learning to predict sleep quality from physical activities in the awake time. Phat *etal.* used Fitbit smartwatches to collect data from 39 students. Volunteers requested to wear Fitbit for the duration of 106 days. They used activity features - burned calories, steps, distance, sedentary, light/fairy/very active times to feed the model to classify good or not good sleep. Their prediction model can reach 62.2% accuracy. From the above literature, we see the possibility of user activity information during the awake time to predict sleep quality. However, people's sleep quality reacts very differently to the change in activities. Therefore, in [21] Iman Azimi proposed a personalised prediction model to assess maternal sleep quality. The monitoring data of a person is used to predict future outcomes. In this work, Bayesian Replicator Neural Networks were exploited to reduce the poor performance of small data. This adaptive model is utilised to evaluate an abnormal score of sleep attributes. The abnormal sleeping score is calculated by the distance between the actual sleep attribute and the predicted attribute. The result of the research proves that the sleep quality of women during pregnancy time reduces significantly.

3 Problem Statements

This section introduces two problems we want to solve in this paper. The first problem is the essence of our framework, which intends to seek the essential features and eliminate the duplicated features to construct the master prediction model with high correlation and low computational complexity. The second question is to find a model for a client that can reuse the knowledge from public data.

1) Problem 1 (Finding optimized feature set): Given a set of variables $V = \{v_i\}_{i=1...N}$, the Multivariate Linear Correlation Analysis ($MLCA$) function $MLCA(V) \rightarrow \hat{V}$ is conducted to select a subset of variables $\hat{V} = \{\hat{v}_j\}_{j=1...\hat{N}}$ of V, where $\hat{N} \leq N$ and $\hat{v_j} \in \hat{V}$ has $\hat{V}$ has minimum correlation while remaining enough information.

2) Problem 2 (Build adaptive client models for different data sets): Given two sets of data, master database $P_p = \{V_{i=1...p}, L_p\}$ and client database $P_c = \{V_{k=1...m}, L_c\}$, where V is the list of features and L is the list of label. Both data sets have m common features and the same type of label. The question is how to transfer the knowledge from the public data to enhance the prediction model in the private data.The detailed solutions for the above problems are presented in the next section

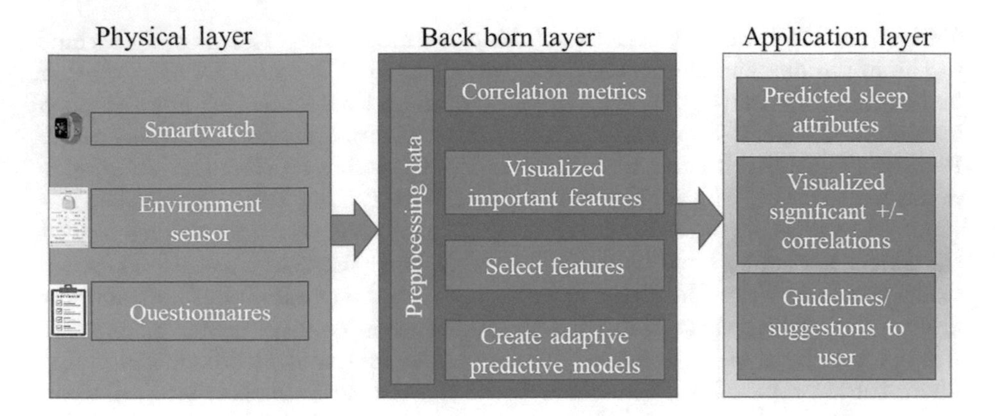

Fig. 1. Block diagram of AE-Sleep system

4 Materials and Methods

4.1 AE-Sleep: System Overview

The block diagram of the proposed system is illustrated in Fig. 1. The system consists of three layers with open containers to make it easy to integrate new blocks into the system. Normally, the physical layer will collect data from various sources to send to the Back born layer via communication links like Bluetooth or Wi-Fi. At this layer, the raw data is cleaned by several functions. After that, the feature section and correlation calculation functions are activated to find important features, and important group features together. Based on important features, the adaptive prediction model is created. The model is forwarded to the next layer to predict sleep efficiency. In addition, important features with a high correlation with sleep attributes are transferred to the application layer for user visualisation and building guidelines.

4.2 Data Sets

In this study, we consider two data set, PMData and private data from an exercise group in Tokyo, Japan. In the PMData, we have life-logging data of 3 for women and 13 for men for five months (from Nov 2019 to March 2020). PMData collected physical activities and sleeping parameters by $Fitbit\ Versa2$. The data also included food pictures, well-being questionnaires, drinking habits, etc. There are around a total of 50 peculiarities in the PMData set. In this research, we highly concentrate on activities and sleeping to see the possibility of using activities to predict sleeping efficiency. Volunteers wore Fitbit $Versa2$ to collect biological data for the whole day for 5 months. Volunteers participated in various exercises, including running, walking, treadmill, outdoor bike, workouts, swimming, dancing, weights, elliptical, yoga, spinning, hockey, skiing, and tennis.

The private data is a data set about a dancing group in Tokyo, Japan. There are 9 participants in the three-month data campaign. Volunteers use a commercial smartwatch- Fitbit sense to collect activities and sleep data, and environmental sensors were installed in the volunteer's bedroom to get information on temperature and air quality.

4.3 Finding Optimized Feature Set

1) PMData: In this section, we highly concentrate on the correlation between activities and sleep efficiency- an important parameter to evaluate the sleep quality of a person. We first see the Pearson correlation between active types at the individual/ group level. We can eliminate a feature from the group level if it has a very high correlation with another one. Assuming that we have two features vector F_a and F_b corresponding with standard deviation δ_a and δ_b respectively. The definition of Pearson's correlation between two features F_a, F_b is as follows: $Cor_{ab} = \frac{\delta_{ab}}{\delta_a \delta_b}$, where δ_{ab} is the covariance between F_a, F_b.

Based on the correlation metrics between features, the near duplicated features will be removed from the feature list. An example of highly correlated features group is illustrated in Figure 2 as ρ parameter, We see that three features - *calo*, *distance*, and *numsteps* have high correlation, specially correlation between *numsteps* and distance is 0.97. We can eliminate one of them from feature list. Moreover, the correlation between features (activities) and sleep attributes is also considered to find important features to solve sleep attribute prediction problems. In this work, we pay attention to the most important sleep attribute - sleep efficiency.

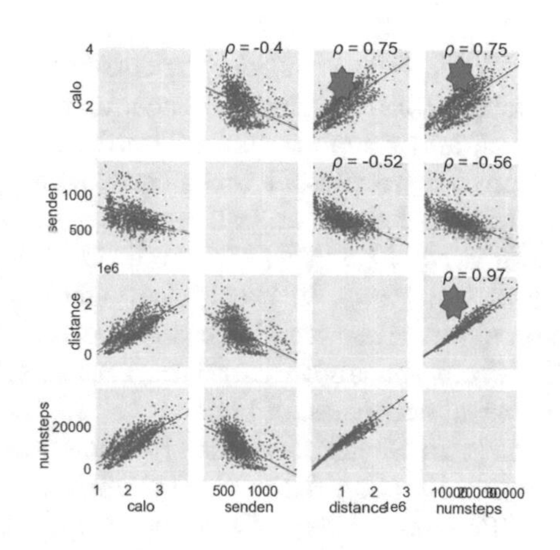

Fig. 2. Illustration of high correlation feature group

Besides doing correlation analysis to find a good feature set, we also compare this kind of analysis between the personal and group levels. From Figure 3, we see from data of 15 people in PMdata that sleep efficiency is highly correlated with light and moderate activities. However, at the personal level, for example, at user1, sleep efficiency does not show a high correlation with moderate activity.

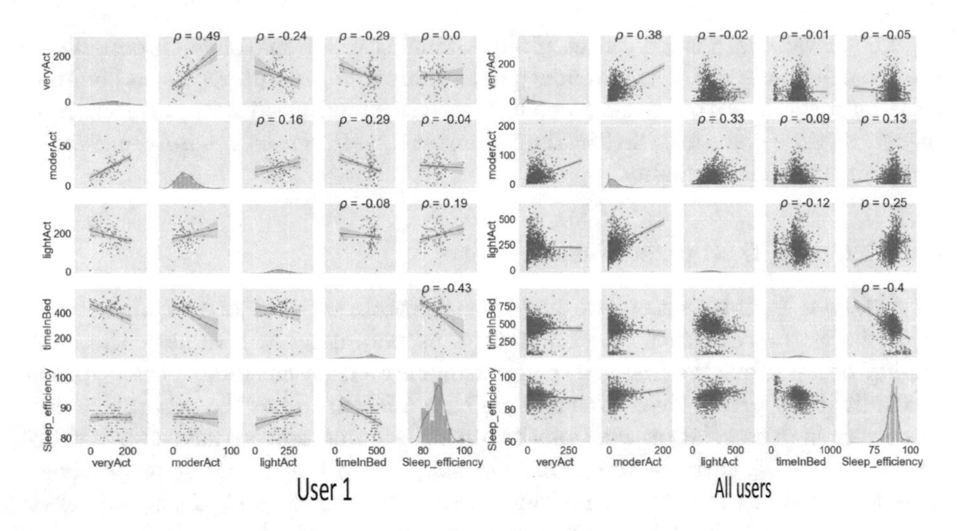

Fig. 3. Correlation between active types and sleep efficiency - one user vs all users

Similarly, we consider the correlation of activities information such as burned calories, total moving distance, number of steps, time of heartbeat in different zones, and sleep attributes (total time in bed, asleep time, sleep efficiency). In Figure 4 we present three correlation tables of user1, user7, and all 15 users. In tables of user7 and all users, we see that sleep efficiency is highly correlated with time of heartbeat in zone1. On the other hand, the data of user 1 does not show this trend. In user1, sleep efficiency correlates to distance and number of steps, but these trends do not happen in user1.

From the above correlation metrics, we can see those correlations between physical activities and choose the high correlation feature group with sleep efficiency. In Sect. 2, we already know the feasibility of using daily physiological information to predict sleep efficiency.

The second step of finding an optimised feature list is to apply an algorithm to calculate the feature score. We consider all PMdata ($N = 1634$) to find the essential features to predict sleep efficiency. We then apply the Random Forest algorithm to see the important feature score of the dataset. Figure 5 presents the feature importance score. Eight important features occupied 84% of total scores.

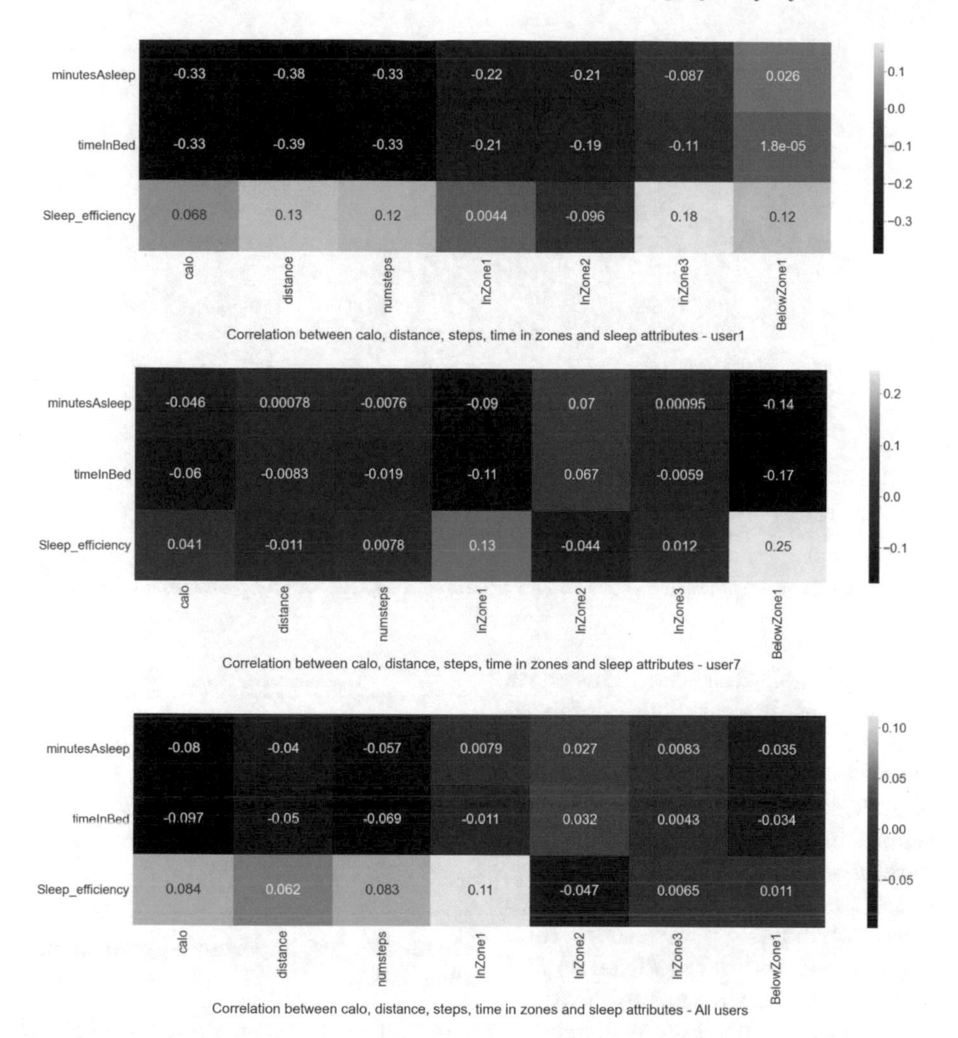

Fig. 4. Correlation between burned calories, distance, number steps, time in different heartbeat zones and sleep efficiency, time in bed, minutes of asleep - one user vs all users

When it comes to personal data, for example, user 1, the importance feature list is different from the group data since the sleep efficiency of people reacts differently with the changing of activities. Figure 6 compare importance feature scores in group of data and user1 data. We see the similar scores of two data sets in *numsteps*, *InZone*1, *Senden*, *veryAct*, *lightAct*, and *InZone*2. The feature time of heartbeat below zone 1 (*BelowZone*1) occupies a high score in the group data but shows a very less score in the user1 data. The *moderAct* (total time of moderate activities) feature is scored just under 20% in the user1 data while this value is only 8% in the group data set.

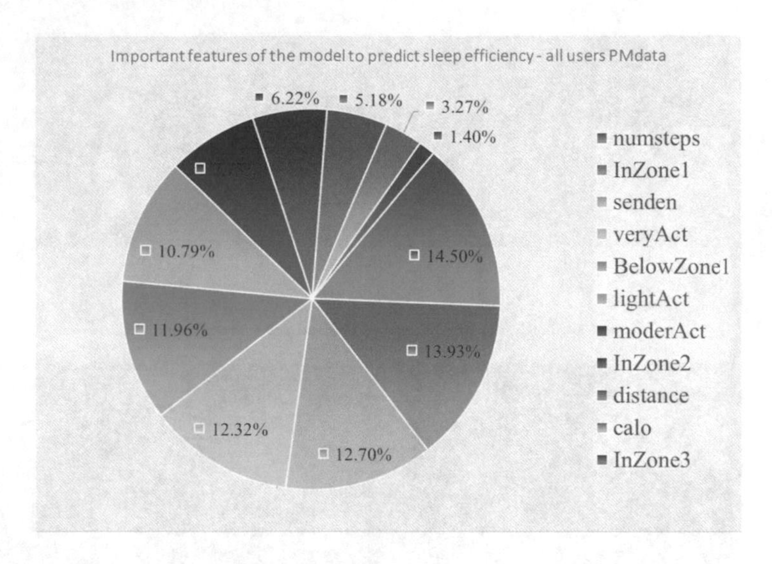

Fig. 5. Feature importance score in the prediction model for sleep efficiency (PMdata)

2) private data set: This data set contains basic activity information of a person from daily life like distance, sedentary time, number of moving steps, and environmental reports in their bedroom such as temperature, carbon dioxide, humidity, etc. The correlation metrics between sleep attributes (asleep time, total time in bed, and sleep efficiency) and activity and environmental features are in Fig. 7. This Figure displays the correlation between data variables at the personal and group levels. Data of User3 shows the high correlation between sleep efficiency and humidity, but this circumstance does not happen in the data of User7 and all user data. This difference means that sleeping attributes of people might react variously to the environment and active levels. In general, the sleeping time (asleep time/ total time in bed) negatively correlates with $CO2$ level and humidity in the bedroom.

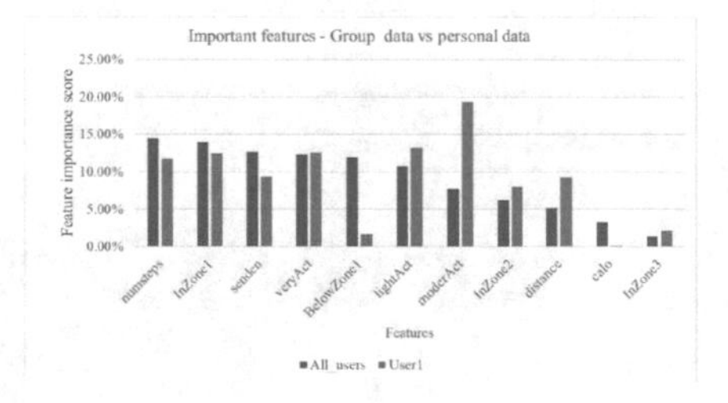

Fig. 6. Feature importance score in personal data

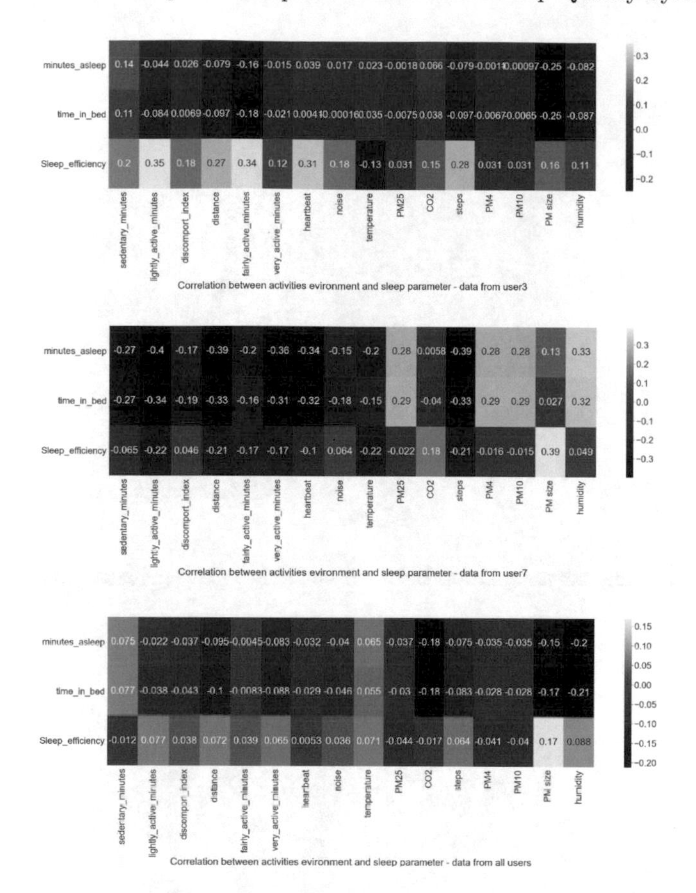

Fig. 7. Correlation between activity,environmental information and sleep efficiency, time in bed, minutes of asleep - one user vs all users

We will consider all the significant correlations between environmental, sleeping, and activity variables in the private data set. From these relationships, we know the factors which can affect sleep attributes. In Figure 8 presents the strong association (i.e. correlation between two variables is larger than 0.1) between features. Deep sleep shows a strong positive relationship with fair/light active levels. On the other hand, we also consider a negative correlation between features. From this analysis (Fig. 9), we can see the negative impacts on sleep attributes. For instance, very active, humidity, and $CO2$ negatively impact deep sleep. In addition, the limitation of correlation value will be set closely to $-1 and 1$ to remove near duplicated features.

In the private data, we again see important features score for the model to predict sleep efficiency. In this data, we have activity and environmental features. The feature importance scores are illustrated in Fig. 10. In the environmental feature group, particulate matter size (PMsize), temperature, humidity, noise, and activities (sedentary, number of steps, fair/light activities) play important

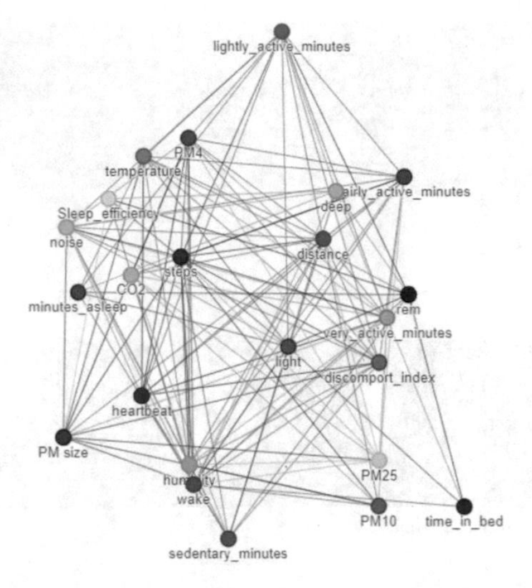

Fig. 8. Significant association between variable in the private group of data

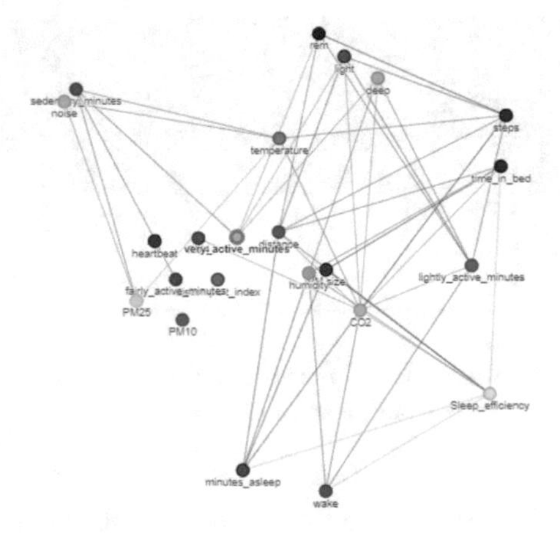

Fig. 9. Significant negative correlation between features

roles in predicting sleep efficiency. By analysing two data sets, we see the possibility of predicting sleep efficiency (SE) from the activities and environmental information of a person. When it comes to prediction models, there are several approaches to predict SE value. In Sect. 2.3, we see that the monitoring data of each person can be used to predict the future values for them-self. This kind of model is called a personalized model. In different circumstances, transfer learning is applied to predict the SE of a person. We consider both the above accesses and present our proposed method in the next section.

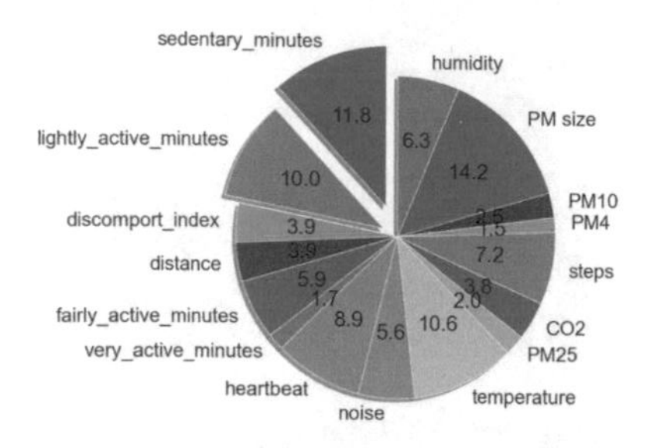

Fig. 10. Feature importance score in the prediction model for sleep efficiency (Private data)

4.4 Models to Predict Sleep Efficiency

The proposed framework to predict sleep efficiency is illustrated in Fig. 11. This framework can utilise the knowledge from global data and apply it to personal data to make the SE predictor more reliable. Assuming that personal and global data have several common features to predict SE. The first step is to apply feature extraction to the master database. Based on the master feature, the master model is created by a random forest regression algorithm. The next step is to find common features between the client and master database, then transfer the master model to get the global model. At the client/personal database, a similar feature selection process is applied to build the client model. Two models, global and client, are fusion to create a prediction model. Moreover, we use monitoring data with all selected features in the client. This data is then augmented to enhance the train data set. The augmented data is used to build a client model.

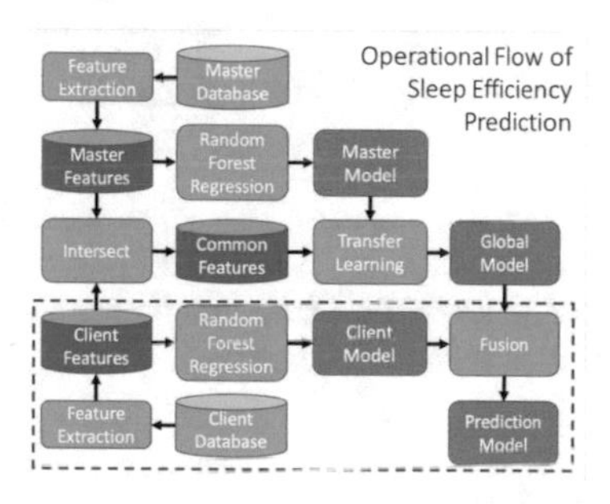

Fig. 11. Operational Flow of Sleep Efficiency Prediction

More detail of preparing data to train client model are interpreted in Fig. 12 and Fig. 13. In Fig. 12, $px7$ days of data will be used before the data augmentation step to predict the future SE values. In this work, we test the performance of the proposed framework with p from $1week$ to $4week$. We also try a personalised adaptive model that learns from previous data to predict the next day's outcome. The interpretation for this process is given in Fig. 13.

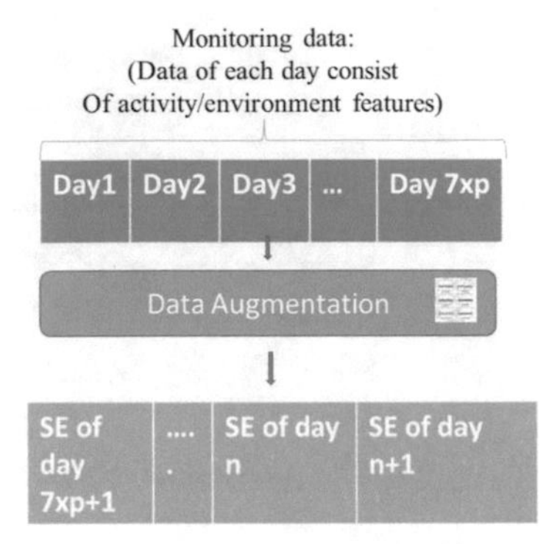

Fig. 12. Illustration of steps to process data to prediction the SE with data augmentation

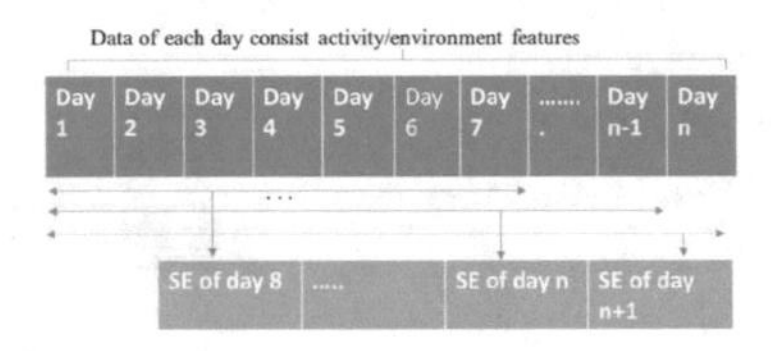

Fig. 13. Illustration of steps to process data to prediction the SE - use previous data to predict future SE values

5 Results

5.1 SE Prediction on PMdata

By applying the framework in Sect. 4.4 on PMdata, we put one person's data separately as client data and use the rest as master data. In the master data, we choose 6 important features (based on analysing results in Fig. 5) to build the master model. The client model is built by all features. Client and master models are constructed by tree-based ensemble learning techniques to produce better prediction performance. We compare our proposed framework prediction results with state-of-art models, which are recently exploited for sleep efficiency prediction. The comparison results in term of $RMSE$ is reported in Fig. 14. We applied our framework with five types of monitoring data (one to four weeks and all previous days). Two deep learning model1 and model2 with structures of $(30, 3, 1)$ and $(30, 4, 1)$ respectively are applied on PMdata to predict SE. Moreover, Decision Tree and Bayesian RNN are employed to estimate the SE values. The $RMSE$ of the proposed models is around 3.5 while other models show a larger error.

Our prediction results suggest that even with small monitoring data (just one week), the proposed framework can produce a good prediction result by combining the knowledge from global/public data with personal data.

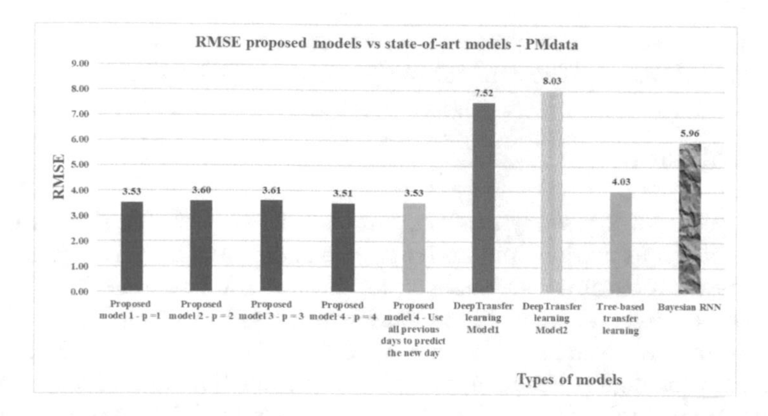

Fig. 14. Comparison between proposed models and current state-of-art model for sleep efficiency prediction of PMdata

5.2 SE Prediction on Private Data

After analysing two data sets, we see that both have some common features like steps, very/light/moderate active time, distance, and sedentary. The critical aspect of the proposed framework is to reuse the knowledge which model learned from public data. Therefore, we used 6 common features of PMdata to train the model to predict SE. This model is then applied to private data as the master model. Both SE prediction results of Pmdata and private data share the same framework as mentioned in Sect. 4.4. The prediction performance of SE values is presented in Fig. 15. Taken together with our proposed models and state-of-art models, these results show fewer root mean square errors in our models than in others. Our model error is around 3.5. In contrast, recently published models have larger than 4 RMSE. The similar prediction results of the proposed

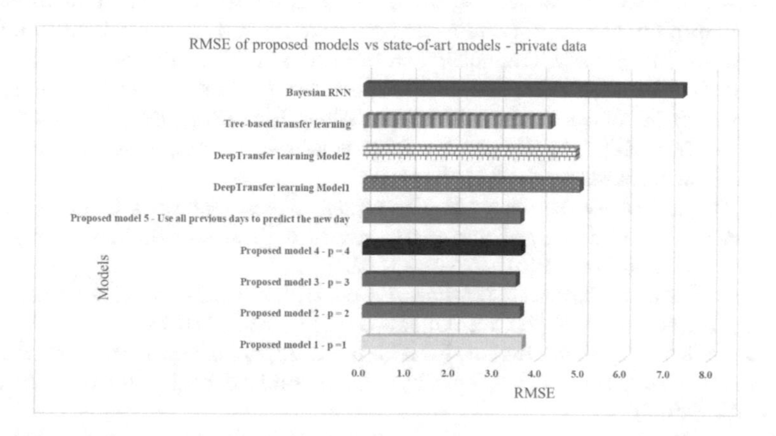

Fig. 15. Comparison between proposed models and current state-of-art model for sleep efficiency prediction of private data

framework on two data sets again approve our method's stability and lead to future applications to solve comparable problems in digital health.

6 Conclusion

The evidence from this study confirms the association between sleep attributes, environmental information, and activity. The significant negative/ positive correlation between features is visualised for further studies. Based on analysing results, we proposed a personalised model which can learn from public data and its data to predict future sleep efficiency. This proposed scheme provides excellent support for digital health applications. We also introduced a system that can learn from public data and make an adaptive model to predict sleep efficiency. Moreover, the proposed method can inform users of the analysed results and give guidelines to users. Generally, the system is designed openly to easily accommodate new data inputs, models and areas.

Acknowledgment. We would like to thank Green Blue Corporation (https://greenblue.co.jp/) and TAOS Institute, Inc. (http://www.taos.tokyo/) for providing the experimental dataset.

References

1. Mcsharry, D.G., Ryan, S., Calverley, P., Edwards, J.C., Mcnicholas, W.T.: Sleep quality in chronic obstructive pulmonary disease. Respirology **17**(7) 1119–1124 (2012)
2. Umemura, G.S., Noriega, C.L., Soares, D.F., Forner-Cordero, A.: Biomechanical procedure to assess sleep restriction on motor control and learning. In: 39th Annual International Conference of the IEEE Engineering in Medicine and Biology Society (EMBC), IEEE 2017, pp. 1397–1400 (2017)
3. Germain, A., Buysse, D.J., Shear, M.K., Fayyad, R., Austin, C.: Clinical correlates of poor sleep quality in posttraumatic stress disorder. J. Traumatic Stress: Official Publication Int. Society Traumatic Stress Studies **17**(6), 477–484 (2004)
4. Colleen, E., Carney, J.D.E., Meyer, B., Lindman, L., Istre, T.: Daily activities and sleep quality in college students. Chronobiology Int. **23**(3), 623–637 (2006)
5. Shim, J., Wan Kang, S.: Behavioral factors related to sleep quality and duration in adults. J. Lifestyle Med. **7**(1), 18 (2017)
6. Yang, P.-Y., Ho, K.-H., Chen, H.-C., Chien, M.-Y.: Exercise training improves sleep quality in middle-aged and older adults with sleep problems: a systematic review. J. Physiother. **58**(3), 157–163 (2012)
7. Mustian, K.M.: Multicenter, randomized controlled trial of yoga for sleep quality among cancer survivors. J. Clin. Oncol. **31**(26), 3233 (2013)
8. Najafabadi, M.T., Farshadbakht, F., Abedi, P.: Impact of pedometer-based walking on menopausal women's sleep quality: A randomized controlled trial. Maturitas **100**, 196 (2017)
9. Wang, F., Boros, S.: The effect of physical activity on sleep quality: a systematic review. Europ. J. Physiother. **23**(1), 11–18 (2021)

10. Vanderlinden, J., Boen, F., Van Uffelen, J.G.Z.: Effects of physical activity programs on sleep outcomes in older adults: a systematic review. Int. J. Behav. Nutr. Phys. Act. **17**(1), 1–15 (2020)
11. Baldursdottir, B., Taehtinen, R.E., Sigfusdottir, I.D., Krettek, A., Valdimarsdottir, H.B.: Impact of a physical activity intervention on adolescents' subjective sleep quality: a pilot study. Glob. Health Promot. **24**(4), 14–22 (2017)
12. Xie, Y., et al.: Association between residential greenness and sleep quality in chinese rural population. Environ. Int. **145**, 106100 (2020)
13. Yu, H., Chen, P., Paige Gordon, S., Yu, M., Wang, Y.: The association between air pollution and sleep duration: a cohort study of freshmen at a university in beijing, china. Int. J. Environ. Res. Public Health **16**(18), 3362 (2019)
14. Nguyen, T.P.V., Nguyen, D.V., Zettsu, K.: Models to predict sleeping quality from activities and environment: Current status, challenges and opportunities. In: Proceedings of the 2021 Workshop on Intelligent Cross-Data Analysis and Retrieval, pp. 52–56 (2021)
15. Stippig, A., Hübers, U., Emerich, M.: Apps in sleep medicine. Sleep Breathing **19**(1), 411–417 (2015)
16. O'Neill, C., Nansen, B.: Sleep mode: Mobile apps and the optimisation of sleep-wake rhythms. First Monday (2019)
17. Hosszu, A., Rosner, D., Flaherty, M.: Sleep tracking apps' design choices: A review. In: 2019 22nd International Conference on Control Systems and Computer Science (CSCS). IEEE, 2019, pp. 426–431 (2019)
18. Topalidis, P., Florea, C., Eigl, E.-S., Kurapov, A., Leon, C.A.B., Schabus, M.: Evaluation of a low-cost commercial actigraph and its potential use in detecting cultural variations in physical activity and sleep. Sensors **21**(11), 3774 (2021)
19. Bhagat, Y.A.: Clinical validation of a wrist actigraphy mobile health device for sleep efficiency analysis. In: 2014 IEEE Healthcare Innovation Conference (HIC), pp. 56–59 (2014)
20. Jaworski, D.J., Roshan, Y.M., Tae, C.-G., Park, E.J.: Detection of sleep and wake states based on the combined use of actigraphy and ballistocardiography. In: 41st Annual International Conference of the IEEE Engineering in Medicine and Biology Society (EMBC). IEEE, pp. 6701–6704 (2019)
21. Azimi, I., et al.: Personalized maternal sleep quality assessment: an objective iot-based longitudinal study. IEEE Access **7**, 93433–93447 (2019)
22. Hsu, C.-Y., Ahuja, A., Yue, S., Hristov, R., Kabelac, Z., Katabi, D.: Zero-effort in-home sleep and insomnia monitoring using radio signals. Proceedings of the ACM on Interactive, mobile, wearable and ubiquitous technologies **1**(3), 1–18 (2017)
23. Ibáñez, V., Silva, J., Navarro, E., Cauli, O.: Sleep assessment devices: types, market analysis, and a critical view on accuracy and validation. Expert Rev. Med. Devices **16**(12), 1041–1052 (2019)
24. Landry, G.J., Best, J.R., Liu-Ambrose, T.: Measuring sleep quality in older adults: a comparison using subjective and objective methods. Frontiers Aging Neurosci. **7**, 166 (2015)
25. Sathayanarayana, A., et al.: Sleep quality prediction from wearable data using deep learning. JMIR Mhealth Uhealth **4**(4), e125 (2016)
26. Palotti, J., et al.: Benchmark on a large cohort for sleep-wake classification with machine learning techniques. NPJ Digital Med. **2**(1), 1–9 (2019)
27. Phan, D.V., Chan, C.L., Nguyen, D.K.: Applying deep learning for prediction sleep quality from wearable data. In: Proceedings of the 4th International Conference on Medical and Health Informatics, pp. 51–55 (2020)

Capability Evaluation of Photothermal Frequency Modulation Technique for Screening Defective of Surface-Mounted Capacitor

Atsushi Yarai(✉)

Department of Electrical, Electronic and Information Engineering, Osaka Sangyo University, Nakagaito, Daito, Osaka 574-8530, Japan
yarai@osaka-sandai.ac.jp

Abstract. Previously, we proposed a new non-destructive evaluation technique for surface mounted capacitor (SMC) based on photothermal frequency modulation. This article is concerned with this technique, involving such the evaluation method as the estimation of equivalent circuit parameter of capacitor and IR camera diagnostic. It is demonstrated that measured capacitance is within the error tolerance even if the stress with a limit of 100 N is applied to SMCs artificially, on the other hand, a proposed technique can be detected "defect symptom" in spite of the stress of 10 N. In addition, it was difficult to detect the symptom even if IR camera is applied. Through these evaluation techniques, it is emphasized that our technique functions as one of the high-fidelity inspection method in spite of a low cost and simple operability in the factory use.

Keywords: Non-destructive evaluation · Surface-mounted capacitor · Photothermal evaluation technique · Infrared inspection

1 Introduction

Surface-mounted devices have been widely used in electronics, prepared as components to be automatically assembled onto electronic circuit boards. Among them, the surface-mounted capacitor (SMC) is one of the extremely important passive devices, especially for reducing the size of a circuit board. Incidentally, it is widely known that a loaded mechanical stress results in damage such as cracking and peeling inside of an SMC even if the force is slight [1, 2]. Nevertheless, the measured capacitance is within the error tolerance. Accordingly, it is extremely important to be able to easily diagnose such damage with non-destructive means. Previously, evaluation methods that use beams for propagating inside of opaque solids such as infrared (IR), X-ray and ultrasonics have been applied. Each has specific advantages and fatal disadvantages. For example, to feed the ultrasonics to an SMC (e.g., scanning acoustic microscope), it is necessary to use water because of sonic coupling, i.e. the SMC must be immersed in a liquid. It is also widely known that dielectric materials containing barium, such as $BaTiO_3$, absorb X-ray, thereby not obtaining a clear contrast image. On the other hand, the evaluation technique

N. K. Suryadevara et al. (Eds.): ICST 2022, LNEE 1035, pp. 48–54, 2023.
https://doi.org/10.1007/978-3-031-29871-4_6

using an IR camera is very useful, however, the cost for this technique is exceedingly high. Consequently, these methods are not necessarily adequate for non-destructive screening at the factory mainly due to instrument cost and operability. Previously, we proposed a new non-destructive evaluation technique for SMC based on photothermal frequency modulation (hereinafter referred as PTFM) [3]. This technique has the following features.

1) A thermal wave is generated on the inside of an SMC by irradiating a periodical laser beam from the upper surface of the SMC. 2) The capacitance is slightly modulated by thermal wave because of the temperature dependence of the dielectric constant. 3) Because of this, the oscillation frequency of the LC oscillation circuit, which is composed of this SMC, results in a frequency modulation [3, 4]. 4) Finally, the change of capacitance of SMC ΔC could be obtained by FM demodulation of the oscillation signal. We believe that the thermal wave is immediately reflected and is dramatically decayed by cutting-off the transmission due to the crack on the inside of SMC, thereby decreasing the change of capacitance. Furthermore, the internal condition of the SMC can be known with a depth-profile by varying the thermal diffusion length of the thermal wave, which is inversely proportional to the square root of the laser modulation frequency.

This articles describes briefly at first the measurement system of PTFM and its experimental results. Next, to verify and to check the results mutually, I demonstrate the change of capacitance and isolated resistance for SMC and the dispersion of upper-surface temperature of SMC using IR camera. Finally, the advantage of our PTFM technique is certified as the results.

2 PTFM Technique

2.1 Measurement System

Figure 1(a) and (b) show a block diagram of a proposed apparatus and a photograph of its sensing head (mount jig), respectively. The cw power of the laser with a wavelength of 806 nm is approximately 0.5 W (at fiber-out). The periodically-modulated laser beam is focused to approximately 50 μm on the upper surface of the SMC, as shown in Fig. 1(b). The universal LC Colpitts RF oscillator designed with the oscillation frequency of 80 MHz is prepared. The oscillation signal is fed to the double band mixer (DBM), converting to the IF signal by mixing with the local oscillation signal. After the IF signal whose frequency is approximately 100 kHz is amplified and it passes through the band pass filter for rejecting spurious signals, the signal is fed to the FM demodulator to convert to a voltage change from a frequency change. After that, the demodulated signal is fed to a lock-in amplifier for signal recovery with the modulation signal of a heating beam. Furthermore, this demodulator is composed of a commercially available integrated circuit for an analog phase locked loop (PLL), enabling the detection of slight frequency change with high sensitivity and high linearity. The experimental results subsequently demonstrate the capability. Furthermore, due to $C_0 \gg \Delta C$, the difference of oscillation frequency Δf can be approached to [3]

$$\frac{\Delta f}{f_0} \cong -\frac{\Delta C}{2C_0} \tag{1}$$

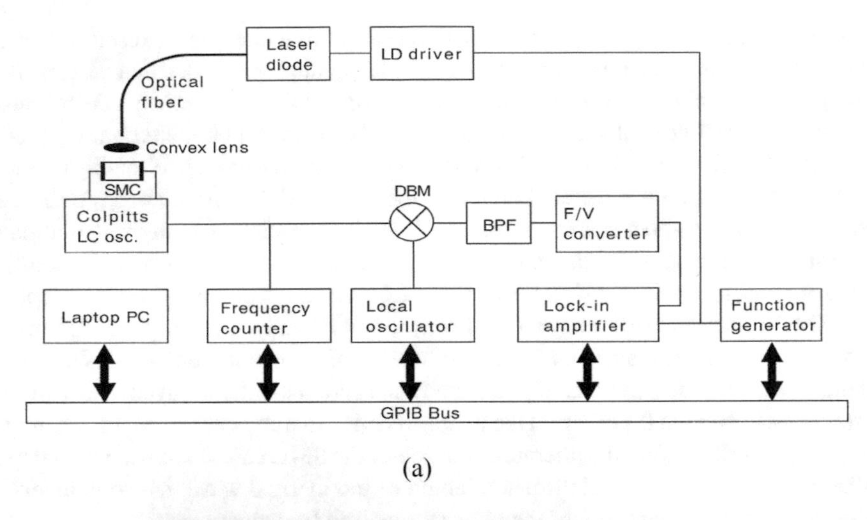

(a)

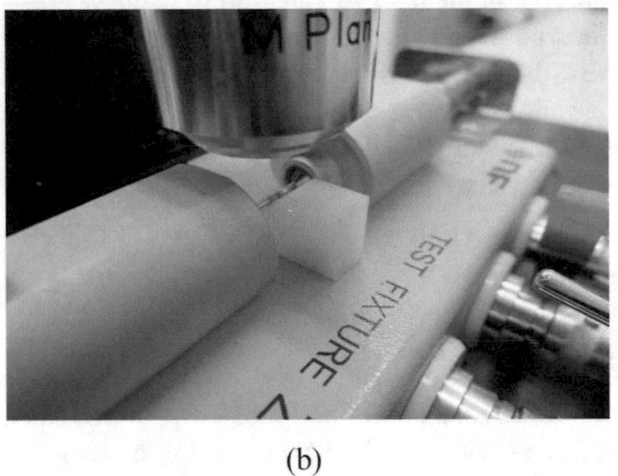

(b)

Fig. 1. (a) Block diagram of experimental apparatus, and (b) photograph of sensing head.

2.2 Prepared Samples

First, a commercially available SMC (whose size is 0.8 mm in width, 0.8 mm in depth and 1.6 mm in length, and whose capacitance *Co* is 100 pF, manufactured by Murata Co. Ltd., Japan) was prepared as the normal test sample. Next, we created defective samples artificially in our laboratory from the normal SMC by loading a mechanical stress to the SMC with the following procedure. The SMC was mounted on the center of the back surface of a universal board for electronic circuits by lead-free solder. After that, the point at the center of this board was pressed using a digital-display force meter on the front surface. This board was fixed to a ring with a diameter of 50 mm so that it sagged elastically like the head of a drum. Hereafter, the value indicated by the force meter is defined as the loaded stress. Loaded stress ranged from 0 to 100 N at about 10-N intervals, and ten samples at each stress level were prepared. Note that the capacitance changes of the SMC measured by an LCR meter fell within the range of 3% at this stress range, i.e. it was within the error tolerance. Accordingly, it was confirmed that test samples made by this method cannot be selected by the different capacitances, either defective or non-defective.

2.3 Measurement Results

Figure 2 shows the modulation frequency f_m of laser beam dependence on the changing ratio of capacitance $\Delta C/C_0$ as a function of loaded stress. Here, the error bars drawn in this figure symbolize the dispersion of ten samples. As can be seen in Fig. 2, it is confirmed that $\Delta C/C_0$ is decreased as the loaded stress increases. The reason for this is believed to be diffusion and expansion inside the crack. It is also demonstrated in Fig. 2 with all test samples that $\Delta C/C_0$ is suppressed as the laser modulation frequency increases. This is because of a reduction of thermal diffusion length by increasing the modulation frequency. In other words, it is possible to obtain a depth profile of the inside of the SMC by sweeping the modulation frequency. Furthermore, it was observed that the error-bar was enlarged and/or average vales of $\Delta C/C_0$ was reached at "limit of detection" when the stress was loaded more than 50 N. Consequently, these experimental data were not plotted in Fig. 2 with a viewpoint of reproducibility.

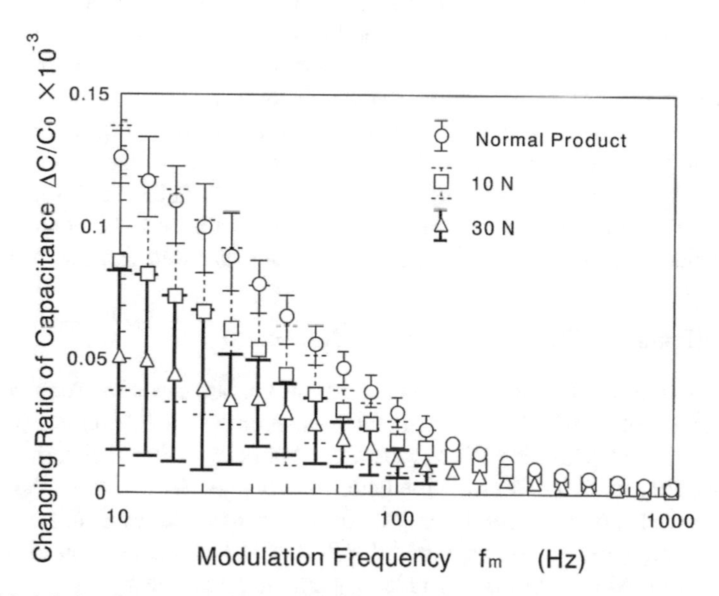

Fig. 2. Modulation frequency of laser dependence of changing ratio of capacitance $\Delta C/C_0$.

3 Reliability Verification of PTFM Diagnostic Technique Compared with Other Techniques

3.1 Measurement of Capacitance and Isolated Resistance

Figure 3 shows the applied stress dependence on the changing of resistance R and capacitance C_0. An impedance gain phase analyzer (HP4194A) which has a function of

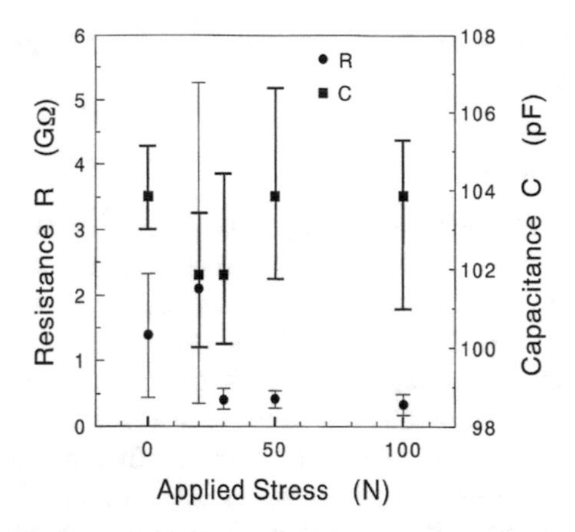

Fig. 3. Applied stress dependence of changing capacitance R and C_0.

frequency sweep until 50 MHz was used for this measurement, enabling the estimation of C and R associated with equivalent circuits. Here, error bars drawn in this figure symbolize the dispersion of measured capacitance and resistance at ten same samples used, prepared for above PTFM. As can be seen in Fig. 3, it was confirmed that the isolated R is decreased slightly as the loaded stress increases. The reason for this is believed to be diffusion and expansion inside the mild distortion of layer structure of capacitor. On the other hand, the change of capacitance due to the applied stress was not observed obviously. In other word, the results were within the error tolerance. Here, as the error tolerance of capacitance of commercially-available SMC is normally 5%, I cannot conclude the quality of SMC and defects products based on these results.

3.2 Deviation of Emitted IR Detected by IR Camera

Figure 4 shows the applied stress dependence on a standard deviation for signal magnitude of all pixel detected by IR camera. In this measurement, SMC samples (which were prepared for above Sects. 2.3 and 3.1) were heated directly by irradiating the modulated laser beam (830 nm, 500 mW) from the bottom-surface of SMC using a 0.6-mm-core-diameter step index optical fiber. IR emitted from upper-surface of SMCs was observed by an IR camera (Electrophysics, model PV-320) that has a resolution of 320 × 240 pixels associated with NTSC analog video signal. Note that the signal averaging of 16 times with each pixel by synchronizing a laser modulation signal and a video signal (which has a vertical synchronization frequency of 60 Hz) was performed to enhance the signal-to-noise ratio of detected signals. As can be seen in Fig. 4, it was observed that the dispersion of brightness among all pixels is increased mildly as the loaded stress increases. The reason for this tendency is believed to be that the temperature distribution irregularity at the upper-surface of SMC was arisen due to the abnormal transmission of thermal wave generated by modulating a heating laser source on the inside of SMC, influencing phenomena such as reflection, diffraction and cut-off at the transmission of thermal wave. Incidentally, it was widely recognized at a research field of photothermal phenomena that these physical phenomena arise due to the inside defects such as a crack.

Previously, a non-destructive diagnostic method using IR camera for quality control of SMC have been widely used, however, the cost for this is extremely high. On the other hand, it is concluded that a proposed PTFM technique in this article is easy and suitable method for screening the quality of SMC.

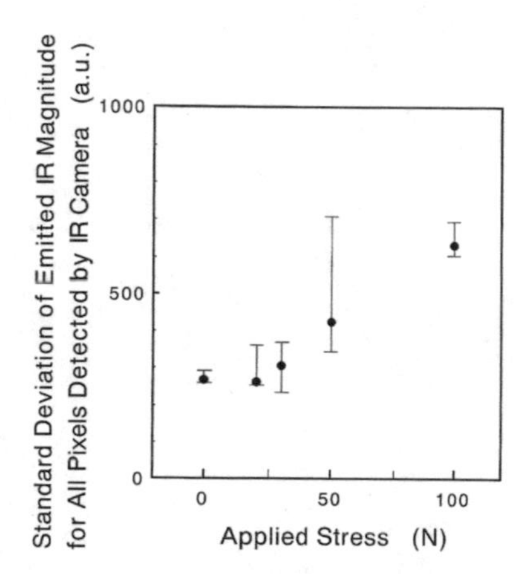

Fig. 4. Applied stress dependence of standard deviation of pixel magnitude for IR camera image.

4 Summary

The advantage of a non-destructive evaluation technique for SMC based on a photothermal frequency modulation was confirmed compared with other technique. This technique features the assistance of a temperature dependence of dielectric constant. The change of capacitance due to the applied stress could not be observed obviously when error tolerance was considered. In case of a use of IR camera, a symptom of the defects on the inside of SMCs could be observed, but it is not obvious tendency. In future work, it is necessary to study in detail the accuracy of the experimental results through internal observation by actually opening in destructive the SMCs. Finally, I believe that our technique will be useful for screening SMCs at the manufacturing stage as the first step toward practical application.

References

1. Krieger, V., Wondrak, W., Dehbi, A., Bartel, W.: Defect detection in multilayer ceramic capacitors. Microelectron. Reliab. **46**(9), 1926–1931 (2006)
2. Wunderle, B., Braun, T., May, D., Mazloum, A., Bouazza, M., Walter, H.: Non-destructive failure analysis and modeling of encapsulated miniature SMD ceramic chip capacitors under thermal and mechanical loading. In: Proceedings on Thermal Investigation of ICs and System 2007 (THERMINIC 2007), pp. 104–109. IEEE, Budapest (2007)

3. Yarai, A., Sato, Y.: Non-destructive evaluation technique of surface-mounted capacitor based on photothermal frequency modulation technique. In: Proceedings on IEEE Int. Symp. on the Applications of Ferroelectrics, pp. 1–4. IEEE, Darmstadt (2016)
4. Cho, Y., Kumamaru, T.: Photothermal dielectric spectroscopy microscope. Rev. Sci. Instum. **67**(1), 19–28 (1996)

Microwave-Assisted Generation of Secondary Nanoparticles and Flame-Assisted Generation of an Amorphous Layer for Improving NO_2 Gas Sensing Behaviors: A Mini Review

Sukwoo Kang[1], Ka Yoon Shin[1], Wansik Oum[1], Dong Jae Yu[1], Eun Bi Kim[1], Hyeong Min Kim[1], Ali Mirzaei[2], Jin-Young Kim[3], Myung Sung Nam[3], Tae Un Kim[3], Myung Hoon Lee[3], Somalapura Prakasha Bharath[1], Krishna Kiran Pawar[1,4], Sang Sub Kim[3](✉), and Hyoun Woo Kim[1,4](✉)

[1] Division of Materials Science and Engineering, Hanyang University, Wangsimni-ro, Seongdong-gu, 222 Seoul, Republic of Korea
hyounwoo@hanyang.ac.kr
[2] Department of Materials Science and Engineering, Shiraz University of Technology, Shiraz, Iran
[3] Department of Materials Science and Engineering, Inha University, Incheon 402-751, Republic of Korea
sangsub@inha.ac.kr
[4] The Research Institute of Industrial Science, Hanyang University, Seoul 133-791, Republic of Korea

Abstract. This mini review is comprised of two approaches of the enhancement of gas sensor response: microwave-assisted and flame-assisted strategies. To achieve enhancement, defects and surface modifications must be addressed. When a SnO_2-graphene mixture is irradiated with microwaves, SnO_2 acquires oxygen vacancies because carbon takes oxygen away from its surroundings. An oxygen vacancy, a type of defect, creates free electrons, increasing the response [1]. In addition, decoration with amorphous carbon, which is a type of surface modification, establishes a heterojunction in the main substance [2]. The heterojunction leads to rectification; hence, electrons flow in one direction to balance the electron concentration. A change in the concentration of electrons affects electron mobility. The gas sensor response is affected by the mobility and concentration of electrons. Oxygen vacancies create electrons according to the Kröger-Vink equation, and heterojunctions accelerate electrons. The sensor response changes when the metal oxide semiconductor gas sensor is exposed to oxidizing and reducing gases. When a substance is oxidized, surface functional groups lose their electrons to remain in equilibrium. Oxygen, a surface functional group, loses electrons and traps them as ions on the surface, resulting in band-bending [3]. Therefore, defects and decorations increase gas sensor response.

Keywords: Gas sensor · NO_2 sensing · Graphene · Amorphous carbon · Tin oxide

N. K. Suryadevara et al. (Eds.): ICST 2022, LNEE 1035, pp. 55–62, 2023.
https://doi.org/10.1007/978-3-031-29871-4_7

1 Introduction

MOS gas sensors have many applications due to their unique characteristics, including fast response speed, wide detection range, and good response. Generally, the gas response improves as the grain size decreases [3]. However, there is a lower limit to grain size. As a result, many studies have aimed to enhance the response without reducing the grain size; for metal oxide gas sensors, the formation of defects and introduction of surface decorations are frequently used to increase mobility in metal oxides [1, 2]. Therefore, we describe herein the fabrication of gas sensors using defects and decorations by microwave (MW) irradiation and amorphous carbon decoration.

2 Two Approaches for Enhancing the NO_2 Sensing

2.1 MW-Assisted Synthesis of graphene-SnO_2 Nanocomposites (NPs)

SnO_2 is a reputable material for gas sensors owing to their high electron mobility and superior stability. to operate the gas sensor, A high temperature of about 300 °C is required. Most metal oxides are sintered at 300 °C, resulting in large grain sizes. When MWs are used to irradiate graphene-SnO_2, free electrons increase the system temperature to above 800 °C. This temperature creates secondary particles around tin oxide nanoparticles. Figure 1 is a schematic of the changes resulting from MW irradiation. [4].

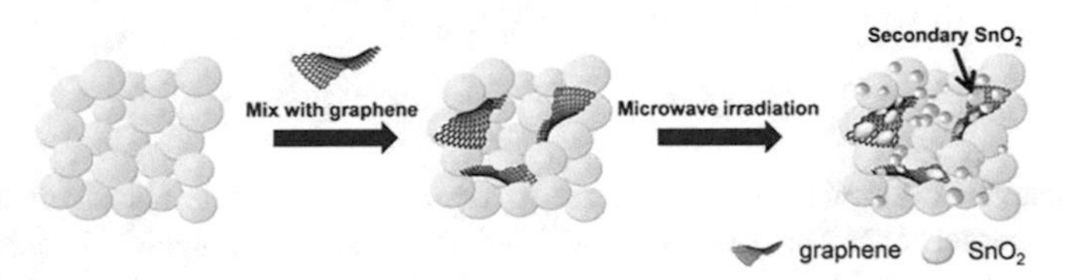

Fig. 1. Schematic of the morphology of graphene-SnO_2 nanocomposites produced by MW irradiation. Reproduced from [4] with the permission of ACS publications.

Upon microwave heating, SnO_2 undergoes a phase transformation that creates SnO_x nanoparticles in the SnO_2. As is evident from the phase diagram, SnO_x decomposes to $Sn_3O_4 + SnO_2$ at about 200 °C [5], which diminishes the number of free electrons. Figure 2 shows TEM images of synthesized secondary SnO nanoparticles.

As shown in the phase diagram, a phase transformation occurs at 200 °C. So, the optimal temperature is 150 °C, where the sensor response is 72.6 at 5 ppm of NO_2 (Fig. 3).

The graphene-ZnO composite synthesis process is the same as that of graphene-SnO_2, with only a variation in the main substance [6]. The different material elevates the optimal temperature of the graphene-ZnO composite to about 300 °C (Fig. 4). The higher is the temperature, the higher is the electron mobility, because the secondary particles of ZnO are more stable at higher temperatures than those of SnO_2. Microwave irradiated graphene-ZnO nanocomposites have heterojunctions that lead to modulation of resistance. In addition, as electrons move to achieve equilibrium, heterojunctions increase electron mobility. Thus, the nanocomposite response and recovery time is much faster than that of a general metal oxide gas sensor.

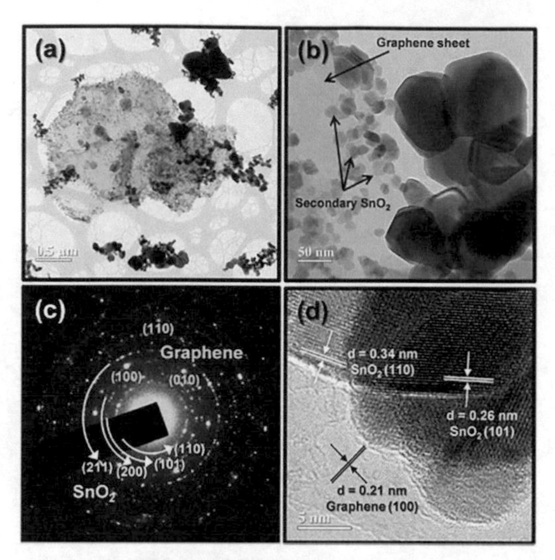

Fig. 2. (a) and (b) TEM images of graphene/SnO_2 nanocomposites at different magnifications and the (c) SAED pattern. (d) HRTEM image of the same material. From [4] with the permission of ACS publications.

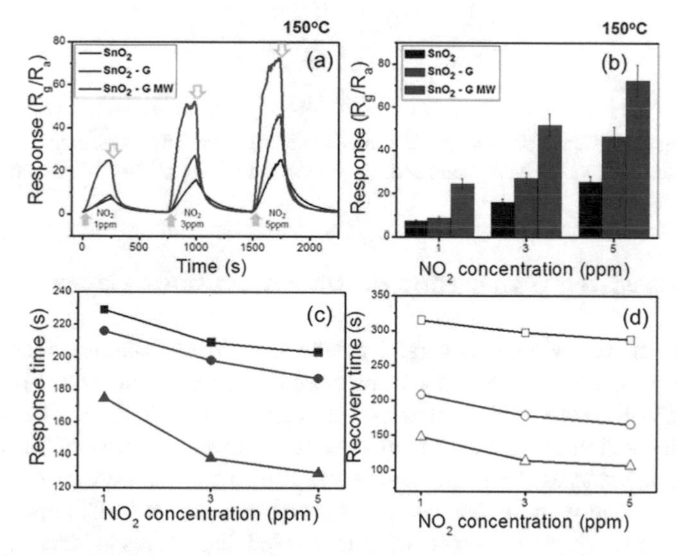

Fig. 3. (a) Transient response curves, (b) response versus NO_2 concentration, (c) response time, and (d) recovery time of pristine SnO_2 and mixtures of SnO_2 and graphene (0.5 wt%), before and after MW irradiation. The temperature was fixed at 150 °C. Reproduced from [4] with the permission of ACS publications.

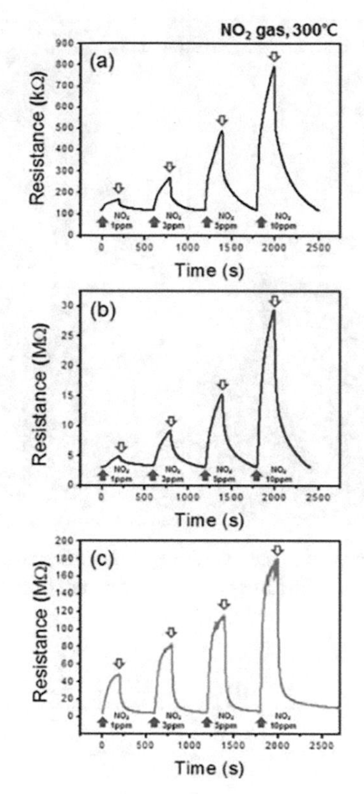

Fig. 4. Transient resistance plots of (a) pristine ZnO sensor and ZnO NPs/graphene sensors that are (b) unirradiated and (c) MW-irradiated, in response to NO_2 at 300 °C. From [6] with the permission of Elsevier.

3 Flame-Assisted Generation of the Amorphous Layer

The response of nanowires is generally greater than that of nanoparticles due to the high nanowire surface-volume ratio. Nanowires (NWs) are synthesized by the vapor-liquid-solid (VLS) route, which relies on heterogeneous nucleation. VLS is a method that entails deposition of Au catalyst onto the substrate and increase of the temperature of the gold catalyst to agglomerate the Au thin film. Then, tin powder that encounters oxygen grows along with the Au catalyst. Amorphous carbon (a-C) deposition, called FCVD, was conducted on the surface of as-fabricated SnO_2 NWs to increase the surface area [7]. Figure 5 shows a schematic of this process. The material exhibits a change in its work function after deposition of a-C (Fig. 6).

The thickness of the a-C layer is a variable and needs to be controlled by deposition time. The appropriate amount of a-C must be deposited to avoid insulating properties (Fig. 7).

TeO_2 is p-type material [8]. Because of hole movement, the response of positive type material is weaker than that of negative type materials [9]. TeO_2 nanowires have a lower growth temperature than SnO_2 nanowires because they easily undergo evaporation

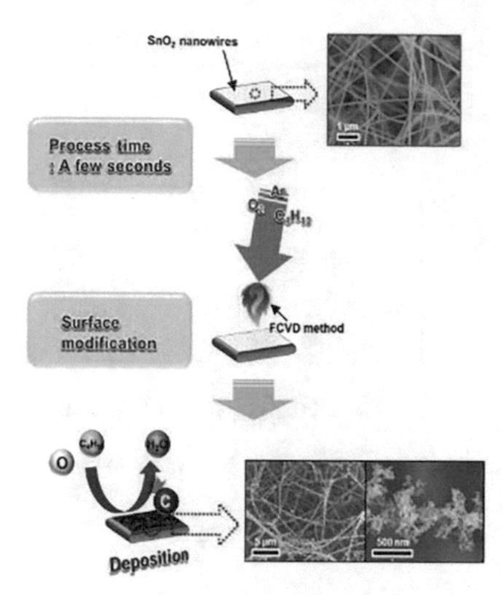

Fig. 5. The steps of the synthesis of an a-C-decorated SnO_2 nanocomposite. The a-C shell is formed on the SnO_2 NWs by FCVD for 5 s. From [7] with the permission of Elsevier.

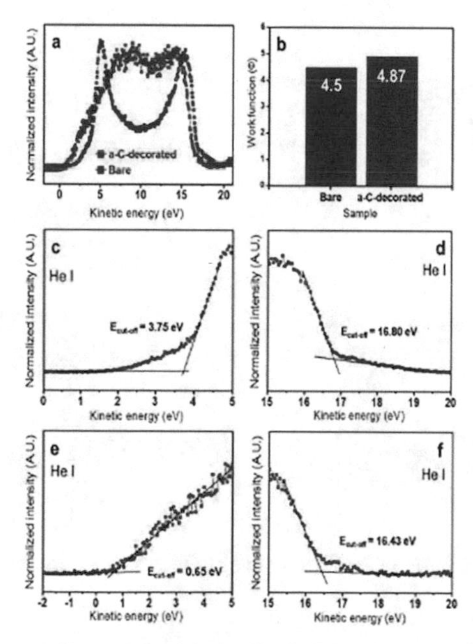

Fig. 6. (a) UPS spectra and (b) work functions of pristine SnO_2 and a-C decorated SnO_2. Characteristic $E_{cut\text{-}off}$ values and valence band edges of (c, d) pristine SnO_2 and (e, f) a-C-decorated SnO_2 samples. From [7] with the permission of Elsevier.

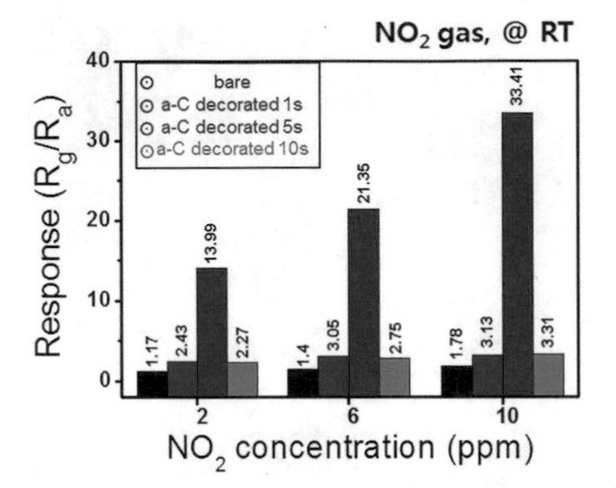

Fig. 7. Comparison of the responses of pristine and a-C decorated SnO_2 NW gas sensors at different deposition times (1, 5, and 10 s) to NO_2 gas at 25 °C. From [7] with the permission of Elsevier.

(Fig. 8). After TeO_2 nanowire synthesis, a-C can be deposited. Figure 9 indicates a TEM image of TeO_2 nanowires with deposited a-C and the accompanying changes to the work function (Fig. 10).

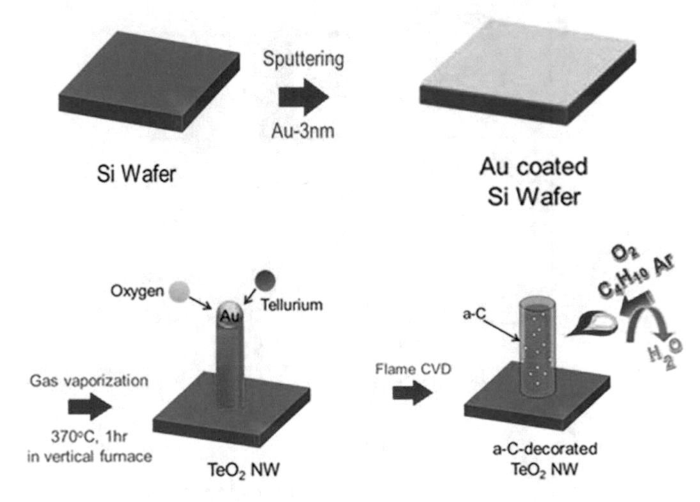

Fig. 8. Schematic of synthesis of a-C-decorated TeO_2 NWs. From [8] with the permission of Elsevier.

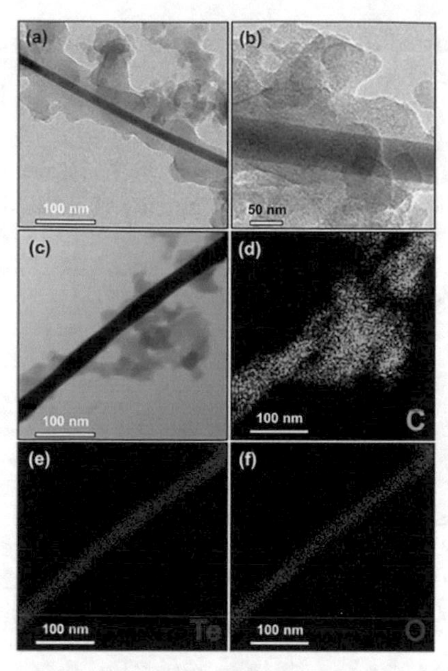

Fig. 9. (a–c) TEM images of a-C-decorated TeO_2 NWs. EDS elemental maps of (d) C, (e) Te, and (f) O. From [8] with the permission of Elsevier.

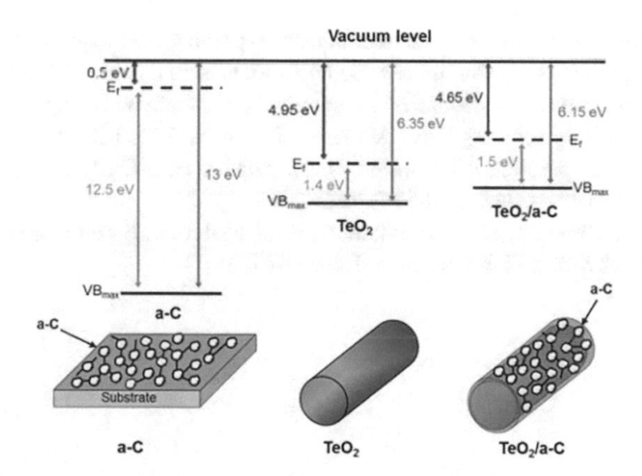

Fig. 10. Energy bands for a-C, TeO_2, and TeO_2/a-C. Reproduced from [8] with the permission of Elsevier.

4 Conclusion

In this mini review, we present two method of increasing MOS gas sensor response. The concentration and mobility of electrons greatly affect the sensitivity of metal oxide gas sensors. The response increases when the main sensing material is deprived of its charge carriers. The formation of heterojunctions and defects was used as a method to easily change the concentration and mobility. Among the techniques that can be employed to do so, we introduce microwave irradiation and FCVD.

Acknowledgments. This research was supported by the Basic Science Research Program through the National Research Foundation of Korea (NRF) funded by the Ministry of Education (2016R1A6A1A03013422). This work was supported by the National Research Foundation of Korea (NRF) grant funded by the Korea government (MSIT) (2019R1A2C1006193). This work was supported by the Technology Innovation Program (20013726) funded By the Ministry of Trade, Industry & Energy (MOTIE, Korea).

References

1. Al-Hashem, M.: Role of oxygen vacancies in nanostructured metal-oxide gas sensors: a review. Sens. Actuators B: Chem. **301**, 126845 (2019)
2. Wang, W., Kumta, P.N.: Nanostructured hybrid silicon/carbon nanotube heterostructures: reversible high capacity lithium-ion anodes. ACS Nano **4**, 2233–2241 (2010)
3. Dey, A.: Semiconductor metal oxide gas sensors: a review. Mater. Sci. Eng. B **229**, 206–217 (2018)
4. Kim, H.W.: Microwave-assisted synthesis of graphene−SnO_2 nanocomposites and their applications in gas sensors. ACS Appl. Mater. Interfaces **9**, 31667–31682 (2017)
5. Batzill, M., Diebold, U.: The surface and materials science of tin oxide. Prog. Surf. Sci. **79**, 47–154 (2005)
6. Kim, H.W.: Synthesis of zinc oxide semiconductors-graphene nanocomposites by microwave irradiation for application to gas sensors. Sens. Actuators B: Chem. **249**, 590–601 (2017)
7. Choi, M.S.: SnO_2 nanowires decorated by insulating amorphous carbon layers for improved room-temperature NO_2 sensing. Sens. Actuators B: Chem. **326**, 128801 (2021)
8. Oum, W.: Room temperature NO_2 sensing performance of a-C-decorated TeO_2 nanowires. Sens. Actuators B: Chem. **363**, 131853 (2022)
9. Callister, W.D., Rethwisch, D.G.: Fundamentals of Materials Science and Engineering, 5th edn. John Wiley & Sons., Hoboken, New Jersey (2012)

A Cognitive Digital Twin Architecture for Cybersecurity in IoT-Based Smart Homes

Sandeep Pirbhulal[1](✉), Habtamu Abie[1], Ankur Shukla[2], and Basel Katt[3]

[1] Norwegian Computing Center, Blindern, P.O. Box 114, 0314 Oslo, Norway
sandeeep@nr.no
[2] Department of Risk and Security, Institute for Energy Technology, 1777 Halden, Norway
[3] Department of Information Security and Communication Technology, Norwegian University of Science and Technology, 2815 Gjøvik, Norway

Abstract. Cognitive Digital Twin (CDT) is an extension of Digital Twin with cognitive capabilities to monitor and analyse complex and unforeseen behaviours and to ensure critical reasoning and decision-making. Thus, CDT has a potential for enhancing cybersecurity for the Internet of Things (IoT)-based applications such as smart homes. In this paper, we developed a conceptual CDT architecture for improving cybersecurity in smart homes with dynamic threat detection and mitigation capabilities. The proposed approach applies closed feedback loops between cognitive process and cybersecurity using artificial intelligence and machine learning techniques. This will allow continuous monitoring of security-related information and analytics with complex behaviours within a virtual environment. The developed approach allows security testing and simulation in the virtual world for prediction and anticipation of dynamic security threats. It also facilitates dynamic updates to the physical world of attack prevention strategies for the dynamic optimization of smart homes security. Finally, this paper discusses the applicability of the developed CDT architecture in other IoT-based critical sectors.

Keywords: Cybersecurity · Cognitive Digital Twins · IoT · Smart Homes

1 Introduction

The Internet of Things (IoT) has been widely used in various sectors, including telemedicine, smart homes, smart cities, etc. [1]. IoT-based smart homes and healthcare [2] facilitate a unique way to support people with special needs, suffering chronically ill, elderly, disabled, and even everyone in pandemic situations such as Covid-19 [3]. Millions of embedded devices use new software stacks that significantly increase the threat level of these IoT devices and the likelihood of turning previously-unexploitable vulnerabilities into actively exploitable vulnerabilities. These challenges were also seen during Covid-19, thus raising the necessity for innovative security solutions in IoT-based applications [4, 5].

In recent times, the digital twin (DT) and cognitive digital twin (CDT) concepts have been increasingly used in different critical sectors. DT and CDT are used to develop a virtual systems to collect real-time data from IoT devices to allow insights into performance

N. K. Suryadevara et al. (Eds.): ICST 2022, LNEE 1035, pp. 63–70, 2023.
https://doi.org/10.1007/978-3-031-29871-4_8

and cyber threats [6–10]. CDT is the advancement of DT that will assist in achieving the goal of industry 4.0 [11]. CDT includes cognitive features that brings the critical aspects of cognition such as attention, perception, learning, memory, decision making, etc. [8, 11]. There exist several studies in the areas of cognitive twins such as enhancing cognition [12], adapted model of CDT [13], providing different levels of self-awareness [14], personalized system for smart cities [15], automatizing cognitive science for cyber-security [16], cognitive cybersecurity analysis [17], etc. Al Faruque et al. [11] developed a CDT framework that focuses on the impact of twin technology on the performance of the product life cycle. Zheng et al. [6] developed a reference architecture for CDT for different applications. However, the above studies did not highlight how CDT can be useful for monitoring and preventing cyber threats. Nguyen, et al. [18] presented an approach which emphasizes the applicability of CDT for cybersecurity using ontology concepts. The main limitation of their approach is that it does not highlight how dynamic updates of the cognitive cycle will be analysed and used for offering adaptive measures to secure systems.

The key limitations of existing CDT approaches can be summarized as follows:

- There is not any systematic CDT framework to enhance cybersecurity which can be used for IoT-based smart home applications.
- There is a lack of automated cyber threats mitigation approach using CDT by analysing complex behaviours, system architectures, and interdependencies of their components and processes.
- Existing CDT approaches for cybersecurity do not apply a feedback loop between cognitive process and cybersecurity, and do not periodically share dynamic updates about security information in a virtual environment, which plays a crucial role in predicting cyber threats.

The main contribution of this study is a provision of dynamic and systematic solution for automated cybersecurity using CDT in IoT-based smart homes. The secondary objectives are:

- To develop a CDT-based conceptual architecture that can monitor, analyse and predict cyber threats.
- To apply a dynamic closed-loop solution within a virtual environment using complex behaviours to address varying dynamic cyber security threats in changing environments.
- To present a systematic way of preventing cyber threats within IoT-based smart homes (physically) by allowing simulation and experimentation in CDT (virtually) using the developed architecture.

This paper is organized as follows: Comparisons between DT and CDT are discussed in Sect. 2. Section 3 describes on the proposed architecture for enhancing cybersecurity in smart homes using CDT. CDT applications, conclusions and future directions are discussed in Sect. 4.

2 Digital Twin vs Cognitive Digital Twin

A DT is a virtual representation of a physical world (system or infrastructure) that captures the system's real-time information, features, and behaviours via physical-virtual world synchronization [10]. The main objective of DT is to empower simulations and testing in the virtual environment to monitor and control the physical system. The advancements in data analytics, i.e., artificial intelligence (AI) and machine learning (ML) offer more capabilities to DT. CDT is self-learning and proactive by incorporating DT and cognitive process (analysing, learning, critical reasoning, planning and decision making) for better understanding and experimenting prediction of unforeseen events of the physical system. An illustration of the comparisons between DT and CDT for smart homes is shown in Fig. 1.

There is a need to comprehend the fact that the CDT technology is not developed to substitute DT. However, CDT is an evolved version of the existing DT technology. DT and CDT technologies have their respective benefits and applicabilities. Before implementing any of them for physical systems, their requirements, specifications, and scenarios must be considered. The empowering technologies of DT are more established, and several case studies are also available, which may serve as references. Nevertheless, CDT requires complex hierarchies to understand relationships between entities and components through lifecycle stages to detect complex behaviours and unforeseen situations [6]. There is not any generic framework or architecture of CDT for enhancing cybersecurity in smart applications. Hence, efforts in that direction are required.

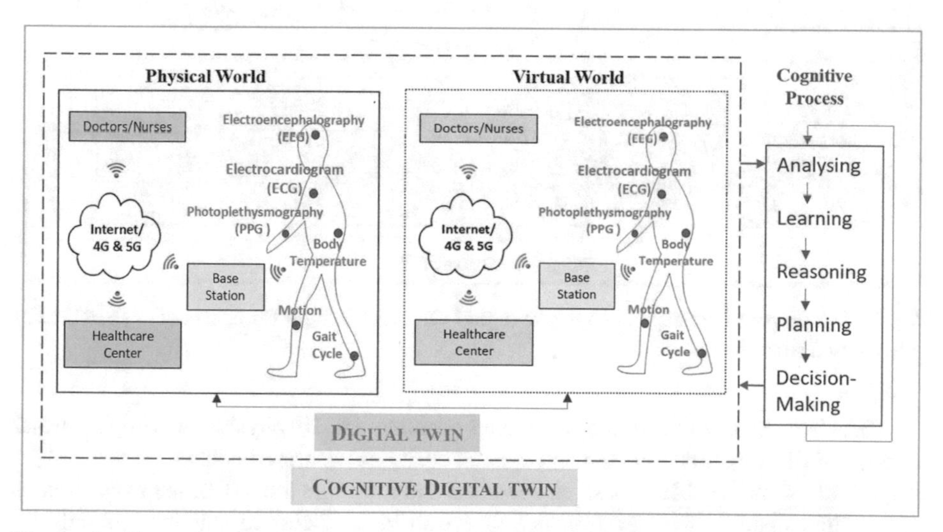

Fig. 1. An illustration of comparisons between Digital Twin and Cognitive Digital Twin of Smart Homes

3 Proposed Cognitive Digital Twin Architecture for Cybersecurity

This study presents a CDT architecture for cybersecurity. It continuously collects security-related information such as vulnerabilities, security threats from the IoT devices of smart homes in the physical world and integrates dynamic updates from the virtual world. It allows analytics, prediction and anticipation of cyber threats by performing simulation and experimentation virtually, and responses of prevention measures autonomously to the physical world. Figure 2 depicts our proposed conceptual CDT architecture for enhancing cybersecurity in IoT-based smart homes. The proposed approach uses a closed feedback loop of AI/ML techniques for analysing complex behaviours and adapting to security variations in the cognitive process of smart homes.

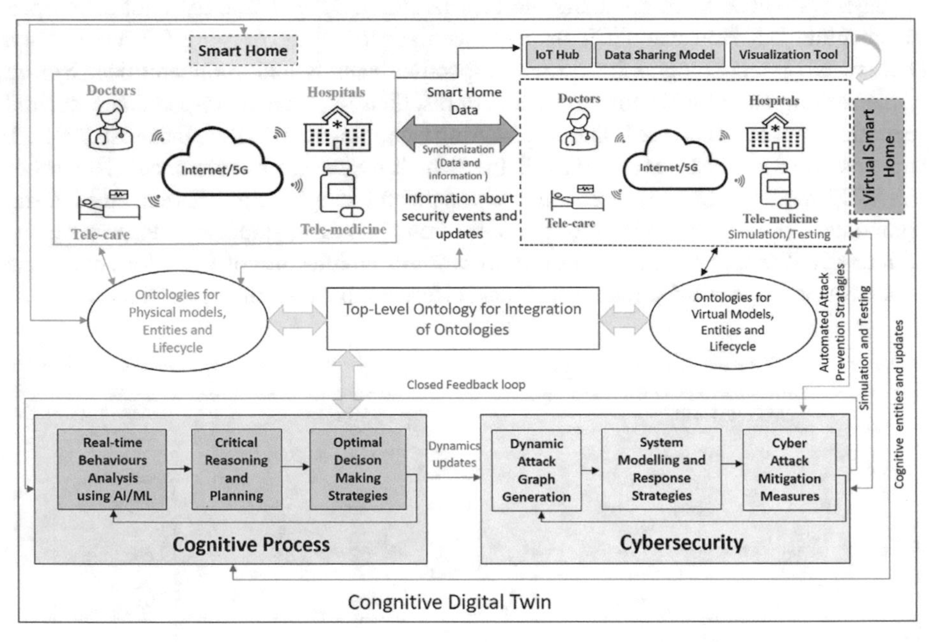

Fig. 2. The proposed conceptual cognitive digital twin architecture for enhancing cybersecurity in IoT-based smart homes

There are three stages in the developed architecture, (i) physical world, (ii)virtual world, and (iii) cognitive process and cybersecurity as described below:

Stage I-Physical World: This stage develops and monitors the IoT-based smart homes having the capability to offer guidance or prescribe medicine to patients remotely by collecting and analysing health data. The health data from medical IoT devices is stored in the cloud, which can be accessed by the concerned authorities (doctors, nurses and hospitals) remotely using wireless communication (5G or Wi-Fi). Hackers can eavesdrop and modify the communication between source and destination devices which can lead to illegal collection and misuse of secret medical information. Hence, security is one of the major concerns in IoT-based smart homes.

Stage II-Virtual World: At this stage, a virtual replica of smart home (physical world) with cognitive properties is developed and visualized to deal with cybersecurity threats efficiently. As depicted in Fig. 2, CDT integrates multiple models to attain more analytical abilities and incorporates expert knowledge to find the optimal solution for avoiding security threats. In our previous research work [19], we developed a framework for reinforcing the cybersecurity in IoT-based healthcare using DT technology. In [20], we discussed the significance of cognitive cybersecurity for anticipating and responding to new and emerging cybersecurity and privacy threats for IoT based critical infrastructures. From this experience, we realized that the cognitive capabilities could learn the behaviour of digital and physical entities to provide better decision-making tool for real operational systems.

Therefore, we focused on CDT for enhancing cybersecurity due to the advantages acquired from integrating human behaviours with the DT models for critical reasoning, planning and decision-making strategies. Our CDT approach will connect the real-time smart home to IoT Hub and cloud to use data models for IoT devices and digital twins using visualization tools, i.e., 2D or 3D to visualize and understand CDT data. Once the virtual setup is visualized and connected with the physical setup, it is essential to check that both worlds have synchronization so that smart home data (i.e., bio-signals or physiological features) or information about vulnerabilities and cyber security events can be shared. In CDT, the integration of human behaviour with twin models can be ensured in numerous ways, such as semantic technologies (category tagging, ontologies, knowledge graphs, natural language processing, etc.), fuzzy logics, and lifecycle methodologies. Ontologies provide a clear understanding of entities, knowledge-sharing capabilities, and reusability. Thus, in this study, the ontology concepts are used for developing interrelationships between entities, models and lifecycle for monitoring heterogeneous smart home data and cyber security information.

Stage III-Cognitive Process and Cybersecurity: This stage discusses how a closed feedback loop between cognitive process and cybersecurity within a virtual environment brings an efficient way to monitor, control, analysis, predict and anticipate security threats in smart homes. Figure 2 shows the outcomes of the top-level integrated ontologies (physical and virtual) are input to the cognitive process. Initially, the AI/ML models analyse the health data and security information to comprehend potential security threats. After that, careful planning and critical reasoning are done to understand the dynamic environments and the plan is carefully monitored and checked for sustainable accuracy. The specific decision is taken by observing security threats in varying conditions. The cognitive updates are sent to the virtual world and communicated with the physical world via synchronization. This helps monitoring dynamic security threats and understanding entities' relationships and updating ontologies. Secondly, the self-learning feature is achieved through the internal closed feedback loop of cognitive process. The outer closed feedback loop between cognitive process and cybersecurity module enables the dynamic detection of security threats by sharing varying updates between two modules. Thirdly, twin models are used to perform simulation and testing in the virtual world. The dynamic attack graphs, response strategies and attack mitigation measures are generated using internal closed feedback loop. Finally, the response strategies and attack mitigation measures are shared with the physical world so that actions against potential threats can be taken sharply.

4 Conclusions, Applicability and Future Directions

Robust and accessible home care is a critical service in today's society. To improve the efficiency of such services, the adoption of IoT to collect patients' data and analysis of collected data are becoming more prevalent amongst healthcare providers and citizens. Unfortunately, IoT devices can be seriously compromised that can breach the privacy of patients' medical data and violate the security of the infrastructures they connect to. To detect cyber-attacks in the IoT-based smart homes and to take quick prevention measures against them are difficult, because security testing and experimentation cannot be performed on functioning real-time medical devices and services since stopping them may affect tele-monitoring of patients. Therefore, our study provides CDT architecture for automated cybersecurity for IoT-based homecare which allows simulation and security testing using twin models and cognitive features via the closed feedback loops. This study can be extended and applied to different critical sectors of Norwegian Centre for Cybersecurity in Critical Sectors (NORCICS) [21] such as smart cities, smart healthcare, smart electricity systems, and industry 4.0. NORCICS is working towards pushing the boundaries of research-based innovation to enhance the capability of private and public sector stakeholders to respond to the current and future cybersecurity risks by developing, validating, and operationalizing innovative socio-technical solutions [21]. It is believed that the proposed architecture will offer new insights into the critical sectors because the virtual representations of real-world systems or products can be innovative backbone for analysing and predicting future events even before they happen. There is a lot of interest in conducting research on the cyber threat landscape of IoT [22] and structuring AI/ML-based security techniques [23] for intelligent applications. However, there is a lack of an efficient approach which not only offers automated detection of threats for IoT networks but also offers threat prevention strategies and measures. The real strength of the proposed CDT solution lies in its ability to learn, anticipate, detect, identify, and protect automatically against cyber threats over time.

The developed CDT architecture can be extended and incorporated with our existing platforms and methodologies to be applied to critical sectors in the NORCICS project:

Smart Healthcare: In our previous research, we developed adaptive security for smart Internet of Things in eHealth (ASSET) platform [24] which can monitor the health data remotely collected from the Raspberry Pi and Shimmer motes. The health data from the medical IoT devices is stored in the cloud, which can be securely accessed by the concerned authorities remotely using wireless communication. In [20], we developed the DT of the ASSET platform [25] which includes the steps for creating virtual system, executing and examining the data sharing model, configuring the function app, analysing and implementing twin models, and performing real-time synchronisation between the physical and virtual systems. Our real-time DT platform can be extended to CDT for enhancing cybersecurity in smart healthcare.

Smart Cities: In [26], we developed resource-efficient approach for data transmission in IoT-enabled smart cities. The resource-efficiency was measured regarding power drain, standard deviation, battery lifetime, delay, and packet loss ratio. The resource-efficient solution can be integrated into CDT for providing balance energy-efficiency and security in smart cities.

Industry 4.0 and Cyber Physical Electrical Systems (CPES): In [20], we developed cognitive cybersecurity framework for ensuring security and privacy in CPS-IoT enabled critical systems based on cognitive cycle of observe and orient, learn, plan, decide and act. This framework can be one of the main pillars to our CDT approach to enhance security in critical sectors (industry 4.0 and CPES).

In the future, we plan to implement the developed CDT architecture on our existing e-healthcare platform [27] and developed prototype [25] to ensure performance analysis regarding automated cyber-attack monitoring and prevention. Moreover, to research how our proposed approach will prevent new cyber threats that may arise due to the advancement of technologies such as 5G and beyond, Information Technology (IT) and Operational Technology (OT) integration.

Acknowledgments. This work has received funding from the Research Council of Norway through the SFI Norwegian Centre for Cybersecurity in Critical Sectors (NORCICS), project no. 310105, and basic institute funding at Norwegian Computing Center (Norsk Regnesentral), RCN grant number 194067.

References

1. Nimmy, K., Dilraj, M., Sankaran, S., Achuthan, K.: Leveraging power consumption for anomaly detection on IoT devices in smart homes. J. Ambient Intell. Hum. Comput. 1–12 (2022)
2. Pirbhulal, S., Pombo, N., Felizardo, V., Garcia, N., Sodhro, A.H., Mukhopadhyay, S.C.: Towards machine learning enabled security framework for IoT-based healthcare. In: 2019 13th International Conference on Sensing Technology (ICST), pp. 1–6. IEEE (2019)
3. Li, W., Su, Z., Zhang, K.: Security solutions for IoT-enabled applications against the disease pandemic. IEEE Internet Things Mag. **4**(4), 100–106 (2021)
4. Pappot, N., Taarnhøj, G.A., Pappot, H.: Telemedicine and e-health solutions for COVID-19: patients' perspective. J. Telemed. e-Health **26**(7) (2020)
5. Hollander, J.E., Carr, B.G.: Virtually perfect? Telemedicine for Covid-19. New Engl. J. Med. (2020)
6. Zheng, X., Lu, J., Kiritsis, D.: The emergence of cognitive digital twin: vision, challenges and opportunities. Int. J. Prod. Res. 1–23 (2021)
7. Pokhrel, A., Katta, V., Colomo-Palacios, R.: Digital twin for cybersecurity incident prediction: a multivocal literature review. In: Proceedings of the IEEE/ACM 42nd International Conference on Software Engineering Workshops, pp. 671–678 (2020)
8. Nguyen, T.N.: Toward human digital twins for cybersecurity simulations on the metaverse: ontological and network science approach. JMIRx Med. **3**(2), e33502 (2022)
9. Zhang, J., Tai, Y.: Secure medical digital twin via human-centric interaction and cyber vulnerability resilience. Connect. Sci. **34**(1), 895–910 (2022)

10. Unal, P., Albayrak, Ö., Jomâa, M., Berre, A.J.: Data-driven artificial intelligence and predictive analytics for the maintenance of industrial machinery with hybrid and cognitive digital twins. In: Curry, E., Auer, S., Berre, A.J., Metzger, A., Perez, M.S., Zillner, S. (eds.) Technologies and Applications for Big Data Value, pp. 299–319. Springer International Publishing, Cham (2022). https://doi.org/10.1007/978-3-030-78307-5_14
11. Al Faruque, M.A., Muthirayan, D., Yu, S.Y., Khargonekar, P.P.: Cognitive digital twin for manufacturing systems. In: 2021 Design, Automation and Test in Europe Conference and Exhibition (DATE), pp. 440–445. IEEE (2021)
12. Eirinakis, P., et al.: Enhancing cognition for digital twins. In: 2020 IEEE International Conference on Engineering, Technology and Innovation (ICE/ITMC), pp. 1–7 (2020). https://doi.org/10.1109/ICE/ITMC49519.2020.9198492
13. Yitmen, I., Alizadehsalehi, S., Akıner, İ, Akıner, M.E.: An adapted model of cognitive digital twins for building lifecycle management. Appl. Sci. **11**(9), 4276 (2021)
14. Zhang, N., Bahsoon, R., Theodoropoulos, G.: Towards engineering cognitive digital twins with self-awareness. In: 2020 IEEE International Conference on Systems, Man, and Cybernetics (SMC), p. 3891 (2020). https://doi.org/10.1109/SMC42975.2020.928335
15. Du, J., Zhu, Q., Shi, Y., Wang, Q., Lin, Y., Zhao, D.: Cognition digital twins for personalized information systems of smart cities: proof of concept. J. Manag. Eng. **36**(2), 04019052 (2020)
16. Andrade, R.O., et al.: An exploratory study of cognitive sciences applied to cybersecurity. Electronics **11**, 1692 (2022)
17. Jiang, Y., Atif, Y.: A selective ensemble model for cognitive cybersecurity analysis. J. Netw. Comput. Appl. **193** (2021). https://doi.org/10.1016/j.jnca.2021.103210
18. Nguyen, T.N.: Cybonto: Towards Human Cognitive Digital Twins for Cybersecurity. arXiv preprint arXiv:2108.00551 (2021)
19. Pirbhulal, S., Abie, H., Shukla, A.: Towards a novel framework for reinforcing cybersecurity using digital twins in IoT-based healthcare applications. In: 2022 IEEE 95th Vehicular Technology Conference (VTC2022-Spring), pp. 1–5. IEEE (2022)
20. Abie, H.: Cognitive cybersecurity for CPS-IoT enabled healthcare ecosystems. In: 2019 13th International Symposium on Medical Information and Communication Technology (ISMICT), pp. 1–6. IEEE (2019)
21. https://www.ntnu.edu/norcics. 21.09.2022
22. Mavroeidakos, T., Chaldeakis, V.: Threat landscape of next generation IoT-enabled smart grids. In: Maglogiannis, I., Iliadis, L., Pimenidis, E. (eds.) AIAI 2020. IAICT, vol. 585, pp. 116–127. Springer, Cham (2020). https://doi.org/10.1007/978-3-030-49190-1_11
23. Memos, V.A., Psannis, K., Lv, Z.: A secure network model against bot attacks in edge-enabled industrial internet of things. IEEE Trans. Ind. Inform. (2022)
24. Berhanu, Y., Abie, H., Hamdi, M.: A testbed for adaptive security for IoT in e-health. In: Proceedings of the International Workshop on Adaptive Security, pp. 1–8 (2013)
25. Orlauskis, V., Pirbhulal, S.: Real-Time Implementation of Digital Twin for IoT based Smart Homes, NR-Notat, DART/14/22 (2022)
26. Sodhro, A.H., Pirbhulal, S., Luo, Z., De Albuquerque, V.H.C.: Towards an optimal resource management for IoT based green and sustainable smart cities. J. Clean. Prod. **220**, 1167–1179 (2019)
27. Abie, H., Balasingham, I.: Risk-based adaptive security for smart IoT in eHealth. In: Proceedings of the 7th International Conference on Body Area Networks, pp. 269–275 (2012)

A Solution to Polynomial Unit Commitment Problems Using Nonlinear Programming

Takehito Azuma(✉)

Utsunomiya University, 7-1-2 Yoto, Utsunomiya, Japan
tazuma@cc.utsunomiya-u.ac.jp

Abstract. In this paper, polynomial unit commitment problems are discussed which are defined as nonlinear optimization problems with some constraints and polynomial performance indices. The optimization problems are described as a class of mixed integer nonlinear programming with polynomial cost functions. A recursive method is derived to reduce the optimization problems with integer variables to nonlinear programming with no integer variables. The proposed recursive method is demonstrated in a small-scale numerical example.

Keywords: Control engineering · Electrical power systems · Unit commitment · Nonlinear programming problems · Mixed integer programming problems · Methods of Lagrange multiplier

1 Introduction

In October 2020, the Japan government declared carbon neutrality to reduce greenhouse gas emissions to zero in Japan by 2050 [1]. In response to this declaration, the sixth basic energy plan was approved by the Japanese cabinet in October 2021. The amount of renewable energy will become almost double and the amount of fossil energy will become almost half by 2030 in comparison with 2019 [2]. Since fossil energy means electrical power by fossil fuel, thermal power generation will be reduced by 2030.

Though electrical power generation from renewable energy has the advantage of a low environmental impact, the disadvantage is that the amount of electrical power generation greatly depends on the amount of solar radiation. Because of the great dependence of weather, it is difficult to control electrical power generation in power systems based on renewable energy. In addition, a reduction of the amount of electrical power generation based on thermal power generation seems to lead to a significant reduction of the adjustment capacity in power systems. General power systems achieve high stability of the balance between the power supply and the demand based on minimizing power generation costs. The adjustment of electrical power generation in thermal power units is achieved as unit commitment problems [3]. In the future unit commitment, it is expected that operation with both stability and economy will become severe. The unit commitment problems will become nonlinear in the future.

Unit commitment problems are considered as optimization problems with some constraints. If the considered performance indices are linear, then the optimization problems

N. K. Suryadevara et al. (Eds.): ICST 2022, LNEE 1035, pp. 71–81, 2023.
https://doi.org/10.1007/978-3-031-29871-4_9

are described as mixed integer linear programming which is called as MILP [4]. If the performance indices are nonlinear, piecewise linearization approaches are applied but complicated [5]. For nonlinear performance indices, the Lagrange method [6] is effective and the quadratic cost function is considered [3]. However solutions to the unit commitment problems for the general polynomial cost functions is not given. Moreover this method based on the Lagrange method is not efficient because analytical solutions are derived.

In this paper, the problem of polynomial unit commitment is considered and an iterative method based on nonlinear programming is derived. If the balance equation of demand power and supply power is considered as a constraint and the polynomial cost function is considered, the unit commitment problem is described as a constrained mixed integer nonlinear programming problem because each state of generators can be considered as an integer variable. Moreover it is difficult to solve the constrained mixed integer nonlinear programming problems. Since it is possible to determine firstly integer-valued variables by using the Lagrange method for this problem, the unit commitment problem is formulated as a nonlinear programming problem that does not include binary variables. In the derived method, the Lagrange multiplier is updated based on the steepest descent method and the unit commitment is reduced to an iterative nonlinear programming problem. The efficacy of the proposed recursive method is demonstrated in a small-scale example.

2 Power Generation Costs and Optimization Problems in Polynomial Unit Commitment

2.1 Generator Cost Functions and Overall Performance Indices

A cost of the generator i at the time k is defined as follows,

$$C_i(P_i(k), U_i(k)) = F_i(P_i(k))U_i(k),$$

where

$$F_i(P_i(k)) = a_{im}P_i^m(k) + a_{i(m-1)}P_i^{m-1}(k) + \ldots + a_{i0},$$
$$U_i(k) \in \{0, 1\}.$$

$P_i(k)$ is the amount of electrical power of the generator i at the time k. $F_i(P_i(k))$ is the power generation cost function of the generator i expressed by a m-degree polynomial. $U_i(k)$ is a binary variable. The state 1 represents the starting state of the generator i at the time k. The state 0 represents the stopped state. Thus $F_i(P_i(k))U_i(k)$ denotes a cost of the generator i at the time k. The cost is $F_i(P_i(k))$ if $U_i(k) = 1$ and 0 if $U_i(k) = 0$.

Furthermore, the sum of the cost functions for the total time $k = 1, 2, \ldots, T$ and all generators $i = 1, 2, \ldots, N$ is defined as follows,

$$J(P, U) = \sum_{k=1}^{T}\sum_{i=1}^{N} C_i(P_i(k), U_i(k)), \tag{1}$$

where P is $P_i(k)$, $i = 1, 2, \ldots, N$, $k = 1, 2, \ldots, T$ and U is $U_i(k)$, $i = 1, 2, \ldots, N$, $k = 1, 2, \ldots, T$.

2.2 Supply-Demand Restraints and Upper and Lower Limits

Taking the power demand in the considered electrical power system as $P_{load}(k)$ at the time k, the balance equation of the demand power and the supply power is described as follows,

$$P_{load}(k) = \sum_{i=1}^{N} P_i(k)U_i(k). \tag{2}$$

For all time $k = 1, 2, \ldots, T$, the above equation must be satisfied to achieve stable power supply.

In addition, the amount of power generation $P_i(k)$ is constrained to the following limits.

$$\text{If } U_i(k) = 0,\ P_i(k) = 0.$$

$$\text{If } U_i(k) = 1,\ P_i^{min}(k) \le P_i(k) \le P_i^{max}(k).$$

Here $P_i^{min}(k)$ and $P_i^{max}(k)$ represent the minimum value and the maximum value of electrical power supplied by the generator i at the time k, and are constant values given in advance. These constraints are described as the following form.

$$P_i^{min}(k)U_i(k) \le P_i(k) \le P_i^{max}(k)U_i(k). \tag{3}$$

In case of $U_i(k) = 0$, $P_i(k)$ becomes zero since both the upper limit and the lower limit are zero.

Here the cost function $F_i(P_i(k))$ is assumed as the next property.

Assumption 1: For the constraint $P_i^{min}(k) \le P_i(k) \le P_i^{max}(k)$, the cost function $F_i(P_i(k))$ is positive, convex downward and monotonically increasing. Therefore the following three conditions are satisfied.

$$\begin{aligned} F_i(P_i(k)) &> 0, \\ \frac{dF_i(P_i(k))}{dP_i(k)} &\ge 0, \\ \frac{d^2F_i(P_i(k))}{d^2P_i(k)} &> 0. \end{aligned}$$

2.3 Unit Commitment as Optimization Problems

From the above definitions, the unit commitment problem in this paper is defined as the following optimization problem.

Optimization problem 1:

$$\min_{P,U} J(P, U),$$

subject to

$$P_{load}(k) = \sum_{i=1}^{N} P_i(k)U_i(k), k = 1, 2, \ldots, T,$$
$$P_i^{min}(k)U_i(k) \leq P_i(k) \leq P_i^{max}(k)U_i(k), k = 1, 2, \ldots, T, i = 1, 2, \ldots, N,$$
$$U_i(k) \in \{0, 1\}, k = 1, 2, \ldots, T, i = 1, 2, \ldots, N.$$

The considered unit commitment problem is to find the amount of power generation $P_i(k)$ and the binary state $U_i(k)$ of all generators $i = 1, 2, \ldots, N$ and all time $k = 1, 2, \ldots, T$ which satisfy the constraints in Eqs. (2) and (3) and minimize the performance index in Eq. (1). We call the problem as polynomial unit commitment because the performance index is the sum of polynomials.

This problem is one of mixed integer nonlinear programming problems since the power generation costs are given as m-degree polynomials of the variable $P_i(k)$. It is difficult to solve this problem computationally as general nonlinear programming problems because the problems contain binary variables. Especially the existence of the binary variable $U_i(k)$ makes this problem complex.

3 A Solution to Polynomial Unit Commitment Based on Nonlinear Programming

3.1 Lagrange Functions and min-max Problems

Considering the constraint equation in Eq. (2), the Lagrange function is defined as follows,

$$L(P, U, \lambda) = J(P, U) + \sum_{k=1}^{T} \lambda_k (P_{load}(k) - \sum_{i=1}^{N} P_i(k)U_i(k)),$$

where λ_k is a Lagrange multiplier and λ denotes $\lambda_k, k = 1, 2, \ldots, T$. The following min-max problem is considered for the above Lagrange function.

Optimization problem 2:

$$\min_{P,U} \max_{\lambda} L(P, U, \lambda),$$

subject to

$$P_i^{\min}(k)U_i(k) \leq P_i(k) \leq P_i^{\max}(k)U_i(k), k = 1, 2, \ldots, T, i = 1, 2, \ldots, N,$$
$$U_i(k) \in \{0, 1\}, k = 1, 2, \ldots, T, i = 1, 2, \ldots, N.$$

The optimal solution P^*, U^*, λ^* of the min-max problem (Optimization problem 2) is equal to the optimal solution P^*, U^* of Optimization problem 1.

3.2 A Solution U of min-max Problems

Transforming the min-max problem (Optimization problem 2) and considering the following function,

$$q(\lambda) = \min_{P,U} L(P, U, \lambda) \tag{4}$$

subject to

$$\begin{gathered} P_{load}(k) = \textstyle\sum_{i=1}^{N} P_i(k)U_i(k), k = 1, 2, \ldots, T, \\ P_i^{\min}(k)U_i(k) \le P_i(k) \le P_i^{\max}(k)U_i(k), k = 1, 2, \ldots, T, i = 1, 2, \ldots, N, \\ U_i(k) \in \{0, 1\}, k = 1, 2, \ldots, T, i = 1, 2, \ldots, N, \end{gathered}$$

the following maximization problem for Lagrange multiplier λ can be obtained.

$$\max_{\lambda} q(\lambda).$$

Note that the optimal solution to this maximization problem P^*, U^*, λ^* is equal to the optimal solution P^*, U^*, λ^* of the min-max problem (Optimization problem 2).

Since it is difficult to solve analytically the solution of this maximization problem, the Lagrange multiplier λ is fixed as $\overline{\lambda}$, then the following equation is given.

$$\overline{L}(P, U) = J(P, U) + \sum_{k=1}^{T} \overline{\lambda}_k (P_{load}(k) - \sum_{i=1}^{N} P_i(k)U_i(k)).$$

By substituting the Eq. (1), the following equation is derived.

$$\overline{L}(P, U) = D + \sum_{k=1}^{T} \sum_{i=1}^{N} (F_i(P_i(k)) - \lambda_k P_i(k))U_i(k),$$

where

$$D = \sum_{k=1}^{T} \lambda_k P_{load}(k),$$

and D does not depend on the variables P, U and is constant. Therefore the equation in Eq. (4) can be written as

$$\overline{q} = D + \min_{P,U} \sum_{k=1}^{T} \sum_{i=1}^{N} (F_i(P_i(k)) - \lambda_k P_i(k))U_i(k).$$

Focusing on the minimization problem in the above equation, the solution P, U is reduced to the solution of the following simple minimization problem under the constraints in Eqs. (2) and (3).

By defining a new function $G(P_i(k))$ and a new constant M_G, the following lemma and theorem are satisfied.

$$\begin{aligned} G(P_i(k)) &= F_i(P_i(k)) - \lambda_k P_i(k), \\ M_G &= \min_{P_i^{\min}(k) \le P_i(k) \le P_i^{\max}(k)} G(P_i(k)). \end{aligned}$$

Lemma 1: For the constraint $P_i^{min}(k) \leq P_i(k) \leq P_i^{max}(k)$, the function $G(P_i(k))$ is convex downward.

Proof: Based on Assumption 1, the following condition is satisfied for the constraint $P_i^{\min}(k) \leq P_i(k) \leq P_i^{\max}(k)$.

$$\frac{d^2G(P_i(k))}{d^2P_i(k)} = \frac{d^2F_i(P_i(k))}{d^2P_i(k)} \geq 0.$$

Theorem 1: The minimization problem for given numbers i, k

$$\min_{P_i(k),U_i(k)} G(P_i(k))U_i(k)$$

subject to

$$P_i^{\min}(k)U_i(k) \leq P_i(k) \leq P_i^{\max}(k)U_i(k),$$
$$U_i(k) \in \{0, 1\},$$

has solutions as follows,

- $U_i(k) = 0$ if $M_G \geq 0$.
- $U_i(k) = 1$ if $M_G < 0$.

Proof: For the constraint $P_i^{\min}(k) \leq P_i(k) \leq P_i^{\max}(k)$, the function $G(P_i(k))U_i(k)$ satisfies the following condition.

$$G(P_i(k))U_i(k) \geq M_G U_i(k).$$

Thus if $U_i(k) = 0$ and $M_G \geq 0$,

$$\min_{P_i(k),U_i(k)} G(P_i(k))U_i(k) = 0,$$

and moreover if $U_i(k) = 1$ and $M_G < 0$,

$$\min_{P_i(k),U_i(k)} G(P_i(k))U_i(k) = M_G.$$

3.3 A Solution *P* of min-max Problems Based on Nonlinear Programming

Since the solution U of the maximization problem is obtained, the following performance index

$$J(P) = \sum_{k=1}^{T} \sum_{i=1}^{N} C_i(P_i(k), U_i(k)),$$

is considered instead of the performance index in Eq. (1). By solving the following minimization problem, the solution P of Optimization problem 2 is obtained.

$$\min_{P} J(P)$$

subject to

$$P_{load}(k) = \sum_{i=1}^{N} P_i(k)U_i(k), k = 1, 2, \ldots, T,$$
$$P_i^{\min}(k)U_i(k) \leq P_i(k) \leq P_i^{\max}(k)U_i(k), k = 1, 2, \ldots, T, i = 1, 2, \ldots, N.$$

Remark 1: The above solutions P, U are not optimal because the Lagrange multiplier λ is fixed. It is needed to compute P, U by updating λ properly.

3.4 A Recursive Method Based on Nonlinear Programming

Finally summarizing the above discussion, the solution P, U of the unit commitment problem in Optimization problem 1 described as one of mixed integer nonlinear programming problems is given as the solution P, U, λ of the following recursive nonlinear programming problem. The proposed recursive method is based on the steepest descent method.

Step 0 (Settings of optimization problems):
The performance index in this paper is defined as Eq. (1). Moreover Optimization problem 1 and Optimization problem 2 are defined.
Step 1 (Updating the Lagrange Multiplier λ):
The Lagrange Multiplier λ is updating by using the following steepest descent method.

$$\lambda_k(j) = \lambda_k(j-1) + \alpha \frac{\partial q(\lambda)}{\partial \lambda_k}.$$

The initial value is zero and the second term after the second iteration is computed in Step 2.
Step 2 (Finding a solution U):
For the given λ, $k = 1a+, 2, \ldots, T, i = 1, 2, \ldots, N$, the following nonlinear programming problem is solved computationally.

$$\min_{P_i(k)} a_{im}P_i^m(k) + a_{i(m-1)}P_i^{m-1}(k) + \ldots + (a_{i1} - \lambda_k)P_i(k) + a_{i0}$$

subject to

$$P_i^{min}(k) \leq P_i(k) \leq P_i^{max}(k).$$

By defining the optimal solution as $\overline{P}_i(k)$ and the optimal value as M_{ik}, then the following solution $U_i(k)$ is solved.

- $U_i(k) = 0$ and $\overline{P}_i(k) = 0$ if $M_{ik} \geq 0$,
- $U_i(k) = 1$ if $M_{ik} < 0$.

Moreover the coefficient of the steepest descent method in Step 1 and the performance index are computed.

$$\frac{\partial q(\lambda)}{\partial \lambda_k} = P_{load}(k) - \sum_{i=1}^{N} \overline{P}_i(k)U_i(k),$$
$$q = J(\overline{P}, U) + \sum_{k=1}^{T} \lambda_k \frac{\partial q(\lambda)}{\partial \lambda_k}.$$

Step 3 (Finding a solution P):

A solution P of the following minimization problem is computed by using the solution P in Step 2.

$$\min_{P} \sum_{k=1}^{T} \sum_{i=1}^{N} C_i(P_i(k), U_i(k))$$

subject to

$$P_{load}(k) = \sum_{i=1}^{N} P_i(k)U_i(k), k = 1, 2, \ldots, T,$$
$$P_i^{min}(k)U_i(k) \le P_i(k) \le P_i^{max}(k)U_i(k), k = 1, 2, \ldots, T, i = 1, 2, \ldots, N.$$

Step 4 (Calculation of the performance index):

The performance index

$$L = \sum_{k=1}^{T} \sum_{i=1}^{N} C_i(P_i(k), U_i(k)) + \sum_{k=1}^{T} \lambda_k (P_{load}(k) - \sum_{i=1}^{N} P_i(k)U_i(k)),$$

is computed. If the value of the performance index

$$\frac{L - q}{L},$$

is converged, the iteration is ended. Otherwise step 1 to step 4 is iterated.

4 Numerical Examples

4.1 Parameter Setting

For $T = 3$ of the all time, the following electrical power demand is considered.

$$P_{load}(1) = 2[\text{MW}],$$
$$P_{load}(2) = 5[\text{MW}],$$
$$P_{load}(3) = 9[\text{MW}].$$

For $N = 2$ of the number of generators, the parameters of generators are given as follows.

- Generator 1

Minimum output: 1[MW](Cost 1.0000)
Maximum output: 8[MW](Cost 52.7020)
Cost function:

$$F_1(P_1(k)) = 0.001P_1^5(k) + 0.004P_1^4(k) + 0.005P_1^3(k) + 0.990$$

- Generator 2

Minimum output: 1.5[MW](Cost 2.0000)
Maximum output: 7[MW](Cost 46.8250)
Cost function:

$$F_2(P_2(k)) = 0.150P_2^3(k) - 0.125P_2^2(k) - 0.005P_2(k) + 1.850$$

The computer which is used in this example is Intel Corei7-10710U(1.10GHz) and 16GB memory. The number of the computed valuables is 12 sine the number of P is 6(2 $\times$ 3) and the number of U is 6(2 $\times$ 3).

4.2 Computational Results

Case 1: The parameter α of the steepest descent method in Step 1 is given as follows,

$$\alpha = 0.45 \text{ if} \frac{\partial q(\lambda)}{\partial \lambda_k} \geq 0,$$
$$\alpha = 0.02 \text{ if } \frac{\partial q(\lambda)}{\partial \lambda_k} < 0.$$

Then the recursive nonlinear programming is converged as follows.

- Iteration number: 9
- Computation time: 0.4253 [s]
- Value of the performance index: $\frac{L-q}{L} < 0.010$

Values of the performance indices are computed as $J = 21.7140$, $L = 21.7140$, $q = 210670$. The optimal solutions are computed as follows,

$$U^* = \begin{bmatrix} U_1^*(1) \; U_1^*(2) \; U_1^*(3) \\ U_2^*(1) \; U_2^*(2) \; U_2^*(3) \end{bmatrix} = \begin{bmatrix} 1 \; 1 \; 1 \\ 0 \; 1 \; 1 \end{bmatrix},$$
$$P^* = \begin{bmatrix} P_1^*(1) \; P_1^*(2) \; P_1^*(3) \\ P_2^*(1) \; P_2^*(2) \; P_2^*(3) \end{bmatrix} = \begin{bmatrix} 2.000 \; 3.1065 \; 5.0896 \\ 0.000 \; 1.8935 \; 3.9104 \end{bmatrix}.$$

The first column of U^* denotes that the generator 1 is active and the generator 2 is negative at the time 1 when the electrical load $P_{load}(1)$ is small. This is because the generator 1 is a based generator which has a smaller cost. The second column and the third column of U^* denotes that the electrical power operation with the generator 1 and 2 is economical.

Each element of P^* shows that the amount of electrical power of each generator is included in the constraint of the upper and lower bound. Sums of columns satisfies the constraints of electrical power demands at all times. All columns shows the amount of electrical power of the generator 1 is larger than that of the generator 2 because the generator 1 is a based generator and has a small cost. It is characteristic that the amount of electrical power of the generator 1 is not maximum. This is because of the nonlinear programming approach.

Case 2: The parameter α of the steepest descent method in Step 1 is given as follows,

$$\alpha = 0.001 \text{ if} \frac{\partial q(\lambda)}{\partial \lambda_k} \geq 0,$$
$$\alpha = 0.0002 \text{ if} \frac{\partial q(\lambda)}{\partial \lambda_k} < 0.$$

These parameter is based on the paper [3]. Then the recursive nonlinear programming is converged as follows.

- Iteration number: 3187
- Computation time: 139.9059 [s]
- Value of the performance index: $\frac{L-q}{L} < 0.010$

Values of the performance indices are computed as $J = 21.7140, L = 21.7140, q = 21.4969$. The optimal solutions are computed as follows,

$$U^* = \begin{bmatrix} U_1^*(1) & U_1^*(2) & U_1^*(3) \\ U_2^*(1) & U_2^*(2) & U_2^*(3) \end{bmatrix} = \begin{bmatrix} 1 & 1 & 1 \\ 0 & 1 & 1 \end{bmatrix},$$
$$P^* = \begin{bmatrix} P_1^*(1) & P_1^*(2) & P_1^*(3) \\ P_2^*(1) & P_2^*(2) & P_2^*(3) \end{bmatrix} = \begin{bmatrix} 2.000 & 3.1065 & 5.0896 \\ 0.000 & 1.8935 & 3.9104 \end{bmatrix}.$$

The computed optimal solution in case 2 is same as that in case 1. As the iteration number becomes quite large, the computation time is increasing. Moreover the values J, L are same but q is not same. This is because the converged Lagrange multiplier in case 2 is different from that in case 1 after adjusting multipliers using the steepest descent method. Thus the parameter α is very important in this recursive nonlinear programming method.

Remark 2: The variable $P_1(1)$ is divided by 500 in the constraint $1 \leq P_1(1) \leq 8$. $P_1(2)$ and $P_1(3)$ are divided by 500 in the same way. Thus the variable P_1 is divided by 500^3. The variable P_2 is set by satisfying the balance equation in Eq. (2). For the divided valuables $P,\ U$, a full search is performed. The solutions are computed as follows,

$$U^{500} = \begin{bmatrix} U_1(1) & U_1(2) & U_1(3) \\ U_2(1) & U_2(2) & U_2(3) \end{bmatrix} = \begin{bmatrix} 1 & 1 & 1 \\ 0 & 1 & 1 \end{bmatrix},$$
$$P^{500} = \begin{bmatrix} P_1(1) & P_1(2) & P_1(3) \\ P_2(1) & P_2(2) & P_2(3) \end{bmatrix} = \begin{bmatrix} 2.008 & 3.100 & 5.088 \\ 0.000 & 1.900 & 3.912 \end{bmatrix}.$$

Almost same results are obtained in case 1 and case 2. It can be concluded that the optimal solution is obtained by using the proposed recursive nonlinear programming method.

5 Conclusion

In this paper, by considering cost functions as m-degree polynomial forms of power generation, polynomial unit commitment problems are discussed. The problems are described as one of nonlinear programming problems that include binary variables. The problems are a kind of mixed integer nonlinear programming problems that are difficult to find the optimal solution by using general computational methods. Therefore a method based on a recursive nonlinear programming is proposed to solve the problems with binary variables by using methods of Lagrange multipliers. A numerical example is shown to demonstrate the efficacy of the proposed recursive method.

References

1. National and Local Decarbonization Realization Conferences: Regional Decarbonization Roadmap ~Strategies for Transitioning to the Next Era, Starting with Local Regions~ (2021)
2. Agency for Natural Resources and Energy: Basic Energy Plan (2021)
3. Wood, A.J., Wollenberg, B.F., Sheble, G.B.: Power Generation, Operation, and Control, 3rd edn. Wiley-Interscience, (2013)
4. Chang, G.W., Tsai, Y.D., Lai, C.Y., Chung, J.S.: A practical mixed integer linear programming based approach for unit commitment, In: IEEE Power Engineering Society General Meeting, pp. 221–225(2004)
5. Viana, A., Pedroso, J.P., : A new MILP-based approach for unit commitment in power production planning **44**(1), 997–1005 (2013)
6. Bertsekas, D.P.: Nonlinear programming, 3rd edn. Athena Scientific (2016)
7. Byrd, R.H., Gilbert, J.C., Nocedal, J.: A trust region method based on interior point techniques for nonlinear programming. Math. Program. **89**(1), 149–185 (2000)
8. Waltz, R.A., Morales, J.L., Nocedal, J., Orban, D.: An interior algorithm for nonlinear programming that combines line search and trust region steps. Math. Program. **107**(3), 391–408 (2006)

Quantifying the Known Unknowns: Estimating Water Take with Optical Remote Sensing and High-Resolution LiDAR DEMs

Robert Day(✉) and Ivars Reinfelds

NSW Natural Resources Access Regulator, Parramatta, Australia
Robert.day@nrar.nsw.gov.au

Abstract. Questions such as "where is the water" are commonly asked in newspapers around the country, particularly in dry years when water is scarce and demands from town water supply, irrigators, traditional owners, and the environment all compete for a limited supply. Ensuring the lawful take of water is critical to maintaining fair and transparent access to water resources, environmental outcomes and to maintaining community trust. Answering questions around "where is the water" is complex as they have broad spatial and temporal scales ranging from a single pump through to entire surface and groundwater systems over periods of months and years. In this paper we present a common approach used by the NSW Natural Resources Access Regulator (NRAR) to address one such question. By utilising remote sensing light detection and ranging (LiDAR) and satellite imagery, we investigate the filling and emptying of a large on-farm storage by calculating storage volume to surface area relationships and integrating satellite imagery analysis to quantify storage volume flux through time. Crop water requirements that relate to storage volume flux are calculated using IrriSat, a publicly available irrigation decision support tool. Employing this approach enables investigation of complex water compliance questions at the scale of individual properties and the enforcement of water law in NSW.

1 Introduction

Water apportioned to irrigators, traditional owners, drinking water supply and the environment is all competing for a limited supply. In wet years with high water availability, most users can access enough volume to meet their needs. In dry years, however, the lack of water availability manifested in reduced allocations leads to robust and passionate debate with questions such as "where is the water" being commonly asked within newspapers across the nation.

The concerns that stem from these questions occur at a variety of spatial and temporal scales. At a small scale, the questions commonly focus on water losses related to critical water supply releases from dams over a period of days to weeks. At a larger scale both spatially and temporally, questions focus on who or what may have caused a stream or aquifer to drain, was there water theft, or whether water recovered for the environment

N. K. Suryadevara et al. (Eds.): ICST 2022, LNEE 1035, pp. 82–91, 2023.
https://doi.org/10.1007/978-3-031-29871-4_10

under efficiency programs can be measured or not in our river systems? These questions require a scalable approach to provide insights as to root causes underlying public perceptions of 'missing' or 'stolen' water.

In NSW, the Natural Resources Access Regulator (NRAR) is responsible for answering compliance aspects around these types of questions, notably whether water has been 'stolen' or taken unlawfully. NRAR has adopted the use of remote sensing as the basis for addressing many of these questions. With the high spatial and temporal resolutions available in many modern satellite imaging platforms, remotely sensed data are ideal for addressing multi-scalar questions at the resolution of individual properties and across large areas for detecting change anomalies over periods of days to multiple years.

In this paper, we present a recent case study to answer a question around "where is the water?" on a large cotton producing property in western New South Wales. Remote sensing analyses combined with climatic data are used to identify crop type and calculate irrigation water requirements. A combination of satellite and Light Detection and Ranging (LiDAR) data are used to calculate time series volumetric changes within a large on-farm storage dam (OFS) to assess crop irrigation requirements against water volume flux within the dam. Such analyses are informative with regard to detection of unmetered, and potentially unlawful, water take at the scale of individual farms.

2 Methods

2.1 Study Area

This study focused on a large on farm storage (OFS) in northwestern New South Wales and irrigated crop fields reliant on water from the OFS (Fig. 1).

2.2 Calculation of Storage Volume Curve

Water volume flux within OFS's cannot be calculated from satellite imagery directly (Brown et al. 2022). Rather a relationship between the observable water surface area and volume is required. To develop this relationship, the best available 1 m digital elevation model (DEM) was downloaded from the Foundation Spatial Data Elevation and Depth Portal (ELVIS). The DEM was processed using QGIS to clip to the dam extent AOI (Fig. 2) and then processed into a storage volume table using the 'Raster Surface Tool' between a minimum elevation of 105.5 m AHD and a maximum elevation of 116.7 m AHD (highest point in DEM) at 0.1 m increments.

2.3 Analysis of Satellite Imagery

Under the NSW Government subscription, PlanetScope 3 m true colour imagery was used for the analysis as it has suitable spatial (3 m on-ground pixel size) and temporal (near daily coverage across NSW) resolution to ensure the water surface area changes in the dam can be observed clearly and sufficiently frequently to calculate water volume flux during the period of interest. From the PlanetScope imagery, the water surface area was digitised in QGIS by observing tonal contrasts between non-water (soil) pixels and wet soil boundaries when present (Lillesand, Kiefer and Chapman 2015).

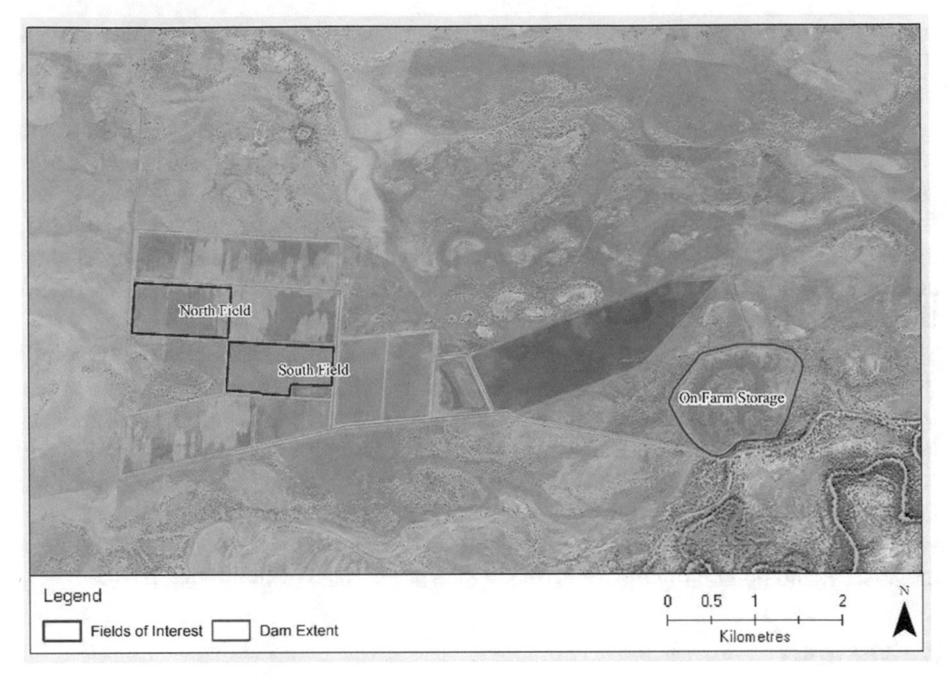

Fig. 1. Study area along the Barwon-Darling River, NSW.

In order to account for uncertainty in volume estimates arising from image interpretation, it was important to understand the magnitude of errors that may have been introduced during the process of digitising the water boundaries (Kimerling 2009; Liro 2015). In a study of the accuracy of digitised riverbank locations from aerial photography, an application fairly similar to digitising water surface areas within dams, Liro (2015) found that multiplying the capture scale of the aerial imagery by 0.25 mm provided a good approximation of digitisation error. Liro (2015) further noted that the accuracy of digitised boundaries is highly dependent on the digitisation scale. For the estimated 1:12,000 capture scale Planet imagery, our experience is that an on-screen digitising scale as low as 1:1800 provides an acceptable balance between observable detail regarding water-soil boundaries (including soil-wet soil boundaries) and image pixelation. Digitising of water surface boundaries for this study was conducted at 1:1800 to 1:2000 scales.

Calculation of Image Capture Scale and Uncertainty Error

The capture scale of the Planet imagery can be estimated using a formula developed by Kimerling (2009) which relates map scale, raster data resolution and map display resolution, (Eq. 1), where pixel ground size (m/pixel), P_{gs}, and pixel density (pixels/m), P_d, such that 1 m on the ground is represented by x m on the map.

$$\frac{1}{x} = \frac{1}{P_{gs} \times P_d} \tag{1}$$

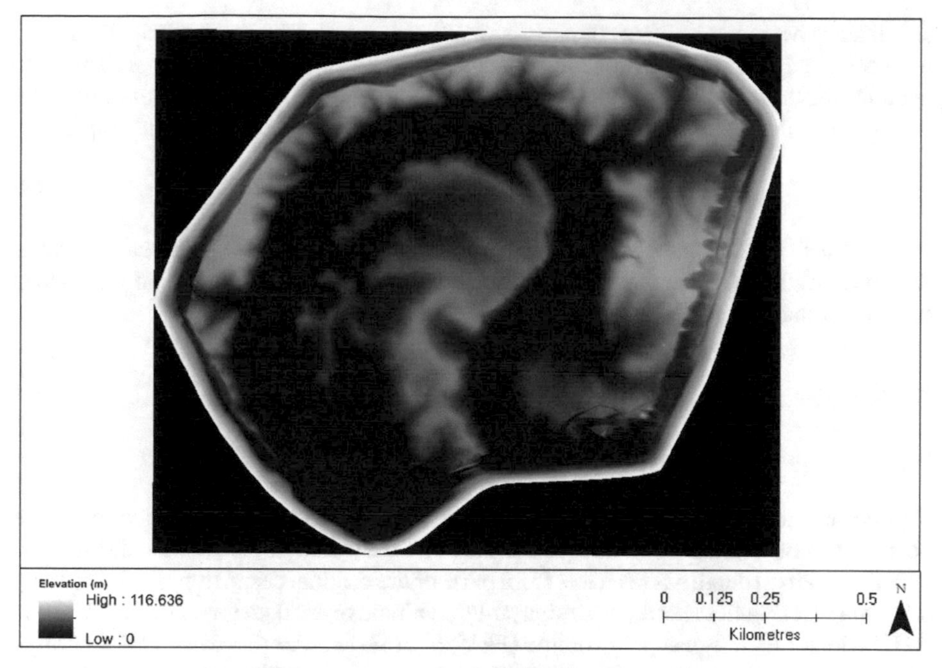

Fig. 2. DEM used for storage volume curve calculation.

At Kimerling's (2009) suggested map pixel density of 4000 pixels per meter, this equation returns an optimal mapping scale of 1:12,000 for Planet (3 m).

Applying Liro's (2015) digitisation error estimate of 0.25 mm multiplied by the 1:12000 optimal mapping scale results in an error of ±3 m (1 pixel). Applying this uncertainty error, the digitised water area polygons were buffered by ±3 m to estimate water volumes within the digitisation margin of error.

2.4 Calculation of Crop Water Usage

For a typical irrigated field, Hornbuckle et al. (2016) suggest that the water balance equation can be simplified as factors such as capillary rise and subsurface flow are negligible. This results in a simplified water balance (Eq. 2) where ΔS is the change in soil water storage (mm), P is the total rainfall (mm), I is the volume of irrigation water (mm) and ET_c is the crop evapotranspiration (the sum of crop transpiration and evaporation from the land surface).

$$\Delta S = P + I - ET_c \tag{2}$$

Raw irrigation water requirement is expressed as the soil water deficit based on an assumption that the crop cycle is initiated with the soil at field capacity (Hornbuckle et al. 2016). The soil water deficit over the crop cycle can therefore be determined as the total crop water requirement less rainfall. A conservative estimate of actual irrigation water required to grow the crop to harvest can be obtained by further subtracting the soil readily available water volume. It is reasonable to assume that an irrigator would not need to restore soil water to field capacity by irrigating before harvest, as there is no need for the crop to continue post-harvest. By subtracting this value from the soil water deficit for each field, the total volume of irrigation water can be determined (Eq. 3):

$$WR = (SWD - RAW) \times Area \tag{3}$$

where WR is the volume of water required (mm per m^2), SWD is the total soil water deficit (mm), RAW is the readily available water for the soil type (mm) and the Area is the area of the field (m^2).

3 Results and Discussion

3.1 Storage Volume Curve

The storage volume curve calculated for the dam is given in Fig. 3. The dam has been designed to have a top of dam level at 116.7 m AHD though the DEM indicates that considerable soil depth has been lost from parts of the original design top level. Applying a 1 m engineering freeboard, full design supply volume of the dam is achieved at 115.7 m AHD with a full design supply volume (FSV) of 5090 Ml. For this dam, a FSV of 5072 Ml had been reported as part of the NSW Government Healthy Floodplains project which characterised OFS volumes across the Northern Basin of NSW. While this volume is 18 Ml lower than the volume of 5072 Ml, this represents a difference of 15 mm in elevation at a surface area of 119 Ha or 0.4%. And is considered negligible.

While the true nature of error within a DEM is difficult to determine due to the randomly distributed nature of error propagated throughout the DEM (Wechsler 1999), the absolute accuracy of LiDAR derived storage volume curves has been assessed as being within 5% of the true volume at full supply level based on limited comparisons with survey data (Morrison and Chu 2018). However, Morrison and Chu (2018 p. 18) also note that "This accuracy assessment is the absolute accuracy when compared to a surveyed storage, whereas the intended use of the stage volume curve is to compare the volume at two points in time, such as before and after a floodplain harvesting event. The relative accuracy along the curve (i.e., the change in volume with elevation) is likely to be considerably higher than the absolute error and therefore these volume changes are likely to be far more accurate than to within the report 5%". In simple terms, this means that the volume differences between two different points along the volume-depth-area curves developed for the FPH project have an accuracy considerably better than 5%. The volume difference of 0.4% was therefore considered an acceptable uncertainty that could be suitably accounted for within the error of digitisation when determining dam water surface area from Planet satellite imagery.

To calculate volumes from this curve, four separate polynomial functions were fitted to the low, middle, and high-volume sections of the SVC based on identifiable shifts in the nature of the surface area to volume relationship (Fig. 4). This approach provided much more precise volume estimates from the polynomial functions, as reflected in the very high coefficients of determination for these equations Fig. 4.

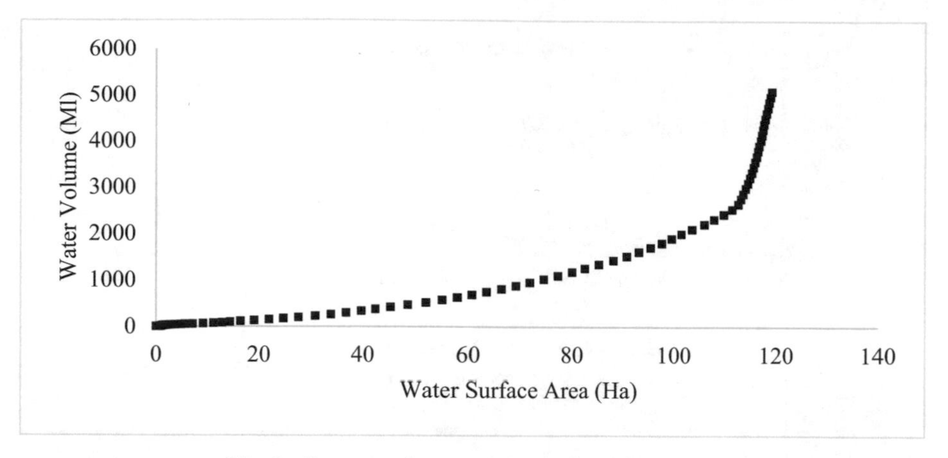

Fig. 3. Storage volume curve calculated for Dam 1

3.2 Volume Calculations

For each of the three datasets, the volume of the dam was calculated using the storage volume curve polynomial functions. Dam volume changes through time reflecting a combination of usage, evaporation, seepage losses and direct rainfall input, and the maximum amount of water held in the storage for each water year, were calculated. Table 1 provides a summary of the calculated volumes for each Planet image analysed and the net change in volume between each image.

From the 'as digitised' volume estimates, a total of 3253 Ml was added to the dam and available for use over the 2016–17 water year (Table 1). While no water was added to the dam during the 2017–18 water year, a total of 1984 Ml was available at the beginning of the 2017–18 water year (Table 1), which was subsequently observed to be released from the dam over the course of the 2017–18 water year.

3.3 Crop Water Usage

From Sentinel false colour imagery, a crop was identified between the 22nd November 2017 and 14th May 2018 across field 1 (54.1 Ha) and field 2 (53.0 Ha) for a total crop area of 107.1 Ha. The date of planting was estimated by subtracting the germination period of between 5 and 14 days (Cotton Australia n.d.). To estimate the highest possible water requirement for the crops, 8 November 2017 was chosen as the planting date and 14 May 2018 as the harvest date.

Crop water usage was estimated using IrriSAT, a publicly available irrigation decision support system. With initial soil water deficit assumed as zero (soil at field capacity at the crop starting date), the total volume of water required by the crops (including rainfall and irrigation) in each field was obtained by multiplying the crop water use (in mm) from IrriSAT by the area of crop (in m^2) and the rooting depth for the crop (assumed to be 1 m) (Fig. 5). This results in a volumetric crop water requirement of 481 Ml for field 1, 394 Ml for field 2, and a total volume of 875 Ml from both rainfall and irrigation as being used by the 107.1 Ha crop area during the 2017–18 water year.

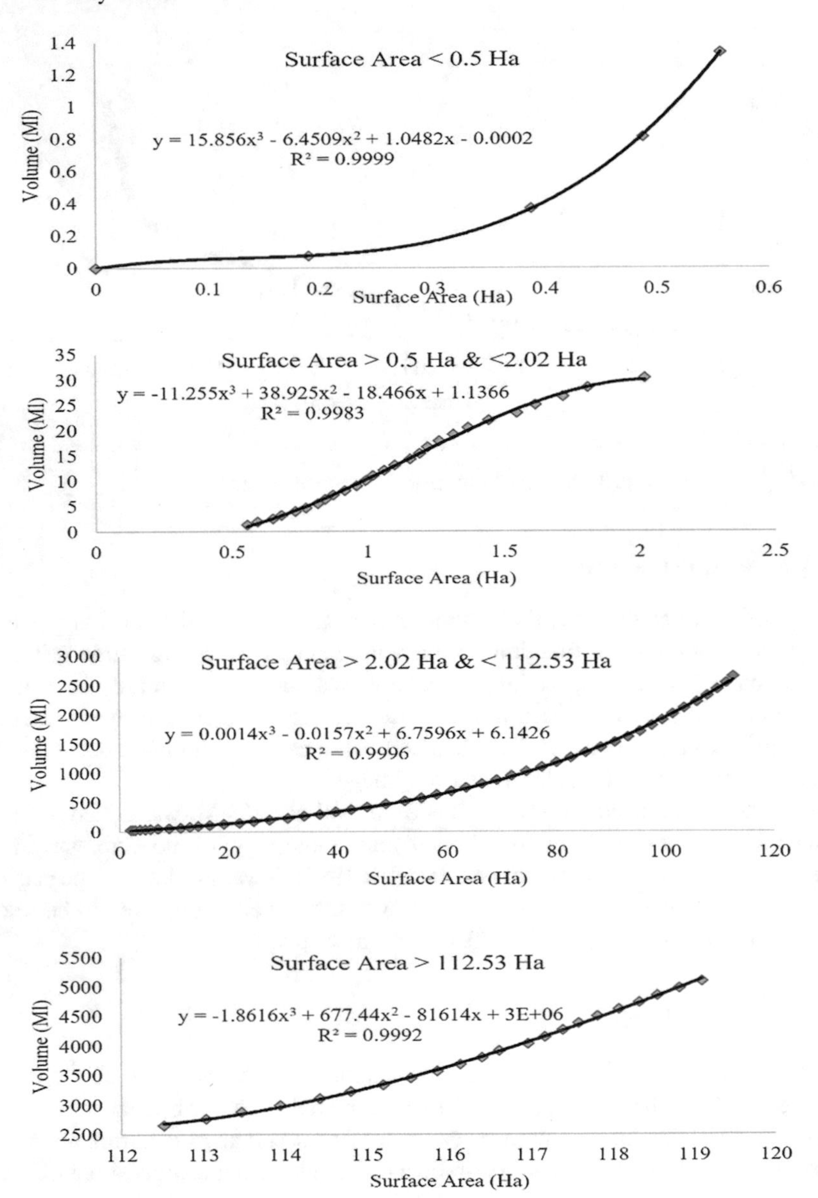

Fig. 4. Polynomial functions applied to the low (sub 2.02 Ha), middle (2.02–112.53 Ha) and high (>112.53 Ha) segments of the storage volume curve for Dam 1.

The aim of irrigation is to keep the plants from wilting by refilling the soil water when the readily available water has been used. Over the lifecycle of a crop, the fields will be topped up several times, as the crop grows and uses water. The sum of this 'topping up' requirement is expressed as the soil water deficit (the total amount of irrigation water needed to keep the plants most productive). It is, however, reasonable to assume that the

Table 1. Calculated Dam Volumes through time

Date (UTC)	+3 m buffer		As digitised		−3 m buffer	
	Volume (ML)	Net change	Volume (ML)	Net change	Volume (ML)	Net change
3/07/2016	869.82		761.66		664.13	
1/08/2016	3108.35	2238.52	2800.01	2038.35	2535.55	1871.42
1/02/2017	2377.40	−730.94	2313.25	−486.76	2250.37	−285.18
12/04/2017	1015.36	−1362.04	918.81	−1394.45	828.62	−1421.74
4/05/2017	2310.37	1295.01	2133.33	1214.53	1958.97	1130.34
1/07/2017	2104.59	−205.78	1984.29	−149.04	1866.00	−92.97

irrigator would not need to restore the soil water to field capacity by irrigating before harvest as there is no need for the crop to continue post-harvest. By doing this, the irrigator would not need to top up the fields before harvest saving the equivalent of the readily available water.

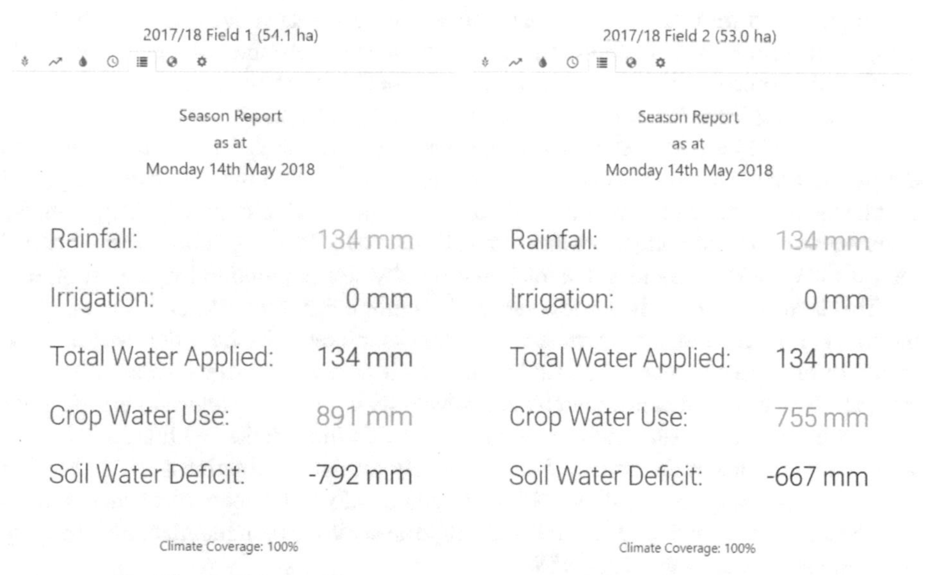

Fig. 5. Polynomial IrriSat crop water usage calculation results

For Field 1 and 2 the IrriSAT total rainfall, crop water use and soil water deficit as total depths (in mm) are shown in 5. By applying Eq. 3 to subtract rainfall and the readily available water for the soil and crop type, the total volume of irrigation water required to meet the crop demand was determined as 374 Ml for field 1, 300 Ml for field 2, and thus a total irrigation volume required over the 2017–18 water year of 674 Ml. In the early stages of crop development, there was excess rainfall, which returned the SWD to zero on several occasions in both fields 1 and 2 of approximately 35 mm (field 1) and 46 mm (field 2) which, under the assumptions of the IrriSAT model, were not available to the crops.

The total water required by the 107.1 Ha crop of 875 Ml describes the volume of water required regardless of its source and includes both irrigation and rainfall. The total irrigation water required (674 Ml) describes just the volume of water required to be supplied to the crops via irrigation assuming 100%. The difference of approximately 200 Ml comes from the 134 mm of rain, which fell at the property during cultivation, and from water savings arising from not refilling the soil profile to field capacity prior to harvest. This approach, which assumes an irrigation efficiency of 100% and an end of crop soil moisture deficit equivalent to the readily available water for the soil and crop type, provides a degree of conservatism in estimating crop irrigation water volumes.

4 Conclusion

In this case study, we sought to understand when water was accessed, what volume was taken and if it was used to support the growth of a crop during any part of the period of interest. Given the broad temporal scale of the question, the lack of instrumentation detailing dam volume flux and crop water usage, and the substantial spatial scale over which the works are located, it was not possible to for NRAR to solely rely on a 'boots on the ground' approach to this and similar complex questions around "where is the water?". We therefore opted to utilise remote sensing approaches to answer these questions.

By utilising LiDAR information, we were able to calculate a storage volume curve for the dam of interest to relate observable water surface areas in satellite imagery to a water volume. This allowed calculation of water flux within the dam over the period of interest, and ultimately, determination of total water available for crop irrigation. By utilising the remote sensing capabilities and climatic data sourcing built-in to the IrriSAT tool, we were able to calculate the total volume of water required to irrigate the crop.

This analysis ultimately led to a successful compliance outcome and demonstrated the value of a remote sensing approach to water compliance. We have demonstrated the ability to successfully relate observable and quantifiable water surface areas on satellite imagery to volumes in dams, thereby allowing tracking of dam water volume changes through time, and the successful combination of this information with crop irrigation water requirements as determined from a publicly available irrigation decision support tool IrriSAT. This approach allows NRAR to objectively assess complex water compliance questions covering broad spatial and temporal scales to help maintain transparency and fairness in water use across NSW.

References

Brown, P., Colloff, M.J., Slattery, M., Johnson, W., Guarino, F.: An unsustainable level of take: on-farm storages and floodplain water harvesting in the northern murray–darling basin, Australia'. Australas. J. Water Resour. 1–16 (2022). https://doi.org/10.1080/13241583.2022.2042061

Cotton Australia: How is cotton grown?. Cotton Australia. https://cottonaustralia.com.au/how-is-cotton-grown (n.d.). 1 Dec 2021

Hornbuckle, J., Vleeshouwer, J., Ballester, C., Montgomery, J., Hoogers, R., Bridgart, R.: IrriSAT Technical Reference (2016)

Kimerling, J.: Mathematical relationships among map scale raster data resolution and map display resolution. https://www.esri.com/arcgis-blog/products/product/imagery/mathematical-relationships-among-map-scale-raster-data-resolution-and-map-display-resolution/ (2009). Last Accessed 15 Jul 2022

Lillesand, T.M., Kiefer, R.W., Chipman, J.W.: Remote Sensing and Image Interpretation, 7th edn. Wiley (2015)

Liro, M.: Estimation of the impact of the Aerialphoto scale and the measurement scale on the error in digitization of a river bank. Zeitschrift Für Geomorphologie **59**(4), 443–453 (2015). https://doi.org/10.1127/zfg/2014/0164

Morrison, T., Chu, C.-T.: Storage Bathymetry Model Update and Application. Department of Industry - Office of Water (2018)

Wechsler, S.P.: Digital elevation model (DEM) uncertainty: evaluation and effect on topographic parameters. https://proceedings.esri.com/library/userconf/proc99/proceed/papers/pap262/p262.htm (1999). 7 Dec2021

Development of a New Line Illumination for Industrial Hyperspectral Imaging Systems

Thomas Arnold[1](✉), Tibor Bereczki[1], and Dirk Balthasar[2]

[1] Silicon Austria Labs GmbH, Villach, Austria
thomas.arnold@silicon-austria.com
[2] TOMRA-Sorting Gmbh, Mülheim-Kärlich, Germany

Abstract. Hyperspectral line scan systems are used in industrial sorting solutions to differentiate materials by subtle spectral features (spectral fingerprint) which are not resolvable by the human eye. Prominent applications are the detection of contaminants in food or the separation of different types of plastic. Hyperspectral line scan systems are suitable for many high throughput applications. A line across the sample, perpendicular to the direction of the relative movement, is projected into an imaging spectrograph. The spectral information for each pixel along this line is projected along the second axis of the two-dimensional detector chip. Thus, it is only necessary to illuminate a narrow line shaped area across the sample. This is usually done with halogen spots which deliver an elliptical illumination spot at the sample. By using several spots in a row and superimposing the illumination spots the area of interest is covered. However, this leads to a non-homogeneous intensity distribution along the scan line and an illumination of a much larger area than necessary. In our approach we use readily available automotive H7 halogen headlight bulbs in combination with an elliptical reflector. We compare different approaches based on mirrors as well as lens arrays to compress the elliptical illumination spot into a line shaped illumination. We compare the different approaches by their intensity distribution of the light along the scanline and the footprint of the light source. Optical simulations and practical measurements of the different light sources are presented.

Keywords: Hyperspectral imaging · halogen illumination · line illumination · automotive H7 headlight bulb

1 Introduction

In many industrial hyperspectral sorting systems halogen spots with aluminum reflectors are used since more than a decade as standard illumination. The halogen spots producing elliptical illumination spots that are superimposed to illuminate an area of interest across the conveyor belt. Due to the beam profile of the halogen spots a much larger area than needed is illuminated and unnecessary heat is introduced. The performance of modern hyperspectral industrial sorting systems is constantly improving. To ensure the high performance the global supply of high-quality halogen spots must be ensured. Therefore, a logic step was taken towards halogen spots that fulfill higher quality requirements

N. K. Suryadevara et al. (Eds.): ICST 2022, LNEE 1035, pp. 92–99, 2023.
https://doi.org/10.1007/978-3-031-29871-4_11

[1]. The automotive H7 headlight bulbs were the perfect candidate. To be able to use the H7 headlight bulbs in industrial sorting systems the emitted light must be collected and reshaped into a line-shaped beam profile. The worldwide change towards circular economy and increasing quality standards lead to new developments in the fields of industrial sorting and quality control solutions. It has already been shown that hyper-spectral imaging (HSI) is a technology that has great potential to solve a variety of inspection tasks. HSI provides spatial and spectral information that can be used to perform chemometric analysis of recycled materials to identify and quantify their chemical composition. The analytical capabilities of the visible (VIS) (400 nm – 700 nm) and the near infrared (NIR) wavelength ranges (900 nm – 2500 nm) are a major advantage enabling non-destructive measurements of chemical parameters. HSI can measure the spectral and spatial information simultaneously and is therefore best suited for industrial applications [2].

2 Hyperspectral Imaging (HSI)

HSI is the combination of machine vision with spectroscopy and comprises the acquisition, processing, and classification of hyper-spectral images [3]. Compared to an RGB camera that captures three broadband channels, a spectral imaging system captures up to several hundred narrow band channels. The increased spectral information makes chemometric analyses possible [4]. The most prominent approaches to acquire hyper-spectral data can be separated into whisk broom or push broom approaches. For high throughput industrial applications both approaches are suitable [5]. To form a line, the whisk broom system uses point spectrometer and a scanning element, and the push broom system uses an imaging spectrograph. A two-dimensional image is formed by a movement of the samples perpendicular to spectral acquisition. The spectral information for each pixel along this observation line is projected along the second axis of the two-dimensional detector chip [6]. The spectral encoding can be provided by an imaging spectrograph (see Fig. 1).

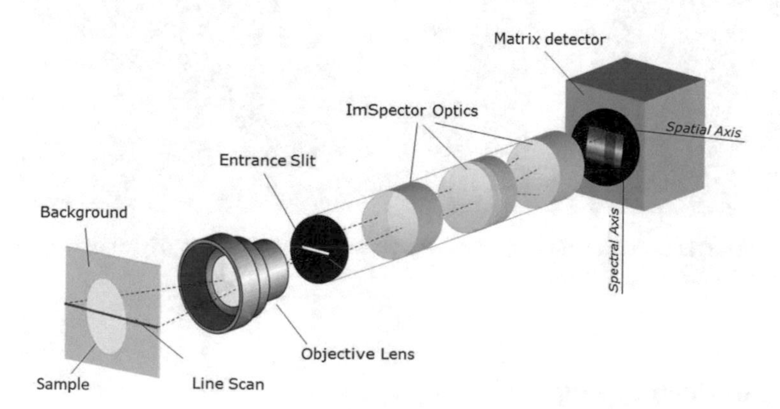

Fig. 1. Imspector imaging spectrograph (Specim, FI)

Since the spatial information along the line is retained, the images contain the spatial information along the first axis and the full spectral wavelength information along the second axis [7]. The spectral and the first spatial dimension are simultaneously acquired, while the second spatial dimension is recorded sequentially due to the movement of the sample relative to the HSI sensor. The HSI data acquisition process produces a three-dimensional dataset with two spatial dimensions and one spectral dimension. For each image pixel a reflectance spectrum is available [8].

3 Optical Simulations

Different approaches based on mirrors as well as lens arrays were simulated using OpticStudio (ZEMAX). The goal was to find a setup that shows high overall illumination intensity that is as homogeneous as possible within a target area of 400x60mm. Mechanical aspects were also taken into consideration to find a setup that is also easy to maintain and retrofit in existing sorting systems. The simulations of the two most promising setups are shown in Fig. 2. Both setups use elliptical reflectors to collect the light from the filament of the H7 headlight bulb and focus it on the imaging plane. However, one setup uses a curved mirror, and the other setup uses an array of lenses to expand the illumination spot along the direction of the scan line of the HSI system. The setup with the lens array is more compact and the lenses also act as cover glass for the illumination system. Due to the fact that both setups concentrate more light in the target area less light sources are needed than compared to a conventional setup without curved mirror or lens array.

Fig. 2. (left) light sources with elliptical reflector and curved mirror, (right) light sources with elliptical reflector and lens array

4 Rapid Prototyping

To be able to verify the results obtained by optical simulations the housings for the new light sources were designed. The design of the housings and mounts was optimized to be suitable for 3D printing. In Fig. 3 the cross-sections of the designed housings and mirror mounts are shown. In Fig. 4 the 3D printed prototypes are shown. Several different lens arrays with different dimensions and number of lenses have been evaluated. Cooling fans and metal heat sinks have been integrated into the housings to perform long term measurements without thermal degradation of the polymer housings.

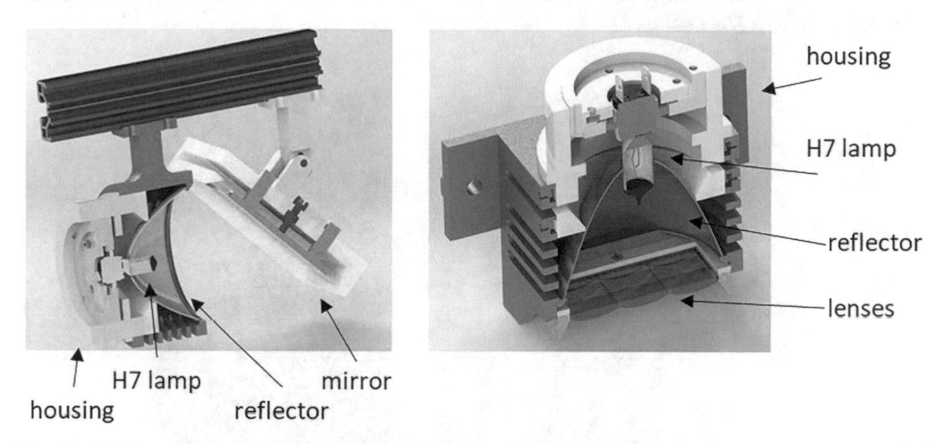

Fig. 3. (left) cross-section of light sources with elliptical reflector and curved mirror, (right) cross-section of light sources with elliptical reflector and lens array

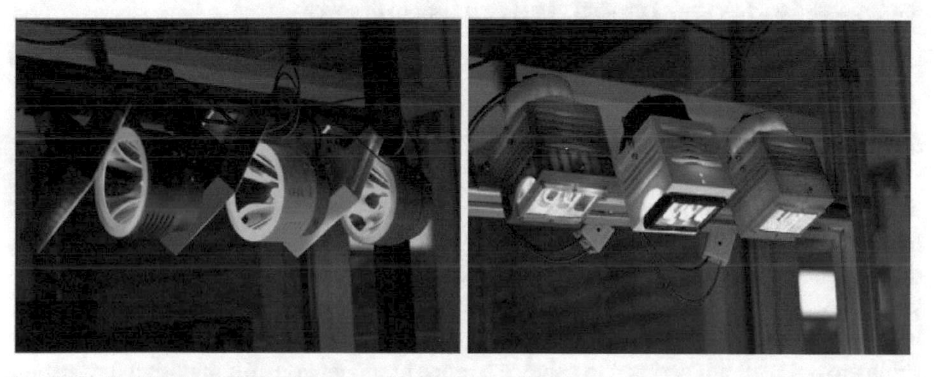

Fig. 4. (left) 3D printed light sources with elliptical reflector and curved mirror, (right) 3D printed light sources with elliptical reflector and lens array

5 Measurements

To evaluate the performance of the new line illumination compared to the initial setup consisting of halogen spots with elliptical reflectors a laboratory setup was established. The setup consists of a 12 Bit monochrome camera (IDS Imaging, GER) with a pixel resolution of 2560 pixel by 1920 pixel and a frame rate of 60 Hz, the light source, and a data acquisition laptop (see Fig. 5). The camera is facing straight down on the target area and the illumination was positioned at an angle of 45°. The geometrical relations are mimicking an industrial sorting system. For evaluating the performance of each light source an area of 400 mm by 60 mm at a working distance of 500 mm was evaluated. The homogeneity of the intensity distribution inside the target area and also the peak intensity were evaluated. Data acquisition and data evaluation were carried out using MATLAB (MathWorks®).

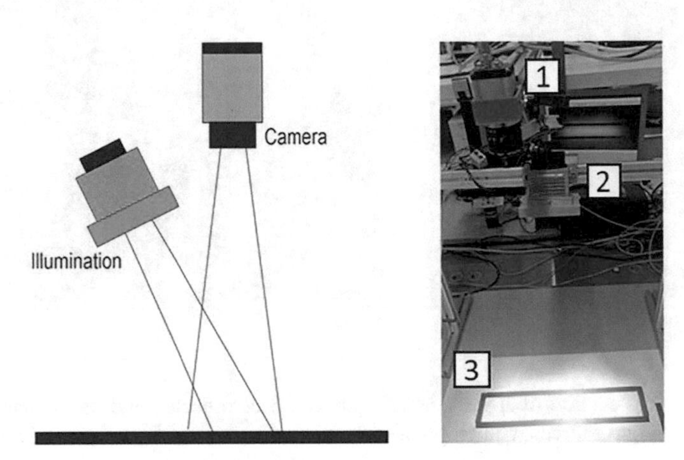

Fig. 5. (left) side view of schematic arrangement of camera (1) and light source (2) relative to the target surface (3), (right) measurement setup used in this work

6 Results

The results of the first measurements show that the new line illumination has a more homogeneous intensity distribution and larger overall intensity. In Fig. 6 the intensity distribution of a light source using a series of halogen spots with elliptical reflectors is shown. Also, the marked target area of 400 mm by 60 mm is visible. It is clearly visible that three elliptical illumination spots are superimposed. However, the illumination spots are not well aligned and to achieve a more homogeneous illumination the spacing between the illumination spots would need to be decreased.

In Fig. 7 the intensity distribution of the new line illumination utilizing the H7 headlight bulb in combination with an elliptical reflector and a lens array is shown. It is visible that the intensity distribution is more homogeneous, and the overall intensity is higher than in the initial setup. The initial setup is using three halogen spots for the illumination and the new line illumination is using two light sources. If three of the new line illuminations are used, then the illumination intensity is even more homogeneous over the whole width of the target area.

The line plot in Fig. 8 shows the horizontal illumination profiles at the middle of the target area for the initial setup and the new line illumination. By changing the lens array the peak intensity and intensity distribution along the scan line can be adjusted. To

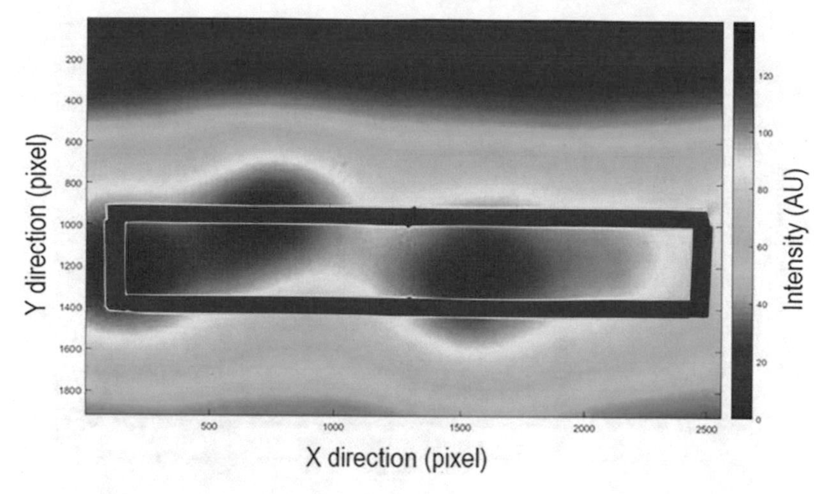

Fig. 6. Spatial intensity distribution of a light source using 4 halogen spots.

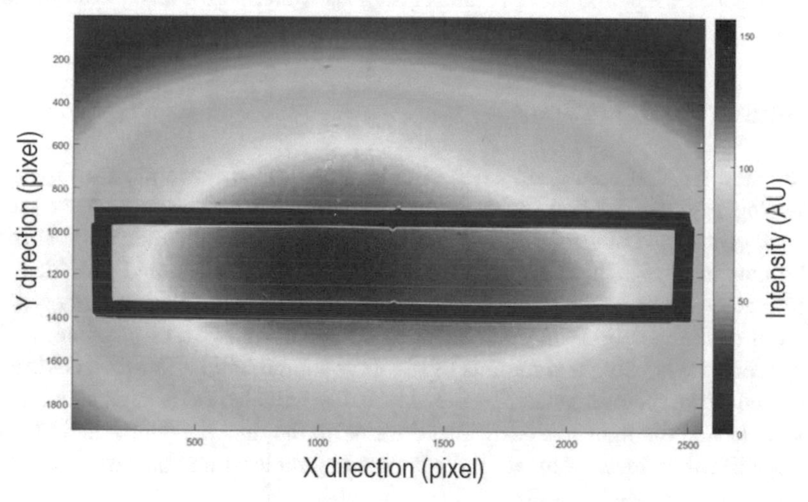

Fig. 7. Spatial intensity distribution of new light source with H7 headlight bulb, elliptical reflector, and lens array.

achieve similar illumination intensity as the initial setup but much more homogeneous illumination one of the new illuminations can replace 3 halogen spots of the initial setup. Due to the standardization of automotive H7 headlight bulbs and the worldwide availability the new light source can reduce energy consumption and shipping costs for spare light sources. Also, H7 headlight bulbs are available in a lot of different variations (spectral distributions) which can benefit industrial applications.

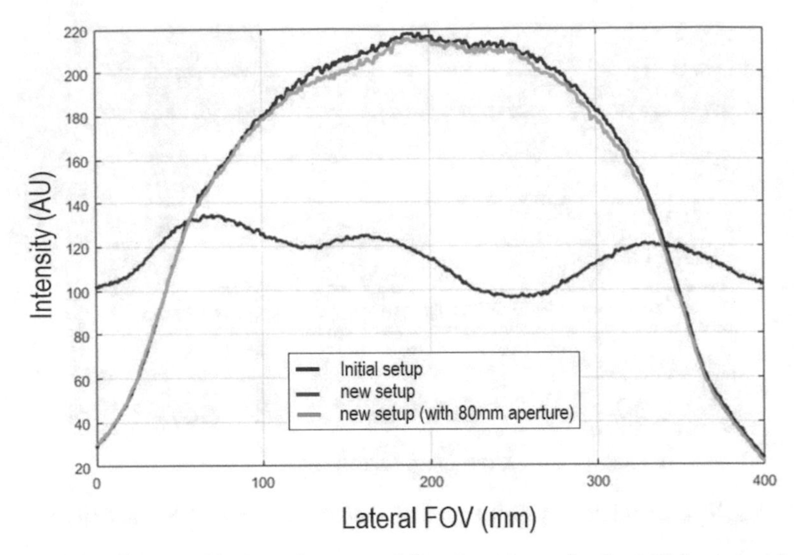

Fig. 8. Horizontal line profile through center of illuminated area for the initial setup and the new light sources.

7 Summary and Outlook

We presented the first results of a new line illumination for an industrial hyperspectral imaging system. The line illumination was evaluated under laboratory conditions with rapid prototyped housings and mirror mounts. Initial results indicate that the new line illumination has a more homogeneous intensity distribution and provides higher overall intensity. The compact dimensions of the version with the lens array will make the retrofit of the new line illumination to existing sorting systems possible. In future investigations the housing and mounts of the light source will be fabricated from metal to fulfill industrial requirements. The metal housing will also act as heat sink and make it possible to seal the light source against dust. With the final design of the light source further qualification measurements and advanced aging tests will be carried out.

Acknowledgement. This work was performed within the COMET Centre ASSIC Austrian Smart Systems Integration Research Center, which is funded by BMK, BMDW, and the Austrian provinces of Carinthia and Styria, within the framework of COMET - Competence Centres for Excellent Technologies. The COMET programme is run by FFG.

References

1. Regulation No 37 of the Economic Commission for Europe of the United Nations (UN/ECE) — Uniform provisions concerning the approval of filament lamps for use in approved lamp units of power-driven vehicles and of their trailers, Official Journal of the European Union (2014)

2. Arnold, T., De Biasio, M., Kammari, R., Sayar-Chand K.: Development of VIS/NIR hyperspectral imaging system for industrial sorting applications. In: Proc. Vol. 11727, Algorithms, Technologies, and Applications for Multispectral and Hyperspectral Imaging XXVII; 117271B (2021) https://doi.org/10.1117/12.2587981, SPIE Defense + Commercial Sensing (2021)
3. Li, Q., He, X., Wang, Y., Liu, H., Xu, D., Guo, F.: Review of spectral imaging technology in biomedical engineering: achievements and challenges. J. Biomed. Opt. **18**(10), 100901 (2013). https://doi.org/10.1117/1.JBO.18.10.100901
4. Groinig, M., Burgstaller, M., Pail, M.: Industrial application of a new camera system based on hyperspectral imaging for inline quality control of potatoes. Proc. OAGM (2011)
5. Bearman, G.H., Nelson, M.P., Cabib, D.: Spectral imaging: Instrumentation, applications, and analysis. In: Proc. SPIE International Society for Optical Engineering (2000)
6. Leitner, R., De Biasio, M., Arnold, T.: High-sensitivity hyperspectral imager for biomedical video diagnostic applications. In: Proc. SPIE: Smart Biomedical and Physiological Sensor Technologies VII (04 2010)
7. De Biasio, T.M., Arnold, R.L.: UAV based multi-spectral imaging system for environmental monitoring. Tech. Mess. **78**(11), 503–507 (2011). https://doi.org/10.1524/teme.2011.0204
8. Amigo, J.M.: (ed.) "Hyperspectral imaging". Elsevier Ltd. book series. Data handling in science and technology, vol. 32, pp/ 0–630 (2019)

Development of a Compact IR-ATR Sensor for Sugar Content Measurement in Liquid Foods

Thomas Arnold[1](✉), Tibor Bereczki[1], Dominik Holzmann[1], Federico Pittino[1], Barbara Oliveira[1], Raimund Leitner[2], Jürgen Holzbauer[2], Frans Starmans[2], and Roland Waldner[2]

[1] Silicon Austria Labs GmbH, Villach, Austria
thomas.arnold@silicon-austria.com
[2] Philips Domestic Appliances Austria GmbH, Klagenfurt, Austria

Abstract. The goal of this development is to develop a compact attenuated total reflection (ATR) sugar sensor which can be integrated in future kitchen appliances like juicers and blenders to track the sugar concentration of the prepared juices and smoothies. Based on the measurement's recommendations can be given to ensure the preferred sugar levels of the users or to support users which need to monitor their sugar intake because of a special diet or health issues. This is an important step towards personalized nutrition. We have evaluated more than 10 different configurations based on the type and arrangement of source, detectors, and ATR crystal. Also, the use of additional lenses and waveguides was evaluated. However, the design showing overall best performance and manufacturability was chosen to build a demonstrator. Optical simulations and practical measurements of different sugar solutions and commercially available fruit juices are presented.

Keywords: IR-ATR spectroscopy · infrared spectroscopy · sugar measurement · diet · nutrition

1 Introduction

In 2015 the world health organization (WHO) published a guideline [1] to provide recommendations on the intake of free sugars to reduce the risk of noncommunicable diseases (NCDs) in adults and children, with a particular focus on the prevention and control of unhealthy weight gain and dental caries. According to this guideline NCDs are the leading causes of death and were responsible for 38 million (68%) of the world's 56 million deaths in 2012 [2]. Modifiable risk factors such as poor diet and physical inactivity are some of the most common causes of NCDs; they are also risk factors for obesity – an independent risk factor for many NCDs – which is also rapidly increasing globally [3]. A high level of free sugars intake is of concern, because of its association with poor dietary quality, obesity, and risk of NCDs [4, 5]. Free sugars contribute to the overall energy density of diets and may promote a positive energy balance [6–8]. Sustaining energy balance is critical to maintaining healthy body weight and ensuring optimal nutrient intake [9]. There is increasing concern that intake of free sugars, particularly in the

N. K. Suryadevara et al. (Eds.): ICST 2022, LNEE 1035, pp. 100–108, 2023.
https://doi.org/10.1007/978-3-031-29871-4_12

form of sugar-sweetened beverages, increases overall energy intake and may reduce the intake of foods containing more nutritionally adequate calories, leading to an unhealthy diet, weight gain and increased risk of NCDs [10–14]. With change in global concern toward food quality over food quantity, consumer concern and choice of healthy food has become increasingly important. This gave rise to the concept of personalized nutrition. Factors including advances in food analytics, nutrition-based diseases, increasing use of information technology in nutrition science and growing consumer concern by better and healthy foods support the theory behind personalization of nutrition [15]. Manufacturers of kitchen appliances support the movement toward personalized nutrition by adding e.g., weight sensors, into an appliance the consumer can be informed about the amount of macronutrients, including sugars. As this information is then based on average values of macronutrients taken from a nutritional database, these values may differ significantly from the actual values. Therefore, they provide software solutions to recommend recipes based on personal preferences or dietary needs. The measurement of food sugar levels is of central importance due to the above-mentioned facts and recommendations. Especially in liquid foods like juices or smoothies it is easy to consume too much sugar because the sugary liquids of fruits get extracted and a single glass of juice can contain more than the daily recommended dose of sugar. In Fig. 1 a demonstrator of an infrared, attenuated total reflection (IR-ATR) sugar sensor is shown that could be integrated in a kitchen appliance like a juicer or blender.

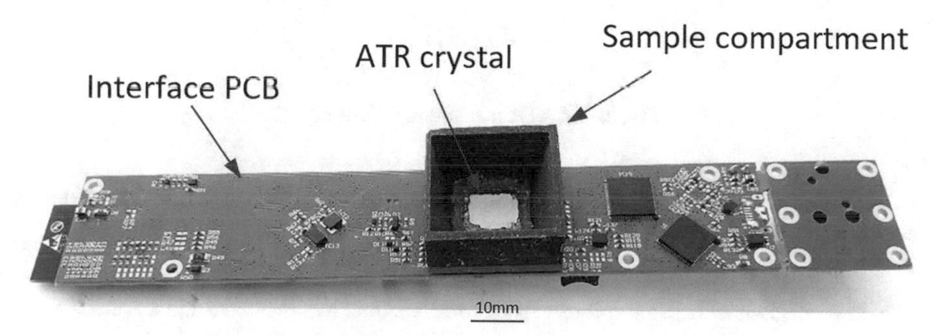

Fig. 1. IR-ATR sugar sensor and readout electronics

2 Infrared Spectroscopy

Infrared (IR) spectroscopy is the measurement of the interaction of infrared radiation with matter by absorption, emission, or reflection. It is used to study and identify chemical substances or functional groups in solid, liquid, or gaseous forms. It can be used to characterize new materials or identify and verify known and unknown samples. The method or technique of infrared spectroscopy is conducted with an instrument called an infrared spectrometer which produces an infrared spectrum like shown in Fig. 3 and Fig. 4. An IR spectrum can be visualized in a graph of infrared light absorption on the vertical axis vs. wavenumber (cm^{-1}) on the horizontal axis. Units of IR wavelength are commonly given in micrometers (μm) which are related to the wavenumber in a reciprocal way. A common laboratory instrument that uses this technique is a Fourier transform infrared (FTIR) spectrometer. A FTIR spectrometer uses an interferometer to acquire the intensity on the detector over the position of the scanning mirror of the interferometer. The resulting signal is called an interferogram which can be Fourier transformed to get a spectrum as shown in Fig. 3 and Fig. 4. The spectroscopic measurements can be performed in different measurement geometries like reflection or transmission. In the case of the measurements in this paper the chosen measurement geometry is attenuated total reflection (ATR). This technique utilizes an ATR crystal to guide the infrared radiation from the source to the detector. Part of the ATR crystal is in direct contact with the liquid sample that is measured and on the surface of the ATR crystal an evanescent wave is present where the IR radiation interacts with the sample. The basic principle is shown in Fig. 2.

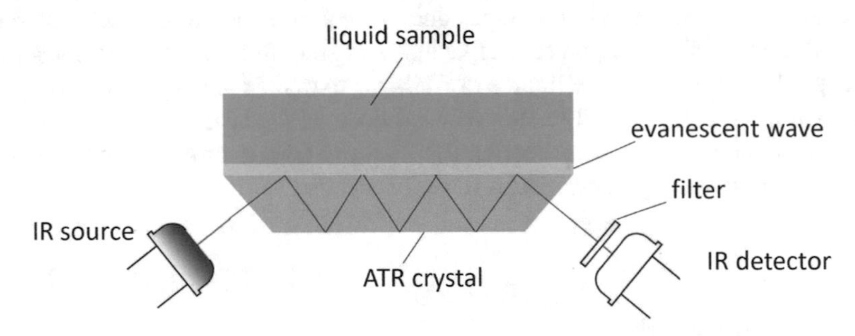

Fig. 2. IR-ATR measurement principle

The infrared part of the electromagnetic spectrum is usually divided into three regions as shown in Table 1.

Table 1. Regions of infrared spectrum with corresponding wavenumbers and wavelengths.

Region	Wavenumber	Wavelength
Near-IR	14000–4000 cm^{-1}	0.7–2.5 μm
Mid-IR	4000–400 cm^{-1}	2.5–25 μm
Far-IR	400–10 cm^{-1}	25–1000 μm

The near-IR can excite overtone or combination modes of molecular vibrations. And the mid-IR is generally used to study the fundamental vibrations and associated rotational–vibrational structure. The far-IR has low energy and may be used for rotational spectroscopy and low frequency vibrations. The infrared spectrum can also be separated into the fingerprint region below 1500 cm^{-1} and the functional group region above 1500 cm^{-1}.

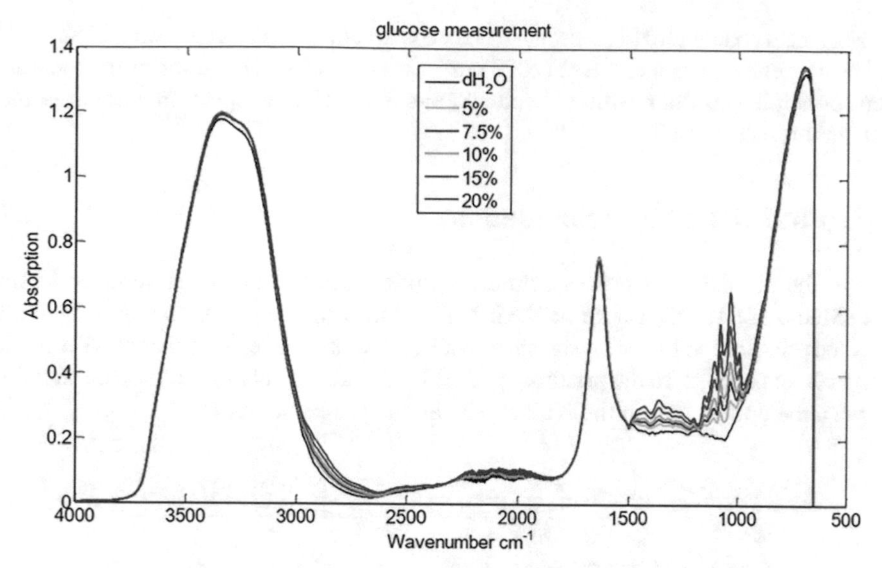

Fig. 3. FT/IR-ATR spectra of aqueous glucose solutions

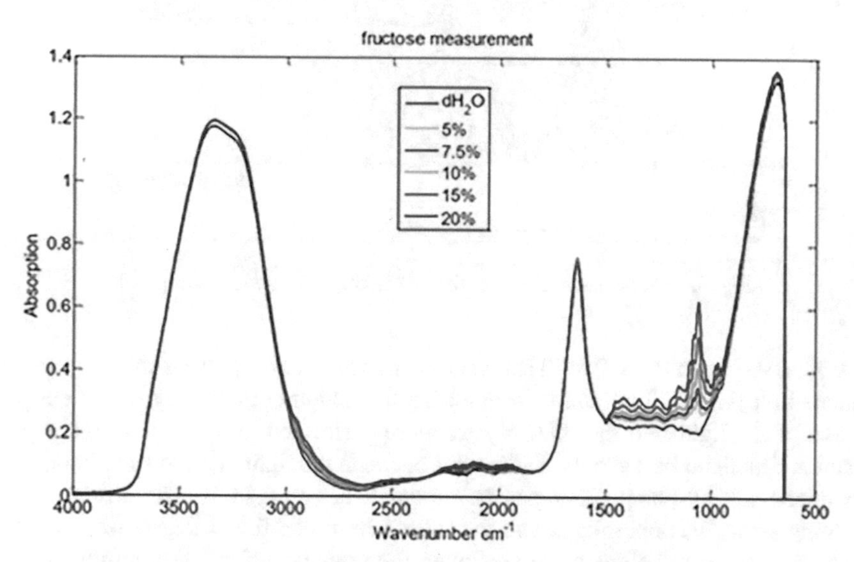

Fig. 4. FT/IR-ATR spectra of aqueous fructose solutions

Pre-studies were performed to investigate different sensing principles and technologies. Spectroscopic technologies like NIR and FTIR spectroscopy as well as hyperspectral sensors were investigated. Due to its high sensitivity and high integration potential IR-ATR spectroscopy was chosen. In Fig. 3 and Fig. 4 FTIR spectra of aqueous glucose and fructose solutions are shown. From these measurements it is visible that the main absorption due to glucose or fructose happens around 1052 cm^{-1}. For this reason, the measurement channel for glucose and fructose was selected to be at 9.5 μm (1052 cm^{-1}) and as a reference channel 5 μm (2000 cm^{-1}) was selected. The proof of the measurement principle was done with a Nicolet Nexus 870 FTIR lab spectrometer based on a germanium ATR crystal.

3 Optical Design and Simulations

During the development process, different optical setups have been simulated using OpticStudio (ZEMAX). Based on the initial test measurements and the optical simulations the prism shaped ATR crystal shown in Fig. 5 was replaced by a planar ATR crystal with Fresnel gratings. In the prism shaped ATR crystal, we only had 3 interaction points between the optical path in the ATR crystal and the sugar solution.

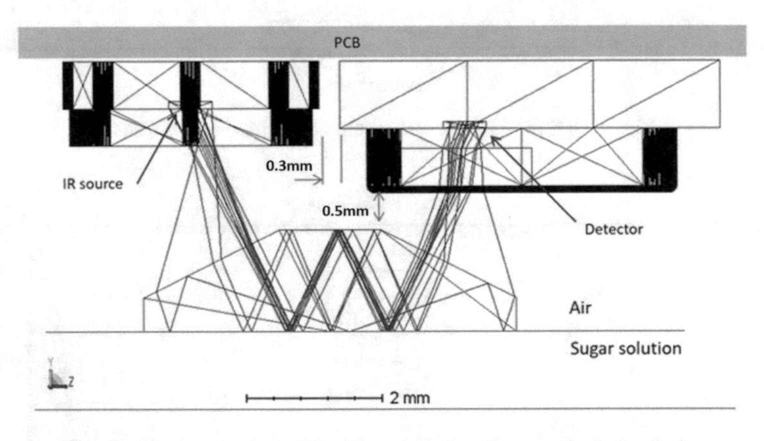

Fig. 5. Geometrical model with ray tracing from optical simulation

With a 0.65 mm thick flat ATR crystal using the Fresnel grating structure we have 6 interaction points. Simulations to optimize the coupling efficiency of the entry and the exit of the light into the ATR crystal were performed. The setup with the planar crystal is beneficial because the interaction between the light and the sample liquid are maximized and all components can be mounted flat on a PCB. The arrangement of the components was optimized, and the influence of additional lenses was evaluated. The goal of the simulations was to estimate the influence of positioning tolerances of the components on the system efficiency. We can increase the signal strength using an IR reflecting waveguide at the IR source side and micro-lenses at the detector side as shown in Fig. 6. However, this results in a more complicated setup with more optical components that have to be positioned precisely. To benefit from the more complicated design the assembly needs active alignment and very tight tolerances for the placement of the individual components. Thus, a simpler design without waveguide and micro lens was selected for the demonstrator.

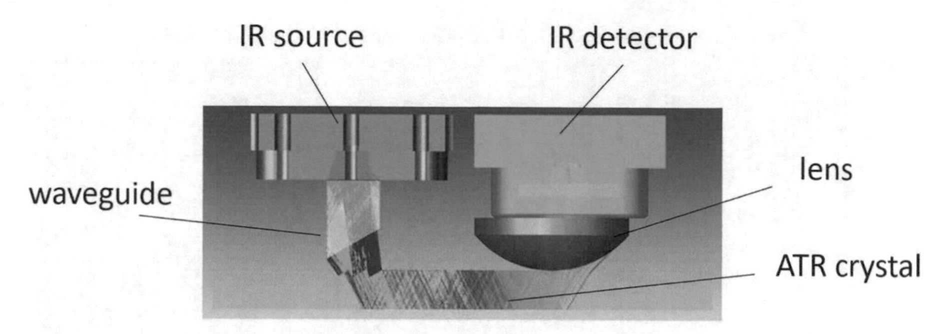

Fig. 6. Setup with flat ATR crystal aluminum waveguide and micro lens.

4 Measurements

For initial tests of the new sugar sensor aqueous glucose and fructose solutions were used. The sugar concentrations ranged from 1 g/100 ml to 10 g/100 ml. The results of the measurements were used to generate calibration models. Also, mixtures of glucose and fructose were measured.

The next step was to measure commercially available fruit juices and smoothies. The values of the nutrition tables of the commercially available juices and smoothies were taken as ground truth for the measurements. The juices were also diluted in the same scheme as the glucose and fructose solutions due to fit the same range of sugar content. The step from sugar solutions to more complex liquids is vital for the testing of the new sensor to evaluate effects from the more complex sample matrix. Also, cross sensitivity tests with e.g. NaCl solutions were carried out and the temperature stability of the sensor was optimized.

5 Results

After optical simulations measurements were performed on a silicon crystal and a simplified setup built and characterized. The following results are from the third generation of the ATR crystal with Fresnel grating and measurement setup (IR source and detectors). A picture of the demonstrator is shown in Fig. 1. In Fig. 7 measurements from fructose solutions with fructose concentrations ranging from 1 g to 10 g per 100 ml of water are shown. Due to the high number of measurements point that the sensor acquires a moving average filter with a kernel size of 20 can be used. The decrease of the sensor signal due to the higher fructose content is clearly visible. These measurements and measurements from glucose samples were used to derive calibration models to link the sensor signal intensity to sugar concentrations.

In Fig. 8 the measurement results from 6 different commercially available fruit juices and smoothies are shown. Due to the high sugar content of up to 12 g/100 ml the samples were diluted to simulate lower sugar contents. For each dilution 3 samples were measured, and the measurement results clearly show the different sugar levels. It is also visible that juices with higher fiber content (e.g., orange juice) show a larger measurement error for higher concentrations.

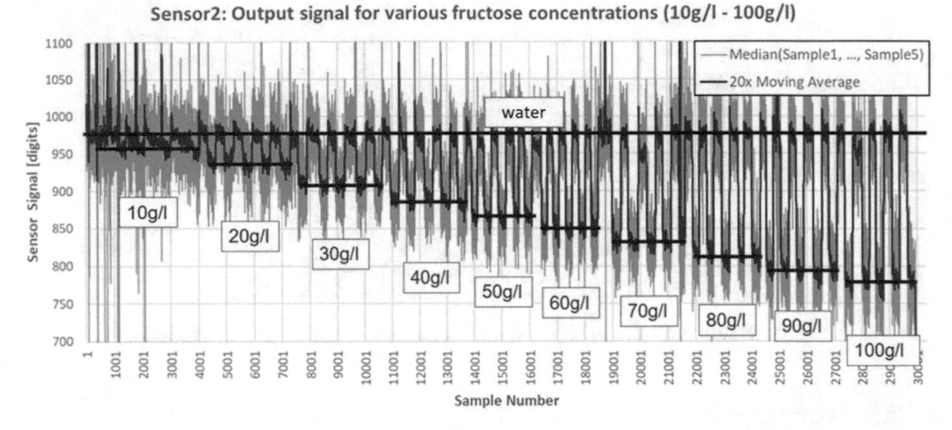

Fig. 7. Sensor signal of the new sugar sensor with respect to different fructose concentrations.

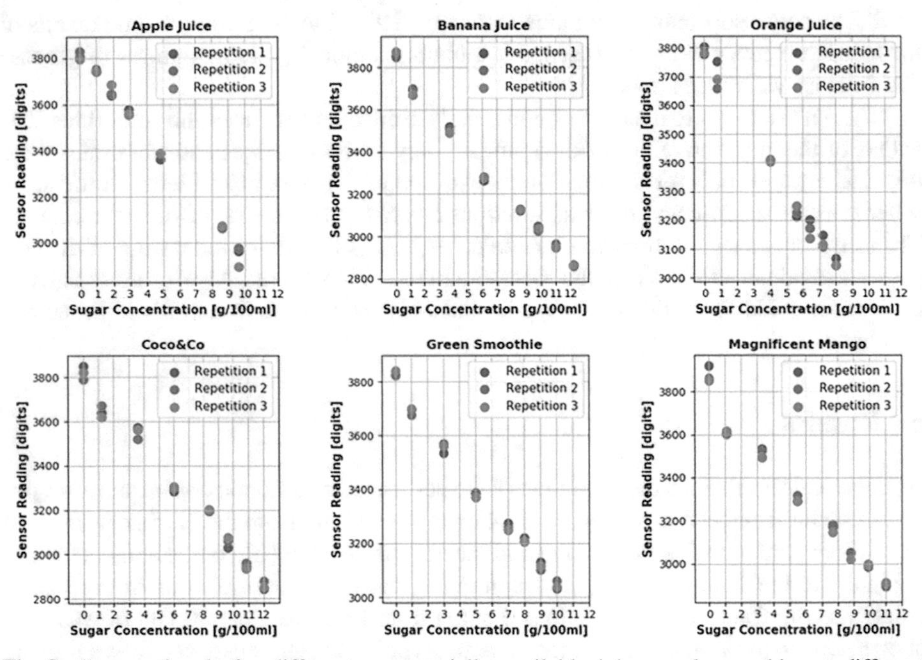

Fig. 8. Sensor signals for different commercially available juices and smoothies at different dilutions.

6 Summary and Outlook

We presented the first results of a new IR-ATR sugar sensor. Calibration measurements of glucose and fructose solutions and measurements of commercial fruit juices and smoothies were presented. The results show that the sensor is capable of covering a wide range of sugar concentrations. The sensor shows great potential to be integrated in a kitchen appliance like a juicer or blender. A sugar sensor for liquid foods can be used to moderate the intake of sugar and help maintaining blood glucose levels which is key to reduce the risk of noncommunicable diseases. This can be achieved through lifestyle changes and the use of technology, which can help make more informed decisions. With the final design of the sugar sensor further measurements on complex samples will be carried out and a combination with other sensors to measure macro nutrients like fat, protein and fiber will be explored.

Acknowledgement. This work was performed within the COMET Centre ASSIC Austrian Smart Systems Integration Research Center, which is funded by BMK, BMDW, and the Austrian provinces of Carinthia and Styria, within the framework of COMET – Competence Centres for Excellent Technologies. The COMET programme is run by FFG.

References

1. WHO guideline on sugars intake for adults and children [cited April 29, 2018]. http://www.who.int/nutrition/publications/guidelines/sugars_intake/en/
2. Invisible numbers: the true extent of noncommunicable diseases and what to do about them. World Health Organization (2022)
3. World health statistics 2022: Monitoring health for the SDGs, sustainable development goals, Geneva (2022)
4. Diet, nutrition and the prevention of chronic diseases: report of a Joint WHO/FAO Expert Consultation. WHO Technical Report Series, No. 916. World Health Organization, Geneva (2003)
5. Global nutrition policy review 2016–2017. Country progress in creating enabling policy environments for promoting healthy diets and nutrition. World Health Organization, Geneva (2018)
6. Johnson, R.K., Appel, L.J., Brands, M., Howard, B.V., Lefevre, M., Lustig, R.H., et al.: Dietary sugars intake and cardiovascular health: a scientific statement from the American Heart Association. Circulation **120**(11), 1011–1020 (2009)
7. World Cancer Research Fund/American Institute for Cancer Research (WCRF/AICR). Food, Nutrition, Physical Activity, and the Prevention of Cancer: A Global Perspective. AICR, Washington, D.C. (2007)
8. Elia, M., Cummings, J.H.: Physiological aspects of energy metabolism and gastrointestinal effects of carbohydrates. Eur. J. Clin. Nutr. **61**(Suppl 1), S40-74 (2007)
9. Fats and fatty acids in human nutrition: report of an expert consultation. FAO Food and Nutrition Paper 91. Food and Agricultural Organization of the United Nations, Rome (2010)
10. Hauner, H., Bechthold, A., Boeing, H., Bronstrup, A., Buyken, A., Leschik-Bonnet, E., et al.: Evidence-based guideline of the German Nutrition Society: carbohydrate intake and prevention of nutrition-related diseases. Ann. Nutr. Metab. **60**(Suppl 1), 1–58 (2012)
11. Malik, V.S., Pan, A., Willett, W.C., Hu, F.B.: Sugar-sweetened beverages and weight gain in children and adults: a systematic review and meta-analysis. Am. J. Clin. Nutr. **98**(4), 1084–1102 (2013)
12. Malik, V.S., Popkin, B.M., Bray, G.A., Despres, J.P., Willett, W.C., Hu, F.B.: Sugar-sweetened beverages and risk of metabolic syndrome and type 2 diabetes: a meta-analysis. Diabetes Care **33**(11), 2477–2483 (2010)
13. Malik, V.S., Schulze, M.B., Hu, F.B.: Intake of sugar-sweetened beverages and weight gain: a systematic review. Am. J. Clin. Nutr. **84**(2), 274–288 (2006)

14. Vartanian, L.R., Schwartz, M.B., Brownell, K.D.: Effects of soft drink consumption on nutrition and health: a systematic review and meta-analysis. Am. J. Public Health. **10**(4), 120 (2007)
15. Chaudhary, N., et al.: Personalized Nutrition and -Omics. Comprehensive Foodomics, pp. 495–507 (2021)

Development of a Loop Animated Full-Color Metasurface Hologram with High Color Reproduction

Hiroki Saito, Masakazu Yamaguchi, Satoshi Ikezawa, and Kentaro Iwami(✉)

Department of Mechanical Systems Engineering, Graduate School of Engineering, Tokyo University of Agriculture and Technology, Koganei 184-8588, Tokyo, Japan
h-saito@st.go.tuat.ac.jp, k_iwami@cc.tuat.ac.jp

Abstract. This study reports a looped full-color holographic movie using dielectric metasurface that exhibits high color reproducibility. Calculations were made to separate the colors in the target image into those represented by the three de-sign wavelengths (443 nm, 532 nm and 633 nm). Then, by reproducing and superimposing their intensity distributions, we succeeded in demonstrating high color reproducibility. Silicon nitride (SiN) was selected as a metasurface mate-rial for its low absorption with a reasonably high refractive index at the visible region. Compared to conventional color holograms made of single-crystal silicon (c-Si), SiN has excellent transparency, although its refractive index is lower than that of c-Si. Therefore, fabrication with a very high aspect ratio is required. This study solves the difficulty and reports the successful color meta-surface hologram cinematic animation. In order to align the viewing angle of the RGB image, the projected images for each component were superimposed by keeping the ratio between the projected wavelength and pixel periods constant. Furthermore, Three-color metasurface holograms were arranged in a square ring to loop the cinematic animation. The average transmittance in each frame image was 84.9%, which was higher than in previous studies. Finally, a loop-moving full-color holographic image consisting of 30 frames was successfully reproduced at a maxi-mum speed of 5 frames per second.

Keywords: Metasurface hologram · Full-color animation · High transmittance metasurface · Microelement projection technology

1 Introduction

In recent years, the advent of 3D technologies such as virtual reality (VR) and 3D glasses has made it possible to convey information that cannot be expressed in 2D. However, these technologies require the wearer to wear equipment, which places a heavy burden on the body, and only certain people are able to view 3D images. Holograms are optical elements that can record and reproduce 3D images and are expected to be applied to very thin planar materials used in 3D displays and wearable sensors [1]. Holograms are gentle to the eyes and brain because there is no need to wear a device, and they have

N. K. Suryadevara et al. (Eds.): ICST 2022, LNEE 1035, pp. 109–118, 2023.
https://doi.org/10.1007/978-3-031-29871-4_13

the advantage of sharing a three-dimensional image with many people at once. Among these holograms, metasurface holograms with nano-planar structures using meta-atoms, which are artificial optical phaseons, are attracting attention. Metasurface holograms can be projected compactly and with a wide viewing angle by controlling light transmission, phase, and wavefront, and the resulting images are characterized by high resolution and low noise. Among other things, it is desired to control the color of reproduced holograms at the design stage to reproduce colors faithfully. Previous studies have reported full-color metasurface holograms designed from HSB color space and multi-color holograms that successfully reproduce animations represented by the three primary colors of RGB [2–5]. However, there are no examples of combining them. In this research, intensity distributions calculated from xyY parameters based on the CIE 1931 color space are reproduced as holograms for each design wavelength (443 nm, 532 nm, and 633 nm) and stacked to achieve full color with high color reproducibility [6]. In addition, the new system combines full color and animation by combining an animation function inspired by the cinematograph method, which uses the afterimage of the human eye to continuously project time-lapse still images into a smooth moving image. Polarization-independent metasurface holograms using silicon nitride octagonal column meta-atoms, which exhibit low absorption and high transmittance, were fabricated, and cinematographic full-color metasurface holograms consisting of 30 frames were successfully reproduced at a maximum speed of 10 frames per second. All projected frame images faithfully reproduced the target image.

2 Material and Methods

Figure 1 shows an overview of the metasurface holograms used in this study. Figure 1(a) shows a unit cell on a periodic metasurface, where one column represents one pixel of the image. Figure 1(b) shows a schematic of a metasurface hologram for animation, inspired by a cinematic approach. Metasurface holograms consisting of 30 frames each of red (633 nm), green (532 nm), and blue (433 nm) main components are arranged in a rectangular shape on a glass substrate, which are sequentially reproduced. Each frame is animated by moving the substrate using an external 2-axis stage.

2.1 Calculation of Transmittance and Phase Delay

Figure 2 shows the results of the calculation of the phase delay depending on the width w by electromagnetic field analysis using the finite element analysis software COMSOL 5.6. Based on this, a column width w of 70 to 170 nm was selected for the wavelength of incident light at 443 nm, 80 to 190 nm at 532 nm, and 70 to 240 nm at 633 nm, respectively, for the design. The average transmittance at each design wavelength was 91.3% at 445 nm, 97.7% at 532 nm, and 97.3% at 633 nm.

2.2 Calculation of Intensity Distribution

The intensity distribution corresponding to each design wavelength using xyY parameters based on CIE 1931 color space was calculated by the following method. First, the target

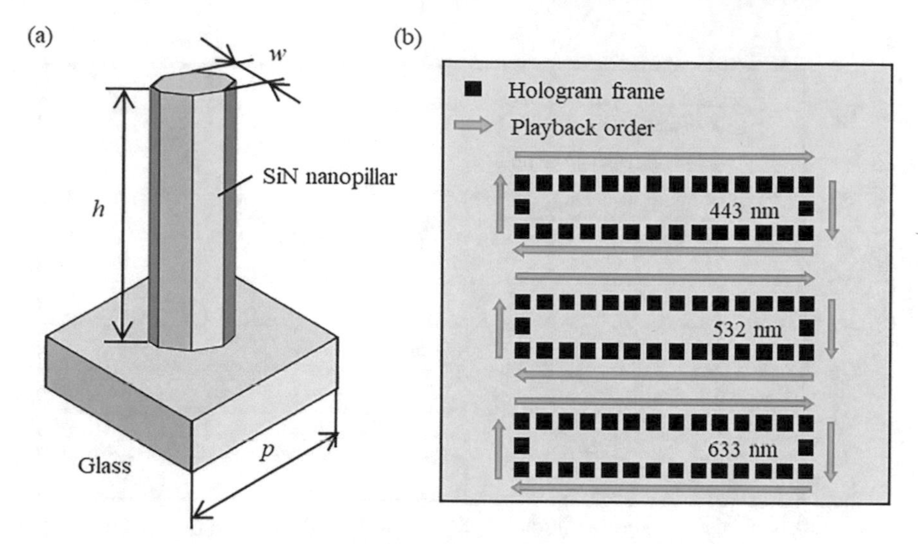

Fig. 1. (a) Schematic diagram of a unit cell on a meta-structured surface. (b) The layout of the metasurface. Holograms consisting of 30 frames were arranged in a looping rectangular shape for wavelengths of 443 nm, 532 nm and 633 nm.

image is separated into RGB components. Next, convert from *RGB* color space to *XYZ* color space from Eq. (1).

$$\begin{pmatrix} X \\ Y \\ Z \end{pmatrix} = \begin{pmatrix} 0.49000 & 0.31000 & 0.20000 \\ 0.17697 & 0.81240 & 0.01063 \\ 0.00000 & 0.01000 & 0.99000 \end{pmatrix} \begin{pmatrix} R \\ G \\ B \end{pmatrix} \tag{1}$$

Since light color mixing calculations are computed in the *xyY* color space, they are converted from the *XYZ* color space from Eq. (2). However, the *Y* value is used as it is because it is an important parameter directly related to intensity.

$$x = \frac{X}{X+Y+Z},\ y = \frac{Y}{X+Y+Z},\ Y = Y \tag{2}$$

Based on the *xyY* values of the full-color image calculated in Eq. (2) and the standardized *xy* values for each design wavelength, the *Y* values at each design wavelength are calculated from Eq. (3) using a simultaneous equation [7].

$$x_m = \frac{\sum\left(\frac{Y_i x_i}{y_i}\right)}{\sum\left(\frac{Y_i}{y_i}\right)},\ y_m = \frac{\sum Y_i}{\sum\left(\frac{Y_i}{y_i}\right)},\ Y_m = \sum Y_i \tag{3}$$

In the above equation, let i be the parameter indicating the design wavelength and m be the parameter indicating after color mixing. That is, x_i denotes the x-value at wavelength i nm, and x_m denotes the x-value of the color mixture (the color to be reproduced). In this study, $i = 443$, 532, and 633. Optical reflectance is a reflectance that takes into account that the brightness perceived by the eye varies with color (wavelength of light). It is

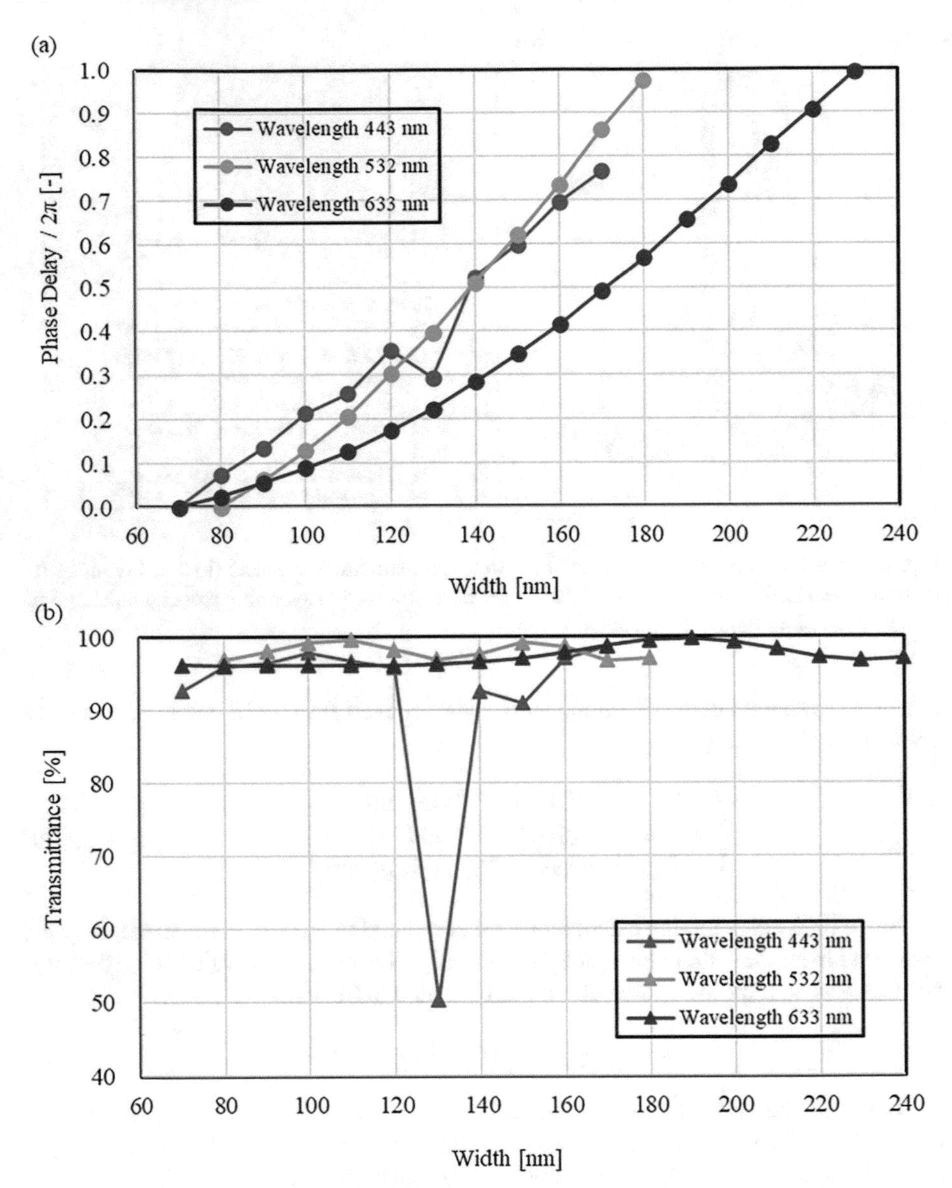

Fig. 2. The numerical simulation results. (a) Simulated phase difference of the SiN nano pillar width when the wavelength is 443 nm, 532 nm and 633 nm, respectively. (b) Simulated transmittance difference of the SiN nano pillar width.

not an intensity value to be reproduced. To obtain the intensity value V from the visual reflectance Y Eq. (4) and Eq. (5) below are used to obtain the intensity value V from the visual reflectance Y [8].

$$10V_i = 2.49268Y'^{\frac{1}{3}}_i + \frac{0.0133}{Y'^{2.3}_i} + 0.0084\sin\left(4.1Y'^{\frac{1}{3}}_i + 1\right) + \frac{0.0221}{Y'_i}\sin(0.39Y'_i - 0.78) - \frac{0.985}{(0.1073Y'_i - 3.084)^2 + 7.54} - \frac{0.0037}{0.44Y'_i}\sin(1.28Y'_i - 0.6784) - 1.5614 \quad (4)$$

$$10V_i = 0.87445Y'^{0.9967}_i \quad (5)$$

2.3 Calculation of the Phase Distribution

The phase distribution in the hologram plane can be obtained by the iterative Fourier transform method (IFTA) shown in Fig. 3. The optimal phase distribution by this method can be obtained by repeating the following steps.

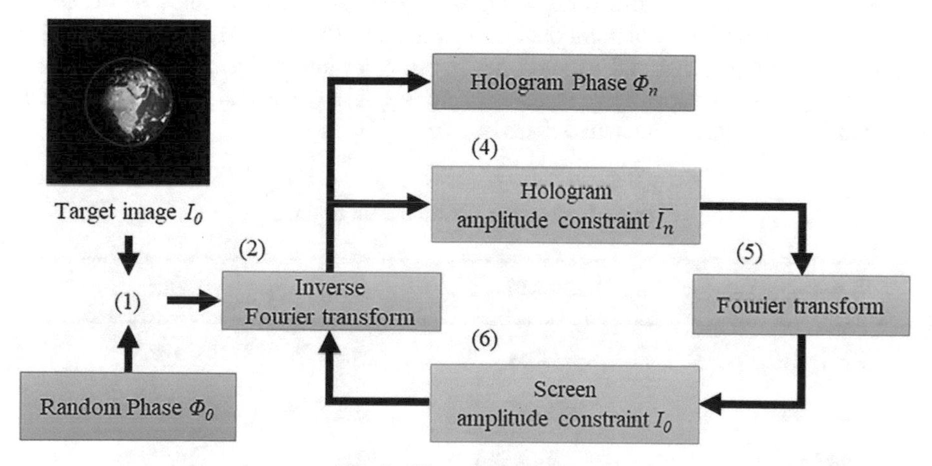

Fig. 3. Flowchart of IFTA.

1). Combines the intensity distribution of the target image with a random phase.
2). Inverse Fourier transform the value obtained in 1).
3). The phase obtained by 2) is the phase distribution reproduced by the hologram.
4). Since the intensity of light incident on the hologram is constant at any plane, the intensity distribution obtained in 2) is replaced by the average value.

5). Fourier transform the combined values of 3) and 4).
6). Since the intensity distribution obtained in 5) is the intensity distribution reproduced on the screen, we replace it with the intensity distribution of the target.

The phase distribution reproduced by the metasurface hologram is the phase distribution when a sufficient projected image is obtained by repeating 1)~6).

Figure 4 shows a frame image of the target movie reproduced in this study. In order to superimpose images, the product of pixel period and resolution is set to be the same at any designed wavelength. In this study, the resolutions of the target images are 2800 × 2800 (443 nm), 2330 × 2330 (532 nm), and 1960 × 1960 (633 nm), respectively. The frame size was thus maintained at 792 μm × 792 μm. Table 1 summarizes the design dimensions of the metasurface determined.

2.4 Fabrication

Metasurface holograms were fabricated using an electron beam lithography (EBL) system based on the design dimensions shown in Table 1. Details of the fabrication process are shown in Fig. 5. SiN film is deposited on a 20 mm × 20 mm glass substrate (a), resist is applied, and the structural pattern is drawn by EBL (b). Chromium (Cr) is then deposited as a mask material (c), and the resist is removed and lifted off (d). Next, unnecessary SiN is removed by reactive ion etching RIE (e). Finally, Cr is removed by wet etching to complete the metasurface hologram (f).

Table 1. Summary of hologram design.

Wavelength [nm]	443	532	633
Pitch p [nm]	283	340	404
Resolution	2800 x 2800	2330 x 2330	1960 x 1960
Width w [nm]	70 − 180	80 − 180	70 − 230
Height h [nm]	1500	1500	1500

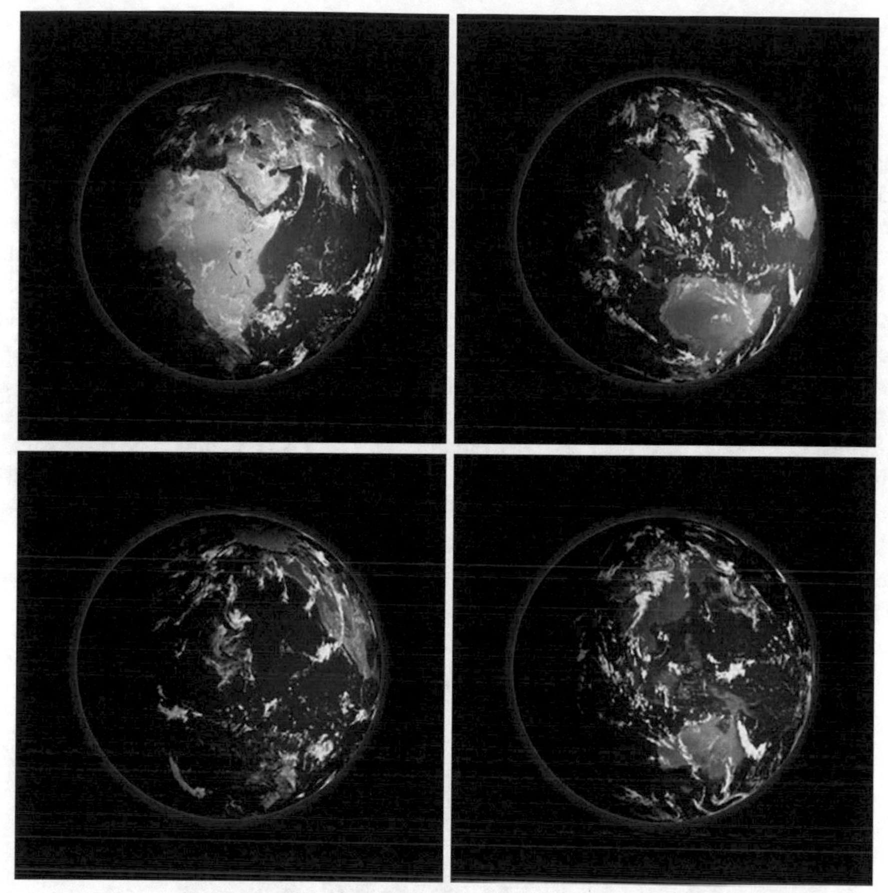

Fig. 4. Frame images in target animation. Among the 30 frames, the frame images shown in the figure are No. 7, No. 14, No. 21 and No. 28.

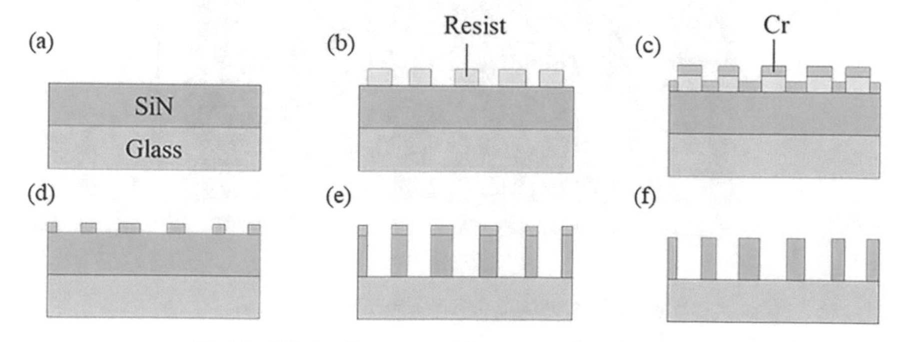

Fig. 5. Fabrication process for metasurface hologram.

3 Results

Images of the produced metasurface holograms are shown in Fig. 6. Rectangular-shaped metasurface holograms were fabricated for the design wavelengths of 443 nm, 532 nm, and 633 nm, respectively. Using the optical system shown in Fig. 7, the laser diameter was narrowed with an iris and the laser intensity corresponding to each component was irradiated parallel to the metasurface hologram. Figure 8 shows the projected image obtained when three wavelengths were input. Compared to Fig. 4, it can be seen that the color holograms are reproduced close to the target frame. As for color reproduction, the brightness is balanced because the characteristics of the designed wavelengths are taken into account. Therefore, we can say that we have succeeded in developing full-color metasurface holograms with multiple wavelengths. The transmittance and diffraction efficiency of the holograms at each designed wavelength were measured, and the transmittance was 89.6% (443 nm), 85.1% (532 nm), and 81.7% (632 nm). The results were adequate, although not as good as those obtained by calculation. The diffraction efficiencies were 7.7%, 12.6%, and 13.8% at 445 nm, 532 nm, and 633 nm, respectively, which were higher than those of previous studies. Finally, the maximum speed of 10 frames/second was successfully achieved by moving the stage and capturing moving images.

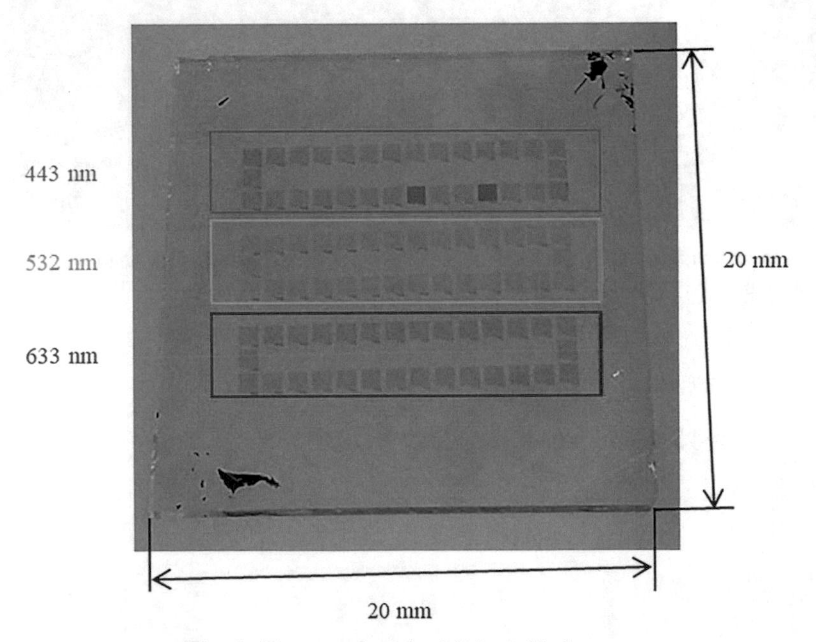

Fig. 6. Photograph of the fabricated hologram.

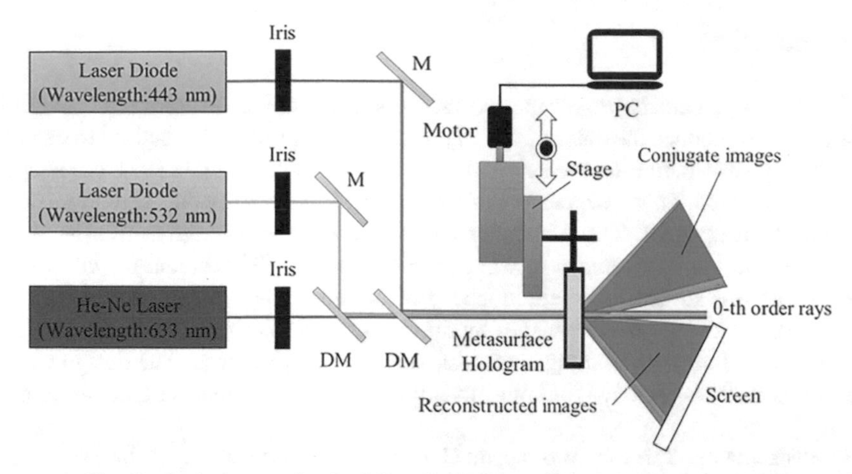

Fig. 7. Optical setup for the full color holographic movie reconstruction.

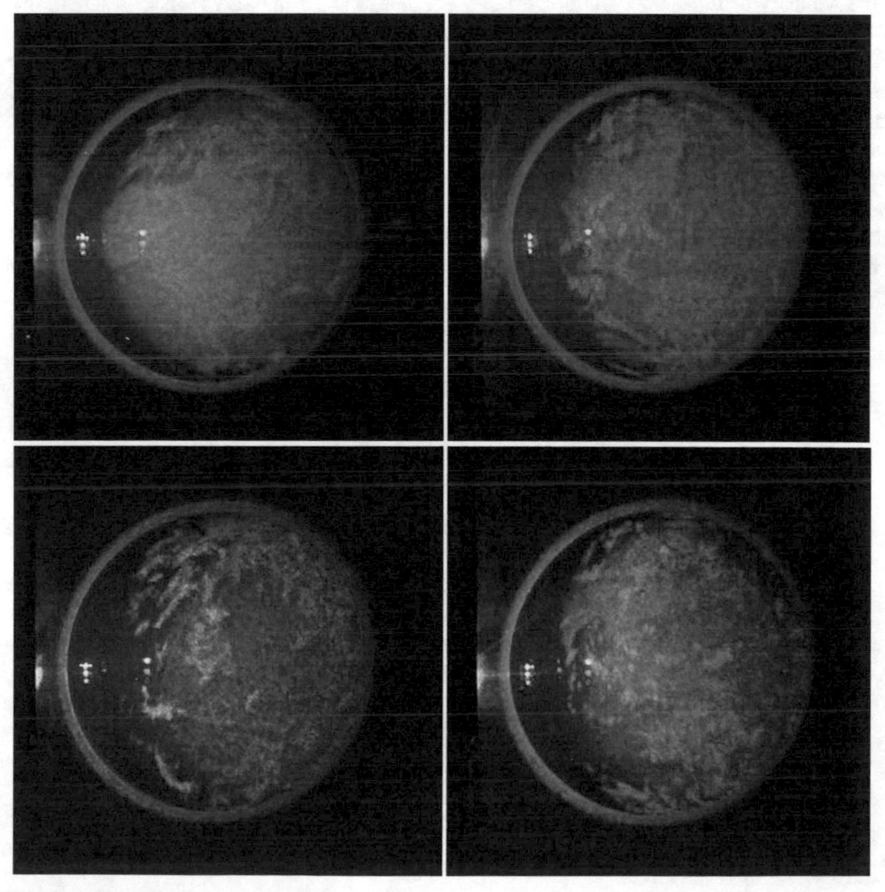

Fig. 8. Pictures of the selected frames corresponding to Fig. 4 of the multicolor holographic movie consisting of 30 frames.

4 Conclusion

We developed a cinematographic metasurface hologram with high color reproducibility. To faithfully reproduce the color of the target image, we proposed a method to calculate the intensity distribution at each design wavelength using the CIE1931 color space. Based on this method, we succeeded in fabricating holograms using silicon nitride as the meta-atom material. The fabricated metasurface holograms were irradiated with lasers at the design wavelengths of 443 nm, 532 nm, and 633 nm, and the color of the target image was successfully reproduced. Furthermore, the substrate was driven by a stage to project a series of still images by the holograms. As a result, a 10-fps animated projection of a full-color rotating earth was successfully produced. The design method used in this study is effective for both full-color and animated metasurface holograms.

Acknowledgements. This work was supported in part by the Japan Society for the Promotion of Science (JSPS) KAKENHI under Grant 21H01781 and Grant 22K04894; in part by the Takeda Sentanchi Supercleanroom, University of Tokyo, through the "Nanotechnology Platform Program" of the Ministry of Education, Culture, Sports, Science, and Technology (MEXT), Japan, under Grant JPMXP09F11857. The authors would like to thank Prof. Y. Mita, Dr. A. Higo, Dr. E. Lebrasseur, and Mr. M. Fujiwara (The University of Tokyo) for their support during sample fabrication, also would like to thank Prof. Lucas Heitzmann Gabrielli (University of Campinas) for developing and maintaining gdstk, a Python library for creating and manipulating GDSII layout files, and also would like to thank a TSUBAME3.0 supercomputer at the Tokyo Institute of Technology for the numerical calculations. This work was supported in part by the Japan Society for the Promotion of Science (JSPS) KAKENHI under Grant 21H01781 and Grant 22K04894; in pard by "Advanced Research Infrastructure for Materials and Nanotechnology in Japan (ARIM)" of the Ministry of Education, Culture, Sports, Science and Technorogy (MEXT), Grant Number JPMXP1222UT1040.

References

1. Kim, L., et al.: Holographic metasurface gas sensors for instantaneous visual alarms. Sci. Adv. **7**(15), 1–9 (2021)
2. Wan, W., Gao, J., Yang, X.: Full-color plasmonic metasurface holograms. ACS Nano **10**(12), 10671–10680 (2016)
3. Bao, Y., et al.: Full-color nanoprint-hologram synchronous metasurface with arbitrary hue-saturation brightness control. Light-Sci. Appl. **8**(95), 1–10 (2019)
4. Izumi, R., Ikezawa, S., Iwami, K.: Metasurface holographic movie: a cinematographic approach. Opt. Express **28**(16), 23761–23770 (2020)
5. Yamada, N., Saito, H., Ikezawa, S., Iwami, K.: Color animation by dielectric metasurface hologram made of silicon nanopillar. Opt. Express **30**(10), 17591–17603 (2021)
6. Fairman, S, Brill, H.M., Hemmendinger, H.: how the CIE 1931 color-matching functions were derived from the Wright–Guild data. Color Res. Appl. **22**(1), 11–23 (1997)
7. LEDs MAGAZINE: https://www.ledsmagazine.com/smart-lighting-iot/whitepoint-tuning/article/16695431/understand-color-science-to-maximize-successwith-leds-part-2-magazine. Last accessed 18 Jul 2012
8. McCamy, S.C.: Munsell value as explicit functions of CIE luminance factor. Color Res. Appl. **17**(3) 205–207 (1992)

Admittance-Type Multifunctional Sensor Based System and Its Controller Design

Bansari Deb Majumder[1(✉)], Sanghamitra Layek[1], and Joyanta Kumar Roy[2]

[1] Narula Institute of Technology, Kolkata, West Bengal, India
bansarideb.research@gmail.com
[2] Eureka Scientech Laboratory, Kolkata, West Bengal, India

Abstract. A multifunctional sensor provides multiple measurements, reducing the cost of using multiple sensors and making the system compact. In multifunction sensor systems, it is very important to have adequate and reliable controls. The system needs to be so designed that an unexpected occurrence in the system can be monitored, controlled, and eliminated. The paper aims to provide an overview of the controller design principle for Admittance-type multifunction Sensor-based process applications. PID controller design, stability, and fragility analysis of the PID controller have been discussed in this paper. Experimental results validate the theoretical concepts.

Keywords: Multifunction Sensor · Admittance-type level sensor · PID controller · Stability · Fragility

1 Introduction

In modern-day Instrumentation systems, different types of sensors are used with separate functions. Each Sensor can measure a single parameter independently. A signal processing algorithm combines all the outputs of the independent Sensor to provide a complete measurement. The generated composite measurement output infers all the individual measurements. Such a type of system is termed a multi-sensor system [1]. In a multi-sensor system, coordination of all the available sensors is important to achieve the system's objectives. The multi-sensor system is classified as a multi-mode and multifunction Sensor. The uniqueness of a multifunction sensor lies in measuring multiple physical parameters from a single measurand. However, such multifunction Sensor demands adequate and reliable controls. An unexpected occurrence in the system needs to be monitored, controlled, and eliminated by the control system. In [2], the researchers have developed a multifunction Sensor capable of simultaneously measuring the liquid level and temperature. Maintaining a constant level in the experimental tank to validate the measurement method is very important. Therefore, an attempt has been made to design a controller and implement in the experimental setup of Admittance-type multifunctional Sensors and instrumentation.

Process control applications are nonlinear, and the complexity of the application increases as the number of parameters increases. Mathematical modeling of process

N. K. Suryadevara et al. (Eds.): ICST 2022, LNEE 1035, pp. 119–127, 2023.
https://doi.org/10.1007/978-3-031-29871-4_14

control applications is time-consuming and computationally expensive. Therefore, the system identification principle is used to frame the mathematical model of any application from experimental input and output data [3]. There are different methods of system identification techniques. Relay-feedback based method is one of the system identification techniques where relays are used to estimate the transfer function [4]. In [5], the authors have used the relay-feedback method to determine the transfer function of a real-time level control system set up where a relay with a hysteresis band is used to excite the process, and transfer function parameters are deduced from the sustained oscillations. In [6], an asymmetrical relay is used to induce sustained oscillations in the system, and the state-space model is estimated. Parametric system identification and controller design for temperature control of heat exchangers have been discussed in [7], where the authors used different time series models and prediction error methods to estimate the dynamics of the heat exchanger system. A laboratory-based experiment to teach the closed-loop system identification scheme has been discussed in [8]. A comparative analysis of different transfer function estimation schemes has been presented in [9].

Further, this paper establishes the formulation of the problem related to the control of process variable needs. This paper provides a detailed procedure to implement a controller for Admittance-type Multifunctional Sensor-based system.

2 Methodology

Figure 1 shows the block diagram of a computerized industrial-level control system with an admittance-type multifunction sensor. The liquid is stored in an industrial storage tank with a fixed datum line. The liquid is pumped into the container made of an HDPE tube from the reservoir. A double-electrode admittance type sensor is used to measure Level and temperature. The Sensor is dipped in the experimental tank. It is connected to the trans-impedance amplifier, and further, the signal is conditioned and fed to the controller using a data acquisition system. The detailed specification of this experimental setup is reported in [2].

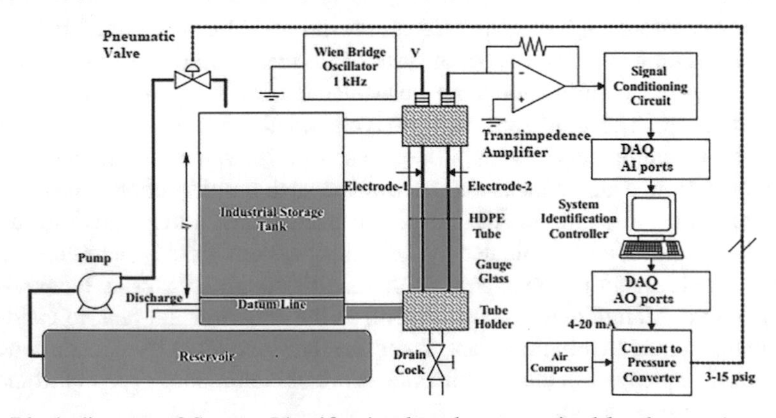

Fig. 1. Block diagram of System Identification based computerized level control system with admittance-type level sensor.

2.1 Admittance-Type Multifunction Sensor

Admittance type sensor consists of double electrodes of cylindrical shape constructed with stainless steel grade-316 (SS-316). The electrodes are separated by a distance. The Sensor can be partially immersed in the liquid. In this case, the impedance between the electrode is calculated by considering both the impedance in the liquid and the impedance in the air. The AC voltage excitation is given to the admittance-type multifunction sensor electrodes. The potential difference between electrodes 1 and 2 is

$$V = \frac{Q}{\pi \epsilon h} \ln \frac{D - r}{r} \tag{1}$$

where V is the voltage across the electrode and, Ɛ is the Conductivity of the liquid, h is the height of the liquid level, r is the radius of the electrode, D is the distance between two electrodes.

The equivalent Admittance, Y, between electrodes dipped in a conducting liquid is obtained from Eq. 2.

$$Y = |Y_P| + |Y_0| = kh + |Y_0| \tag{2}$$

Here, the $|Y_o|$ Now, the decomposition of the admittance Y is performed by considering the effect of temperature and ionic concentration on the value of Conductivity and permittivity [2]. is constant since the fringe resistance, fringe capacitance, and frequency values are constant. The Admittance is linearly proportional to change in the liquid Level above the datum line. However, the value of k depends on the variation of Conductivity and permittivity. It is constant for a particular temperature and ionic concentration of the liquid. Further Conductivity and permittivity of liquid change with the variation of temperature and ionic concentration [10, 11].

The equation of decomposition of parameters is given by

$$Y = \left[\underbrace{\left\{ \frac{\pi h}{\ln\left(\frac{D-r}{r}\right)} \sqrt{(4\sigma^2 + \omega^2 \varepsilon^2)} \right\} h}_{\text{Level Factor}} \pm \underbrace{\left\{ \frac{\pi}{\ln\left(\frac{D-r}{r}\right)} \sqrt{(4{\sigma_T}^2 + \omega^2 {\varepsilon_T}^2)} \right\} h}_{\text{Temperature Factor}} \pm \underbrace{\left\{ \frac{\pi}{\ln\left(\frac{D-r}{r}\right)} \sqrt{(4{\sigma_c}^2 + \omega^2 {\varepsilon_c}^2)} \right\} h}_{\text{Ion Concentration Factor}} \right] \tag{3}$$

The parameter separation of Level, temperature, and ionic concentration are very cross-correlated and complex. A trans-impedance amplifier circuit measures Admittance between two electrodes. Amplitude and frequency stabilized sinusoidal excitation of 1 kHz are provided to the operational amplifier circuit. The output of the trans-impedance amplifier is then provided to the signal conditioning circuit. Data acquisition system NI-DAQ-6211 has been used to interface the signal conditioning circuit's output to the PC for measurement purposes. LabVIEW software is used for measurement purposes. In the literature, cross-sensitivity of the admittance sensor is performed [12]. Also, a LabVIEW program is developed for the cross-sensitivity analysis of the Sensor [13].

Further, this program helps to decompose the composite admittance signal into level and temperature. However, the ionic concentration of the liquid is kept constant. For control purposes, the PC-based controller's command is provided to the data acquisition system and then passed through a current-pressure converter which actuates a pneumatic valve connected to the inlet to the experimental tank.

From the experimental setup, measurement data related to the water level is considered for control action, neglecting other parameters. To design the controller for the process (experimental setup), mathematical modeling must be performed to get the required transfer function. The derived transfer function is essentially required to design a suitable control system.

2.2 System Identification

The methodology for building mathematical models of dynamic systems using measurements of the system's input and output signals is called System Identification. System identification requires measurement of the input and output signals in the time or frequency domain. The Block diagram of the system identification process is shown in Fig. 2.

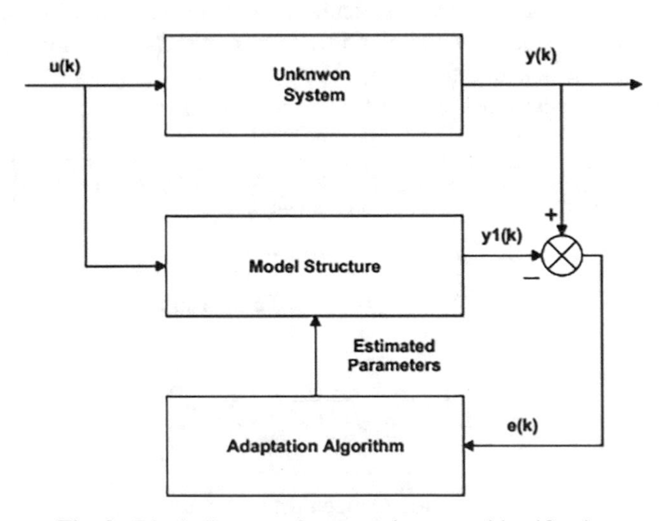

Fig. 2. Block diagram of parametric system identification

Here the process dynamics and process parameters are unknown. Only the input stimulus and output responses are known. From time-varying input and output signals, the dynamics of the process need to be ascertained. Parametric system identification is a time-domain-based approach, whereas non-parametric system identification is a frequency-domain approach. As process control applications are time-dependent, parametric system identification approaches have been chosen to estimate the process's dynamics. In this work, system identification-based methods have been used to estimate the parameters.

3 Controller Design and Analysis

The controller design helps to control a system or a process. There are various toolboxes available for controller design in many simulation platforms. One of the conventional controllers is the PID control, which offers a faster settling time with lesser transients. The traditional PID controller is described as

$$u(k) = K_p\left(e(t) + \frac{1}{T_i}\int_0^t e(\tau)d\tau + T_d\frac{de(t)}{dt}\right) \tag{4}$$

where e(t) is the error, Kp is the controller's proportional gain, Ti is the integral time constant, Td is the derivative time constant, and u(k) is the controller's output.

The transfer function of an ideal PID controller is described as

$$G_c(s) = K_p\left(1 + \frac{1}{T_i s} + T_d s\right) \tag{5}$$

To avoid derivative kick due to pure derivative action, a first-order low-pass filter (LPF) is used, which can be represented as

$$G_c(s) = K_p\left(1 + \frac{1}{T_i s} + \frac{T_d s}{1 + T_d s}\right) \tag{6}$$

Series form of PID controller can be realized if both zeros are real, i.e., Ti $\geq$ 4Td. Series form of PID controller is

$$G_c(s) = K_p(\alpha + sT_d)(1 + \frac{1}{\alpha T_i s}) \tag{7}$$

A PID controller has been designed to control the liquid level in the experimental tank. In the experimental setup, there are two control mode operations. One is Auto Mode for programmable through NI LabVIEW, and the other is Manual Mode to control the temperature. Further, various toolboxes are available in many simulation platforms to design the controller strategy of level control, from the front panel switches to, Temperature controller active in manual mode operation. A modified Ziegler-Nichols tuning formula computes the parameters based on the Nyquist frequency response plot for tuning PID controller parameters. After tuning the parameters, the following analysis is required to verify the designed PID controller's performance, i.e., Stability and Fragility Analysis.

The stability analysis of the control system is performed to determine the range of the controller gains. For the PI controller, the stability Gain Margin (GM) and phase margin (PM) will reduce, and the closed-loop system will become oscillatory.

$$\left|1 + \frac{1}{j\omega T_i}\right| = \sqrt{1 + \frac{1}{\omega^2 T_i^2}} > 1, \forall_\omega \tag{8}$$

$$Angle\left(1+\frac{1}{j\omega T_i}\right)=tan^{-1}\left(\frac{-1}{\omega T_i}\right)<0,\forall_\omega \tag{9}$$

PD controller, the derivative term will compensate for the phase lag caused by the integral term and improve disturbance rejection. System robustness is expressed by the stability margin, which is the reciprocal of maximum sensitivity.

$$|1+j\omega T_d|=\sqrt{1+j\omega^2T_d^2}>1,\forall_\omega \tag{10}$$

$$Angle(1+j\omega T_d)=tan^{-1}(\omega T_d)\epsilon\left[0,\frac{\pi}{2}\right],\forall_\omega \tag{11}$$

The derivative term will compensate for the phase lag caused by the integral term and improve disturbance rejection. System robustness is expressed by the stability margin, which is the reciprocal of maximum sensitivity, which can be represented as

$$M_s=\max_\omega|S(jw)|=\max_\omega\left|\frac{1}{1+G_C(jw)G_P(jw)}\right| \tag{12}$$

The fragile, non-fragile, and resilient PID controller concept was introduced [14]. Fragility is the loss of robustness due to perturbation in the controller parameter. For quantitative analysis of Fragility, Delta-Epsilon Fragility Index is used, which is represented as

$$RFI_{\Delta_e}=\frac{M_{s\Delta em}}{M_{so}}-1 \tag{13}$$

where $M_{s\Delta em}$ represents extremum maximum sensitivity and M_{so} represents nominal maximum sensitivity.

The condition for the test of Fragility is

$$\begin{cases} RFI_{\Delta 20}>0.5 & Fragile \\ RFI_{\Delta 20}\le 0.5 & Non-Fragile \\ RFI_{\Delta 20}\le 0.1 & Resilient \end{cases} \tag{14}$$

Microcontroller-based Fuzzy-based intelligent single-loop process controller has been used to control the temperature parameter. And a digital microcontroller-based TDS meter cum controller with a range of 0-1000ppm/2000uS/cm has been used. These two controllers are additional controllers used to control the parameters separately in manual mode in the process.

4 Results and Discussions

The experimental data are collected from the process in the VI platform for the designed controller simulation. The experimental data are collected from the process plant at 1-s intervals. A few samples were initially collected, and the system's water level in (cm) was measured.

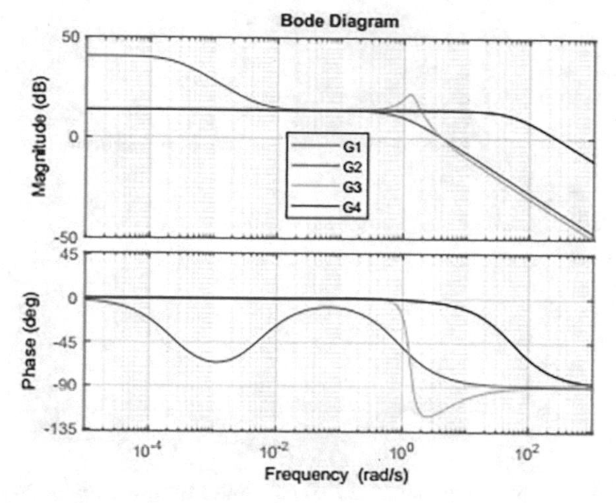

Fig. 3. Comparative analysis of validation results in subspace identification (Frequency Domain)

Table 1. Transient Performance Analysis of Controller

Controller	$\%M_p$	t_s
PID (set-point regulation)	26.0226%	0.0417 s

Further, the subspace identification method's comparison plot has been achieved in the frequency domain, as shown in Fig. 3. The gain margin and phase margin has been computed from the bode plot. Further, the transient performance of the designed controller has been studied (Table 1).

The transient performance shows the PID controller has a maximum overshoot ($\%M_p$) of 26.0226% and settling time, $t_s = 0.0417$ s.

4.1 Transient Response of Controller

The PID controller's transient response analysis considers different parameters such as maximum overshoot (%Mp) and settling time. Figure 3 illustrates the set-point regulation feature of the PID controller.

The response of the PID controller for the admittance sensor used for level measurement is shown in Fig. 4. The step signal is applied to the designed controller, and the response achieved is satisfactory. Moreover, the maximum sensitivity of the controller is also considerably high.

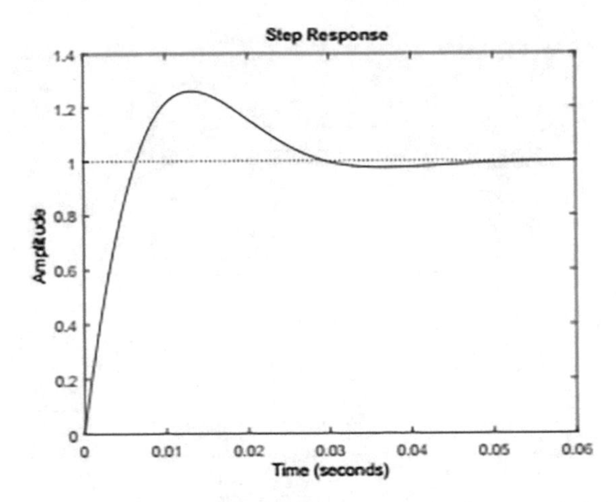

Fig. 4. Set-point regulation of PID controller for level control system

5 Conclusion

The present work provides the principle of controller design for a multifunction sensor-based process. A double electrode admittance type multifunction sensor has been used to measure Level and temperature simultaneously. The time series model and subspace identification concepts have been used to estimate the transfer function model from real-time experimental data from the developed process. PID controller design, stability, and Fragility analysis of the PID controller have been discussed in this paper. The experimental results validate the theoretical concepts. The advantage of the admittance sensor system is that it can continuously measure liquid Level and temperature. The said admittance sensor can measure the boiler's liquid Level and temperature continuously, and the said controller can control the boiler drum level control. In practice, a hydra-step type sensor measures the liquid level in the boiler drum, which performs the discrete measurement. The system identification design method will be executed once to evaluate the controller parameter once. Furthermore, these parameters can be programmed in PLC in industrial control operations.

References

1. Majumder, B.D., Roy, J.K., Padhee, S.: Recent advances in multifunctional sensing technology on a perspective of multi-sensor system: a review. IEEE Sens. J. **19**(4), 1204–1214 (2018)
2. Majumder, B.D., Roy, J.K.: Multifunctional admittance-type sensor and its instrumentation for simultaneous measurement of liquid level and temperature. IEEE Trans. Instrum. Meas. **70**, 1–10 (2021). https://doi.org/10.1109/TIM.2020.3041107
3. Ljung, L.: System Identification: Theory for the users (1987)
4. Liu, T., Wang, Q.-G., Huang, H.-P.: A tutorial review on process identification from step or relay feedback test. J. Process Control **23**, 1597–1623 (2013)
5. Ghorai, P., Majhi, S., Pandey, S.: Dynamic model identification of a real-time simple level control system. J. Control Decis. **3**(4), 248–266 (2016)

6. Ghorai, P., Majhi, S., Pandey, S.: State space approach for identification of real-time plant dynamics. In: 2017 Indian Control Conferrence, pp. 1–6 (2017)
7. Gupta, S., Gupta, R., Padhee, S.: Parametric system identification and robust controller design for liquid-liquid heat exchanger system. IET Control Theor. Appl. **12**(10), 1474–1482 (2018)
8. de Klerk, E., Craig, I.K.: A laboratory experiment to teach closed-loop system identification. IEEE Trans. Educ. **47**(2), 276–283 (2004)
9. Broersen, P.M.T.: A comparison of transfer function estimators. IEEE Trans. Instrum. Meas. **44**(3), 657–661 (1995). https://doi.org/10.1109/19.387302
10. Conductivity Theory & Practice. Technical Manual, Radiometer Analytical SAS, France, pp. 22–23 & 32–35
11. Wang, P., Anderko, A.: Computation of dielectric constants of solvent mixtures and electrolyte solutions. Fluid Phase Equilib. **186**, 103–122 (2001)
12. Roy, J.K., Deb, B.: Investigation of cross sensitivity of single and double electrode of admittance type level measurement. In: Proceedings of the 6th International Conference on Sensing and Technology (ICST), pp. 234–237 (2012)
13. Roy, J.K., Majumder, B.D.: Real time measurement of water level using admittance method and fuzzy based linearizer. In: Proceedings of the 10th International Conference on Sensing and Technology (ICST), pp. 1–5 (2016)
14. Alfaro, V.M., Vilanova, R., Arrieta, O.: Fragility analysis of PID controllers. In: 2009 IEEE Control Applications, (CCA) & Intelligent Control, (ISIC), pp. 725–730. IEEE (2009)

Fabrication of Linear Polarization-Separating Silicon Metalens at Long-Wavelength Infrared

Noe Ishizuka[1(✉)], Jie Li[2], Wataru Fuji[2], Satoshi Ikezawa[1], and Kentaro Iwami[1]

[1] Department of Mechanical Systems Engineering, Tokyo University of Agriculture and Technology, Fuchu, Japan
n-ishizuka@st.go.tuat.ac.jp

[2] Opto-Science R&D Department, R&D Technology Center, Tamron Co., Ltd., Saitama, Japan

Abstract. A polarization-separating metalens at the long-wavelength infrared was designed, developed, and demonstrated. The silicon quadrangular pillars are adopted as meta-atoms to achieve full 2π phase coverage at the design wavelength of 10.6 μm for both of two orthogonal linear polarizations with high transmittance. Their dimensions were determined by electromagnetic field analysis based on a finite element method, and transmittances of more than 70% were obtained in all meta-atoms. The metalens was fabricated through electron beam lithography and reactive ion etching, and SEM observation showed that the fabrication was done while maintaining good verticality. An imaging test was performed using a commercially available micro-bolometer, and it was confirmed that the light was focused at different positions for each linear polarization. In the case of 45°-polarized illumination, it was confirmed that the light was focused at both locations. The fabricated metalens successfully showed the ability of linear polarization separation of a thermal image.

Keywords: Meta-surface · Metalens · Polarization · Long Wavelength infrared · thermal images

1 Introduction

Recently, metamaterials (holography [1], metalens [2, 3] etc.) have been the subject of various application researches and developments [3, 4], which have properties not found in the conventional optical field. Metalens can control the phase locally, which enables thinner and smaller lens. Metalens have a particular advantage in that the polarization separation function can be realized only by the lens. Metalens have been studied extensively in the visible light range, but there has been little research in the far-infrared [5].

Among the far-infrared radiation, the long-Wavelength infrared radiation represents the temperature change near room temperature according to Wien's displacement law. Long-wavelength infrared light detects temperature and can be used to capture images at night or in poor visibility conditions such as a fire [6].

N. K. Suryadevara et al. (Eds.): ICST 2022, LNEE 1035, pp. 128–134, 2023.
https://doi.org/10.1007/978-3-031-29871-4_15

There are two types of cameras for Long-Wavelength infrared: cooled cameras with high performance and uncooled microbolometers [7], which are relatively inexpensive. However, microbolometers have a MEMS-structure [8], making it difficult to attach a polarizer. Metalens, however, can realize a polarization separation function with the lens itself. Therefore, the implementation of polarization-separating Metalens in long-wavelength infrared light has significant advantages. The implementation of Metalens may make it possible to use polarization separation more easily and inexpensively.

In this study, polarization separation is implemented in the long-wavelength infrared. We fabricated a meta-lens with polarization separation at 10.6 um, which is a Long-Wavelength infrared light.

2 Principle

When light passes through a dielectric column, the column acts as a dielectric waveguide. Therefore, by changing the size of the column, a phase change occurs. Therefore, the phase can be controlled locally. The light can be focused by arranging the dielectric so that it is equivalent to the phase equation of a spherical lens as Fig. 1(a).

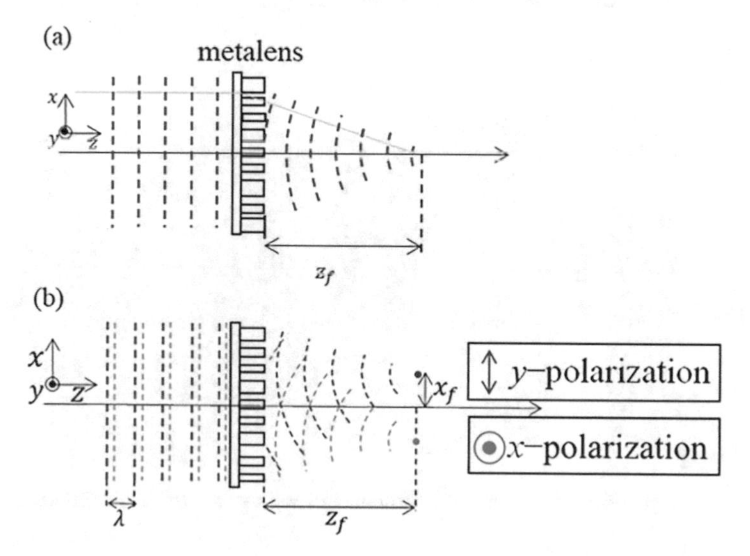

Fig. 1. Principle of metalens (a) Fresnel metalens (b) Polarization metalens

Polarized light is light in which the electric and magnetic fields oscillate regularly in time and space as they travel. Obtaining polarization information can be very useful for obtaining additional visual information. Examples of possible applications include classification of pollutants in the atmosphere and non-contact fingerprint detection [10]. To realize the polarization separation function, a meta-lens that forms images at different locations depending on the direction of polarization is required. To achieve this, an anisotropic waveguide must be used as the dielectric medium. By using waveguides with different aspect ratios, as Fig. 1(b) it is possible to use polarization according to direction. The polarization-separating meta-lens follows the equation below.

$$\varphi = \frac{2\pi}{\lambda}\left(\sqrt{x_f^2 + z_f^2} - \sqrt{\left(x \pm x_f\right)^2 + y^2 + z_f^2}\right) \tag{1}$$

where λ is the wavelength, z_f is focal length in z direction, and x_f is focal length in x direction [11].

3 Result

3.1 Electromagnetic Field Analysis

A dielectric prism made of single-crystal silicon was used as the meta-lens as Fig. 2(a). The design wavelength was set to 10.6 μm. The height of the columns of the meta-lens was set to 20 μm, and the distance between the centers of the columns was set to 3 μm. The design was carried out so that the transmittance would be more than 70% in the entire design area. The meta-atoms were determined so that a phase change of 2π could be obtained for each of the x- and y-polarizations.

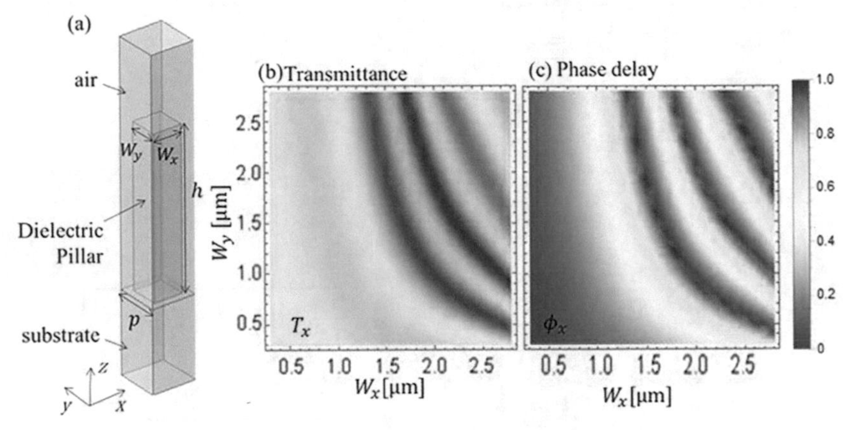

Fig. 2. Electromagnetic field analysis (a) Structure of meta-atom (b) Transmittance (c) Phase delay

3.2 Metalens Design

Based on these results, the meta-lens was designed to satisfy Eq. (1). The designed CAD data is shown in Fig. 3. The focal lengths in the z and x directions were set to 3.96 mm and 1 mm, respectively. The lens diameter is 4.25 mm. This is in accordance with the use of a far-infrared camera using a microbolometer. The meta-atom geometry was arranged by dividing each polarization into eight segments.

In order to verify that polarization separation is actually achieved, a simulation was performed with a model that is 1/20th of the designed model. The same NA and the same meta-atoms were used. The results of the analysis confirmed image formation for both x-polarization and y-polarization. The same level of incident intensity was confirmed for 45°-polarization.

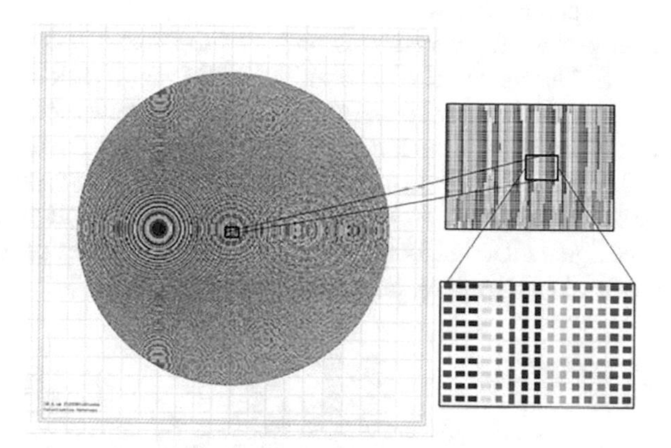

Fig. 3. Design of metalens

3.3 Focusing Confirmation by Electromagnetic Field Analysis

In order to verify that polarization separation is actually achieved, a simulation was performed with a model that is 1/20th of the designed model. The same NA and the same meta-atoms were used. The results of the analysis confirmed image formation for both x-polarization and y-polarization. The same level of incident intensity was confirmed for 45°-polarization (Fig. 4).

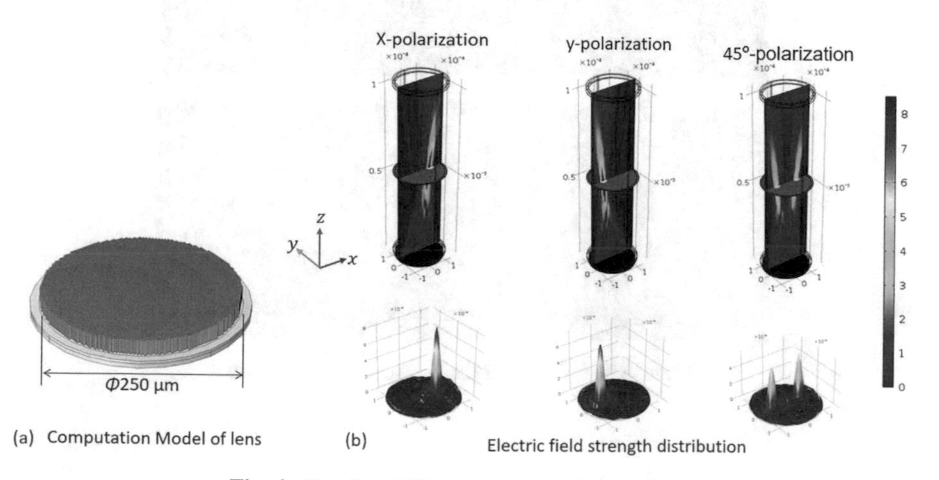

Fig. 4. Results of Electromagnetic field analysis

3.4 Production and Condensed Light Confirmation

In fabrication, the columns were fabricated by electron beam lithography followed by deep silicon etching as Fig. 5. After that, the resist was removed by dry etching. The substrate used was a single crystal silicon substrate. SEM images of the fabricated results are shown below Fig. 6. From the results, it was found that the meta-lens was fabricated while maintaining perpendicularity.

Using the meta-lens we fabricated, we confirmed the focusing of the polarization-separating meta-lens. The camera used was the LW10F90-E (TAMRON) The lens uses an uncooled microbolometer as an image sensor. Figure 7 shows the focused light. A soldering iron was used to check the focusing. The figures show the image taking with attached spherical lens as Fig. 7(a). It's the focusing results for x-polarized light as Fig. 7(b), y-polarized light as Fig. 7(c), and light with 45° polarization as Fig. 7(d), by metalens. Since the light collection location differs depending on the polarization, it can be said that a polarization-separating metalens has been realized.

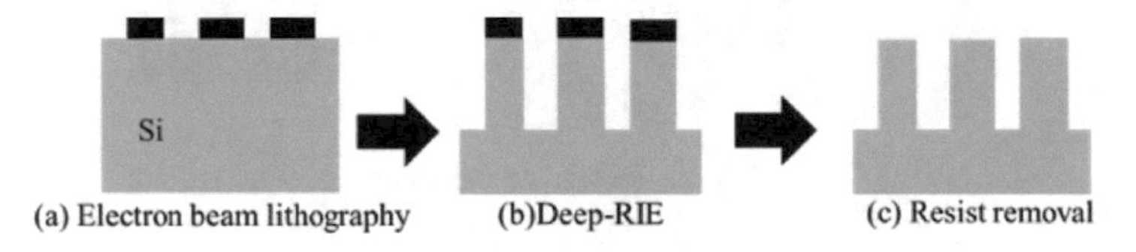

Fig. 5. Manufacturing Engineering

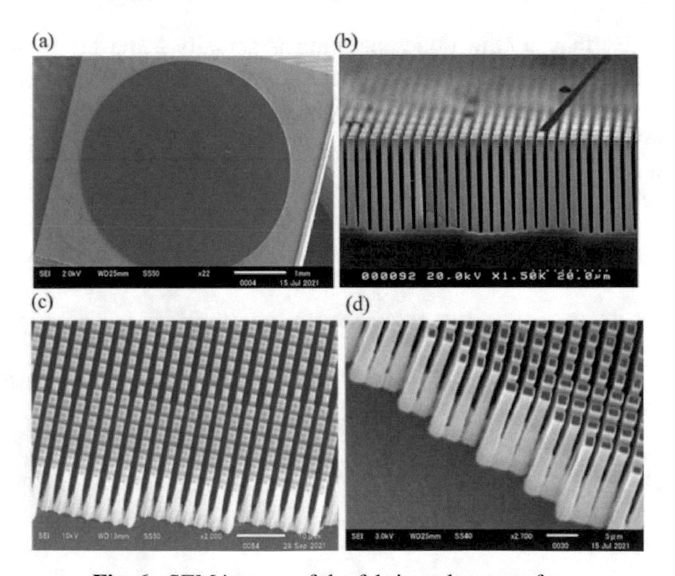

Fig. 6. SEM images of the fabricated metasurface

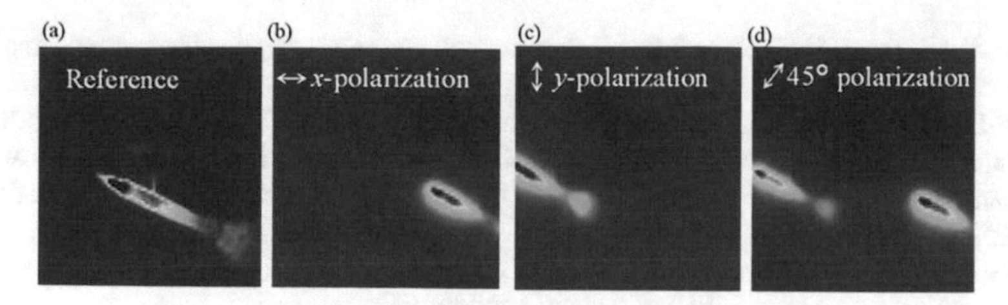

Fig. 7. Thermal image with linearly polarized light

4 Conclusion

In this study, a polarization-separating meta-lens was fabricated at 10.6 μm. The meta-atom achieved a transmittance of more than 70% and a phase of 2π in the entire range. The light-collecting performance was confirmed by detecting thermal images. It was confirmed that the polarization separation function, which focuses light to different positions depending on the direction of polarization.

Acknowledgement. This work was supported in part by the Japan Society for the Promotion of Science (JSPS) KAKENHI under Grant 21H01781 and Grant 22K04894; in part by the Takeda Sentanchi Supercleanroom, University of Tokyo, through the "Nanotechnology Platform Program" of the Ministry of Education, Culture, Sports, Science, and Technology (MEXT), Japan, under Grant JPMXP 1222UT1041. The authors would like to thank Prof. Y. Mita, Dr. A. Higo, Dr. E. Lebrasseur, and Mr. M. Fujiwara (The University of Tokyo) for their support during sample fabrication, also would like to thank Prof. Lucas Heitzmann Gabrielli (University of Campinas) for developing and maintaining gdstk, a Python library for creating and manipulating GDSII layout files, and also would like to thank a TSUBAME3.0 supercomputer at the Tokyo Institute of Technology for the numerical calculations.

We have conducted various studies on dielectric metasurfaces and their optical characterization.

References

1. Yamada, N., Saito, H., Ikezawa, S., Iwami, K.: Demonstration of a multicolor metasurface holographic movie based on a cinematographic approach. Opt. Express **30**(10), 17591–17603 (2022). https://doi.org/10.1364/OE.457460
2. Ogawa, C., Nakamura, S., Aso, T., Ikezawa, S., Iwami, K.: Rotational varifocal moiré metalens made of single-crystal silicon meta-atoms for visible wavelengths. Nanophotonics **11**(9), 1941–1948 (2022)
3. Ikezawa, S., Yamada, R., Takaki, K., Ogawa, C., Iwami, K.: Micro-optical line generator metalens for a visible wavelength based on octagonal nanopillars made of single-crystalline silicon. IEEE Sens. J. **22**(15), 14851–14861 (2022). https://doi.org/10.1109/JSEN.2022.3186060
4. Fan, Q., Liu, M., Yang, C., Le, Yu., Yan, F., Ting, Xu.: A high numerical aperture, polarization-insensitive metalens for long-wavelength infrared imaging. Appl. Phys. Lett. **113**(20), 201104 (2018). https://doi.org/10.1063/1.5050562

5. Bhan, R.K., Dhar, V.: Recent infrared detector technologies, applications, trends and development of HgCdTe based cooled infrared focal plane arrays and their characterization. Opto-Electron. Rev. **27**(2), 174–193 (2019). https://doi.org/10.1016/j.opelre.2019.04.004
6. Niklaus, F., Vieider, C., Jakobsen, H.: MEMS-based uncooled infrared bolometer arrays: a review. In: Proceedings of SPIE 6836, MEMS/MOEMS Technologies and Applications III, 68360D (2008)
7. Gruev, V., Perkins, R., York, T.: CCD polarization imaging sensor with aluminum nanowire optical filters. Opt. Express **18**(18), 19087–19094 (2010)
8. Miyata, M., et al.: Compound-eye metasurface optics enabling a high-sensitivity, ultra-thin polarization camera. Opt. Express **28**, 9996–10014 (2020)

Semantically Processed Sensor Data in Health Care, Legislation Compliant, Ontologies

Ollencio D'Souza(✉), Subhas Mukhopadhyay, and Michael Sheng

Macquarie University, North Ryde, NSW, Australia
ollencio.dsouza@students.mq.edu.au

Abstract. Sensor technologies protect life, property and those that are vulnerable, an ontology of most concern to researchers in public health. Detecting risks, such as intrusion and fire, is as crucial as delivering care to the most vulnerable. Today the sophistication of these multi-spectral sensor systems provides much more than "event alarms", improving the quality of work and life for all. The technology, operational scope and necessary compliance with standards and legislation (the law) make it a challenging environment. We present core concepts in designing sensors and semantically aware systems to enhance security, safety and compliance, especially in assisted living centres, where solutions have evolved to perform in a data-driven domain. The complexity of products, systems and processes ensures that data is always available to analyse events live and forensically. The emergence of machine learning makes sensors "intelligent", improving the approach to risk management. We present options that bring together legislated processes and technology that facilitate compliance with a data-driven focus for the assisted living community.

Keywords: Healthcare · Legislated compliance · real-time · Security · fire safety · Standards · Machine Learning sensors · Data drive

1 Introduction

The physical protection of life and property in assisted living centres is a known ontology driven by sensors that provide the status of many risk-associated criteria. As a risk management practitioner, I have presented in 3rd person my experiences in the field to be clinical in my approach. Most deployments adhere to risk management guidelines developed by licenced experts for that environment. [1] Data collected from sensors establish the risk profile of a site. The sensor design objectives are to create virtual "perimeters" in areas of activity and no activity. Knowledge of occupancy of a room, the temperature, humidity, and airflow in those areas, are some of the "parameters" defined in the risk management plan, guided by Australian Standards and produced by experts in the industry. Sensors detect real-time activity and relevant conditions. For example, excess airflow negates fire detection devices that rely on "aspiration", a process of taking a sample of air, to measure "smoke" particulates, which pass over the sensitive Opto coupled detection technology [2] because the airconditioning airflow was too high.

N. K. Suryadevara et al. (Eds.): ICST 2022, LNEE 1035, pp. 135–148, 2023.
https://doi.org/10.1007/978-3-031-29871-4_16

Deliveries that are rushed and inadvertently placed become a trip hazard. Doors kept ajar when they must remain locked, the ill-timed dispensation of medications and poor response to "assistance calls" are a few glaring examples of how legislated processes are disrupted.

In summary, the interdependence of sensors, the property's physical characteristics, and the risk profile require management systems to monitor essential criteria continuously. [3] The technologies deployed are constantly updated to keep up with performance requirements. The sensor technology must function in a data-driven business domain and must "sense" (detect) and "respond" to alerts generated by the systems installed on the premises. Sensor applications [4], over the years, have also undergone updates in fundamental concepts and technology, making them "intelligent" using Machine Learning.

2 Methods and Technology

In matters related to life, property and assisted living [5] risks, legislated minimum operational standards are enforced. It is essential to choose suitable sensors to ensure mandated operational workflows are followed (Fig. 1). Sensors are ideally suited to operate across this environment to ensure functional compliance is observed [6]. Sensors use a good part of the electromagnetic spectrum from gamma to RF (radio frequency) to ensure risk events are detected, prevented and responded to, as defined by the Australian standard on risk management [7]. Events like a fire, burglary or a loss of a patient with dementia who walked through open doors take only a few minutes to act out but result in disastrous consequences such as the house getting burnt down, the burglar getting away or panic when someone is missing. Every incident has a predictor/precursor to a risk event, e.g., smoke before the fire, e.g. strangers in the neighbourhood, an unattended door open alarm, etc. The general "method" to stay vigilant is better Situational Awareness derived from timely analysis of sensor data [8].

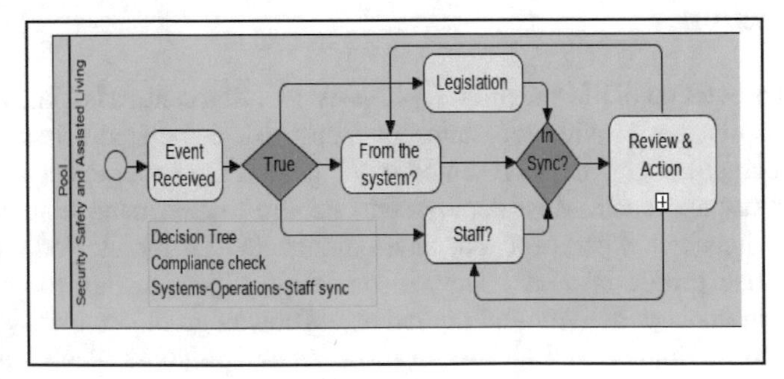

Fig. 1. Workflow that ensures compliance with legislation using data from sensors to direct/correct the process

3 Current Technology and Processes

The standard "security", "fire safety", or "Assisted Living" system has a centralised "panel" to which sensors are wired and programmed to manage the outcomes based on the status of each of the sensor devices connected to it;

1. If sensors do not send an "alarm" signal (or system heartbeat), the panel will interpret that as an "offline" situation. i.e. communications failure.
2. Each sensor is "supervised", so it sends an alert if it goes offline (i.e. does not respond to a heartbeat poll or the impedance of the connection changed).
3. The panel will send that event type to a supervisor via SMS, pager, phone, etc., from where a physical response is initiated.
4. Sensors are never shut off and operate 24x7. The central (programmed) panel decides what to do with the signals from the sensor [9].
5. In some cases, 10-year-old sensor technologies are still used for these critical sensing tasks. To ensure a fundamental change to the adoption of newer base sensor technology, new sensor developments are being tried out to make the shift to a data-driven environment.
6. The user panels can be programmed to present simple push buttons to set or unset operating sensors so they alarm if operating procedures/schedules are bypassed such as keeping a door open with an obstruction.
7. Other "nurse call" and/or triggered calls for assistance are part of the operating procedure because the time to respond is measured. When a response has satisfactorily concluded, the operator or the patient can turn the request off. Assisted living emergency pendant status and location are continuously monitored [10].
8. All sensor "transactions" (start to finish) are recorded. Legislation requires that records are kept (audit trails) also used to analyse process compliance [11].

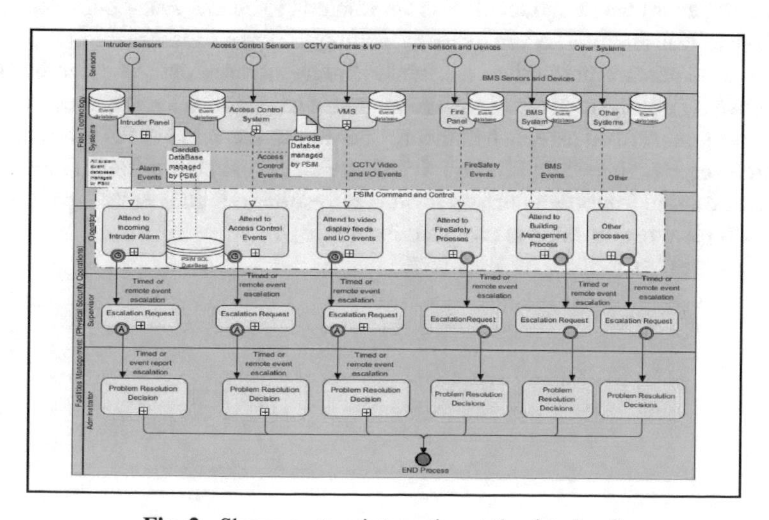

Fig. 2. Shows system integration at the data level)

4 Current Technology and Process Weakness

The current technology weaknesses in hardware and software deployments are critiqued to contrast it with the emerging changes presented in the following chapter.

4.1 Data Science Techniques to Quantify Risk

The records show a high false alarm rate [2] (90% + industry average, [12, 13]) caused mainly by "less featured" sensors or poor procedures [14], which in turn lead to deterioration in behaviour because repeated alarms are ignored. Current trends are moving decision-making to the "edge" (smart sensors) to improve alarm verification and reduce unwanted callouts [15].

The technology used today is explained within the construct of data science classification techniques using the Confusion Matrix;

a) A False-Positive (false alarm) ratio (FAR) is calculated as the number of false alarms (FP) per total number of alarms (FP + TN) in the study. There is a cost associated with every False Alarm, and studies show that it is more cost-effective to keep false positives to a minimum by verification[12, 13, 16].
b) The "Confusion Matrix" provides results such as (a) Accuracy, (b) Precision (c) Recall (d) False Positive Rate – sometimes termed FAR, (e). False Negative Rate (poor risk assessment or prediction) and the (f) Specificity from counts of the key events that occur.
c) Detectors have localised logic and multiple sensor modules, in some cases based on contrasting sensing technologies, to "detect" the event with higher certainty (improve prediction). So contrasting event data is used to confirm the validity of an incident.
d) Other ratios matter and contribute to the overall response strategy adopted. FN (or False Negative) where the incident is "predicted" to be low risk – but turns out to be high risk resulting in a severe incident leading to serious losses.

    In a healthcare predictive risk management process, data is essential. Data is obtained by assessing every event and response to establish systemic weaknesses in delivering prescribed care. In healthcare operations, the value of FN is ordinarily low (increasing Recall = TP/(TP+FN). In other risk domains, a different classification would be more important such as lower FN results in higher email spam received, etc. A typical results table is provided in Figure 3.
e) Results- RECALL 50%, FP Rate 80%

	Actual	
	TP	FP
Predicted	15	4
	15	1
	FN	TN

Accuracy	0.4571
Precision	0.7895
Recall	0.5000
FPR	0.8000
FNR	0.5000
Sp	0.2000

Fig. 3. Typical "Confusion Matrix" Results table

The total cost per "false call out" is estimated to be $1825 [12]. The sample tests show poor accuracy at 45%, a False positive rate at 80%, average Recall at 50% and low specificity at 20%. Methods to improve accuracy include focusing on sensor reliability to improve predictability and reduce false positives.

The system keeps a record of alarms generated, but the response team usually fill out details on the alarm response after investigation. At this point, errors and incorrect representation of facts often distort the recorded workflow.

4.2 Current System Configuration Weaknesses

Existing designs use a panel-based configuration method. Multiple local panels cascade into a Sub-panel. Multiple Sub-Panels cascade into a Master Panel and so on for large sites. All sensors, hardwired to a local panel, generate data that eventually cascade to the master panel. In more extensive cascades, the time taken for the alarm data to reach the end-user (latency) is high, and reference to other relevant linked data from other systems, like CCTV, is lost [17].

4.3 Legacy Protocols, Communications and Support Methods

Existing serial communication protocols over RS485, MODBUS and similar serial comms are very efficient and often reliable locally [18]. But the cabling up the chain towards a master panel introduces data delays, "configuration" and "support" issues because they are cabling-intensive tasks.

5 Converged Network – Services over IP

The new paradigm is based on a converged network (IP) network built to share information from sensors to central control, where data analysis and storage occur, enabling access to stored information by the end-user. Figure 2 shows (grey area) data created by each system is ready to be directly accessed by the end-user, losing none of its references to the source event [19] timing.

Two major reviews by the Government; "Review of Assistive Technology Programs in Australia" and "A Technology Roadmap for the Australian Aged Care Sector [20]" [21] have set the roadmap and a framework for providers of several "technology product services" such as "Nurse Call" and "Access Control." The owner of the premises generally provides these technologies to remain a compliant provider.

The industry adopts the IP configuration because almost every panel manufacturer in security and safety (fire) and assisted care uses a product that primarily works over the ethernet network. Some legacy systems have been "converted" to IP using serial to ethernet converters. This opens up many new configuration opportunities enabling functionality to be deployed anywhere on the premises [22].

6 Sensor Technology Developments

The rapid adoption of Ethernet as the industry-standard communications platform enables many systems to move to Ethernet TCP/IP or UDP protocol-based communications. Some benefits are locating a sensor device anywhere on the premises network (wired or wireless), identifying it with an IP address, and making them flexible, upgradeable, configurable intelligent and multifunctional [23] (Fig. 4).

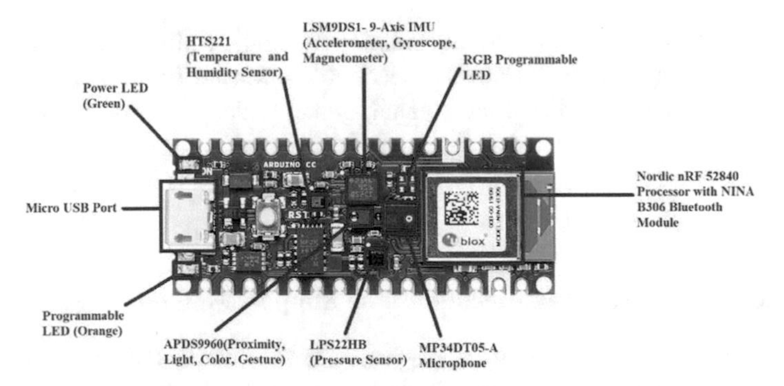

Fig. 4. Example of a multisensor microcontroller device for Sensor Fusion Machine Learning (TinyML)

On board sensors: Refer to Fig.4
Sensor1: Temperature
Sensor1: Humidity
Sensor2: 9axis IMU (AGM)
Sensor3: Proximity (LCG)
Sensor3: Colour (LCG)
Sensor4: Pressure
Sensor5: Sound

15 individual sensors
(in 5 bundled groups)
Each sensor measure contributes to the rich parametric ML data source.

Note:
The usefulness of "TinyML" is often not understood fullyTo train a model in the cloud based on large data sets, explore inter dependencies, confirm functionality and then "squeeze" them down to reside on resource constrained devices to function just as well is a phenomenal benefit. "WakeWords & Triggers" optimse power consumption.
On Board sensors extract vital multispectral information.

6.1 The Current and Evolving State of the Technology

The sophistication of sensors has remained dormant because these sensors, Table 1, have been approved via legislation. Changes will be accepted if newer sensors deliver better outcomes [24].

Table 1. Sensor technology application overview

Sensor Catg.	Types of Sensors		
	Function	*Method*	*In/outdoors*
Security	Detecting movement	MicroWv	In/out
Security	Detecting warm obj movement	PIR	In
Security	Detecting movement	IR beam	In/out
Fire Safety	Detect smoke	Opto/Ion	In
Security/safety & fire	Video analysis & thermal (incl fire) analysis in marked areas	Multi-Spectral vision	In/out
Access Control	Token verified pass through	RF readr + programed card	In/out
Access Control	Biometrics	Video Analysis	In/out
Security	Vibration/break glass/audio	Audio Analysis	In
Assisted Living	Personal tags, duress, geoloc	Tracking/SOS	In/out

The MCU unit with on board sensors provide many more data points to analyse relationships between data these sensors generate which are then explored by Machine Learning [25]. If training using the right "mimic" of the environment is presented to the sensor "FUSION" training procedure excellent results are achieved. A study of different modes of each "on board" sensor is presented in Table 2.

Table 2. On Board Device Sensors and their contribution to domain requirements

SNo	Sensor detection	What would the sensors contribute to the target domains	"Wish List" of outcomes desired
1	Colour	Colour is observed in "equipment", "dress", "flame","device colour", etc. all features that aid "recognition"	The ML Model will learn more as colour will be an added feature during training
2	Brightness	Light intensity is a serious consideration, and sensing ambient light, or lights, turning off and on are some "scene features" of interest that change	This feature will improve "reliability" if the level of "illumination" is managed
3	Proximity	The sensing is based on IR which can also reflects back in total darkness a really important input to detect motion	A triggering mechanism is useful even as a "Wake word" to trigger "analysis
4	Gesture	Guestures can be used to automate functions	Automation in some cases will help
5	Motion	Motion detection is an important sensory input to enable such things as recognising "change"	Again "triggering" ML analysis to turn on will improve analytic outcomes

6.2 The Location

The location of the devices is an important concern. Trials with the following configurations worked best.

1. Rigid housing on the wall – along side the bed – coupled by BLE (blue tooth low energy) Fig. 5.
2. In a housing with sensor board in adjustable flexible strapping – strapped to the bed head. Figure 6.

Model training is done in the cloud. (TinyML – Tensorflow lite) reduced models are uploaded to the MCU to transfer trained "behaviour" models (filters) critical to the risk management process [25]. The sensor table puts in focus most sensor types and functions. Although these functions appear similar, they are delivered differently over an

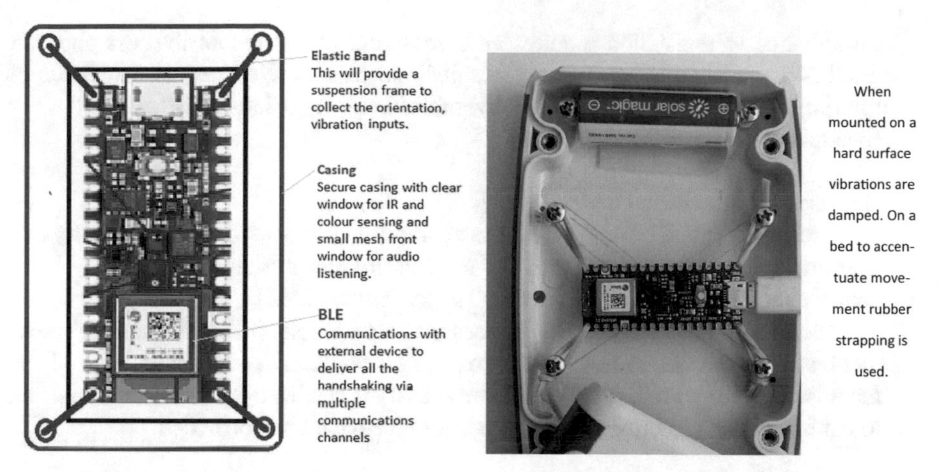

Fig. 5. Construction of strapping

Fig. 6. Image of strapping using rubber

IP network. The converged network environment can now "merge" functions. Adding new types of sensor technologies such as Multi-spectral Video analytics, Biometrics, and Laser scanning [26], other geolocation forms of sensing can be incorporated into the configuration to provide a richer data set for model training.

Sensor analytic outputs are semantically interpreted to create operationally recognisable instructions. Examples: Warning of potential fire (excessive heat no air flow events), or "intruder" (permissionless entry event) or "door left open" event (potential loss of dimentia patient walking out) and many more [27].

An example of a semantic model [19] creation and deployment is explained in the following graphical note (Fig. 7).

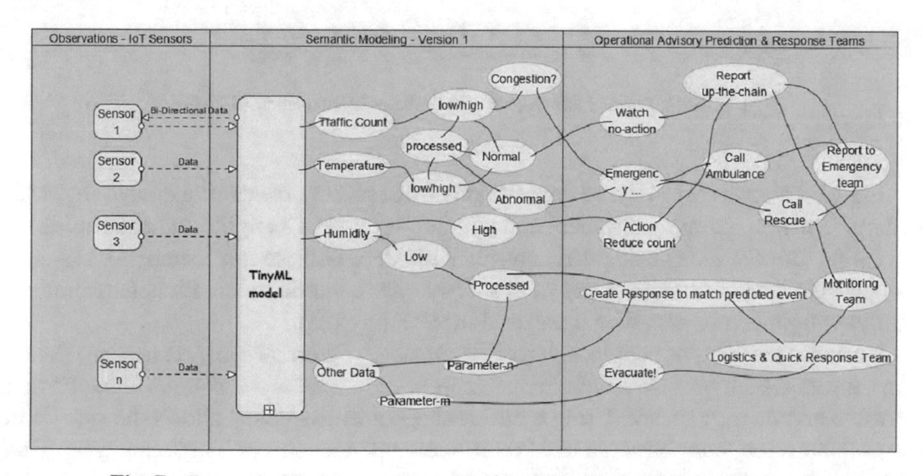

Fig. 7. Semantic Model translates data jargon to "actionable intelligence"

Compliance to legislation requires an assessment of the responsiveness and care provided is established by measuring "results", such as derived results that focus on interpreting "data analysis jargon" as deliverables as per legislation [28].

Examples:

1. Has adequate rest been obtained? [21]

   A response graph (Fig. 8) plotting intrusions, restfullness, day, night activity etc., over an agreed period, is analysed and used to produce a routine report.
2. Has adequate exercise and "recreation" been delivered? [21]

   Analysis of response activity is assessed using graphs (Fig. 10) showing movement into and out of areas such as the room, exercise area or outdoor activity. Movement is recorded with the time of day the activity is carried out. This data is analysed to confirm if the prescribed activity and exercise have been provided.

6.3 Derived Results

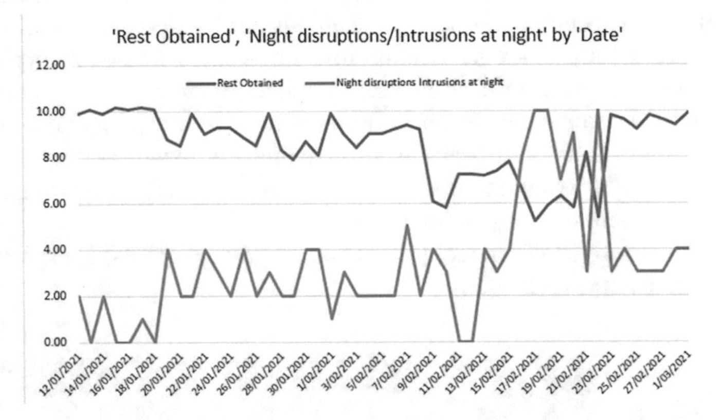

Fig. 8. Rest pattern analysis when there is intrusion or disruption

Figure 8 shows two analysed data streams from sensors measuring everything from bed movement to human movement and other activity such as lights on/off or noise.

As previously indicated, microcontroller clusters with sophisticated MEMS sensors provide data of different types such as movement, inertia, orientation, temperature, humidity, light levels, etc., which are used in the study [25].

Accurate tracking of people activity is done using "thermal"(low resolution) detection sensors which provide a definitive detection of human movement as a "blob" non intrusive and privacy secure. Using a different spectral frequency allows the system to use collaboration, corroboration and "reinforcement" techniques in data analysis. This technique uses more than one "sensing feature" to confirm the activity. Such as the motion sensor and the thermal sensor will confirm that movement has occurred and is identified as human (Fig. 9).

Fig. 9. Low resolution "colourised" thermal image of human in the 8 to 14 μm thermal band

Critical operational aspects such as activity & sleep patterns, assistance requests, service efficiency, medication, illegal or forced entry, and eating frequency can all be tracked over an IP network because services delivered are "visible" and detection "customisable" (Fig. 10).

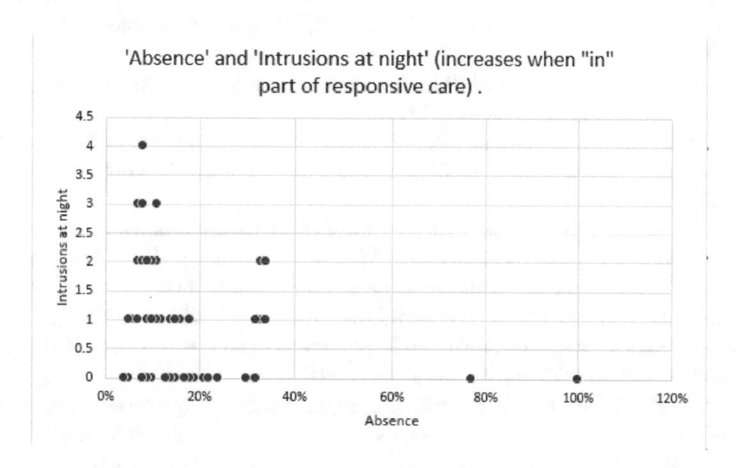

Fig. 10. Scheduled intrusions, care delivery, absence/presence in room

The generation of data using intelligent sensor clusters that use Machine Learning, changes the way legislation-driven process optimisation can be achieved using live and recorded data analysed for workflow compliance. In Fig. 10, sensor data analysis helps qualify the quality of care in terms of the criteria stated in legislation that benchmark the required outcomes [29].

7 Conclusion

Sensor technology is a significant data source to safeguard life, property and assist those most vulnerable in our society, such as the aged. Essential compliance to legislation provides the framework to create supportive functions that provide "early warning" to correct procedures before they become operational issues. Most of this occurs in real time; hence temporal synchronisation is essential. Digital Transformation benefits extend to security, fire safety and assisted living. Custom Sensor technology can guide operational staff on policies and procedure compliance because they mimic operating procedures (like a digital twin), and provide automatic notification of lapses and tools to help the provider to stay compliant.

References

1. Crosweller, M.: Disaster management and the need for a relational leadership framework founded upon compassion, care, and justice. Clim. Risk Manag. **35**, 100404 (2022). https://doi.org/10.1016/j.crm.2022.100404
2. Gladstone, N.: False fire cost taxpayers $100 million a year. The Sydney Morning Herald (2020). https://www.smh.com.au/national/nsw/false-fire-cost-taxpayers-100-million-a-year-20200220-p542j2.html (accessed Jul. 01, 2021)
3. Aguileta, A.A., Brena, R.F., Mayora, O., Molino-Minero-Re, E., Trejo, L.A.: Multi-sensor fusion for activity recognition-a survey. Sensors **19**(17), 3808 (2019). https://doi.org/10.3390/s19173808
4. Mukhopadhyay, S.C.: Intelligent Sensing, Instrumentation and Measurements. Springer Berlin Heidelberg, Berlin, Heidelberg (2013)
5. Sonia, S., Semwal, T.: A multimodal human sensing system for assisted living. EAI Endorsed Trans. Pervasive Health Technol. **6**(24), 167285 (2020). https://doi.org/10.4108/eai.26-11-2020.167285
6. Degerli, M., Ozkan Yildirim, S.: Enablers for IoT regarding wearable medical devices to support healthy living: the five facets. In: Marques, G., Bhoi, A.K., Albuquerque, V.H.C., Hareesha, K.S. (eds.) IoT in Healthcare and Ambient Assisted Living. SCI, vol. 933, pp. 201–222. Springer, Singapore (2021). https://doi.org/10.1007/978-981-15-9897-5_10
7. Risk Management. Standards Australia. https://www.standards.org.au/standards-catalogue/sa-snz/publicsafety/ob-007 (accessed Jul. 01, 2021)
8. Wurthmann, K.: An illustrative example of applying systems engineering tools for risk management when launching new technologies: the case of lifeboat insufficiency on the RMS Titanic. In: Proceedings of the 2019 IEEE Technology Engineering Management Conference (TEMSCON), pp. 1–5 (2019). https://doi.org/10.1109/TEMSCON.2019.8813655
9. Pech, M., Vrchota, J., Bednář, J.: Predictive maintenance and intelligent sensors in smart factory: review. Sensors **21**(4), 1470 (2021). https://doi.org/10.3390/s21041470
10. Casiddu, N., Porfirione, C., Monteriù, A., Cavallo, F. (eds.): LNEE, vol. 540. Springer, Cham (2019). https://doi.org/10.1007/978-3-030-04672-9
11. Leveson, N., Samost, A., Dekker, S., Finkelstein, S., Raman, J.: A systems approach to analysing and preventing hospital adverse events. J. Patient Saf. **16**(2), 162–167 (2020). https://doi.org/10.1097/PTS.0000000000000263
12. Marks, M., He, Y., Buckley, G.: False Alarms and Cost Analysis of Monitored Fire Detection Systems, p. 13

13. Sampson, R.: False burglar alarms (2011)
14. Tilley, N., Thompson, R., Farrell, G., Grove, L., Tseloni, A.: Do burglar alarms increase burglary risk? a counter-intuitive finding and possible explanations. Crime Prev. Community Saf. **17**(1), 1–19 (2015). https://doi.org/10.1057/cpcs.2014.17
15. Witzig, C.S., et al.: When good intentions go bad—false positive microplastic detection caused by disposable gloves. Environ. Sci. Technol. **54**(19), 12164–12172 (2020). https://doi.org/10.1021/acs.est.0c03742
16. Kang, P., Finn, A.M., Gillis, T.M., D'souza, O.: Context-Aware Alarm System (2011)
17. Eifert, T., Eisen, K., Maiwald, M., Herwig, C.: Current and future requirements to industrial analytical infrastructure—part 2: smart sensors. Anal. Bioanal. Chem. **412**(9), 2037–2045 (2020). https://doi.org/10.1007/s00216-020-02421-1
18. Gupta, G.S., Mukhopadhyay, S.C.: Embedded microcontroller interfacing: designing integrated projects, 1. Aufl., 1st ed., vol. 65. Springer-Verlag, Berlin, Heidelberg (2010). https://doi.org/10.1007/978-3-642-13636-8
19. Detro, S.P., et al.: Applying process mining and semantic reasoning for process model customisation in healthcare. Enterp. Inf. Syst. **14**(7), 983–1009 (2020). https://doi.org/10.1080/17517575.2019.1632382
20. review-of-assistive-technology-programs-in-australia-final-report.pdf. https://docs.google.com/viewer?url=https%3A%2F%2Fwww.health.gov.au%2Fsites%2Fdefault%2Ffiles%2Fdocuments%2F2021%2F01%2Freview-of-assistive-technology-programs-in-australia-final-report.pdf (accessed Jul. 01, 2021)
21. Madai, V.I., Higgins, D.C.: Artificial intelligence in healthcare: lost in translation?. ArXiv210713454 Cs (2021). Accessed: Sep. 18, 2021. http://arxiv.org/abs/2107.13454
22. Isa, I.S.B.M., El-Gorashi, T.E.H., Musa, M.O.I., Elmirghani, J.M.H.: Energy efficient fog-based healthcare monitoring infrastructure. IEEE Access **8**, 197828–197852 (2020). https://doi.org/10.1109/ACCESS.2020.3033555
23. Kim, H.J., Lee, U., Kim, M., Lee, S.: Time-synchronization method for CAN–ethernet networks with gateways. Appl. Sci. **10**(24), 8873 (2020). https://doi.org/10.3390/app10248873
24. Brena, R.F., Aguileta, A.A., Trejo, L.A., Molino-Minero-Re, E., Mayora, O.: Choosing the best sensor fusion method: a machine-learning approach. Sensors **20**(8), 2350 (2020). https://doi.org/10.3390/s20082350
25. D'Souza, O., Mukhopadhyay, S.C., Sheng, M.: Health, security and fire safety process optimisation using intelligence at the edge. Sensors **22**(21), 8143 (2022). https://doi.org/10.3390/s22218143
26. Nait-ali, A. (ed.): SB, Springer, Singapore (2020). https://doi.org/10.1007/978-981-13-0956-4
27. D'Souza, O., Mukhopadhyay, S., Akhter, F., Khadivizand, S., Memar, E.: Extracting operational insights from everyday IoT data, generated by IoT sensors over LoRaWAN. In: Mandal, J.K., Roy, J.K. (eds.) Proceedings of International Conference on Computational Intelligence and Computing. AIS, pp. 241–249. Springer, Singapore (2022). https://doi.org/10.1007/978-981-16-3368-3_23

28. Zgheib, R., et al.: A scalable semantic framework for IoT healthcare applications. J. Ambient. Intell. Humaniz. Comput., 1–19 (2020). https://doi.org/10.1007/s12652-020-02136-2
29. Banbury, C.R., et al.: Benchmarking TinyML systems: challenges and direction. ArXiv Prepr. ArXiv200304821 (2020)

A Standalone Millimeter-Wave SLAM System for Indoor Search and Rescue

Huyue Wang(✉) and Kevin I-Kai Wang

The University of Auckland, Auckland, New Zealand
hwan685@aucklanduni.ac.nz, kevin.wang@auckland.ac.nz

Abstract. Real-time location data offers great insight to achieve many intelligent and context-aware applications, such as personalised location-aware services, navigation, surveillance, and search and rescue. While outdoor location can be easily collected by the global positioning system (GPS), indoor location data remains challenging to collect and faces several challenges such as cost, preconfigured infrastructures, and limited long-term accuracy. In this study, a novel Simultaneous Localization and Mapping (SLAM) system is proposed and implemented targeting indoor search and rescue application, using the state-of-the-art millimeter-wave (mmWave) radar sensor. The proposed system is completely self-contained and requires no prior installation or configuration of any other devices. A 2-dimensional map and movement trajectory data is produced as the output of the SLAM system. Typical straight line and L-shape pathway experiments have been conducted. In both cases, the proposed system is capable of achieving submeter accuracy, which are promising results demonstrating its ability to provide accurate SLAM for indoor applications.

Keywords: Indoor navigation · indoor positioning · millimeter-wave · Simultaneous Localization and Mapping (SLAM)

1 Introduction

Location data is a critical context information for many modern days intelligent services, or the so-called Location Based Services (LBS) [1]. Outdoor location data is typically collected and dominated by the Global Positioning System (GPS) since its worldwide accessibility for civilian applications in the 1980s due to its wide coverage and availability [2, 3]. These location data have demonstrated great values in many outdoor applications such as navigation, tracking, and search and rescue. However, these existing systems mainly focuses on outdoor applications due to the limited and interfered indoor reception of satellite and cellular signals, and cannot offer seamless indoor location tracking, which limit the scope of its applicability.

Similar to outdoor environment, location data is equally valuable for indoor applications for the purposes of tracking, navigation, personalised intelligent services, and emergency services. For example, the position information is frequently acquired as it offers valuable insights that can assist our daily activities. The indoor positioning

N. K. Suryadevara et al. (Eds.): ICST 2022, LNEE 1035, pp. 149–161, 2023.
https://doi.org/10.1007/978-3-031-29871-4_17

system is highly beneficial to Internet of Things (IoT) and home automation applications by offering real-time location information to the gateway devices. For instance, the researches in [4] and [5] show the investigation toward the aspect of mobile robot localisation, and the collection and analysis of walking pattern in an indoor environment. The researches in [6] and [7] also illustrate how the positioning information can be used to assist the elderly and telecare for home automation.

The indoor environment is, however, more challenging than outdoor environment. For example, various obstacles (walls, furniture, and occupants) result in a more complex and hostile environment which leads to issues such as multipath fading of wireless signal [8]. Indoor positioning applications also typically demand higher precision than outdoor applications. Over the last decade, there are many researches on different indoor positioning systems using wireless technologies such as Wi-Fi [9, 10], Bluetooth [11, 12], Infrared [13, 14], and ultrasound [15, 16]. These radio-based technologies typically require pre-installed infrastructures, which is very often environment dependent and may be too costly or even impossible to retrofit for certain applications.

Further, positioning problem can be further extended into localisation and mapping, where localisation refers to the process of determining one's position, and mapping refers to the process of establishing a view of the surrounding environment with the knowledge of walking path [17, 18]. Such systems are usually referred to as the Simultaneous Localization and Mapping (SLAM) systems, which offers great value in the fields such as robotics, unmanned autonomous vehicles (UAV), and search and rescue applications, where the systems need to operate in an unknown environment without relying on any prior installation of sensing devices.

In this study, we aim to investigate and propose a SLAM system targeting an indoor search and rescue application, such as for firefighters and rescue robots. The system aims to offer the ability to track a person or a robot in an unknown environment while at the same time map out the surrounding physical environment for others to follow through the same path. The SLAM problem was first introduced in 1991 and an Extended Kalman Filter (EKF)-based solution was proposed [17, 19]. The recent trend focuses more on graph-based SLAM algorithms rather than the traditional filter-based solutions, which aim to solve the SLAM problem using graphical model such as in [17, 20].

In order to address the challenges of lack of pre-installed and pre-configured infrastructure and lack of visibility in such hostile environments, the designed system needs to be fully self-contained without requiring any additional devices, and not impacted by any dust particles in the air. As such, the state-of-art millimeter-Wave (mmWave) radar has been chosen as the base solution for designing the indoor positioning system due to its unique features that can overcome the aforementioned limitations.

Based on the application requirements, a unique SLAM system is proposed and implemented, where the system is completely self-contained without requiring any infrastructure to be pre-installed. This is essential for exploring unknown environment, where retrofitting devices is impossible. Different to existing approaches, the proposed system is also designed solely based on mmWave radar data, without needing any other sensor inputs.

The rest of the paper is organised as follows. Section 2 provides an overview of the existing mmWave-based localisation technologies. Section 3 explains the design of the proposed mmWave-based SLAM system with the actual system prototype and its algorithmic components, followed by the experiment results presented in Sect. 4. Section 5 concludes the study with ongoing future tasks.

2 Related Work

Millimeter-Wave (mmWave) refers to radio frequency signals that operate between 30 GHz to 300 GHz spectrum, it is a wide spectrum that is less occupied [15]. The mmWave spectrum can support a wide range of applications. For instance, the fifth generation of cellular networks (5G) is utilizing 28 GHz, 38 GHz, and 70–80 GHz bands [21]. Moreover, the recently developed IEEE 802.11ad standard allows Wi-Fi to operate at the 60 GHz band [15].

For indoor positioning scenario, 60 GHz is a commonly used spectrum because the high atmospheric absorption and high path loss nature of this spectrum makes it an ideal option for positioning due to its ability to reduce self-interference. Technically, the double-bounce reflection is the maximum multipath signal that a mmWave channel can take because the higher-order bouncing signals suffer from severe attenuation at this frequency range, which means the line of sight propagation path can be easily distinguished [15]. Moreover, the beamwidth ranges between 7 and 10°s and has less energy scattering with reflection, which takes advantage of angular-based positioning method, such as angle of arrival (AoA) [15].

As mentioned, the mmWave spectrum can also be used in future Wi-Fi and cellular networks, so there is an overlap between those technologies, which offers great potential in the aspect of indoor positioning. For instance, the research in [22] suggested that 1.32 m of accuracy can be achieved through a RSSI-Fingerprint based indoor positioning method using a 60 GHz Wi-Fi system.

Regarding the passive method for mmWave technology, the interest of combining mmWave with Simultaneous Localization and Mapping (SLAM) has attracted some attention due to the short wavelength of mmWave offers the ability to integrate a large number of antenna elements into a small device [15]. Therefore, there is a potential to bring mmWave based radar technology into personal applications and indoor environment to provide the ability of target positioning along with mapping the environment [23]. Moreover, the short wavelength and the wide bandwidth can also increase the capability of handling multi-target scenarios which are very challenging using existing technologies.

3 Standalone mmWave SLAM System

The SLAM system designed in this study is illustrated in the block diagram shown in Fig. 1 which contains four major stages, namely the data collection, pose calculation, 2D conversion, and position optimisation stage. In the first stage, the raw mmWave-radar data is collected and a 3-dimensional (3D) point cloud data is generated as the output which captures the radar measurements of the surrounding environment. The 3D point cloud data is processed using the Iterative Closest Point (ICP) algorithm to estimate the current pose of the tracked user (i.e. the location and heading angle) using

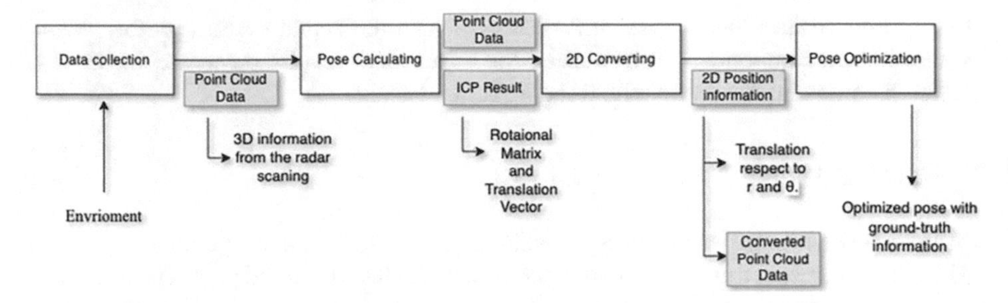

Fig. 1. mmWave-based SLAM system overview.

only mmWave radar data. During the second stage, the adjacent point cloud data will be sorted based on the sequence of frames in chronological order, then the rotation and translation between each frame will be calculated to estimate the current pose. Moreover, in this stage, numbers of frames will be merged to ensure a satisfying amount of data for performing the pose calculation. The outputs of the second stage (i.e. the ICP algorithm) are the displacement and orientation information between two adjacent frames. This information can be used to trace the moving trajectory. In this study, a 2-dimensional (2D) trajectory is targeted to fulfil the requirements of an indoor search and rescue application. Therefore, in the third stage the output of the ICP algorithm is converted to produce a 2D SLAM output (i.e. a 2D map with movement trajectories). In the last stage, the converted 2D position and mapping information will be used for further optimisation to produce the final SLAM output (i.e. a 2D map with movement trajectories). In the final stage, the iSAM2 algorithm is adopted to perform optimisation to generate a smooth trace of movement trajectory.

In the rest of this section, the first subsection introduces the mmWave radar sensor used in this study to generate the raw point cloud data. The second subsection explains the ICP algorithm that is used to extract pose estimation based on two adjacent point cloud frames. The third subsection describes how the 3D trajectory is mapped to a 2D trajectory and eventually optimised by the iSAM2 algorithm presented in the last subsection.

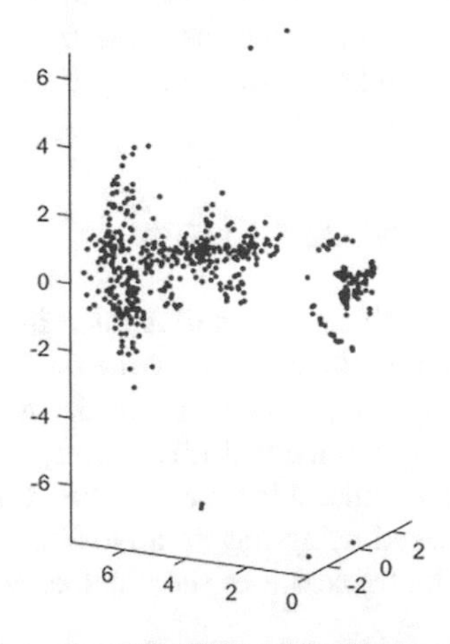

Fig. 2. An example of the IWR-1143 3D cloud point output.

3.1 mmWave Radar Sensor

Based on the application requirements, a mmWave radar sensor is selected in this study for its immunity to low visibility and hostile lighting condition in the physical environment. The mmWave radar also enables the SLAM system to be self-contained without requiring any additional infrastructure deployment as needed by other wireless localisation technologies.

In general, a radar system consists of two parts, the transmitter and the receiver part. The transmitter part is responsible for emitting a particular radio signal to the physical environment, whereas the receiver consists of an antenna that is used to sense the signal that reflects from the obstacle or the echo signal. In addition, the receiver is also responsible for revealing and analysing the echo signal, and the result will be used to estimate the angle and distance of an obstacle depending on the type of signal that the system used. Ideally, the distance and the angle of the object could be workout by sampling the time delay or phase difference between the transmitted signal and the echo signal.

In this study, the Texas Instruments IWR-1443 mmWave sensor module is used. It is a frequency modulated continuous wave (FMCW) based mmWave sensor and operates between 76 to 81 GHz bandwidth [24]. In the FMCW modulation, a sinusoidal wave that cycles through a complete range of frequency increments is considered as a "chirp". For IWR-1443, one chirp refers to a signal that has a linearly increased frequency from 77 GHz to 81 GHz as it has a 4 GHz bandwidth. The signal has a period of 40 us and the rate of increase is at 100 MHz/us [24].

In our current implementation, the collected point cloud data is transmitted to a host computing device for further data processing (i.e. the data from the radar device will be the input for both ICP and iSAM2 algorithm). Overall, the data comes from the IWR-1443 to the host computer is a point cloud data of the sensed environment. An example of one point cloud data frame collected by IWR-1443 is shown in Fig. 2. The sensed points that belong to the same frame are sent to the host with a prescribed packet structure and header information such as the number of frames and number of detected obstacles, etc.

3.2 ICP Algorithm

As discussed previously, the SLAM problem can be split into the localisation and mapping part, which involve multiple steps of sensing and mapping the physical world and tracing one's position in time. While mmWave radar can be naturally applied to map the surrounding environment, other sensor inputs may be required to estimate the trajectory of the target. Very often, an additional Inertial Measurement Unit (IMU) data is fused to provide additional pose information and to enhance the accuracy of mmWave-based localisation [25–27].

In this research, the Iterative Closest Point (ICP) algorithm is used as a replacement for the IMU to measure the target's movement. It is a dominating algorithm in the aspect of aligning 3D models and is first proposed in 1992 [28]. The process of alignment is about estimating a relative rigid-body transformation by minimising the error metric of a pair of corresponding points on two related data sets [29]. The ICP algorithm is commonly applied to point cloud data which is the output generated by the mmWave radar.

The process of alignment can be used to infer the location and orientation change, which allows the use of mmWave radar data to trace target trajectories without using an additional IMU. The ICP algorithm is an iterative algorithm and each iteration is performed between two consecutive frames of point cloud data, which is sorted in chronological order. One of the frames is used as the geometric *reference*, and the other frame is the data that is to be aligned to the reference, which is usually referred to as the *reading*. Specifically, the ICP algorithm process can be split into five steps, namely selection, matching, weighting, rejection, and minimising, which will be explained in more details in this section.

In the selection stage, one or some of the points are selected for matching both point cloud frames. Some selection strategy can be used to improve the matching accuracy. A random selection is considered as one of the baseline selection methods, which will randomly select the point pairs in each iteration. With random selection, the effect caused by outliers will be reduced [30].

The matching stage aims to find the correlation of those selected points. It is a process of approaching the *reading* point cloud frame to the *reference* frame. In the most direct manner, the distance between the selected point pairs with all of its near neighbours will be calculated, and the shortest one will be kept. However, the nearest-neighbours-based method may also lead to a mismatch when a curvature exists in the data. Alternatively, in order to compensate the uncertainty caused by curvature, the line-surface intersection technique can be used [30]. In this technique, the data point is paired based on the surface intersection indicated in the selection phase.

In each iteration, it is impossible to perfectly match all the selected points in the *reading* and *reference* frame due to the existence of outlier points, and the radar movement will also result in the shift of the sampled points. Therefore, the weighting stage is to evaluate the influence of each pair of matched points. The weight is typically calculated by evaluating the importance of each individual matching pairs. For example, one common weighting function is to work out the ratio of the distance between a point pair against the maximum separation distance between two points as shown in (1):

$$\omega = 1 - \frac{dist(p, q)}{dist_{MAX}} A = \pi r^2 \tag{1}$$

After weighting, the rejection stage is used to eliminate those point pairs which do not fit the criteria, which is then followed by the minimisation performed in the last stge. The aim is to minimise the difference between the *reading* and *reference* frame. At last, the ICP algorithm will output a rotation matrix and translation vector that best fitted the *reading* point cloud frame to the *reference* point cloud frame. Through using the ICP algorithm, the pose (i.e. the rotation matrix and translation vector) can be estimated without using additional sensor information such as an IMU.

3.3 2D Conversion

The process of converting the 3D results into a 2D plane consists of two steps. The first step is to correct the potential orientation errors along the accumulated and mapped trajectory and the second step is the 2D projection of the corrected trajectory. To begin with, the rotation matrix, the translation matrix, and the trajectory matrix built by the ICP algorithm will be used as inputs. The algorithm will loop over every location point in the trajectory, and in each iteration, the Euclidian distance between every successive point will be calculated and then work out an average distance. The trajectory could be evaluated by comparing the length between each point on it. Then, for every successive point, if the distance between two consecutive points is larger than 1.2 times or less than 0.6 times of the average distance value, it will be rearranged. The rotation and translation matrix that resulted in this displacement will be replaced with the ones from the previous points. This step smooths the trajectory by removing extreme points along the trajectory. In the second step, the smoothed trajectory is projected onto a 2D plane by converting the displacement between each pair of location points into an angle with respect to the x-axis and the Euclidean distance between the pair of consecutive points. In addition to the trajectory points, the same process is also applied to the obstacle points in a point cloud frame to rearrange the environment mapping to be aligned with the trajectory.

3.4 Map and Trajectory Optimisation

The last stage of the proposed SLAM system is to perform further optimisation to ensure a smoother SLAM output. In this stage, the iSAM2 algorithm is adopted and its process is based on the 2D information produced in the previous stage. In the iSAM2 algorithm, the factor graph is used to describe the SLAM problem, and it solves the SLAM problem by matrix factorisation, and builds up the graphical model of Bayes tree by factor elimination [31, 32]. The iSAM2 algorithm is designed for incremental updates, which offers the capability to operate in real-time for practical purposes. While the experiment trials discussed in this study are done offline, that means the sampling of the environment is completed before the optimisation stage, the incremental update features are still used by performing the iSAM2 algorithm iteratively.

In practice, the iSAM2 update can be separated into three steps, which are factor adding, setting an initial guess, and calculating the best estimation. In each iteration, the iSAM2 process starts with adding new factors, which are based on the 2D information calculated in the previous 2D conversion stage. The 2D information will then be used to initialise new variables to be added into the iSAM2 algorithm. A fluid linearisation is then performed to handle updates of non-linear data, which can simply be considered as a relinearisation process when needed. The next step is to re-estimate the model, which is a Bayes tree, with the process of updating non-linear factor, and finally update the current estimation.

4 Experiments

4.1 Experimental Design

The proposed SLAM system is evaluated in one of the levels of an actual office building which has a square-shaped corridor, and one segment of the corridor is shown in Fig. 3. All the experiments are completed using this square corridor with an inner wall of

Table 1. The IWR-1143 parameters.

Parameter	Value
Frequency band	77–81 GHz
Frame Rate	15 frames per second (fps)
Range Resolution	0.04 m
Maximum Unambiguous Range	8.19 m
Maximum Radial Velocity	2.67 m/s
Radial Velocity Resolution	0.34
Range Detection Threshold (CFAR)	10 dB
Range Peak & Doppler Peak Grouping	Disabled
Static Clutter Removal	Disabled

approximately 6.8 m long, and 9 m long for the outer wall of each side. Before the actual SLAM experiment, a series of preliminary experiments have been conducted to find out the best sensing parameters for the selected mmWave radar in the target indoor environment. The summary of the radar parameters is listed in Table 1. In general, the ICP algorithm requires a large amount of point cloud data. In order to form a denser point cloud frame that can be used for pose estimation, every set of six consecutive frames were merged into one in our experiment. The IWR-1143 mmWave radar is mounted on a user at a height that is 1.5 m above the ground, and the antenna is fixed facing the forward direction of the user at all times.

In this study, to evaluate the system performance under different experimental trials, the Root Mean Square (RMS) value is used as in (2).

$$RMS = \sqrt{\frac{\sum_{i=1}^{n}(Z_i - X_i)^2}{N}} \tag{2}$$

where N is the number of sample data to be evaluated; Z is the ground truth value; X is the sample value to be evaluated. In a real-world localisation scenario, it is impossible to find the exact ground truth point that matches with the estimated location. Therefore, the linearly projected Z_i on the ground truth walking path is considered as the reference value.

4.2 A Straight-Line Walking Test

A straight-line test has been conducted at the segment (A) illustrated in Fig. 3(a), where the user will start with the origin (0, 0), and stop at the end of the inner wall nearly segment (B). The same experiment has been conducted 3 times to ensure the results are reliable and reproducible. All three trials exhibit similar pattern in their results, and Fig. 4 shows the results of one of the experiments, where the axes are in unit of meters and the starting point is assumed at (0, 0).

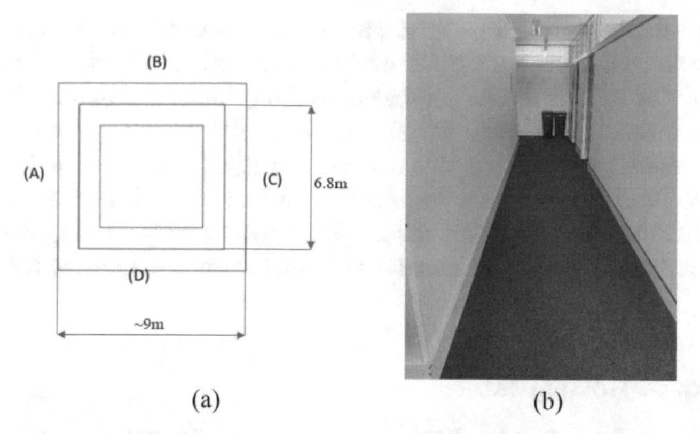

(a) (b)

Fig. 3. (a) An office building with a square-shaped corridor is used as the experimental environment. (b) A segment of the corridor is shown with a concrete wall on the left hand side and office wall on the right hand side.

The solid dots in Fig. 4(a) represent the environment features (or sometimes referred to as the landmarks), which are the walls in this case, and the hollow dots are the movement trajectory of the user. As shown in Fig. 4(a), a linear movement and pathway can be observed in both the trajectory and the sensed environment (i.e. the landmark mapping). More specifically, the RMS value of the estimated trajectory and the reference ground truth (i.e. the blue line) is 0.25 m, and is 0.45 m between the estimated environment features (i.e. the solid red dots) and the wall reference (i.e. the dark lines). This is a fairly promising result, demonstrating submeter level accuracy in a straight-line walking test. However, the total length of the path should be 6.8 m, and the trajectory is shown on the graph only has 4.6 m, which results in a 2.2 m error.

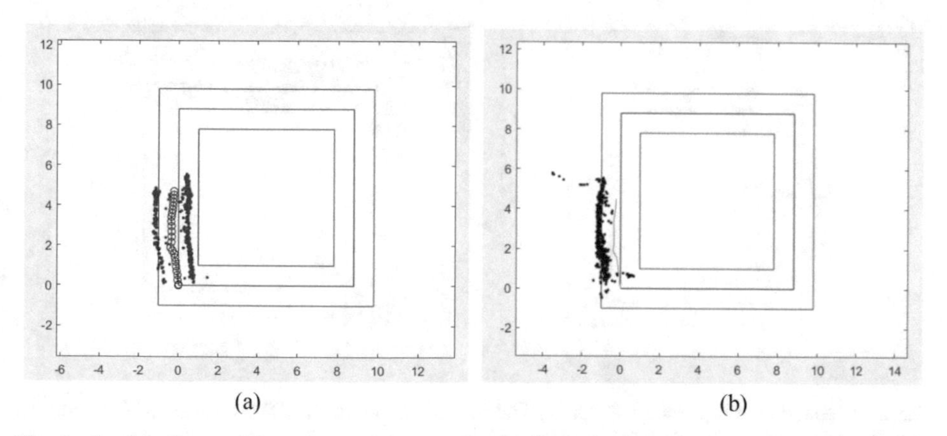

(a) (b)

Fig. 4. Straight line walking test result for (a) The SLAM output of the 2D converted ICP output. (b) The SLAM output of the iSAM algorithm.

As mentioned in Sect. 3, the 2D converted ICP output will be sent to the iSAM2 algorithm for further optimisation. The result is shown in Fig. 4(b), where the green line represents the optimised trajectory, and the dark solid points are the environmental features. The same criteria are used to evaluate the iSAM2 algorithm, which result in an RMS value of 0.26 m in the estimated walking trajectory and an RMS value of 0.29 m in the estimated landmark locations, and a total travelled distance of 4.6 m. In this case, the iSAM2 algorithm is able to achieve a slight improvement in the landmark estimation. While the iSAM2 algorithm seems to offer a slightly better result as the RMS value is smaller than the ICP result, the graphical results shown in Fig. 4(b) shows that some points representing the inner wall have been trimmed out during the optimisation process.

4.3 An L-shape Walking Test

Different to the straight-line test, the L-shape test involves a 90° turn (change in orientation) which makes the localisation and mapping more challenging. This is also a critical test to examine the effectiveness of using the ICP algorithm to replace actual IMU sensor data. The walking path is approximately 13.6 m, including both segment (A) and (B). The user starts from the bottom left corner, and stop at the top right corner. Figure 5 shows the results of the 2D converted ICP algorithm and the iSAM2 algorithm. Similar to Fig. 4, Fig. 5 axes are also in unit of meters and the starting point is assumed at (0, 0).

Referring to Fig. 5(a), both the trajectory and the environmental features are clearly reflecting an L-shape movement and pathway, with the correct turning motion. The same behaviour can be observed in Fig. 5(b) for the iSAM2 result, and the loss of information that occurred in the straight-line test still exists. However, the distribution of the points that represent the environmental features in Fig. 5(b) is more concentrated to give a clearer indication of the landmarks (i.e. the walls), and the features are less overlapped with the trajectory after optimisation.

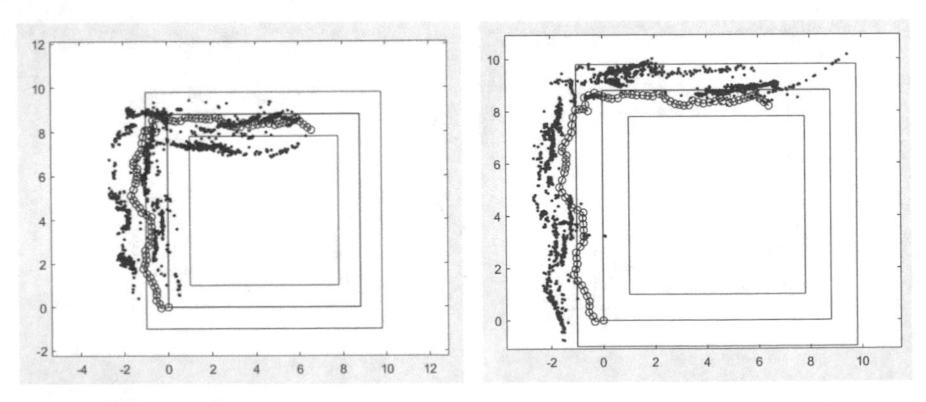

Fig. 5. L-shape walking test result for (a) The SLAM output of the 2D converted ICP output. (b) The SLAM output of the iSAM algorithm.

In comparison, the RMS results of the L-shape test is higher than the straight line test result, which is to be expected as the total travelled distance and the spatial complexity (the change in orientation is involved) of the path has increased. The ICP algorithm has achieved an RMS value of 0.81 m for the movement trajectory and an RMS value of 0.66 m for the environmental features. The difference in the final destination point is 1.1 m. In comparison, the iSAM2 algorithm achieved an RMS value of 0.81 m for the movement trajectory and an RMS value of 0.81 m for the environmental features, with a slightly better end point mismatch of 1 m. This could be explained as the ICP results contain both the inner and the outer wall, whereas the iSAM2 results contain mostly just the outer wall. This could lead to the issue where some inner wall points are actually matching to the outer wall points, which can lead to a misleading RMS result.

In the L-shape test, the proposed SLAM system also achieved submeter level accuracy, and the results are fairly consistent with what has been observed in the straight line test. This demonstrates that the proposed SLAM is also able to handle change of orientation, which is a promising result for more complex indoor environment.

5 Conclusion

Over the last decade, positioning data has demonstrated great value in various intelligent services. However, the same outcome has not been achieved for indoor applications due to lack of pre-installed or pre-configured infrastructure, high cost, and relatively low long-term accuracy for indoor positioning technologies. In this study, a new SLAM system is proposed to address these challenges by providing a completely self-contained mmWAVE SLAM system, targeting indoor search and rescue applications. The proposed SLAM system makes use of the ICP algorithm to replace the need of additional IMU data. The system has been evaluated on typical straight line and L-shaped pathway and has demonstrated submeter level accuracy with correctly detected change in orientation. These preliminary results are promising. Further studies will be conducted to further improve the accuracy and reliability of the SLAM system, and to examine the performance in more complex indoor environments.

References

1. Lee, G., Yim, J.: A review of the techniques for indoor location based service. Int. J. Grid Distrib. Comput. **5**(1), 1–22 (2012)
2. Youssef, M.: Indoor Localization. Encyclopedia of GIS, pp. 547–552 (2008)
3. Jochen, S., Agnès, V.: Location-Based Services, 1st edn. Elsevier, USA (2004)
4. Wang, H., et al.: Mobile robot indoor positioning system based on K-ELM. J. Sens. **2019** (2019)
5. de Cillis, F., et al.: Indoor positioning system using walking pattern classification. In: Proceedings of the 22nd Mediterranean Conference on Control and Automation, pp. 511–516 (2014)
6. Tabbakha, N.E., Tan, W.H., Ooi, C.P.: Indoor location and motion tracking system for elderly assisted living home. In: Proceedings of the 2017 International Conference on Robotics, Automation and Sciences, pp. 1–4 (2018)
7. Santoso, F., Redmond, S.J.: Indoor location-aware medical systems for smart homecare and telehealth monitoring: state-of-the-art. Physiol. Meas. **36**(10), R53 (2015)
8. Al-Ammar, M.A., et al.: Comparative survey of indoor positioning technologies, techniques, and algorithms. In: Proceedings of the 2014 International Conference on Cyberworlds, pp. 245–252 (2014)

9. Liu, F., et al.: Survey on WiFi-based indoor positioning techniques. IET Commun. **14**(9), 1372–1383 (2020). https://doi.org/10.1049/iet-com.2019.1059
10. Kim, J., Han, D.: Passive WiFi fingerprinting method. In: Proceedings of the 9th International Conference on Indoor Positioning and Indoor Navigation (2018)
11. Satan, A., Toth, Z.: Development of bluetooth based indoor positioning application. In: Proceedings of the 2018 IEEE International Conference on Future IoT Technologies, pp. 1–6 (2018)
12. Satan, A.: Bluetooth-based indoor navigation mobile system. In: Proceedings of the 19th International Carpathian Control Conference, pp. 332–337 (2018)
13. Lai, K.C., Ku, B.H., Wen, C.Y.: Using cooperative PIR sensing for human indoor localization. In: Proceedings of the 27th Wireless and Optical Communication Conference, pp. 1–5 (2018)
14. Yang, D., Bin, Xu., Rao, K., Sheng, W.: Passive infrared (PIR)-based indoor position tracking for smart homes using accessibility maps and a-star algorithm. Sensors **18**(2), 332 (2018). https://doi.org/10.3390/s18020332
15. von Zabiensky, F., Kreutzer, M., Bienhaus, D.: Ultrasonic waves to support human echolocation. In: Antona, M., Stephanidis, C. (eds.) UAHCI 2018. LNCS, vol. 10907, pp. 433–449. Springer, Cham (2018). https://doi.org/10.1007/978-3-319-92049-8_31
16. Hoeflinger, F., Saphala, A., Schott, D.J., Reindl, L.M., Schindelhauer, C.: Passive Indoor-Localization using Echoes of Ultrasound Signals. In: Proceedings of the 2019 International Conference on Advanced Information Technologies, pp. 60–65 (2019)
17. Khairuddin, A.R., Talib, M.S., Haron, H.: Review on simultaneous localization and mapping (SLAM). In: Proceedings of the 5th IEEE International Conference on Control System, Computing and Engineering, pp. 85–90 (2015)
18. Azril, N., Zaman, B., Abdul-Rahman, S., Mutalib, S., Shamsuddin, R.: Applying Graph-based SLAM Algorithm in a Simulated Environment. In: Proceedings of the 6th International Conference on Software Engineering & Computer Systems. IOP Science, Malaysia (2019)
19. Leonard, J.J., Durrant-Whyte, H.F.: Mobile robot localization by tracking geometric beacons. IEEE Trans. Robot. Autom. **7**(3), 376–382 (1991)
20. Kaess, M., Ranganathan, A., Dellaert, F.: iSAM: fast incremental smoothing and mapping with efficient data association. In: Proceedings of the IEEE International Conference on Robotics and Automation, pp. 1670–1677 (2007)
21. Yassin, A., Nasser, Y., Awad, M., Al-Dubai, A.: Simultaneous context inference and mapping using mm-Wave for indoor scenarios. In: Proceedings of the 2017 IEEE International Conference on Communications, pp.1–6 (2017)
22. Wei, Z., Zhao, Y., Liu, X., Feng, Z.: DoA-LF: a location fingerprint positioning algorithm with millimeter-wave. IEEE Access **5**, 22678–22688 (2017)
23. Gualda, D., et al.: Coverage analysis of an ultrasonic local positioning system according to the angle of inclination of the beacons structure. In: Proceedings of the 10th International Conference on Indoor Positioning and Indoor Navigation (2019)
24. IWR1443 Single-Chip 76-to 81-GHz mmWave Sensor 1 Device Overview. https://www.ti.com/lit/ds/symlink/iwr1443.pdf. last accessed: 2022/01/23
25. He, C., Tang, C., Yu, C.: A federated derivative cubature Kalman filter for IMU-UWB indoor positioning. Sensors **20**(12), 3514 (2020)
26. Poulose, A., Han, D.S.: Hybrid indoor localization using IMU sensors and smartphone camera. Sensors **19**(23), 5084 (2019). https://doi.org/10.3390/s19235084
27. al Mamun, M.A., Rasit Yuce, M.: Map-aided fusion of IMU PDR and RSSI fingerprinting for improved indoor positioning. In: Proceedings of the 2021 IEEE Sensors, pp. 1–4. IEEE, Sydney (2021)
28. Besl, P.J., McKay, N.D.: A method for registration of 3-D shapes. IEEE Trans. Pattern Anal. Mach. Intell. **14**(2), 239–256 (1992)

29. Kjer, H.M., Wilm, J.: Evaluation of surface registration algorithms for PET motion correction. https://lucidar.me/fr/mathematics/files/icp_bscthesis.pdf. last accessed: 2022/03/27
30. Godin, G., Rioux, M., Baribeau, R.: Three-dimensional registration using range and intensity information. In: Proceedings of the Photonics for Industrial Applications, pp. 279–290. SPIE (1994)
31. Kaess, M., et al.: ISAM2: incremental smoothing and mapping using the Bayes tree. Int. J. Robot. Res. **31**(2), 216–235 (2012). https://doi.org/10.1177/0278364911430419
32. Aldroubi, A., Hamm, K., Koku, A.B., Sekmen, A.: CUR decompositions, similarity matrices, and subspace clustering. Front. Appl. Math. Stat. **4**, 65 (2019)

Integrated Compound-Eye Alvarez Metalens Array for Apposition Image Acquisition

Hyo Adegawa, Katsuma Aoki, Satoshi Ikezawa, and Kentaro Iwami(✉)

Department of Mechanical Systems Engineering, Graduate School of Engineering, Tokyo University of Agriculture and Technology, Fuchu, Japan
Hyo-adegawa@st.go.tuat.ac.jp

Abstract. In this study, Alvarez metalenses, which is advantageous for miniaturization, with a compound-eye arrangement are designed, fabricated, and characterized. We designed three different lenses with different focal length designs, one of which varies focal length from 0.75 mm to 3.75 mm with substrate displacements from 10 μm to 50 μm. The compound-eye lens consists of a 4 × 4 array of 0.5 mm square lenses with a height of 400 nm, which is thinner compared to the conventional refractive micro-lens array (several dozen micrometers height). Alvarez metalenses were developed using the dielectric metasurface consisting of amorphous silicon nanopillars. Pillars were fabricated by electron beam lithography and reactive ion etching. Observation of the fabricated lens structure showed that the nanopillars of specific dimensions were missing. However, the image was obtained with the lenses we fabricated. The resolution obtained from the smallest identifiable pattern in the captured image was 6.35 lp/mm. Focal length measurements confirmed that the measured values agree well with the theoretical values.

Keywords: Metalens · Alverez lens · Compound-eye lens · Amorphous silicon · Image processing

1 Introduction

Varifocal lenses installed in various devices are required to be thinner as those devices become smaller. Metalenses [1–4] are lenses with sub-wavelength scale structures and are characterized by their extremely thinness compared to refractive lenses. Therefore, by using metalens technology, significant thinning can be expected even in varifocal lenses. Arbabi et al. fabricated and demonstrated a tunable metalens realized by implementing vertical MEMS actuation [5]. This lens requires reserve space between lenses for actuation and is not suitable for overall miniaturization.

In this study, we focused on two technologies: the Alvarez lens, a type of variable focus lens, and the compound-eye lens. Alvarez lenses [6, 7] are relatively advantageous for thinning because the focal length is adjusted by displacing the lenses perpendicular to the optical axis. Compound-eye lenses have a shorter focal length than monocular lenses of the same numerical aperture. We have combined these two technologies to produce compound-eye Alvarez metalenses that operate at a wavelength of 900 nm.

N. K. Suryadevara et al. (Eds.): ICST 2022, LNEE 1035, pp. 162–169, 2023.
https://doi.org/10.1007/978-3-031-29871-4_18

2 Methods

2.1 Design of Metaatoms for Metalens Fabrication

We have conducted various studies on dielectric metasurfaces and their optical characterization [4, 8–11]. This study fabricated metalenses by arranging amorphous silicon (a-Si) pillars in a hexagonal lattice. The a-Si pillars act as a waveguide, and light transmitted through them is phase-delayed relative to light that is not. The amount of phase delay varies with the dimensions of the pillars. Therefore, if the wavefront of transmitted light is controlled to have a phase distribution like that of a refractive lens by arranging pillars of different dimensions on the substrate, these pillar arrays can function as an ultra-thin lens.

Figure 1 shows the relation between the width of the pillar and the transmission and phase delay of light. The inset of the graph in Fig. 1 is a schematic of the metaatom. The a-Si octagonal pillars are aligned periodically on a hexagonal lattice on the glass substrate with a height of $h = 400$ nm and a lattice constant $p = 360$ nm. The analysis revealed that a pillar with a width of 110 nm to 250 nm could be used to achieve both high transmittance and a phase delay of 0 to 2π, which is necessary to reproduce the lens.

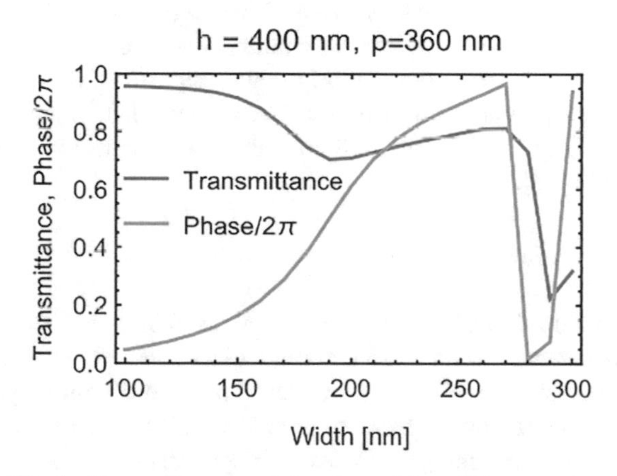

Fig. 1. Transmittance and the phase delay of the metaatom.

2.2 Design of Alvarez Metalens

Alvarez lenses are a type of varifocal lens. These lenses are composed of a pair of two lenses. The focal length can be continuously changed by displacing both lenses perpendicular to the optical axis direction. The Alvarez metalens is a metalens that realizes the functions of the Alvarez lens [6, 7].

Figure 2a shows schematics of the Alvarez metalens. The phase delays ϕ_{reg}, ϕ_{inv} that the two lens substrates impart to the transmitted light are given by Eq. (1) using the in-plane coordinates with the lens center as the origin.

$$\phi_{\text{reg}}(x, y) = -\phi_{\text{inv}}(x, y) = A\left(\tfrac{1}{3}x^3 + xy^2\right) \tag{1}$$

A is a constant that determines the focal length of the lens and is expressed as in Eq. (2).

$$A = \frac{\pi}{2\lambda f d} \cong \frac{\pi \text{NA}}{L\lambda d} \tag{2}$$

In Eq. (2), NA, f, λ, L are the numerical aperture, the focal length, the wavelength, and the lens width, respectively. When two lens substrates are superimposed with a mutual displacement of $\pm d$ in the x x-direction the phase across the lens is expressed by Eq. (3).

$$\begin{aligned}\phi(x, y) &= \phi_{\text{reg}}(x + d, y) + \phi_{\text{inv}}(x - d, y) \\ &= 2Ad\left(x^2 + y^2\right) + \tfrac{2}{3}Ad^3 \\ &\cong 2Ad\left(x^2 + y^2\right)\end{aligned} \tag{3}$$

The addition of ϕ_{reg} and ϕ_{inv} corresponds to stacking two lens substrates. This phase is equivalent to that of a spherical lens, indicating that the Alvarez metalens has the same imaging action as a spherical lens. We designed three types of Alvarez metalenses and determined the constants A to have NA values of 0.3, 0.4, and 0.5 at the reference displacement of $d = 30$ μm at the working wavelength of 900 nm. The compound-eye Alvarez metalens produced consists of a 4 × 4 array of 0.5 mm square monocular Alvarez metalenses (Fig. 2b). This lens is thinner compared to the conventional refractive micro-lens array [12, 13].

2.3 Fabrication

Figure 3 shows schematics of the Alvarez metalens fabrication process. A 400 nm thick a-Si layer was deposited on a 525 μm thick glass substrate by sputtering. Alvarez metalenses were patterned on the resist by electron beam lithography and developed. After depositing Al and removing the resist with a stripping solution, the pillar structure was formed by reactive ion etching the Si layer using the remaining Al as a mask. Finally, Al was removed by wet etching. The electron beam lithography process is known to cause a systematic error of + 30 μm in the width of the fabricated pillars, which was addressed using an inverse design method.

Figure 4 shows a CAD drawing of the designed Alvarez metalens and an SEM image of the fabricated Alvarez metalens. Comparing the drawings and observed images revealed that pillars with widths of 110 nm and 120 nm are missing. This is due to the loss of the Al mask during the resist removal process.

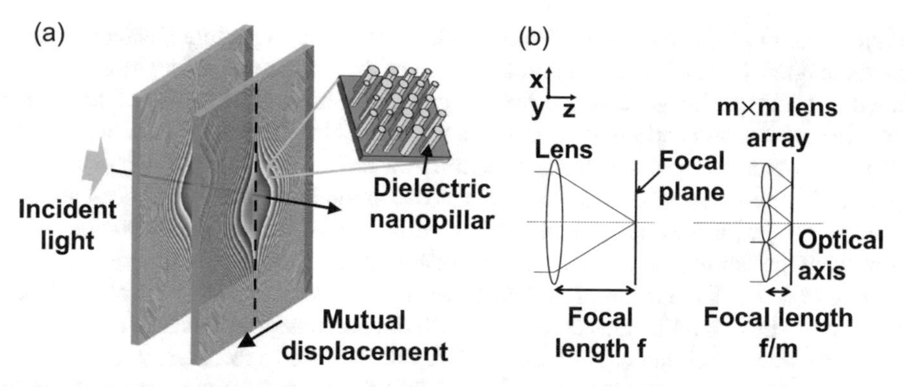

Fig. 2. (a) Schematics illustrations of Alvarez metalenses. (b) Compound-eye configuration of the fabricated Alvarez metalens.

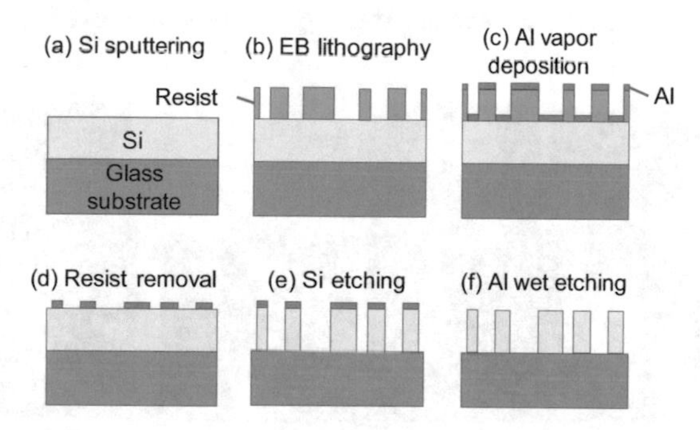

Fig. 3. Fabrication process of Alvarez metalens.

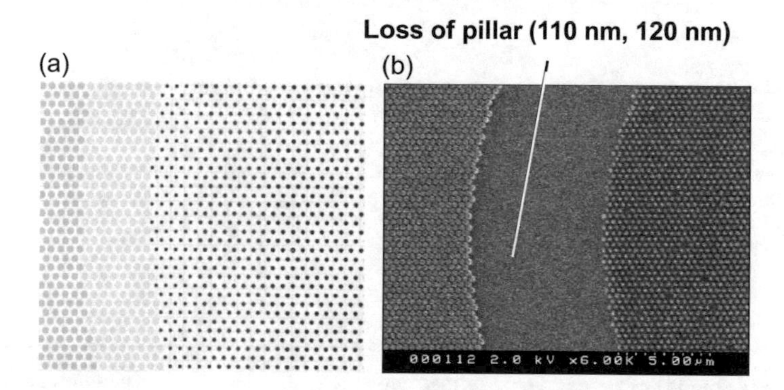

Fig. 4. (a) CAD drawing of designed Alvarez metalens. (b) SEM image of fabricated Alvarez metalens.

3 Results and Discussion

To evaluate the imaging performance of fabricated lenses, we illuminated a 1951 USAF resolution test chart with 900 nm LED in transmission and imaged the pattern onto an image sensor of a CMOS camera. Figure 5 shows the images captured using fabricated compound-eye lenses. Images captured by each of the 4 × 4 monocular lenses are combined into a single image. Inverted real images were obtained when the Alvarez metalens was convex, and upright imaginary images were obtained when it was concave. The resolution obtained from the smallest identifiable pattern of the target image was 6.35 lp/mm. This result may be related to the fact that some transmitted light does not contribute to image formation due to the lack of metaatoms. Higher resolution is expected by improving the fabrication process and producing metalenses without structural defects.

Figure 6 shows the result of the focal length measurement. The fabricated Alvarez metalenses show focal lengths agree well with the theoretical estimation, especially in the region where $|d|$ is large. The lens with NA = 0.2 showed a focal length of 0.576 mm at $d = 70$ μm, which corresponds to NA = 0.398. On the other hand, the 2 mm square monocular lens achieved a focal length of 2.28 mm at $d = 120$ μm, which corresponds to NA = 0.401. Therefore, in comparing monocular and compound-eye lenses of 2 mm, the compound-eye lens succeeded in shortening the focal length by 1/4.

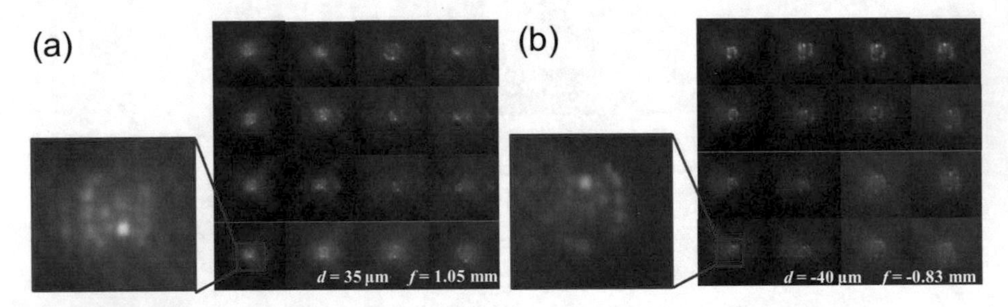

Fig. 5. Captured images on CMOS camera, (a) convex state, (b) concave state.

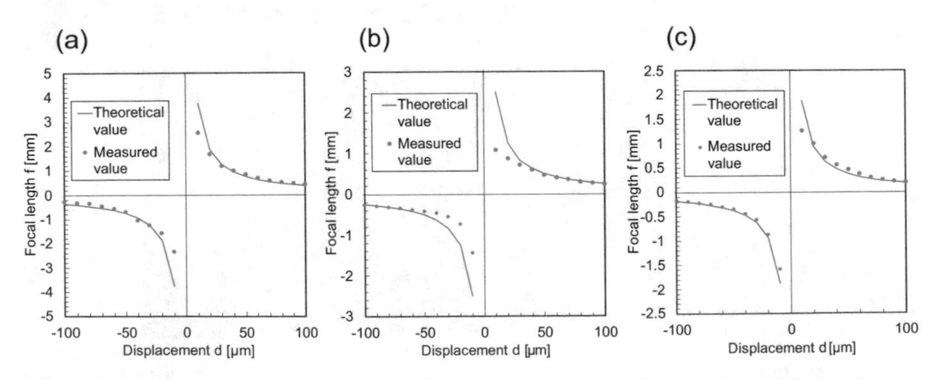

Fig. 6. Measurement results of focal length, (a) NA = 0.2, (b) NA = 0.3, (c) NA = 0.4.

Considering the production and evaluation, we changed the design and fabrication process. In the design, we added lens arrays which consist of 50 μm square monocular lens (1/10th the size of the previous fabrication) to the sample layout we produce. This change is to verify smaller lens arrays can be manufactured and work. In the fabrication process, we changed the type of resist from FEP-171D (FUJIFILM Electronics Materials Co., Japan) to ZEP 520A-7 (ZEON Co., Japan) used in the electron beam lithography process. This enables the fabrication of pillars with a width of 120 nm or less, and lenses with no pillar defects can be expected.

Figure 7 shows a CAD drawing and SEM images of Alvarez metalens after changes to the design and fabrication process. Comparison of CAD drawing and SEM images confirmed the smallest width of 100 nm and the largest width of 240 nm pillars were fabricated among those used in the design. Since the lens was successfully fabricated with no defects in the pillar structure, the process modification was effective in fabricating the microstructure.

Figure 8 shows the result of demonstration of focusing action. For three different lens arrays with different NA, a spot of light was observed in the center of each individual lens arranged in a 4 × 4 configuration.

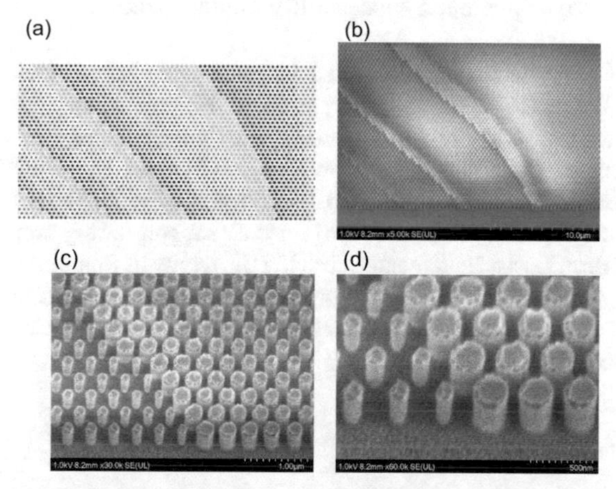

Fig. 7. (a) CAD drawing of nanopillars of Alvarez metalens, (b), (c), (d), SEM image of fabricated nanopillars at 5000x, 30,000x, and 60,000x, respectively.

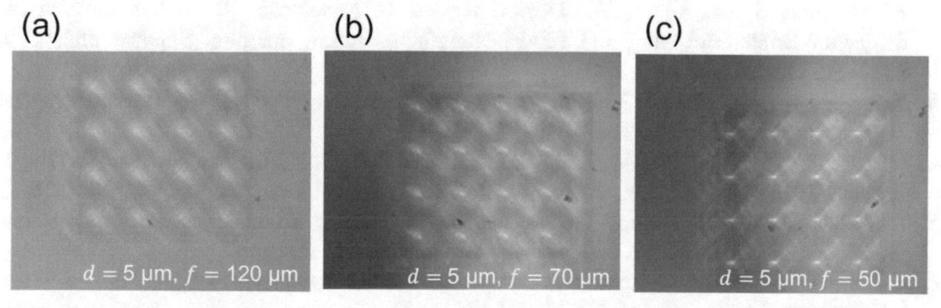

Fig. 8. Measurement results of focusing action, (a) reference NA = 0.1, (b) reference NA = 0.2, (c) reference NA = 0.3.

4 Conclusion

We have designed, fabricated, and characterized compound-eye Alvarez metalens arrays. Although some of the pillars were missing, observation of the captured image and measurement of the focal length showed that the fabricated lenses functioned as both concave and convex lenses, and that the focal length varied continuously. The resolution obtained from the smallest identifiable pattern in the target image is 6.35 lp/mm. This is expected to be improved by optimizing the process and producing lenses with no metaatom defects. In addition, compound-eye metalenses showed focal length reduction by 1/4 of that of a monocular lens. To further reduce the size of the compound meta-lens array and to produce a lens with no pillar defects, we modified the design and process and produced another metalens array. As a result of the process changes, it is possible to fabricate pillars with smaller widths than before and lens arrays with no structural defects. The demonstration results also show the manufactured metalens arrays have focusing action and the metalens arrays have been successfully miniaturized.

Acknowledgement. This work was supported in part by the Japan Society for the Promotion of Science (JSPS) KAKENHI under Grant 21H01781 and Grant 22K04894; in pard by "Advanced Research Infrastructure for Materials and Nanotechnology in Japan (ARIM)" of the Ministry of Education, Culture, Sports, Science and Technorogy (MEXT), Grant Number JPMXP1222UT1056. The authors would like to thank Prof. Y. Mita, Dr. A. Higo, Dr. E. Lebrasseur, and Mr. M. Fujiwara (The University of Tokyo) for their support during sample fabrication, also would like to thank Prof. Lucas Heitzmann Gabrielli (University of Campinas) for developing and maintaining gdstk, a Python library for creating and manipulating GDSII layout files, and also would like to thank a TSUBAME3.0 supercomputer at the Tokyo Institute of Technology for the numerical calculations.

References

1. Arbabi, A., Horie, Y., Ball, A., et al.: Subwavelength-thick lenses with high numerical apertures and large efficiency based on high-contrast transmitarrays. Nat. Commun. **6**, 7069 (2015)
2. Khorasaninejad, M., Chen, W., Devlin, R., et al.: Metalenses at visible wavelengths: diffraction-limited focusing and subwavelength resolution imaging. Science **352**(6290), 1190–1194 (2016)
3. Wang, S., Wu, P.C., Su, V.C., et al.: A broadband achromatic metalens in the visible. Nat. Nanotechnol. **13**, 227–232 (2018)
4. Ogawa, C., Nakamura, S., Aso, T., et al.: Rotational varifocal moiré metalens made of single-crystal silicon meta-atoms for visible wavelengths. Nanophotonics **11**(9), 1941–1948 (2022)
5. Arbabi, E., Arbabi, A., Kamali, S.M., et al.: MEMS-tunable dielectric metasurface lens. Nat. Commun. **9**, 812 (2018)
6. Colburn, S., Zhan, A., Majumdar, A.: Varifocal zoom imaging with large area focal length adjustable metalenses. Optica **5**, 825–831 (2018)
7. Han, Z., Colburn, S., Majumdar, A., et al.: MEMS-actuated metasurface Alvarez lens. Microsyst. Nanoeng. **6**, 79 (2020)
8. Ikezawa, S., Yamada, R., Takaki, K., Ogawa, C., Iwami, K.: Micro-optical line generator metalens for a visible wavelength based on octagonal nanopillars made of single-crystalline silicon. IEEE Sens. J. **22**(15), 14851–14861 (2022)

9. Yamada, N., Saito, H., Ikezawa, S., Iwami, K.: Demonstration of a multicolor metasurface holographic movie based on a cinematographic approach. Opt. Express **30**(10), 17591 (2022)
10. Iwami, K., Ogawa, C., Nagase, T., Ikezawa, S.: Demonstration of focal length tuning by rotational varifocal moiré metalens in an ir-A wavelength. Opt. Express **28**(24), 35602–35614 (2020)
11. Izumi, R., Ikezawa, S., Iwami, K.: Metasurface holographic movie: a cinematographic approach. Opt. Express **28**(16), 23761 (2020)
12. Kangsen, L., Xinfang, H., Qiang, C., et al.: Flexible fabrication of optical glass micro-lens array by using contactless hot embossing process. J. Manuf. Process. **57**, 469–476 (2020)
13. Sohn, I., Choi, H., Noh, Y., et al.: Laser assisted fabrication of micro-lens array and characterization of their beam shaping property. Appl. Surf. Sci. **479**, 375–385 (2019)

Measuring Mueller Matrices of 'Zesy002' Kiwifruit Peel and Pericarp Slices

Damenraj Rajkumar[1,2,3](✉), Rainer Künnemeyer[2,3], Jevon Longdell[1,3], and Andrew McGlone[2]

[1] Department of Physics, The University of Otago, Dunedin, New Zealand
damen.rajkumar@plantandfood.co.nz
[2] The New Zealand Institute for Plant and Food Research Limited, Ruakura, New Zealand
[3] The Dodd-Walls Centre for Photonic and Quantum Technologies, Dunedin, New Zealand

Abstract. A transmission-based polarized light system was used to make polarized light measurements on 20 eating ripe, yellow-fleshed kiwifruit (*Actinidia chinensis* var. *Chinensis* 'Zesy002') in the 650–1000 nm Near-Infrared (NIR) region. Mueller matrices of kiwifruit outer pericarp slices and skin peel were calculated from measurements made using a rotating retarder fixed polarizer (RRFP) polarimeter. The Lu-Chipman decomposition was used to decompose the filtered experimental matrices into depolarizer, retarder and diattenuation matrices, with the depolarization coefficient, linear retardance and circular retardance extracted from those decomposed matrices. A strong linear relationship was observed between the de-polarization coefficient of the pericarp slices and the penetrometer firmness of the fruit, the firmer kiwifruit showing a higher depolarization coefficient than softer fruit. The depolarization coefficients were spectrally of only moderate or low slope, lower at higher wavelengths, and those for the skin peel were generally higher than those of the pericarp slices. The latter observation is likely due to the more complex light scattering morphology of the skin peel. The linear retardance was spectrally flat and varied greatly by sample, possibly due to variation in the presence of birefringent material such as cellulose. Although circular retardance was negligible for the pericarp slices, spectrally flat around zero, that was not the case for skin peel, which was significantly non-zero and with a strong spectral trend of decreasing circular retardance with wavelength. These results provide some new insights into understanding the measurable responses of polarized light in fruit.

Keywords: Polarization · Near-Infrared · Fruit · Mueller Matrix · Spectroscopy

1 Introduction

The New Zealand horticulture industry boasts earnings of more than NZ$6.6 billion with kiwifruit exports exceeding NZ$2.5 billion in 2020 [1]. To reliably maintain demand and standards for premium quality fruits, optical techniques such as near-infrared spectroscopy (NIRS) are sometimes used to non-destructively grade fruit in terms of internal quality parameters, such as the consumer taste proxies of soluble solids content (SSC),

N. K. Suryadevara et al. (Eds.): ICST 2022, LNEE 1035, pp. 170–182, 2023.
https://doi.org/10.1007/978-3-031-29871-4_19

dry matter content (DMC) and firmness. However, NIRS commercial systems are far from perfect, and there is much room for useful technological advances to be made [2]. A property of NIRS light that is currently not used for quality assessment is the polarization state of light.

Studies involving the use of polarized light to determine internal quality of horticultural produce are few, but two recent studies have demonstrated potential for useful non-destructive assessment. One study examined the polarized light response of three green 'Conference' pears over a ten-day period, monitoring the degree of linear and circular polarization (DOLP and DOCP) of the diffusely reflected light as the fruit softened [3]. There was an observed decrease in both DOLP and DOCP on the monitored pears with time. A fairly similar monitoring study on apples ('Red Star' and 'Golden Delicious') over a period of two weeks revealed a slow increase of DOLP [4].

Such contrasting differences in results from two similar fruit monitoring studies, one demonstrating increasing and the other decreasing DOLP with time, suggests more certain or detailed knowledge regarding the polarization response of fruit would be useful. A knowledge gap is the relative response of polarized light on the thin outer skin or peel layer, and the near-surface underlying fruit tissue (outer pericarp layer). These are quite different tissue types and, for instance, on yellow-fleshed kiwifruit the very thin skin peel comprises a layer of dead cells with a hypodermis consisting of two to three layers of closely packed cells compared with the outer pericarp tissue, which has small and large parenchyma cells [5]. It can be expected the two tissue types have different polarization responses and that the overall non-destructive fruit response to polarized light will depend on the combined effect of the skin peel and the underlying outer pericarp layers that are traversed by the measured light.

The aim of this experimental study was to separately measure the spectroscopic polarized light response of the skin peel and outer pericarp slices of yellow-fleshed kiwifruit (*Actinidia chinensis* var. *Chinensis* 'Zesy002'). A static rotating retarder fixed polarizer (RRFP) polarimeter is used to analyze, using a Mueller matrix decomposition methodology, the polarized light transmission characteristics in the NIR range commonly used for NIRS fruit grading (650–1000 nm). The polarization response was interpreted in terms of differences observed with independently measured light scattering and fruit firmness properties.

By measuring the Mueller matrix of kiwifruit skin peel and outer pericarp tissue, properties such as depolarization, retardance and diattenuation can be determined. These properties might correlate with parameters describing internal quality of kiwifruit and might lead to alternative measurement methods for non-destructive evaluation of fruit.

2 Materials and Methods

2.1 Kiwifruit Samples and Preparation

Twenty 'Zesy002' kiwifruit were harvested from The New Zealand Institute for Plant and Food Research Limited (Plant & Food Research), Ruakura Orchard (Hamilton, New Zealand), stored in a cool room (2–3 °C) for 2 days before being placed in a 22 °C laboratory to soften to eating firmness. The decay in fruit firmness, to the eating ripe stage for measurement, was monitored non-destructively using a portable impact probe [6].

2.2 Experimental Procedure

As shown in Fig. 1, an approximately 2-mm outer pericarp slice and one skin peel sample was excised from each side of each kiwifruit using a commercial meat slicer (WFS30MGB3, Wedderburn, New Zealand). Preparation of the outer pericarp sample for measurement involved first adding four drops of 6% solution of citric acid to the exposed surfaces, to prevent oxidative browning, and then gently sandwiching the slice between two 100 × 100 mm sheets of borosilicate glass of 1.12 mm thickness. Optical properties of the sandwiched sample were then measured using the Plant & Food Research laboratory's bespoke IAD (Inverse Adding-Doubling) integrating sphere method [7]. Experimental polarization measurements on the sandwiched sample, to generate the required Mueller matrix, were completed immediately afterwards.

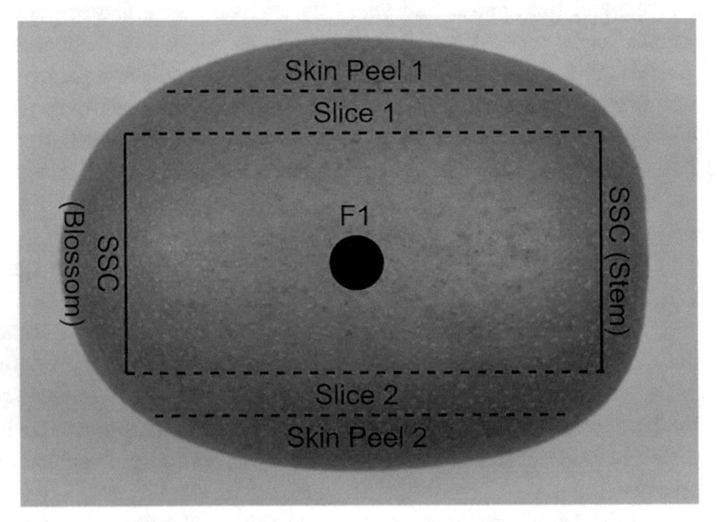

Fig. 1. Positions where outer pericarp slice and skin peel samples are excised from a kiwifruit. Soluble solids content (SSC) is taken as an average of the stem and blossom end. F1 shows the position where one penetrometer firmness measurement is taken with F2 (not shown) being measured on the opposite side.

Preparation of the skin peel sample, after removal from the fruit, first required careful cleaning away of any internal fruit tissue from the inner surface, achieved by gently scraping with a scalpel blade until only the clean and dry skin layer remained. The skin peel layer was also sandwiched between glass slides before measured directly by the RRFP. Due to the large port size (40 mm diameter) of the IAD system, optical coefficients were not measured for smaller skin peel samples (approximately 20 mm diameter).

The measurement sequence was to remove the skin peel and outer pericarp layers from one side of the fruit, prepare and measure the outer pericarp sample, and then clean and measure the skin peel sample. The process was then repeated on the other side of the fruit. Excised sections of the kiwifruit not immediately used at any stage were placed back on the cut kiwifruit surfaces and the whole fruit wrapped in cling wrap (Gladwrap®, Plant Based Cling Wrap, Sydney, Australia) to reduce moisture loss.

Penetrometer firmness measurements were then taken on the remaining two intact surfaces (see Fig. 1) using a GUSS fruit texture analyzer (Fruit Texture Analyser GS-20, GUSS Manufacturing Limited, South Africa). SSC measurements were acquired by squeezing juice from the stem and blossom ends into a refractometer (PAL-1, Atago Co. Ltd, Tokyo, Japan). Summary statistics for penetrometer firmness and SSC are given in Table 1.

Table 1. Penetrometer firmness and SSC distribution of kiwifruit used (N = 20).

	Mean	Standard Deviation	Minimum	Maximum
Firmness (kgf)	4.39	2.29	0.40	7.51
SSC (°Brix)	14.6	2.8	8.2	17.8

2.3 Rotating Retarder Fixed Polarizer (RRFP) Polarimeter

Mueller matrices of kiwifruit pericarp slices and skin peel were measured using a RRFP (Fig. 2). The polarization state generator (PSG) onsists of a feedback-stabilized halogen lamp (Newport Oriel Instruments, Irvine, California, USA). The output light is collimated using an achromatic fiber collimator (C40SMA-B, Thorlabs Inc., Newton, NJ, USA). A 5 mm aperture is placed before the PSG. A rotatable linear polarizer (LPVIS100-MP2, Thorlabs Inc., Newton, NJ, USA) is placed before and after a Fresnel rhomb (FR600QM, Thorlabs Inc., Newton, NJ, USA) to produce circularly (right and left) and linearly (0°, 45°, 90° and 135°) polarized light in the 650–1000 nm range as shown below,

$$\boldsymbol{S}_{Circular(\theta=\pm 45^\circ)} = \boldsymbol{M}_{\lambda/4}\boldsymbol{M}_{pol}\boldsymbol{S}_{unpolarized},$$

$$\boldsymbol{S}_{Circular(\theta=\pm 45^\circ)} = \begin{bmatrix} 1 & 0 & 0 & 0 \\ 0 & 1 & 0 & 0 \\ 0 & 0 & 0 & -1 \\ 0 & 0 & 1 & 0 \end{bmatrix} \frac{1}{2} \begin{bmatrix} 1 & cos(2\theta) & sin(2\theta) & 0 \\ cos(2\theta) & cos^2(2\theta) & sin(2\theta)cos(2\theta) & 0 \\ sin(2\theta) & sin(2\theta)cos(2\theta) & sin^2(2\theta) & 0 \\ 0 & 0 & 0 & 0 \end{bmatrix} \begin{bmatrix} 1 \\ 0 \\ 0 \\ 0 \end{bmatrix}, \quad (1)$$

$$\boldsymbol{S}_{\mathrm{Linear}(\theta=0^\circ,\pm 45^\circ,90^\circ)} = \boldsymbol{M}_{\mathrm{pol}}\boldsymbol{M}_{\lambda/4}\boldsymbol{S}_{unpolarized},$$

$$S_{Linear(\theta=0°,\pm45°,90°)} = \frac{1}{2}\begin{bmatrix} 1 & cos(2\theta) & sin(2\theta) & 0 \\ cos(2\theta) & cos^2(2\theta) & sin(2\theta)cos(2\theta) & 0 \\ sin(2\theta) & sin(2\theta)cos(2\theta) & sin^2(2\theta) & 0 \\ 0 & 0 & 0 & 0 \end{bmatrix}\begin{bmatrix} 1 & 0 & 0 & 0 \\ 0 & 1 & 0 & 0 \\ 0 & 0 & 0 & -1 \\ 0 & 0 & 1 & 0 \end{bmatrix}\begin{bmatrix} 1 \\ 0 \\ 0 \\ 0 \end{bmatrix},$$

where $\boldsymbol{S}$ is the input Stokes vector before interacting with the sample. $\boldsymbol{M}_{\lambda/4}$, $\boldsymbol{M}_{pol}$ and $\boldsymbol{S}_{unpolarized}$ are the quarter waveplate and rotatable linear polarizer Mueller matrices, and Stokes vector for an unpolarized source, respectively. θ represents the angles that the polarizer is rotated at to produce linear and circular polarization states.

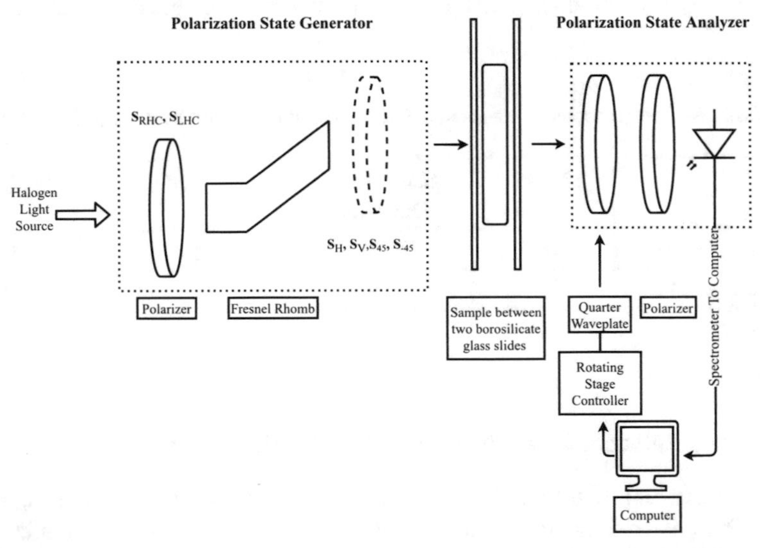

Fig. 2. Experimental set up of rotating retarder fixed polarizer (RRFP) polarimeter. A Zeiss spectrometer records spectra obtained after each quarter waveplate rotation.

Light collected after interacting with the sample is processed by the polarization state analyzer (PSA), which consists of a plano-convex lens (LA1986-B, Thorlabs Inc., Newton, NJ, USA), an achromatic quarter waveplate (AQWP10M-980, Thorlabs Inc., Newton, NJ, USA) and a fixed linear polarizer (LPVIS100-MP2, Thorlabs Inc., Newton, NJ, USA). The quarter waveplate is rotated to six angles (±30.4°, ±53.2°, ±79.9°) using a motorized mount (KPRM1E/M, Thorlabs Inc., Newton, NJ, USA), which minimizes the condition number of the data matrix of the PSA [8, 9]. Spectra for each quarter waveplate angle is collected by a fiber collimator package (F260SMA-B, Thorlabs Inc., Newton, NJ, USA) coupled to a 550 μ m multimode fiber (M37L02, Thorlabs Inc., Newton, NJ, USA) and is recorded by a Zeiss spectrometer (MMS-1, Zeiss, Jena, Germany).

Stokes vectors ($\boldsymbol{S}(\lambda)$) measured by the RRFP system are related to the six intensities measured by the spectrometer through the equation,

$$\boldsymbol{S}(\lambda) = \boldsymbol{W}(\lambda)\boldsymbol{I}(\lambda), \tag{2}$$

where $\boldsymbol{W}(\lambda)$ is the data reduction matrix (See Appendix A) for each wavelength and $\boldsymbol{I}(\lambda)$ are the spectra measured for each quarter waveplate angle.

Data Acquisition and Analysis

The Mueller matrix is then calculated using the equation,

$$\boldsymbol{M}(\lambda)_{sample} = \boldsymbol{S}(\lambda)_{sample} \cdot \boldsymbol{S}(\lambda)^{-1}{}_{input}, \tag{3}$$

where $\boldsymbol{M}(\lambda)_{sample}$, $\boldsymbol{S}(\lambda)_{sample}$ and $\boldsymbol{S}(\lambda)_{input}$ are the sample's Mueller matrix, measured Stokes vectors and input Stokes vectors, respectively, for each wavelength λ (nm).

Experimentally measured Mueller matrices may contain random noise that can produce non-physically realizable Mueller matrices. To ensure that they correspond to physically realizable ones, these are filtered using a method outlined in [10]. An experimental Mueller matrix $\boldsymbol{M}_{exp}$ is first converted to a coherency (covariance) matrix $\boldsymbol{C}$,

$$\boldsymbol{C} = \frac{1}{4}\sum\nolimits_{i,j=1}^{4} M_{i,j}\boldsymbol{A}\left(\boldsymbol{\sigma}_i \otimes \boldsymbol{\sigma}^*{}_j\right)\boldsymbol{A}^{-1}, \tag{4}$$

where $M_{i,j}$ are elements of the experimental Mueller matrix, $\boldsymbol{A}$ is the unitary matrix and $\boldsymbol{\sigma}_i$ are the Pauli matrices with the 2×2 identity, multiplied as outer products. The unitary matrix $\boldsymbol{A}$ is given by

$$\boldsymbol{A} = \begin{bmatrix} 1 & 0 & 0 & 1 \\ 1 & 0 & 0 & -1 \\ 0 & 1 & 1 & 0 \\ 0 & i & -i & 0 \end{bmatrix}, \tag{5}$$

and Pauli matrices as in [11],

$$\boldsymbol{\sigma}_1 = \begin{pmatrix} 1 & 0 \\ 0 & 1 \end{pmatrix}, \boldsymbol{\sigma}_2 = \begin{pmatrix} 1 & 0 \\ 0 & -1 \end{pmatrix}, \boldsymbol{\sigma}_3 = \begin{pmatrix} 0 & 1 \\ 1 & 0 \end{pmatrix}, \boldsymbol{\sigma}_4 = \begin{pmatrix} 0 & -i \\ i & 0 \end{pmatrix}. \tag{6}$$

Negative eigenvalues associated with the coherency matrix $\boldsymbol{C}$ results in a non-physically realizable Mueller matrix. We can subtract the eigenvectors associated with the negative eigenvalues to 'filter' the experimental Mueller matrix.

The Lu-Chipman decomposition is used to decompose the filtered matrices into depolarizer ($\boldsymbol{M}_{\Delta}$), retarder ($\boldsymbol{M}_R$) and diattenuation matrices ($\boldsymbol{M}_D$) [12, 13],

$$\boldsymbol{M}_{filtered} = \boldsymbol{M}_{\Delta}\boldsymbol{M}_R\boldsymbol{M}_D. \tag{7}$$

The depolarization coefficient, linear retardance and circular retardance are then calculated from those decomposed matrices using the equations defined in [13],

$$\Delta = 1 - \frac{|tr(\boldsymbol{M}_{\Delta}) - 1|}{3}, \tag{8}$$

$$\delta = cos^{-1}\left(\sqrt{\left(\boldsymbol{M}_{R,22} + \boldsymbol{M}_{R,33}\right)^2 + \left(\boldsymbol{M}_{R,32} - \boldsymbol{M}_{R,23}\right)^2} - 1\right), \tag{9}$$

and

$$\psi = tan^{-1}\left(\frac{\boldsymbol{M}_{R,32} - \boldsymbol{M}_{R,23}}{\boldsymbol{M}_{R,22} + \boldsymbol{M}_{R,33}}\right), \tag{10}$$

where Δ is the net depolarization coefficient, δ is the linear retardance and ψ is the circular retardance. $\boldsymbol{M}_{R,ij}$ refers to elements of the retardance Mueller matrix after applying the Lu-Chipman decomposition. These polarization properties were averaged across the measurements made separately on the two sides of each kiwifruit.

3 Results and Discussion

3.1 IAD Method – Reduced Scattering Coefficient

The reduced scattering coefficient of the pericarp slices decreases steeply as the fruit soften from 7.5 kgf to 4 kgf, but then flattens considerably in the relationship below 2 kgf (Fig. 3). In kiwifruit, softening is closely associated with degradation and swelling of the cell wall [14, 15]. As kiwifruit decrease in firmness, the cell wall layer, which maintains the rigidity of structures such as parenchyma cells, breaks down into weaker structures that swell with water. This means less light scattering, due to less variation in refractive index in the tissue, and longer light pathlengths between scattering events.

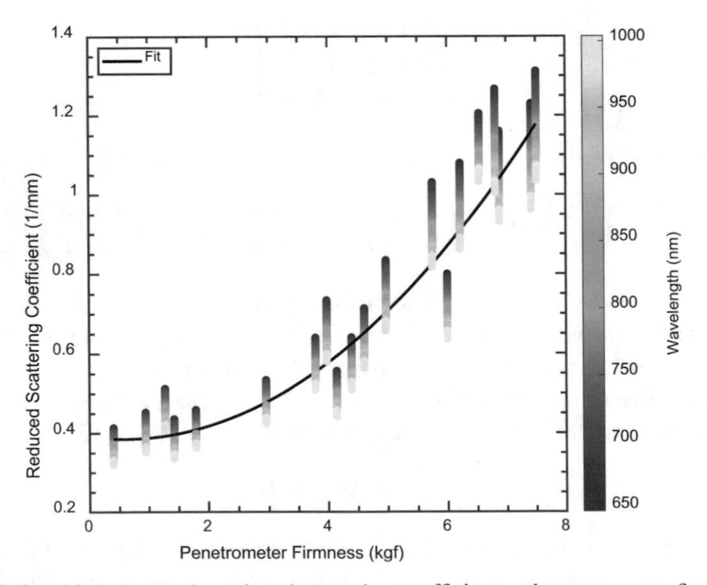

Fig. 3. Relationship between the reduced scattering coefficient and penetrometer firmness of the outer pericarp slices for 20 kiwifruit. Each vertical line represents an average scattering coefficient of one kiwifruit across the wavelength range 650 nm to 1000 nm (color bar). The black line represents a best fit curve through the data at 827 nm.

3.2 Depolarization Coefficient

A stronger linear relationship with fruit firmness is observed for the depolarization coefficient (0 – non-depolarizing, 1 – completely depolarizing) of the pericarp slices (Fig. 4). In particular, there is no flattening in the relationship below 2 kgf, as was the case for the scattering coefficient.

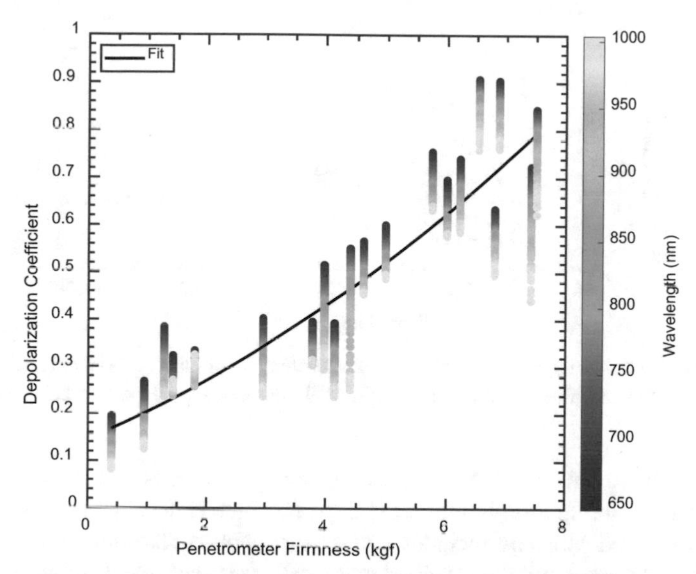

Fig. 4. Relationship between depolarization coefficient and penetrometer firmness for kiwifruit pericarp slices. Each vertical line represents an average depolarization coefficient of one kiwifruit for the wavelength range of 650 nm to 1000 nm (shown by the color bar). The black line represents a fit for 827 nm.

As kiwifruit get softer, the depolarization coefficient tends towards 0 but it is unlikely to reach it because of the inhomogeneous structures that remain while the fruit is recognizably still a whole fruit (i.e., not completely senesced or collapsed into a decomposed state). Inhomogeneous structures such as cell walls and various phase boundaries, for example around the intercellular spaces and the cytoplasm, present hard refractive index changes that will scatter the light photons as they propagate through the biological material [5, 16].

The depolarization coefficients were higher for shorter wavelengths, harder fruit and for skin peel relative to the pericarp slice, as shown by examples in Fig. 5. The light scattering data (not shown) were similar, which is expected since depolarization is largely predicated on multiple light scattering [17, 18]. A feature of the skin peel data, not observed in the pericarp data, is a notable dip in the depolarization coefficient at c.680 nm, the approximate position of a chlorophyll absorption peak [19]. The skin peel of the yellow-fleshed kiwifruit can have a considerable amount of chlorophyll, sometimes forming a thin green line immediately under the skin peel, but the internal pericarp of a fully ripe, yellow-fleshed kiwifruit will not.

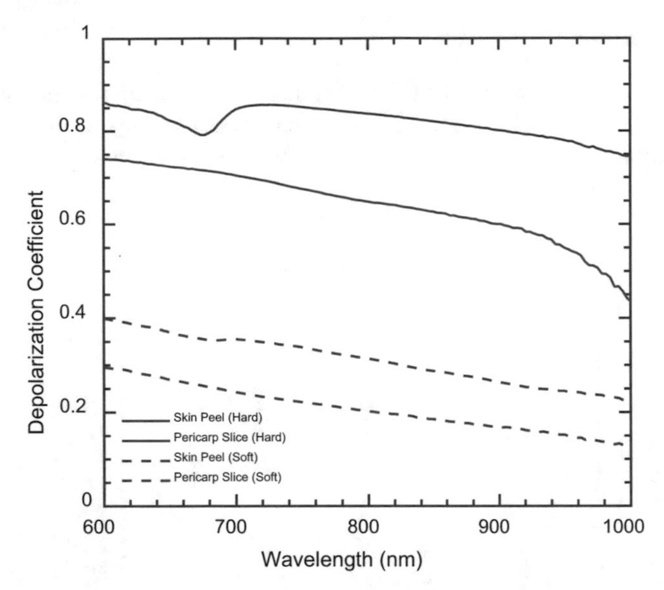

Fig. 5. Example of depolarization coefficient vs wavelength for hard (solid line) and soft (dashed line) kiwifruit. Red and blue lines represent skin peel and pericarp slices, respectively.

The differences between the skin peel and pericarp slice is probably related to the complex light scattering morphology of skin peel, which is dominated by small closely packed parenchyma cells and includes lenticels, trichomes (hair structures), collapsed (dead) cells and some small thick-walled stone cells (sclerenchyma cells) [5]. The outer pericarp slice contains a less dense mix of both small and large parenchyma cells (diameters of ~100 μm and ~200 μm, respectively), so will have far less light scattering interfaces per volume element compared with skin peel.

3.3 Linear and Circular Retardance

The linear retardance was flat for most of the wavelength range, higher for the skin peel than the slice, but varied considerably from sample to sample as evidenced the relatively large standard deviations and overlap of values (Fig. 6a). The measured linear retardance could possibly be the effect of birefringent material in skin peel and outer pericarp such as cellulose in cell wall, which exhibits different refractive index along and across the axis of these fibrils [20]. The circular retardance was negligible, quite tightly around the zero value, for the pericarp slices but not for skin peel samples, which were significantly higher and exhibited a decreasing trend with wavelength (Fig. 6b). Again, as with the depolarization coefficient, the differences here between skin peel and pericarp slices is most likely driven by the greater light scattering occurring in the skin peel compared with the pericarp. It may be that the circular retardance is the most precise measure for indicating such differences.

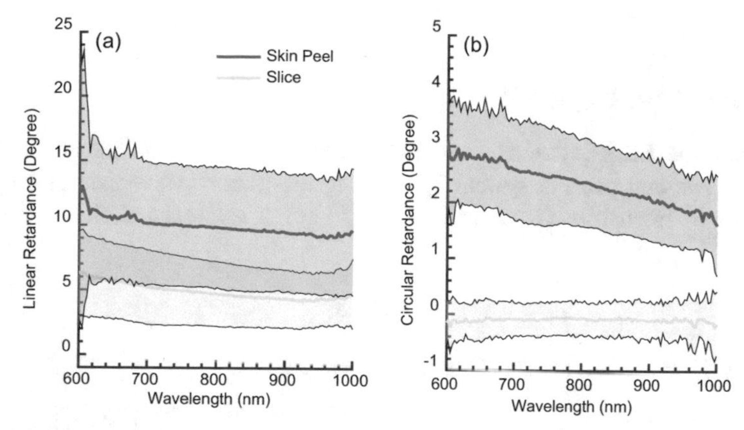

Fig. 6. (a) Linear and (b) Circular retardance vs wavelength for kiwifruit pericarp slice (yellow) and skin peel (brown). Filled areas represent the mean ±1 standard deviation for 40 samples.

4 Conclusion

The depolarization coefficient of the pericarp slices showed a strong decreasing trend with fruit firmness and probably stronger than that observed with the reduced scattering coefficient, particularly for firmness values below 2 kgf. Measurements presented here have highlighted the possibility of using the depolarization coefficient as an optical method to measure kiwifruit penetrometer firmness, which is an important internal quality parameter in terms of fruit storage and consumer preference. In general, the depolarization coefficient of the skin peel was greater than that of pericarp slices, which is probably related to the more complex light scattering structure of skin peel compared with that of the pericarp slices.

The linear retardance was spectrally flat for both the skin peel and pericarp slices, with the skin peel data slightly higher but overlapping considerably with the pericarp slice data. Measurements of linear retardance could be caused by birefringent material such as cellulose because of its appearance in the cell wall. The circular retardance was essentially zero and flat for the pericarp slices but not for the skin peel, which was significantly higher and with a decreasing trend with wavelength. These striking differences in spectral form for the circular retardance of the skin peel, both to the linear retardance spectra and to that of outer pericarp spectra, are unexplained and it would be insightful now to compare it with similar measurements on other fruits that may have different skin peel structures. To investigate the polarization responses due to different skin peel and tissue types, future work will extend these measurements onto other fruit cultivars such as apples and nectarines. These results provide new insights into understanding the measurable responses of polarized light in fruit and yellow-fleshed kiwifruit in particular.

Acknowledgements. Damenraj Rajkumar acknowledges the financial support of a PhD scholarship from Plant & Food Research and funding from the Dodd-Walls Centre for Photonic and Quantum Technologies and the University of Otago.

Appendix A

Calibration of RRFP System

To calibrate the RRFP system, a method similar to that of Boulbry et al. [21] was employed. We calibrate the system by rotating the quarter waveplate and acquiring M number of intensities ($\boldsymbol{I} = [I_1 I_2 \cdots I_M]^{\boldsymbol{T}}$) for a reference Stokes vector ($\boldsymbol{S} = [S_0 S_1 S_2 S_3]^{\boldsymbol{T}}$),

$$\boldsymbol{I} = \boldsymbol{A} \cdot \boldsymbol{S}, \tag{11}$$

where $\boldsymbol{A}$ is the PSA data matrix. This can be rewritten as,

$$\boldsymbol{I} = \begin{bmatrix} A_{11} & \cdots & A_{14} \\ \vdots & \ddots & \vdots \\ A_{M1} & \cdots & A_{M4} \end{bmatrix} \cdot \begin{bmatrix} S_0 \\ S_1 \\ S_2 \\ S_3 \end{bmatrix} \text{Or}\, \boldsymbol{I} = (\boldsymbol{S} \otimes \boldsymbol{eye}(\boldsymbol{M})) \cdot \begin{bmatrix} A_{11} \\ A_{12} \\ A_{13} \\ A_{14} \\ A_{21} \\ A_{22} \\ A_{23} \\ A_{24} \\ \cdot \\ \cdot \\ \cdot \\ A_{M4} \end{bmatrix}, \tag{12}$$

where $\boldsymbol{eye}(\boldsymbol{M})$ and $\otimes$ are the identity matrix and Kronecker product, respectively. We can stack for each reference Stokes vectors ($\boldsymbol{S_N}$) such that

$$\begin{bmatrix} \boldsymbol{I}_1 \\ \boldsymbol{I}_2 \\ \cdot \\ \cdot \\ \cdot \\ \boldsymbol{I}_N \end{bmatrix} = \begin{bmatrix} \boldsymbol{B}_1 \\ \boldsymbol{B}_2 \\ \cdot \\ \cdot \\ \cdot \\ \boldsymbol{B}_N \end{bmatrix} \cdot \mathit{flatten}(\boldsymbol{A}), \tag{13}$$

where $\boldsymbol{I}_N$ are intensities measured for each Stokes vector $\boldsymbol{S}_N$, with $\boldsymbol{B}_N = \boldsymbol{S}_N \otimes \boldsymbol{eye}(\boldsymbol{M})$ and $\mathit{flatten}(\boldsymbol{A})$ being the single column form of the PSA data matrix $\boldsymbol{A}$. We can find $\mathit{flatten}(\boldsymbol{A})$ by minimizing the following function,

$$\sum_{i=1}^{N} \left\| \begin{bmatrix} \boldsymbol{I}_1 \\ \boldsymbol{I}_2 \\ \cdot \\ \cdot \\ \cdot \\ \boldsymbol{I}_N \end{bmatrix} - \begin{bmatrix} \boldsymbol{B}_1 \\ \boldsymbol{B}_2 \\ \cdot \\ \cdot \\ \cdot \\ \boldsymbol{B}_N \end{bmatrix} \cdot \mathit{flatten}(\boldsymbol{A}) \right\|^2 . \tag{14}$$

We can simply get $\boldsymbol{A}$ by multiplying the pseudoinverse of $\overline{\boldsymbol{B}}\left([\boldsymbol{B}_1\boldsymbol{B}_2\cdots\boldsymbol{B}_N]^T\right)$ with $\overline{\boldsymbol{I}}\left([\boldsymbol{I}_1\boldsymbol{I}_2\cdots\boldsymbol{I}_N]^T\right)$,

$$flatten(\boldsymbol{A}) = pinv\left(\overline{\boldsymbol{B}}\right)\cdot\overline{\boldsymbol{I}}. \tag{15}$$

The matrix $\boldsymbol{A}$ can be used to calculate a set of intensity measurements $\boldsymbol{I}$ from a Stokes vector $\boldsymbol{S}$ but by taking the pseudoinverse of $\boldsymbol{A}$, we can determine the data reduction matrix $\boldsymbol{W}$ which can calculate a Stokes vector $\boldsymbol{S}$ from a set of intensity measurements,

$$\boldsymbol{S} = pinv(\boldsymbol{A})\cdot\boldsymbol{I}\text{ or }\boldsymbol{S} = \boldsymbol{W}\cdot\boldsymbol{I}. \tag{16}$$

References

1. Aitken, A.G., Warrington, I.J.: Fresh Facts New Zealand Horticulture. The New Zealand Institute for Plant and Food Research Ltd. https://www.freshfacts.co.nz/files/freshfacts-2020.pdf. Accessed 27 Feb 2022
2. Walsh, K., McGlone, V., Han, D.: The uses of near infra-red spectroscopy in postharvest decision support: a review. Postharvest Biol. Technol. **163**, 111139 (2020)
3. Nassif, R., Pellen, F., Magné, C., Le Jeune, B., Le Brun, G., Abboud, M.: Scattering through fruits during ripening: laser speckle technique correlated to biochemical and fluorescence measurements. Opt. Express **20**(21), 23887–23897 (2012)
4. Sarkar, M., Gupta, N., Assaad, M.: Monitoring of fruit freshness using phase information in polarization reflectance spectroscopy. Appl. Opt. **58**(23), 6396–6405 (2019)
5. Hallett, I., Sutherland, P.: Structure and development of kiwifruit skins. Int. J. Plant Sci. **166**(5), 693–704 (2005)
6. Scalisi, A., O’Connell, M.G., McGlone, A., Langdon-Arms, S.: Evaluation of a portable impact probe for rapid assessments of flesh firmness in peaches and nectarines. Acta Horticulturae (in press). The Proceedings of the XII International Symposium on Integrating Canopy, Rootstock and Environmental Physiology in Orchard Systems (2021)
7. Wang, Z., Künnemeyer, R., McGlone, A., Burdon, J.: Potential of Vis-NIR spectroscopy for detection of chilling injury in kiwifruit. Postharvest Biol. Technol. **164**, 111160 (2020)
8. Tyo, J.S.: Design of optimal polarimeters: maximization of signal-to-noise ratio and minimization of systematic error. Appl. Opt. **41**(4), 619–630 (2002)
9. Sabatke, D., Descour, M., Dereniak, E., Sweatt, W., Kemme, S., Phipps, G.: Optimization of retardance for a complete Stokes polarimeter. Opt. Lett. **25**(11), 802–804 (2000)
10. Cloude, S.R.: Conditions for the physical realisability of matrix operators in polarimetry. In: Polarization Considerations for Optical Systems II, Proc. SPIE, vol. 1166, pp. 177–187 (1990)
11. Kuntman, E.: Mathematical work on the foundation of Jones-Mueller formalism and its application to nano optics (2019)
12. Lu, S.-Y., Chipman, R.A.: Interpretation of Mueller matrices based on polar decomposition. JOSA A **13**(5), 1106–1113 (1996)
13. Ghosh, N., et al.: Mueller matrix decomposition for polarized light assessment of biological tissues. J. Biophotonics **2**(3), 145–156 (2009). https://onlinelibrary.wiley.com. https://doi.org/10.1002/jbio.200810040
14. Redgwell, R.J., MacRae, E., Hallett, I., Fischer, M., Perry, J., Harker, R.: In vivo and in vitro swelling of cell walls during fruit ripening. Planta **203**(2), 162–173 (1997)

15. McGlone, V., Abe, H., Kawano, S.: Kiwifruit firmness by near infrared light scattering. J. Near Infrared Spectrosc. **5**(2), 83–89 (1997)
16. Tuchin, V.V.: Polarized light interaction with tissues. J. Biomed. Opt. **21**(7), 071114 (2016)
17. Ahmad, M., Alali, S., Kim, A., Wood, M.F.G., Ikram, M., Vitkin, I.A.: Do different turbid media with matched bulk optical properties also exhibit similar polarization properties? Biomed. Opt. Express **2**(12), 3248–3258 (2011). https://doi.org/10.1364/boe.2.003248
18. Ghosh, N., Gupta, P.K., Patel, H.S., Jain, B., Singh, B.N.: Depolarization of light in tissue phantoms – effect of collection geometry. Opt. Commun. **222**(1), 93–100, (2003). https://doi.org/10.1016/S0030-4018(03)01567-0
19. Qin, J., Lu, R.: Measurement of the optical properties of fruits and vegetables using spatially resolved hyperspectral diffuse reflectance imaging technique. Postharvest Biol. Technol. **49**(3), 355–365 (2008)
20. Vignolini, S., et al.: Pointillist structural color in Pollia fruit. Proc. Natl. Acad. Sci. **109**(39), 15712–15715 (2012). https://www.ncbi.nlm.nih.gov/pmc/articles/PMC3465391/pdf/pnas.1210105109.pdf
21. Boulbry, B., Ramella-Roman, J.C., Germer, T.A.: Improved method for calibrating a Stokes polarimeter. Appl. Opt. **46**(35), 8533–8541 (2007)

Detecting Sensor Faults and Outliers in Industrial Internet of Things

Anuroop Gaddam(✉) and Vinodha Govender

School of Information Technology, Deakin University, Geelong, VIC, Australia
Anuroop.Gaddam@deakin.edu.au

Abstract. The Internet of Things (IoT) is rapidly growing in nearly every industry. People are adopting and using IoT technology in every facet of their lives, ranging from civilian use (e.g., smart homes, hobbies, fitness, and health monitoring), to military (e.g., battlefield surveillance systems) to industrial (i.e., industrial automation, plant and equipment monitoring) etc. More and more companies are introducing IoT technology innovations to increase efficiency. In the resources industry specifically, IoT can technically be used to imitate human behaviour and separate or even eliminate human and operational equipment interaction, hence keeping people safer. Another vital IoT offering in the mining space is equipment reliability. Maintenance and engineering teams are switching from the traditional time-based maintenance approach to predictive maintenance using IoT. Therefore, a key deliverable from IoT technology in mining is to have accurate and reliable data to assist management with making informed decisions. In the IoT process chain, there are several influences that can cause sensor faults and result in erroneous data. The data can also contain anomalies or outliers. This paper aims to describe IoT sensor faults, anomalies or outliers, fault detection using Deep anomaly detection techniques and mitigation or correction strategies specific to the Industrial Internet of Things within the resources industry.

Keywords: Outlier detection · Industrial Internet of Things (IIoT) · Internet of Things (IoT) · Federated Learning · Equipment monitoring · Resources industry · Sensor anomaly · Deep Anomaly Detection · Temperature monitoring · Wireless sensor network

1 Introduction

Internet of Things is a convergence of different innovations that assist physical devices with internet-based applications to gather information by using heterogeneous sensors for monitoring purposes [1]. IoT has the dynamic capability to sense, communicate, compute, and even control their environment [2]. A sensor is a device that responds to a physical stimulus and transmits a result for measurement or operating a control system [3]. The heterogeneous smart sensors used for monitoring equipment are essential and crucial components as they are the

N. K. Suryadevara et al. (Eds.): ICST 2022, LNEE 1035, pp. 183–200, 2023.
https://doi.org/10.1007/978-3-031-29871-4_20

front end of the IoT process and capture raw data from the physical world. In the resources industry, these IoT sensors are often found to be deployed or installed on equipment in environments that are harsh and prone to high noise, contaminants, high vibration, earth movement, severe weather conditions, etc. [4].

This inevitably means that the sensors are imperfect and prone to failure, accelerated wear and tear, tampering and theft, etc. All these conditions cause the sensors within IoT to produce unusual and erroneous readings, often known as outliers. There is no standard confined definition for sensor outliers, however in the context of IoT the sensor outlier is commonly known to be an irregularity or a divergence in sensor behaviour during the process of cataloguing particular parameters or events when compared to its previous behaviour or readings [5].

1.1 Industrial IoT's in Action

Worsley Alumina Pty Ltd (WAPL) is a resources company in Collie, Western Australia that mines bauxite and processes white alumina powder to aluminium smelters around the world [6]. WAPL employs Bayer's process to extract alumina from bauxite in a commercially efficient process using hot caustic slurry [7], as shown in Fig. 1.

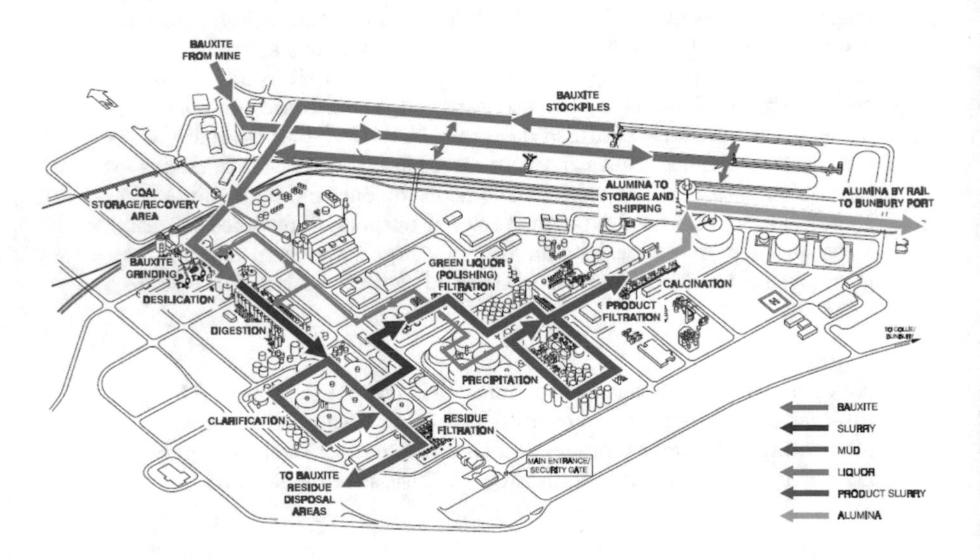

Fig. 1. The Worsley Alumina refinery process flow. [21]

The extraction process involves the following five basic steps,

- Dissolving alumina from crushed and ground bauxite ore using caustic liquor at high temperature and pressure to make an alumina-rich solution. When the crude bauxite is digested, sodium aluminate and silicates are formed.

- Settling and filtering to remove the sodium aluminate, silicates, iron oxide and solid residue or red mud, leaving alumina-rich liquor, also called green liquor.
- Seeding the sodium aluminate in caustic solution with crystalline hydrated alumina and cooling. Most of the alumina precipitates out as hydrate crystals which are washed and dried to produce almost pure alumina.
- The Worsley refinery process extends the original Bayer Process to include the additional step of calcination, where the three molecules of water bound into the tri-hydrated alumina crystals are driven off to make the alumina product.
- Concentrating the caustic liquor and recycling back into the basic Bayer Process.

Worsley realised the commercial benefits of commissioning the Internet of Things and Wireless Sensor Networks in the process plant since the past 10 years.

Worsley has IoT sensors deployed on critical process plant and equipment at the mine and refinery [8,9], these are:

- The piezometer is a sensor used to measure the strength of tailing dam walls through the variables of frequency and thermal resistance.
- The OBDBox receiver is a fleet tracking sensor that monitors vehicles for speed and GPS location [10].
- Honeywell and Emerson instruments measure pressure and temperature in process pipes.
- Bently Nevada sensor instruments measure vibration and temperature for the milling process plant.
- Rockwell and ABB Variable Speed Drive (VSD) smart sensors measure and control speed and power on motors all-around the site.

In recent years, there has been a need to monitor the health and performance of critical electrical and mechanical equipment such as pumps, blowers, cooling fans and other rotating equipment. In addition to equipment failure strategies, such as maintenance plans and critical spares availability, the Asset Management teams also require real-time data for predictive analysis of equipment failures i.e., moving from a time-based maintenance strategy to a predictive maintenance strategy. Maintenance performed re-actively on critical plant operational equipment assets in the mining industry can impact production yield and result in unplanned costs.

A McKinsey Global Institute study shows that IoT offers a potential economic impact of between 1.2 and 3.7 trillion dollars for factory operations management and predictive maintenance by 2025 [11,12]. The McKinsey report also showed that IoT-based predictive maintenance helps to reduce maintenance costs of factory equipment by up to 40 percent. It can also reduce equipment downtime by up to 50 per cent and reduce equipment capital investment by 3 to 5 percent by extending the useful life of machinery" [13]. This paper presents Industrial IoT's sensor faults detection techniques in the operational environment.

2 Collecting Sensor Data for Outlier Detection and Correction

This section presents current research conducted at Worsley Alumina and comprises data obtained through a series of interviews with various stakeholders in the Maintenance and Engineering teams and physical observation of the sensor as it goes through the operational life-cycle for rotables and other electrical and mechanical equipment. Some sensor faults or outlier causes are easier to detect than others. Figure 2 depicts a classification of sensor outlier detection and correction ranging from easy to more complex. Intrinsic, operator discipline and pseudo-sensor outliers are detected through manual or physical visualisations by humans. While the extrinsic, intermittent, known, and sporadic sensor outlier events require machine learning to be detected in the operational application layer as shown below in the Fig. 3. These are the primary sensor fault or outlier causes observed by visually inspecting the sensor hardware:

2.1 Sensor Intrinsic Faults and Outliers

These are intrinsic faults that occur internally to a sensor. IoT sensors are made of electronic components like amplifiers, batteries, transducers, analogue filters, excitation control and compensation sensors. [4] These components can often fail without prior warning. Sensor intrinsic faults are fairly easy to detect and easy to

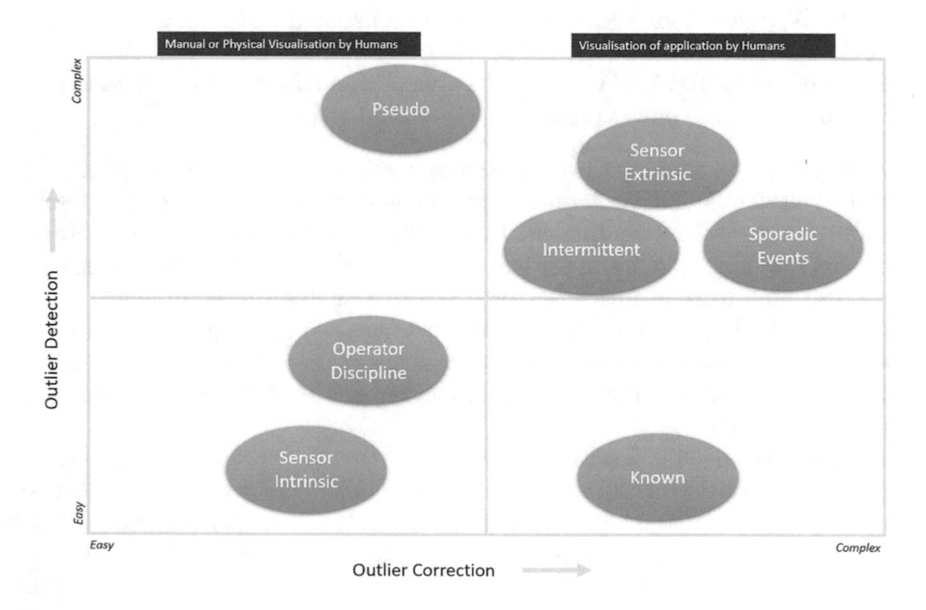

Fig. 2. Levels of detection and correction for sensor outliers.

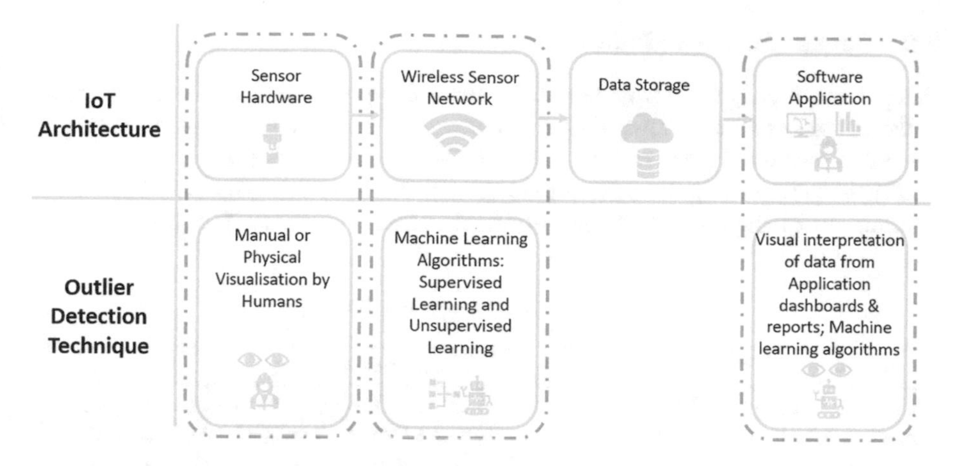

Fig. 3. IoT architecture & outlier detection techniques relevant to the respective layer.

correct (see Fig. 4). They typically result in a null reading value from the sensor. The null value could be interpreted as an equipment failure. The immediate response is for the Operator to investigate both the sensor and equipment to confirm the situation. In most cases, the quick solution fix is to simply replace the sensor. The sensor may be run for forensics to evaluate diagnosis. Depending on the outcome, the sensor may be fixed and re-used or decommissioned if it cannot be repaired.

Fig. 4. Melted sensor.

2.2 Sensor Intrinsic Failure Scenario

WAPL conducted a comparison between solar-powered sensors and battery-powered sensors and found that solar-powered sensors were not as robust as battery powered. Furthermore, some of the sensors were placed in areas that had little to no light and could not charge up efficiently. Table 1 and Table 2 give a summary of the correction strategy during intrinsic sensor failures.

Table 1. Scenario of a flat battery.

Cause	Effect	Correction strategy
A sensor battery that has reached the end of its lifespan. A sensor battery that experiences ultra-low temperature spikes e.g., very cold weather	Flat (dead) battery(sensor reading is 0v)similar to the sensor having no battery. There won't be equipment data available	*Know the approximate sensor battery failure or warranty expiry time and implement a maintenance plan for battery replacement. For sensor battery failures that occur off-cycle setup alarm notifications for low battery alerts. Where possible, consider using alternate power sources, such as wired sensors to a fixed power source or use solar power*

Table 2. Scenario of internal components hardware failure

Cause	Effect	Correction strategy
The accelerometer standard design for the sensor is to measure 0 - 25 mm/s for movement. Excessive vibration and temperature changes from the equipment can cause a data deviation. Loose wiring connection within the sensor. Poor sensor quality checks from the equipment manufacturer	Internal sensor components hardware failure for an accelerometer measuring x, y, z returns no readings	*Ensure operational equipment is maintained and does not exceed excessive vibration thresholds. Ensure the sensor equipment manufacturer has a stringent quality control process*

2.3 Pseudo Outliers

A type of extrinsic outliers, pseudo outliers are interesting as they are caused by external factors (see Fig. 5 that imitate the sensor readings and are generally perceived as being accurate. The nature of pseudo-outliers is that they are extremely difficult to detect but once detected the correction is easy to action, shown in Fig. 2. Table 3, Table 4 and Table 5 provide a summary of the correction strategy during pseudo outliers and sensor failures.

Table 3. Scenario of the impact of environmental conditions - debris.

Cause	Effect	Correction strategy
Due to environmental conditions over time, the magnet can accumulate particles of steel, rust, and dust	The debris collected on a sensor embeds itself on the magnet poles and gives false readings	*Replace magnetic sensor mounts with stable fixed-mounted quick-disconnects*

Table 4. Scenario of a melted sensor.

Cause	Effect	Correction strategy
When the equipment that the sensor is placed on fails e.g., a bearing ceased due to lack of lubrication where the auto-lube dispenser failed. The equipment was not maintained for a period of time, and this caused a fire on the bearing.	Slow melting of the sensor could spike temperature readings; eventually, there will be no data flow when the sensor perishes.	*Up-skill the equipment monitoring team with an understanding of the maintenance processes for the equipment and sensors. Set up KPI (Key Performance Indicators) to ensure there's operating discipline and accountability from the monitoring teams*

Table 5. Scenario of the impact of environmental conditions - mud and grease.

Cause	Effect	Correction strategy
Harsh mining environment can cause excessive contaminants such as mud and grease to cake on the sensor unit cover	Produces false vibration readings and false alarms are sent out	*Regularly clean and maintain equipment that are subject to contaminants*

Fig. 5. A sensor covered with environmental contaminants.

2.4 Operator Discipline Outliers

Operator discipline outliers are typically caused by the Operators managing the sensor and equipment. These occur from a lack of operator training on sensor management, inadequate procedural documentation that supports process understanding, or negligence. These outliers are fairly easy to detect and correct if the sensor and equipment are checked and maintained on a regular basis. Table 6, Table 7, Table 8 and Table 9 give a summary of the correction strategy during different types of operator discipline outliers and sensor failures.

Table 6. Scenario of the incorrect sensor used.

Cause	Effect	Correction strategy
Sensors mounted on equipment that is not rated for industrial use can fail more quicker than a sensor that is built for industrial use	The incorrect type of sensor used provides false readings	*Ensure industrial-rated quality sensors are tested and implemented*

Table 7. Physical removal of the sensor scenario.

Cause	Effect	Correction strategy
The sensor monitoring equipment is displaced due to the operator's lack of process understanding or negligence. New untrained staff were asked to attend to the equipment without being properly communicated	Equipment won't be monitored i.e., no readings transmitted	*Ensure there is a documented procedure for sensor management. Ensure the operators monitoring and supporting the sensors are trained on sensor management*

Table 8. Incorrect sensor installation.

Cause	Effect	Correction strategy
The procedure for sensor management is not followed and the sensor is not properly installed i.e., the sensor mounting is not done correctly or the quick-disconnect is not bolted down firmly	Causes excessive vibration on the mount instead of the equipment and data is inaccurate	Ensure there is a documented procedure for sensor management. Ensure the operators monitoring and supporting the sensors are trained on sensor management

Table 9. Scenario of sensor maintenance delays.

Cause	Effect	Correction strategy
The procedure for sensor management is not followed. Sensor maintenance is not carried out timely	Sensor reading is null (0v) and equipment is not monitored. Equipment not monitored can fail catastrophically without any monitoring or triggers	*Ensure there is a documented procedure for sensor management. Ensure there is a sensor maintenance plan*

An example of critical equipment experiencing a catastrophic disaster due to Operator discipline outliers is shown in Fig. 6. A pulley bearing caught fire and caused a material loss of approximately AUD 20,000. The sensors on this equipment had flat batteries that had not been replaced in time.

Fig. 6. Critical equipment experiencing a catastrophic disaster due to Operator discipline outliers.

2.5 Extrinsic Sensor Errors

Extrinsic sensor errors occur for position and orientation (pose) with respect to the environment, equipment or another sensor. Extrinsic faults are not easily detected but can be discovered by observing data trends produced by a sensor as explained in Table 10. These kinds of sensor faults, errors or outliers are observed by visually observing the application dashboard or automated reports, and data loggers.

Table 10. Scenario of the impacts of the sensor's location.

Cause	Effect	Correction strategy
Poor network design and incorrect sensor placement i.e., sensors located on the boundary of the repeater network can sometimes miss data packets being transmitted from the sensor to the repeater network	Intermittent data flow between the boundary sensor and the repeater network	*Perform thorough testing of network connection after the sensor is installed*

2.6 Sporadic Sensor Events

When an outlier occurs from an unlikely and unpredictable situation that affects the sensor this is considered to be a sporadic event. Sporadic events are both complex to detect and correct (see Fig. 2). Examples of sporadic events are described below. Another real-world example is exploitation by cybercriminals. IoT is a complex network comprising different components and could therefore provide hackers with more than one point of entry. Cybercriminals can use IoT botnets to infect devices with malware, which allows the attacker to control the connected devices, and IoT application errors, which could generate false alarms to the monitoring team and leave equipment vulnerable. A sensor placed on equipment usually emits a single frequency. The introduction of a second sensor vibrating at the same frequency can cause the second equipment to vibrate - this is called resonance. The current correction process is discussed in Table 11. A sensor placed on equipment usually emits a single frequency or signal wave. A harmonic occurs when there is a series of signals emitted.

Table 11. Scenario of resonance.

Cause	Effect	Correction strategy
A sensor is usually mounted on equipment using a quick disconnect (or *puck*). In some instances, sensors are mounted on a magnet. Due to environmental conditions over time, the magnet can accumulate particles of steel, rust and dust. This embeds itself on the magnet poles	Sensor can move, cause resonance, and shift off its axis causing axis orientation changes	*Replace magnetic sensor mounts with stable fixed-mounted quick-disconnect*

A sensor placed on equipment usually emits a single frequency or signal wave. A harmonic occurs when there is a series of signals emitted. The effect and the current correction process of this type of outlier scenario are discussed in Table 12

Table 12. Scenario of harmonic interference.

Cause	Effect	Correction strategy
Harmonic interference and resonance from adjacent equipment, broken structures, or footings on equipment	There could be an electrical frequency interference. Induced vibration coming from adjacent equipment could provide false readings (higher than actual)	*Ensure equipment and mechanical structures are maintained. As far as possible ensure equipment being monitored do not have interference from adjacent equipment*

2.7 Intermittent Sensor Outliers and Faults in IIoT

Intermittent outliers occur at random times and not for too long. These outliers are usually difficult to detect physically and can be observed through an intimate knowledge of data trends for specific equipment. The scenarios, the effects and the current correction process of this type of outlier scenario are discussed in Table 13 and Table 14.

Table 13. Scenario of the blocked transmission path.

Cause	Effect	Correction strategy
When a battery reaches the end of life, the battery voltage can drop below recommended threshold values (2.6v $<$ 2.85v). A healthy battery is $>$2.85v	Sensor gives erroneous values	*Know the approximate sensor battery failure or warranty expiry time and implement a maintenance plan for battery replacement. For sensor battery failures that occur off-cycle setup alarm notifications for low battery alerts*

Table 14. Scenario of minor decline in battery voltage.

Cause	Effect	Correction strategy
Heavy vehicles (e.g., mobile cranes) can block the transmission path between the sensor and the repeater network	The blocked transmission path creates intermittent data flow	*Expand the repeater network to allow alternate communication paths*

A real-life example of intermittent sensor fault is noticed at the WAPL as shown in Fig. 7, a sensor data showing loss of communication during the night when the battery was very cold and re-establishes communication again in the morning when the day warms up. This battery needs to be replaced immediately to prevent further communication failures.

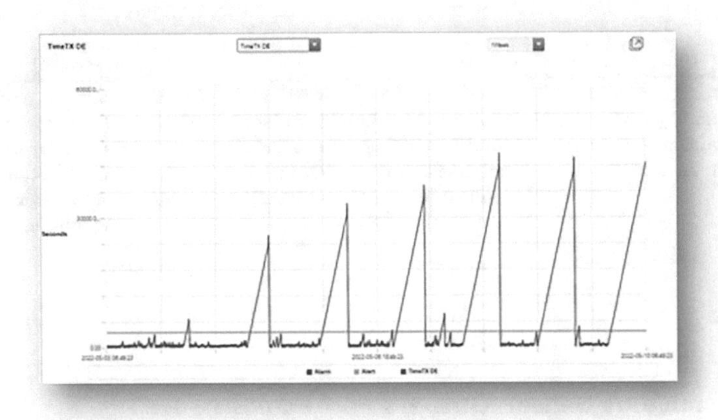

Fig. 7. Critical network failure due to intermittent battery failure.

2.8 Known Sensor Outliers

In some operational situations, equipment may be required to be monitored urgently for anomalies. On these occasions, the Operators hastily implement an IoT sensor solution on the basis that the sensor readings may not be accurate. Known outliers are acceptable and temporary by design. Known outliers are easily detected, however not easily corrected due to time constraints (refer Fig. 1). The scenarios, the effects and the current correction process of this type of outlier scenario are discussed in Table 15 and Table 16.

Table 15. Scenario of incorrect sensor orientation placement.

Cause	Effect	Correction strategy
The original sensor design is for vertical placement on the equipment. However, in some instances due to the equipment design, the sensor needs to be placed horizontally.	The horizontal placement of the sensor can sometimes prevent data flow from the sensor to the repeater network.	*Ensure the sensor placement is always vertical. If the sensor needs to be placed horizontally then this may require an additional repeater located closer to the horizontal sensor*

Table 16. Scenario of installing incorrect sensor mounting types.

Cause	Effect	Correction strategy
Some equipment has an auto-lube dispenser as shown in Fig. 8. A short-term solution design to have sensors monitoring this equipment is to use an elbow adaptor to have the sensor and auto-lube dispenser connected jointly	There are slight variations in temperature data.	*Separate the auto-lube dispenser mount adaptor and the sensor*

Fig. 8. sensor installed on equipment whilst connected to an auto-lube dispenser.

Some key learnings that have surfaced as improvement opportunities from this research for optimal sensor fault mitigation are:

- Establish an automated sensor monitoring mechanism.
- Develop procedural hardware and machine learning approach for IoT hardware component management i.e., sensors, batteries, repeater boards, Modbus boards etc.
- Develop a real-time sensor monitoring hardware and software for their health and durability in an operational environment.
- Introduce Key Performance Indicators based on Machine learning based automated alerts to measure the support team's alignment with the processes and identify improvement gaps.

3 Automated Deep Anomaly Detection Techniques

Machine learning based techniques could be employed to ensure the sensor faults and outliers automated detection in real-time throughout the IoT cycle. The Deep Anomaly Detection (DAD) is always a popular choice in the IIoT context as the DAD serves as an underpinning method in detecting anomalies! [14]. Existing research about the DAD broadly classified into four categories,

3.1 Statistical

These methods use past data readings to approximate a model of the correct behaviour of the sensor. When a new data reading is registered, it is compared to the model. If the data is incompatible with the data model, then it is regarded as an anomaly [15].

3.2 Probabilistic

Classification of anomalies based on a probabilistic model is performed by measuring the probability of reading with respect to the probability model. If the probability falls below a predefined threshold, then it is considered to be an anomaly. Examples of these are Hidden Markov Models (HMMs) or Bayesian Networks (BNs) [16].

3.3 Proximity or Nearest-Neighbour Based

This is a technique used to analyse the sensors' data point in relation to its neighbours i.e., by using the distances between sensor data measurements to differentiate between abnormal and correct readings [17].

3.4 Clustering-Based

The cluster-based analysis is a technique that groups related data instances into clusters of similar behaviour [18]. The initial readings from a sensor are used as a baseline to create the clusters and then the new sensor measurements.

3.5 Machine Learning Algorithms to Detect Sensor Anomalies in IIoT

Some Machine Learning algorithms to detect sensor anomalies specific to IIoT [19] are:

- Supervised Learning
  - KNN (K-Nearest Neighbour)
  - SVM (Support Vector Machine)
  - CESVM (Centred hyper ellipsoidal)

- Unsupervised Learning
  - DBSCAN (Density-based anomaly detection)
  - K Means
  - PCA (Principal Component Analysis)

4 Federated Learning Based IIoT Outlier Detection

In this section, we briefly discuss about outliers and introduce federated deep learning, and gradient compression as follows.

4.1 Outliers

In the data science context, outliers (also referred to as anomalies or abnormalities) are a phenomenon where the captured data points are substantially divergent from other captured observations [20]. If we consider R_1, R_2 and R_3 are regions that comprised most of the data observations, therefore they are considered to be normal data instance regions. When captured data points S_1 and S_2 are far from the R_1, R_2 and R_3 thereby classified as outliers. In order to formally define the outliers we assume that an

n-dimensional data set $\boldsymbol{y}_i = (y_{i,1}, y_{i,2}, y_{i,3} \ldots, y_{i,n})$ follows a normal distribution and its mean μ_j and variance σ_j for each dimension where $i \in \{1, \ldots, m\}$ and $j \in \{1, \ldots, n\}$. Specifically, for $j \in \{1, \ldots, n\}$, under the assumption of the normal distribution, we have,

$$\mu_j = \sum_{i=1}^{m} y_{i,j}/m, \sigma_j^2 = \sum_{i=1}^{m} (y_{i,j} - \mu_j)^2 / m \tag{1}$$

If there is a new vector $\boldsymbol{y}$, the probability $p(\boldsymbol{y})$ of anomaly can be calculated as follows:

$$p(\boldsymbol{y}) = \prod_{j=1}^{n} p\left(y_j; \mu_j, \sigma_j^2\right) = \prod_{j=1}^{n} \frac{1}{\sqrt{2\pi}\sigma_j} \exp\left(-\frac{(y_j - \mu_j)^2}{2\sigma_j^2}\right) \tag{2}$$

We can then conclude whether vector $\boldsymbol{y}$ belongs to an outlier based on the probability value.

5 Proposed Federated Learning Based IIoT Sensor Outlier Detection

To detect sensor outliers within the IIoT, we aim to design our framework using a collaboratively distributed deep learning paradigm. Conventional distributed deep learning techniques require a certain amount of sensor time series data to be aggregated and analysed at central servers during the model training phase. Such a training process needs additional network infrastructure requirements and data-intensive communication between the IIoT's that exist in an industrial setting and there are possibilities for potential data privacy leakage risks for IIoT devices.

To address these challenges, a collaboratively distributed deep learning paradigm, called Federated Learning (FL), is proposed for IIoTs to train a global model while keeping the training data sets locally without sharing raw training data. Sensor outlier detection using FL has three phases, the initialisation phase, the aggregation phase, and the update phase.

During the initialisation phase, FL with N edge devices and a parameter aggregator, i.e., a cloud aggregator, distributes a pre-trained global model ω_t on the existing sensor data sets. Then, each IIoT device creates a local data set $\mathcal{B}_k$ of size B_k to train and improve the existing global model ω_t in each iteration. During the aggregation phase, the cloud aggregator amasses all the local data uploaded by the individual IIoT devices. During this process, the local loss function needs to be optimised as below,

$$\min_{x \in \mathbb{R}^d} F_k(x) = \frac{1}{B_k} \sum_{i \in B_k} \mathbb{E}_{z_i \sim B_k} f(x; z_i) + \lambda h(x) \tag{3}$$

where $f(x; z_i)$ is the local loss function for the single IIoT device $k \forall \lambda \in [0,1], h(x)$ is a regulariser function for individual IIoT device k, and $\forall i \in [1, \ldots, n], z_i$ is sampled from the local data set $\mathcal{B}_k$ on k device.

$$\omega_{t+1} \leftarrow \omega_t + \frac{1}{n} \sum_{n=1}^{N} F_{t+1}^{n} \tag{4}$$

where $\sum_{n=1}^{N} F_{t+1}^{n}$ denotes model updates aggregation and $(1/n)\sum_{n=1}^{N} F_{t+1}^{n}$ denotes the average aggregation (i.e., the FedAVG algorithm).

Both the cloud aggregator or the central server and the local IIoT devices repeat the above process till the global model reaches convergence. The proposed framework considers a generic setting for on-the-device DAD within the IIoT for sensor outlier, fault detection. The cloud aggregator (central server) and IIOT devices work collaboratively to train a DAD model by using a training algorithm for detecting the sensor outliers and identifying the type of outlier. Individual IIoT devices train a shared global model locally (i.e., on the device) on their own sensor time-series data set and upload their model updates to the cloud aggregator. By implementing the FedAVG algorithm [?], the cloud aggregator incorporates and aggregates these model updates to obtain a new global model. Finally, the individual cloud aggregator sends an updated global model to the individual IIoT devices to achieve accurate and timely outlier detection.

6 Conclusion and Future Work

In this article, we have presented various sensor fault and outlier detection observations, i.e., causes, impacts and correction strategies that had been conducted from manual observation at an operational plant, as well as visual data analysis of IoT application dashboard reports. We have also presented classifications of sensor fault causes and presented the level of complexity of fault detection and correction. This information will be especially useful to help determine sensor faults and outliers and to develop a strategy for future mitigation. Additionally, we have presented the proposed automatic sensor fault and outlier detection strategies that are algorithm-based and can be useful in locating outliers using machine learning algorithms. The proposed FL framework can enable a decentralised, efficient approach in detecting IIoT sensor outliers and further solving the problem of data islands.

In the future, our research will focus on testing the proposed framework and enhancing FL framework to be a more robust sensor outlier detection model.

The findings presented in this paper will be relevant for management at any operational plant within the resources industry, who are keen to explore and deploy IoT solutions within their organisation.

Acknowledgements. We would like to humbly thank the gracious and ever-inspiring staff at Worsley Alumina Private Limited for their generous support of the research conducted at the site and for allowing the inclusion of imagery in this paper.

References

1. Al-Sarawi, S., Anbar, M., Abdullah, R., Al Hawari, A.B.: Internet of things market analysis forecasts, 2020–2030. In: 2020 Fourth World Conference on Smart Trends in Systems, Security and Sustainability (WorldS4). IEEE, pp. 449–453 (2020)
2. Bouras, C., Kokkinos, V., Papachristos, N.: Performance evaluation of LoraWan physical layer integration on IoT devices. In: 2018 Global Information Infrastructure and Networking Symposium (GIIS). IEEE, pp. 1–4 (2018)
3. Javaid, M., Haleem, A., Rab, S., Singh, R.P., Suman, R.: Sensors for daily life: a review. Sensors Int. **2**, 100121 (2021)
4. Gaddam, A., Wilkin, T., Angelova, M., Gaddam, J.: Detecting sensor faults, anomalies and outliers in the internet of things: a survey on the challenges and solutions. Electronics **9**, 511 (2020). https://doi.org/10.3390/electronics9030511
5. Saleem, J., Hammoudeh, M., Raza, U., Adebisi, B., Ande, R.: IoT standardisation: Challenges, perspectives and solution. In: Proceedings of the 2nd International Conference on Future Networks and Distributed Systems, pp. 1–9 (2018)
6. Worsley Alumina [WWW Document], n.d. https://www.south32.net/our-business/australia/worsley-alumina (accessed 1.3.22)
7. Healy, S.J.: Bayer Process Impurities and Their Management, in: Raahauge, B.E., Williams, F.S. (Eds.), Smelter Grade Alumina from Bauxite: History, Best Practices, and Future Challenges, Springer Series in Materials Science. Springer International Publishing, Cham, pp. 375–426 (2022). https://doi.org/10.1007/978-3-030-88586-1
8. Pihnastyi, O., Sytnikova, A.: Construction of Control Systems of Flow Parameters of the Smart Conveyor using a Neural Network (2021)
9. Jonas, R.K.: Digital Transformation in Alumina Refining. In: Martin, O. (Ed.), Light Metals 2018, The Minerals, Metals & Materials Series. Springer International Publishing, Cham, pp. 79–87 (2018). https://doi.org/10.1007/978-3-319-72284-9-12
10. FTP Solutions - FTP Solutions, leading the way in operational technology. [WWW Document], n.d. https://www.ftpsolutions.com.au/ (accessed 1.3.22)
11. Bauer, H., Patel, M., Veira, J., 2014. The Internet of Things: Sizing up the opportunity. Retrieved from: McKinsey at http://www.mckinsey.com/insights/high_tech_telecoms_internet/the_internet_of_things_sizing_up_the_opportunity
12. Dahlqvist, F., Patel, M., Rajko, A., Shulman, J.: Growing opportunities in the Internet of Things. McKinsey & Company 1–6 (2019)
13. A Complex View of Industry 4.0 - Vasja Roblek, Maja Meško, Alojz Krapež, 2016 [WWW Document], n.d. https://journals.sagepub.com/doi/full/10.1177/2158244016653987 (accessed 8.3.22)
14. Deep Learning for Anomaly Detection: A Review: ACM Computing Surveys: Vol 54, No 2 [WWW Document], n.d. https://dl.acm.org/doi/abs/10.1145/3439950 (accessed 8.3.22)

15. Nesa, N., Ghosh, T., Banerjee, I.: Outlier detection in sensed data using statistical learning models for IoT, in: 2018 IEEE Wireless Communications and Networking Conference (WCNC). pp. 1–6 (2018). https://doi.org/10.1109/WCNC.2018.8376988
16. Bhatti, M.A., Riaz, R., Rizvi, S.S., Shokat, S., Riaz, F., Kwon, S.J.: Outlier detection in indoor localization and Internet of Things (IoT) using machine learning. J. Commun. Netw. **22**, 236–243 (2020). https://doi.org/10.1109/JCN.2020.000018
17. Evaluation of Machine Learning-based Anomaly Detection Algorithms on an Industrial Modbus/TCP Data Set — Proceedings of the 13th International Conference on Availability, Reliability and Security [WWW Document], n.d. https://dl.acm.org/doi/abs/10.1145/3230833.3232818 (accessed 8.3.22)
18. Syafrudin, M., Alfian, G., Fitriyani, N.L., Rhee, J.: Performance analysis of iot-based sensor, big data processing, and machine learning model for real-time monitoring system in automotive manufacturing. Sensors **18**, 2946 (2018). https://doi.org/10.3390/s18092946
19. Syafrudin, M., Alfian, G., Fitriyani, N.L., Rhee, J.: Performance analysis of iot-based sensor, big data processing, and machine learning model for real-time monitoring system in automotive manufacturing. Sensors **18**, 2946 (2018). https://doi.org/10.3390/s18092946
20. Guo, Y., Zhao, Z., He, K., Lai, S., Xia, J., Fan, L.: Efficient and flexible management for industrial Internet of Things: a federated learning approach. Comput. Netw. **192**, 108122 (2021). https://doi.org/10.1016/j.comnet.2021.108122
21. Regan, P.: Worsley Alumina Pty Ltd [Review of Worsley Alumina Pty Ltd]. In School of Engineering and Energy, Murdoch University. Process Control Department (2008). https://researchrepository.murdoch.edu.au/id/eprint/763/1/Report.pdf
22. Zhang, P., Hong, Y., Kumar, N., Alazab, M., Alshehri, M.D., Jiang, C.: BC-EdgeFL: a defensive transmission model based on blockchain-assisted reinforced federated learning in iiot environment. IEEE Trans. Industr. Inf. **18**, 3551–3561 (2022). https://doi.org/10.1109/TII.2021.3116037

Thermal Li-Ion Battery Management with Distributed Sensors on Cooling Foil Shields

Valentin Mateev[1](✉), Zhelyazko Kartunov[2], and Iliana Marinova[1]

[1] Technical University of Sofia, 1000 Sofia, Bulgaria
vmateev@tu-sofia.bg
[2] HTS Design Department, Sensata Technologies Inc., 1582 Sofia, Bulgaria

Abstract. Lithium-based batteries operation is related to many safety risks of dangerous flaming, integrity destruction, or even explosion. Battery cell temperature rise is an early and reliable indicator for such irreversible malfunctioning and can be used as a convenient overload indicator even in the initial process stages. Reducing the number of sensing elements, while keeping detection selectivity, is very useful for optimal temperature monitoring systems. Often, accurate heat source location sensing is difficult especially in heavy operational conditions, related to ambient temperature changes, other heating sources, active cooling systems, etc. which are typical for electrical vehicles. In this paper, we propose a measurement approach for detecting the transient overheating of li-ion battery cells in critical overload conditions. Distributed thermal sensors mounted on battery cooling foil shields are used for accurate abnormal heat source location determination by sophisticated numerical reconstruction procedure.

Keywords: Li-Ion batteries · temperature sensing · battery safety · battery management system

1 Introduction

Accumulator batteries for electric and hybrid vehicles are composed of thousands, even tenths of thousands of elements. Direct temperature measurements and long term monitoring of each element is extremely difficult, because of the huge number of required sensors, wiring of sensor arrays, complicated signal routing and multiplexing, multi-channel data processing is another important issue that must be addressed in Battery Management Systems (BMS). Altogether these reasons result in high costs for these monitoring systems. Electrochemical accumulator batteries are critical in modern technologies for transportation, computation, communication, and many more. At present, Li-Ion batteries are characterized by optimal energy and power density, long operational life, high efficiency in static and dynamic modes, etc. Nevertheless, these batteries could become unstable in certain conditions, causing dangerous flaming and even explosions. Because of that, their operational conditions are continually monitored by precise protective controllers and sensors to overcome these issues. Monitoring systems using mainly fast temperature variations during operation [1, 2], pressure distribution changes [3], gas

N. K. Suryadevara et al. (Eds.): ICST 2022, LNEE 1035, pp. 201–208, 2023.
https://doi.org/10.1007/978-3-031-29871-4_21

leakages caused by polymer chemical degradation [4], or battery cell electric impedance changes [1–5]. Protective systems must react simultaneously in case of detected overheating, overvoltage, overcharging, short circuits, and even sudden mechanical impact [6].

Battery temperature mode must be kept in optimal range for electrochemical reactions of operational charge/discharge cycles. Even small deviations from the optimal temperature mode will reduce efficiency or indicate battery cell internal malfunctioning. To control the temperature mode active cooling/heating systems are employed. Here is considered a passive cooling system for battery stacks, which uses high thermally conductive copper foil for intensive temperature equalizing between battery cells. Over the cooling foil, also called Cooling Foil Shield, an array of distributed thermal sensors are mounted and used for accurate abnormal heat source location determination by sophisticated numerical reconstruction procedure.

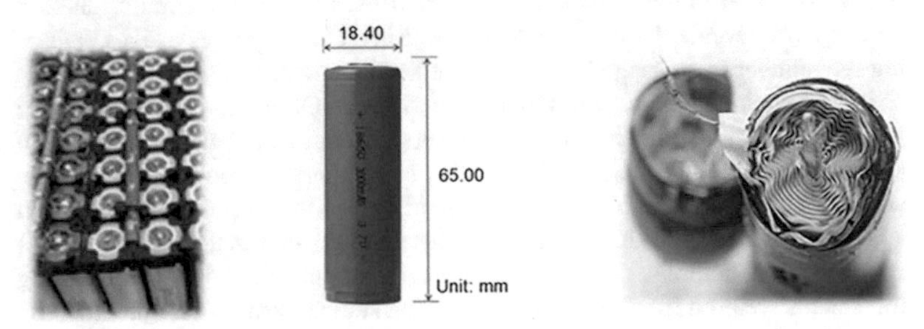

Fig. 1. Lithium-ion battery stack (left), 18650 cell outlook size (center), vertical cross-section of the battery cell 18650 (right).

Vertical cross-section of the 18650 battery cell is shown in Fig. 1, it is a multilayer axisymmetric roll in an aluminum enclosure with an internal volume of 17 284 mm^3 and outer lateral surface area of 3 757 mm^2 without terminals.

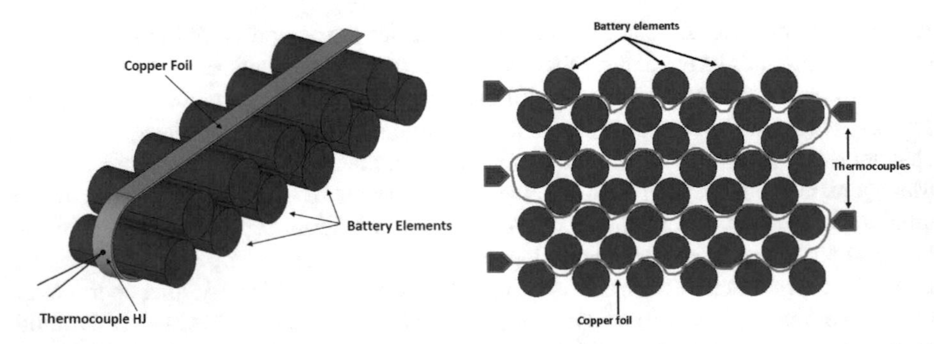

Fig. 2. Cooling copper foil situated in a battery stack with thermal sensors locations in red. 11 elements set (left) and complete battery stack (right) are presented.

2 Design Concept

Each battery element outer enclosure is connected with high thermal conductivity material, e.g. thin copper foil (Fig. 2). This copper foil is insulated from the main battery electric circuit and could help the battery cells cooling. Foil is connected with thermal conductive glue and it goes by each battery in the battery module. At a given distance over the copper foil thermocouple sensors are attached, typically one sensor per 12–20 battery elements can be used.

The thermocouple sensor measures the integral temperature formed by several battery elements depending on their proximity with the sensor. Relatively low temperature, low cost thermocouple types are used (T, J, K) or alternatively, thermoresistors Pt100, Pt1000 could be used.

For the whole battery the thermocouples and copper foil assembly is represented in Fig. 2 (right). It takes into account the relative thermocouple sensor location and proximity to the separate battery elements. The closest elements have a larger impact over the resulting temperature. Each battery element temperature is determined by simple and fast numerical data reconstruction algorithm. Computational cost is low, containing several low order interpolations per element array.

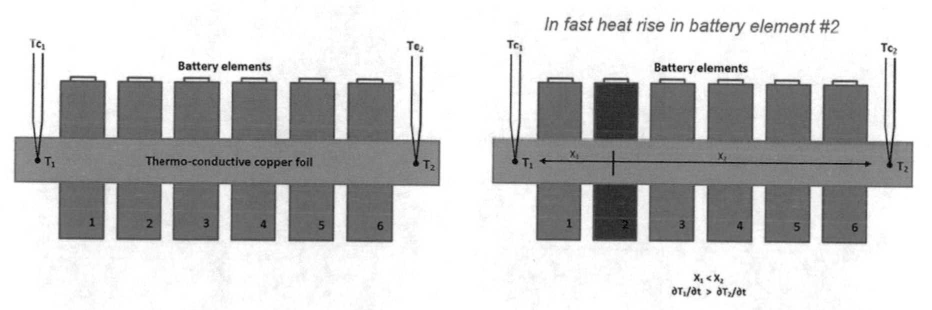

Fig. 3. Unbalance between two temperature changes is used for the location of the overheated element.

3 Reconstruction Method

The proximity to high temperature cell is related to the temperature gradient and its change in time, where temperature is measured by thermocouples. A smaller distance between thermocouple and heat source corresponds to a faster response i.e. faster temperature rising, measured by the closest ti the source thermocouple (Fig. 3).

The temperature reconstruction algorithm is based on the thermal-conductivity equation. The heat transfer, in the solid cell domain, is described by the following transient equation,

$$\rho c_p \frac{\partial T}{\partial t} - \lambda \nabla^2 T = Q, \tag{1}$$

where ρ is material average density, c_p is specific heat capacity of the material, $\mathbf{q}$ is conductive heat flux, Q is power heat source density of the cells.

For each thermocouple (1, 2, …) the temperature changes ∂T over time t are equal to,

$$\left| \begin{array}{l} \frac{\partial T_1}{\partial t} = k\left(\sum_{i=1}^{n} Q_i - \lambda \nabla^2 T\right) \\ \frac{\partial T_2}{\partial t} = k\left(\sum_{i=1}^{n} Q_i - \lambda \nabla^2 T\right) \end{array} \right. \tag{2}$$

where Q is the heat power density of battery elements from one array (n); k is $1/\rho c_p$; and λ – solid material thermal conductivity. Laplacian is calculated as a vector by the relative distance between thermocouples and each battery element.

Thermographic image sequence results are used for heating process time-constant determination. In all experiments, natural convection cooling is adopted with the rate of 8 W/°C·m^2. At 1.9 C (rated current starts from 0.2 C) the estimated heat source volumetric density is 1273 kW/m^3, distributed in 17284 mm^3. The thermal-time constant (T) of the battery cell heating process is temperature dependent, close to the initial temperature 24 °C, T = 10–12 min, after adding 30 °C overheating, it is reduced to T = 5–7 min, because of internal volume heat capacity change [2].

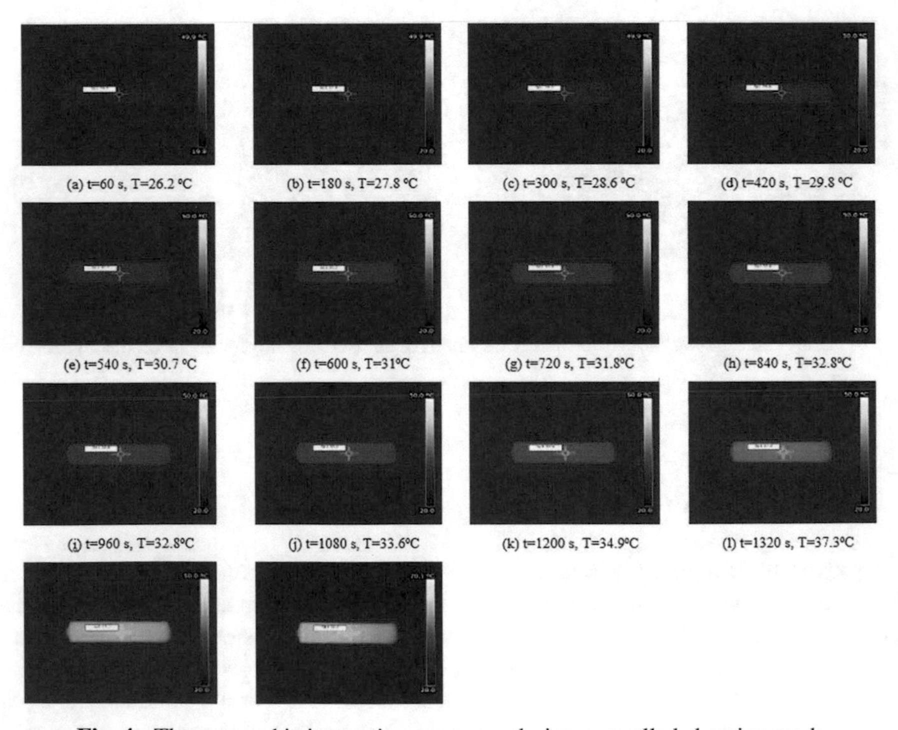

Fig. 4. Thermographic image time sequence during controlled charging mode.

A dataset of thermographic images has been created during discharging of the battery cell module (Fig. 4). On each moment, the picture is shown of the maximum temperature reached on the surface of the battery cell. Charging voltage and current diagrams are shown in Fig. 5. In discharging mode, a constant active power is used P = 22 W, equal to approximately 1.9 C. C is battery cell capacity in (Ah).

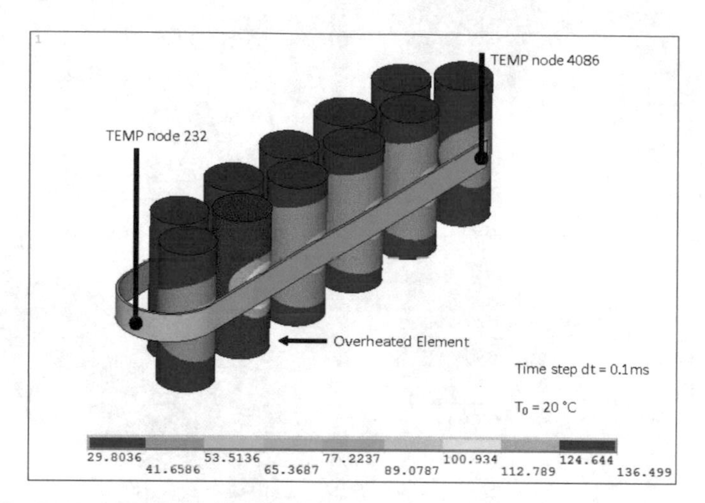

Fig. 5. Transient thermal field FE model of overheated element #2 at 100 s.

Modeling results for forward thermal problem solving (Fig. 5) are confirmed by single cell thermographic imaging. Modeled transient thermal field distribution results (Fig. 7) are used for inverse heat source reconstruction, and accurate battery cell malfunctioning location.

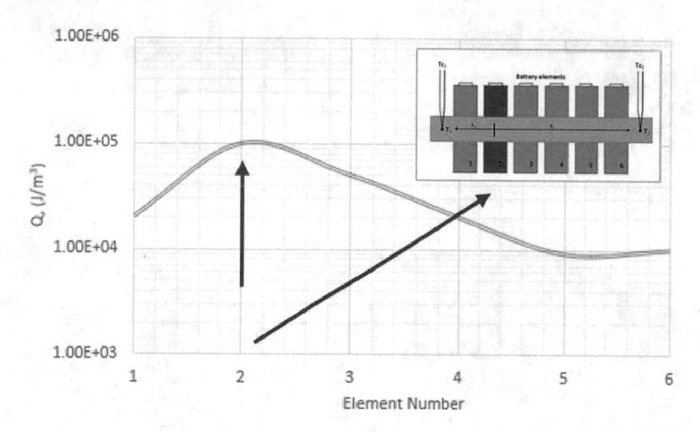

Fig. 6. Reconstructed heat sources in a six element stack, the abnormal heat source is cell #2.

Reconstructed heat sources $Q_{1\text{–}6}$ by system (2), in a six element stack (1–6), is presented on Fig. 6. The abnormal heat source after the reconstruction is found to be cell #2. For this reconstruction are used modeled transient temperature distributions with short-circuited battery. Transient heating is presented at 10 s time-step, for 90s. Short-circuit cells (Fig. 7) are heated faster than normal (Fig. 6).

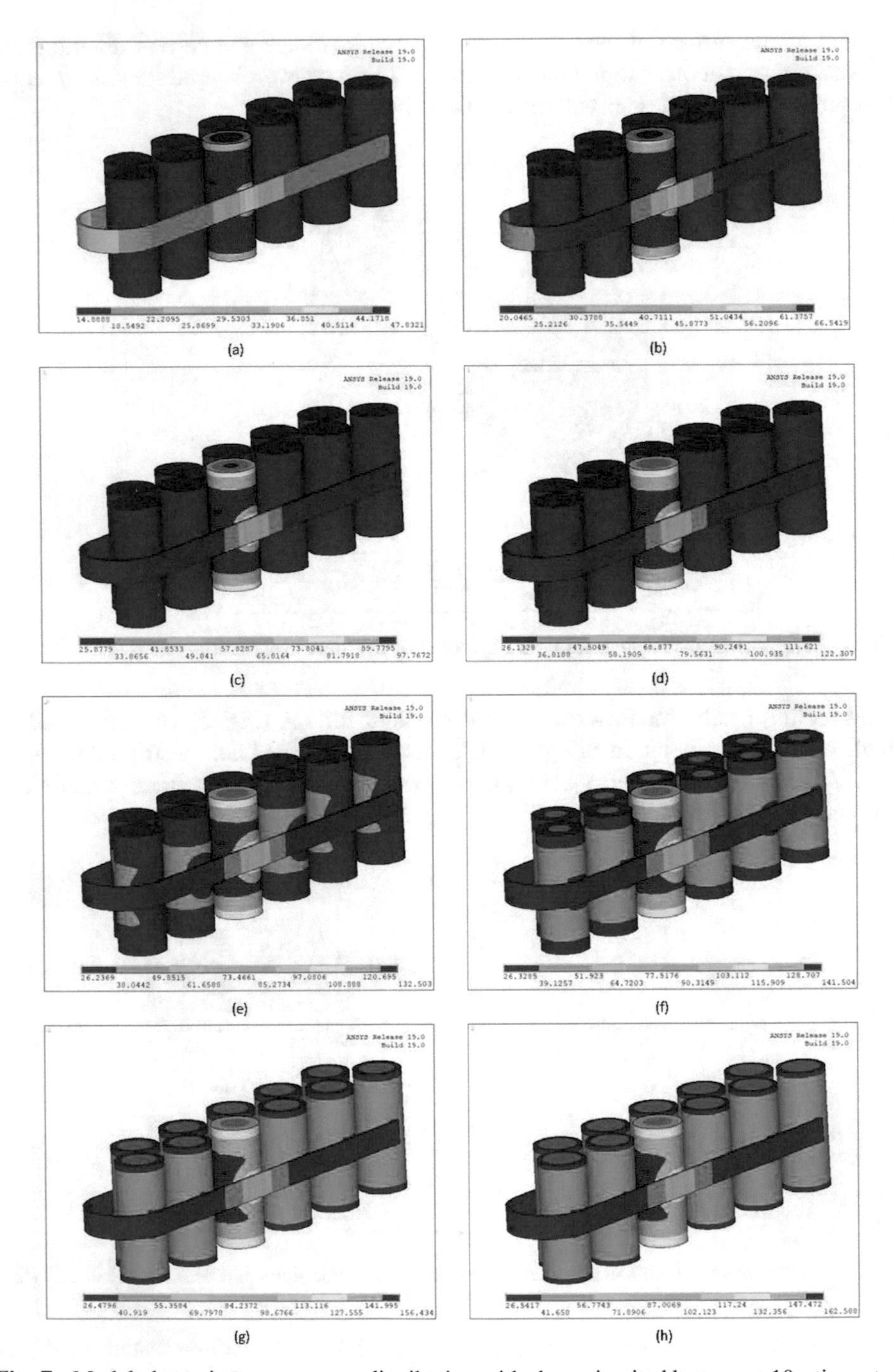

Fig. 7. Modeled transient temperature distribution with short-circuited battery, at 10 s time-step, from 10 s to 90 s.

4 Conclusion

It was found that short-circuit cells are heated faster than normal 100 vs. 1400 s. Heating speed and heat source density are good detecting principles to catch the overheating earlier, before it become irreversible. The passive cooling system for battery stacks, which uses high thermally conductive copper foil for intensive temperature equalizing between battery cells is promising not only for cooling, but for distributed detection. In distributed detection, the number of sensor elements is smaller than the number of battery elements under observation. Over the cooling foil, also called Cooling Foil Shield, an array of distributed thermal sensors are mounted and used for accurate abnormal heat source location determination by sophisticated numerical reconstruction procedure. This Cooling Foil Shield is an excellent place for distributed temperature sensing.

The same Cooling Foil Shield approach could probably be applied to battery connection busses/wires, which have a significant cross-section higher than thermal cooling foils. In that case measuring thermocouples will be place directly over the battery connection busses/wires with a proper electric insulation. Systematic efforts for optimizing both Cooling Foil Shield method and sensor placement optimal configurations must be further made.

Acknowledgement. Some measuring and data processing techniques, used in this research, are developed with the support of the National Science Fund of the Ministry of Education and Science of Republic of Bulgaria within the KP-06-N47/2 contract.

References

1. Mateev, V., Marinova, I., Kartunov, Z.: Automatic system for li-ion battery packs gas leakage detection. In: Proceedings of the 2018 12th International Conference on Sensing Technology (ICST), Limerick, Ireland, 4–6 December 2018, pp. 13–16 (2018)
2. Ralchev, M., Mateev, V., Marinova, I.: Transient heating of discharging li-ion battery. In: 21st International Symposium on Electrical Apparatus & Technologies (SIELA), pp. 1–3 (2020)
3. Dung, L.R.; Li, H.P.: A voltage-gradient based gas gauge platform for lithium-ion batteries. In: Proceedings of the 2018 IEEE International Conference on Applied System Invention (ICASI), Chiba, Japan, 13–17 April 2018, pp. 789–792 (2018)
4. Hannan, M., Hoque, M., Hussain, A., Yusof, Y., Ker, P.: State-of-the-art and energy management system of lithium-ion batteries in electric vehicle applications: issues and recommendations. IEEE Access **6**, 19362–19378 (2018)
5. Wenger, M., Waller, R., Lorentz, V., März, M., Herold, M.: Investigation of gas sensing in large lithium-ion battery systems for early fault detection and safety improvement. In: Proceedings of the IECON 2014–40th Annual Conference of the IEEE Industrial Electronics Society, Dallas, TX, USA, 29 October–1 November 2014, pp. 5654–5659 (2014)
6. Gong, W., Chen, Y., Kou, L., Kang, R., Yang, Y.: Life prediction of lithium ion batteries for electric vehicles based on gas production behavior model. In: Proceedings of the 2017 International Conference on Sensing, Diagnostics, Prognostics, and Control (SDPC), Shanghai, China, 16–18 August 2017, p. 275–280 (2017)
7. Burgués, J., Jiménez-Soto, J.M., Marco, S.: Estimation of the limit of detection in semiconductor gas sensors through linearized calibration models. Anal. Chim. Acta. **1013**, 13–25 (2018)

8. Hutchinson, M., Oh, H., Chen, W.H.: A review of source term estimation methods for atmospheric dispersion events using static or mobile sensors. Inf. Fus. **36**, 130–148 (2017)
9. Marinova, I., Mateev, V.: Inverse source problem for thermal fields. COMPEL Int. J. Comput. Math. Electr. Electron. Eng. **31**(3), 996–1006 (2012)
10. Kishkin, K., Kanchev, H., Arnaudov, D.: Modeling the influences of cells characteristics in battery bank. In: 2022 22nd International Symposium on Electrical Apparatus and Technologies (SIELA),Bulgaria, pp. 1–5 (2022)

Optical Performance and UV Detection Properties of ZnO Nanofilms Using FDTD Simulation

Zachary Stosic[1](✉), Xiaohu Chen[1], David Payne[2], and Noushin Nasiri[1]

[1] NanoTech Laboratory, School of Engineering, Macquarie University, Sydney, Australia
Zachary.stosic@students.mq.edu.au, Jayden.chen@hdr.mq.edu.au, Noushin.nasiri@mq.edu.au

[2] School of Engineering, Macquarie University, Sydney, Australia
David.payne@mq.edu.au

Abstract. Through the use of Lumerical three-dimensional (3D) finite-difference time-domain (FDTD) Electromagnetic Simulator software and the numerous testing parameters, we examined the optical properties, such as the absorbance (%A), transmittance (%T) and the reflectance (%R) of the simulated ZnO nanofilms in relation to the film morphology. This simulation-based experimentation approach enabled effective and efficient testing of a broad range of light wavelengths (λ, 250–650 nm) incident on ZnO thin films with various microstructures, from bulk materials with zero porosity to porous nanostructured films. In this work, the impacts of film thickness (60–320 nm) and particle size (20–40 nm in diameters) on optical properties of the fabricated films were carefully and systematically investigated, to understand the relationship of ZnO nanofilm morphologies and their optical performance. The simulated data was validated using experimental results from a flame-made ultraporous ZnO nanoparticle network (UNN). The finding of this study confirms the capacity of FDTD as a powerful tool to provide insights into the optical phenomenon of nanostructured metal oxide thin films.

Keywords: Nanoparticle · Thin film · Simulation · FDTD · ZnO · Optical properties · Transmittance · Absorbance

1 Introduction

Ultraviolet (UV) radiation is the highest energy component of the solar spectrum reaching the earth's surface [1]. In the skin, UV-light stimulates the synthesis of vitamin D, an essential compound for many metabolic processes [2, 3]. However, prolonged exposure to UV radiation from the Sun can cause severe health issues including erythema, skin pigmentation, DNA damage and skin cancer [1, 4, 5]. There is a growing need for the development of a UV detector that is simple, accurate and robust enough to be utilized as a wearable device to improve sun safety and prevent skin cancer [1].

Nanostructured wide bandgap semiconductors including ZnO, TiO_2, SnO_2, GaN and SiC are commonly reported as potential materials for UV photodetection applications

N. K. Suryadevara et al. (Eds.): ICST 2022, LNEE 1035, pp. 209–222, 2023.
https://doi.org/10.1007/978-3-031-29871-4_22

[6–8]. Among them, ZnO is the most investigated nanostructured material for UV photodetection due to its wide direct band gap of 3.37 eV, large exciton binding energy of 60 meV at room temperature, and ease of fabrication [9–11].

A wide range of nanofabrication techniques have been used to synthesise highly crystalline, pure, and homogeneous ZnO nanofilms including chemical vapour deposition (CVD) [12, 13], flame spray pyrolysis (FSP) [9–11, 14], pulsed laser deposition (PLD) [15, 16], atomic layer deposition (ALD) [17, 18] and sol-gel [19, 20], to name a few. Among them, FPS is one of the most highly promising and versatile techniques for the rapid and scalable synthesis of nanostructured metal oxide films for a wide range of application including gas sensors [21, 22], solar cells [23, 24], functional surfaces [25, 26] and tissue engineering coatings [27, 28]. The high temperature formation in the flame renders high thermal stability of the fabricated particles and films, as compared to other lower temperature wet techniques [29]. In previous works, the FSP synthesis technique were used to fabricate highly porous, crystalline photodetectors made of networks of ultrafine ZnO nanoparticles, with an average particle size of 16–20 nm [9, 11, 14]. The fabricated nanofilms resulted in excellent photosensitivity towards UV radiation with record high milliampere photocurrents and nanoampere dark currents. This strong response is attributed to the unique morphology of the ZnO nanostructures that resulted in the absorption of more than 80% of the incoming UV radiation and transmission of nearly 90% of the visible light [9, 11, 14]. However, further studies are required to investigate the interaction of light with the surface of the nanofilms, and the key roles of particle size and film porosity played in optical performance of these thin nanostructures.

2 Simulation Outline

The Lumerical 3D FDTD Electromagnetic Simulator[1] software uses multi-coefficient models for accurate material modelling over large wavelength ranges [30, 31]. This allows the interactions between materials and electromagnetic fields to be both monitored and visualised. This simulation methodology utilises 'frequency domain power monitors' to capture data before and after the test material (ZnO thin films) to analyse total transmission and absorption of UV and visible light through the films [32]. Additionally, a third monitor behind the light source is used to capture wavelength reflection data.

Figure 1 shows the simulation set-up used in the ZnO films of a porous lattice of spheres to simulate the flame-made nanostructure which consists of a network of vertically stacked nanoparticles (NPs). The self-assembly mechanisms of ultrafine nanoparticle aerosols have been recently modelled [26, 33], indicating the formation of individual surface-bond agglomerates constituted by depositing nanoparticles. Individual components of the simulation are as follows:

FDTD Solver Region: To specify the area/volume that is to be simulated as well as the simulation time, mesh, and the boundary conditions, the FDTD solver region was used as previously reported [34, 35]. Specific boundaries can be set such as Perfectly Matched

[1] More information on Lumerical FDTD simulation software is available at https://www.lumerical.com/products/fdtd/.

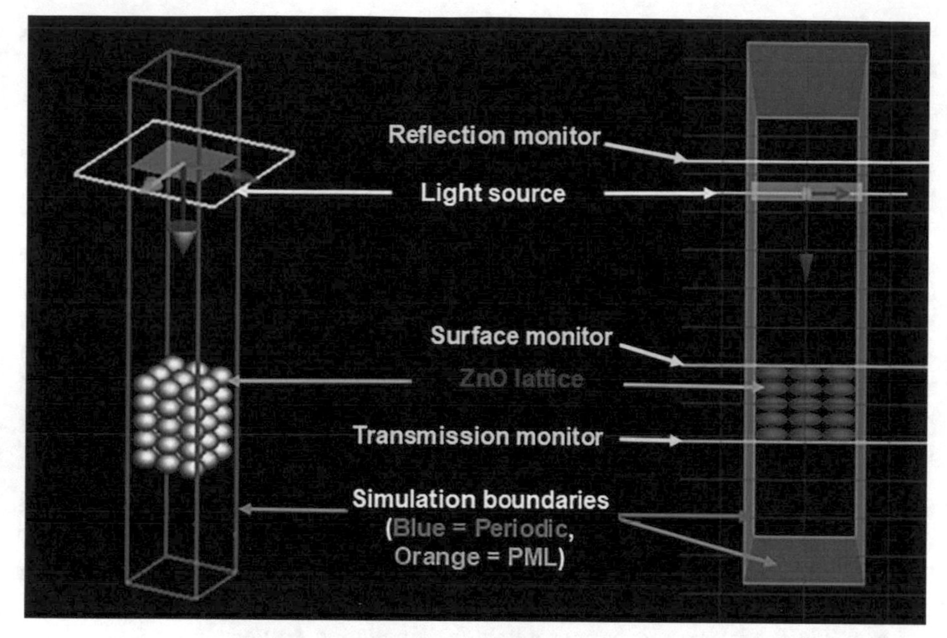

Fig. 1. Schematic of the labelled ZnO spherical lattice simulation set-up.

Layer (PML) boundaries (Fig. 1, highlighted in orange) which act as perfect absorbers to remove any errors from incident light, and Periodic boundaries (Fig. 1), highlighted in blue) which shrink the infinitely wide periodic structure into a small replica unit for effectively numerical calculation.

Plane Wave Source: To test the optical performance of the modelled nanolayers, a plane wave source above it (Fig. 1 highlighted in grey) is used to create uniform electromagnetic energy [36]. This source injects the energy along the direction of propagation indicated by the pink arrow and towards the target layer (Fig. 1).

Frequency Domain Power Monitors: To capture the required data, three frequency domain power monitors were utilised for these simulations [37], namely, the surface, transmission and reflection monitors. These monitors are used to capture an array of data within the FDTD software. However, this experiment focuses on the transmission data measuring the wavelengths crossing the 2D monitor plane.

The highest monitor capturing reflective data is placed above the light source (Fig. 1) ensuring all data is collected from the reflected sources. The next two monitors are placed on either side of the simulated ZnO structures. The total transmission and absorption of light through the nanolayers were measured by comparing the data collected from both monitors.

ZnO Material Data and Geometry: This experiment was carried out using pre-recorded ZnO material data and two geometric structures (bulk film and a lattice of spheres). To properly simulate the optical performance of ZnO nanofilms, the material's refractive index [38] was applied to the thin film structures. This ensures the interaction of light with the film nanostructure as it would under 'real' non-simulated test scenarios.

Initially, the optical characteristics of bulk ZnO thin films with the packing density of 100% (zero film porosity) were investigated for the film thickness of 60–300 nm. This range utilises the simulation methodology and provides the opportunity to test thicknesses that are usually impractical or difficult to produce in the laboratory using the FSP synthesis technique.

A rectangular lattice of spheres was created to simulate the nanoparticles as the building blocks of these nanostructured films [26, 33]. Throughout this experiment, being able to control the diameters and total layer numbers allowed the manipulation of particle size and film thickness. Repeating simulations with these variables provided insights into the relationship between particle size, film thickness and overall optical performance. A visual example of the ZnO lattice of spheres tested in this study is illustrated in Fig. 1.

Movie Monitor: 2D cross sectional 'movie' was created to visually analyse interactions within the simulation. These 'video' monitors work by taking snapshots of the fields over the entire simulation and running them together to create a movie file. This tool while not producing any raw data is useful to visualise complex interactions within simulations and to quickly pick up errors or bugs within the simulation set-up.

3 UNN ZnO Film Fabrication and OpticaL Measuremnts

An UNN ZnO thin film was fabricated via a super-fast and facile FSP method by controlling a spraying time of 50 s while maintaining the other fabrication parameter settings the same as our previous report elsewhere [11]. The optical properties of the film were measured using a Lambda UV/Vis spectrophotometer (L265, Perkin Elmer, USA), where a reflectance module (P/N: N4103003) was equipped for reflectance measurement.

4 Results and Discussion

By taking advantage of the additional opportunities the simulation software offered, the scope of this project allowed a wide array of wavelengths to be tested. Wavelengths ranging from 250–650 nm were used in the simulations allowing data capture for all of the UVA and UVB spectra plus the majority of UVC and visible light.

Optical Properties of the Bulk ZnO Thin Films

Figure 2 shows the optical properties including the absorbance (%A, Fig. 2a), transmittance (%T, Fig. 2b) and reflectance (%R, Fig. 2c) of the bulk ZnO thin films as a function of the film thickness from 60 to 300 nm. For all film thicknesses, a constantly high UV absorption ($\lambda < 370$ nm) but low visible light ($\lambda \geq 400$ nm) absorption was observed (Fig. 2a), depicting a sharp cut-off around 370 nm which is very well in line with ZnO direct bandgap of 3.37 eV [39]. Increasing the film thickness from 60 to 300 nm resulted in an increase in UV light absorbance from 33.6% to 82.7% ($\lambda = 325$ nm), suggesting a

17.45% per 100 nm enhancement in UV absorbance as the ZnO film thickness increases. Similarly, the visible light absorbance increases from 5.5% for film thickness of 60 nm to 25.4% for film thickness of 300 nm ($\lambda = 600$ nm) demonstrating 8.3% per 100 nm enhancement in visible light absorbance. This limited light absorbance at the visible region demonstrates an intrinsic "visible blind" characteristic of ZnO films [11].

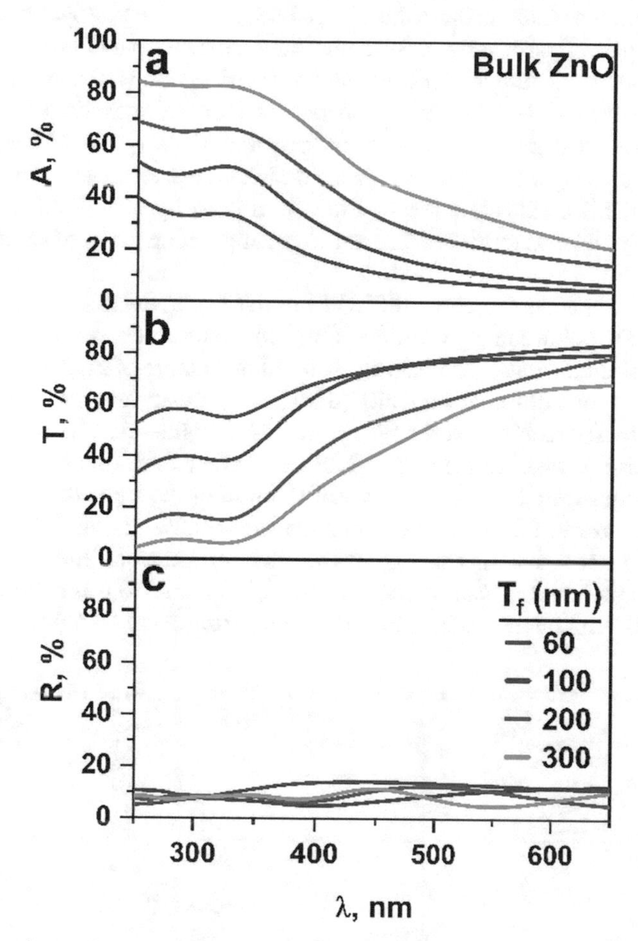

Fig. 2. The absorbance (a), transmittance (b) and the reflectance (c) of the bulk ZnO with different film thicknesses.

The bulk ZnO thin films exhibit high visible light transmission with a %T of over 60% at $\lambda > 600$ nm (Fig. 2b). In the UV region ($\lambda = 325$ nm), the transmittance dramatically decreases from 53.8% to 38.3% by increasing the thickness from 60 to 100 nm. Further increase in the film thickness up to 300 nm reduces the UV transmittance down to 5.6% (Fig. 2b). The selectively high UV absorption and high visible light transmission is in good agreement with previous reports on ZnO nanofilms as visible-blind UV photodetectors [9, 11, 40]. A reflectance of less than 15% was observed of bulk ZnO thin films over the UV-Vis range regardless of their film thicknesses. No explicit correlation was found between %R and the light wavelength for most of the simulated bulk ZnO nanofilms. This could be attributed to the non-porous, highly dense morphology of these bulk films [41, 42].

Optical Properties of Porous ZnO Thin Films Made of Vertically Stacked Nanoparticles (film$_{NP}$).
To simulate building blocks within a thin film, a lattice of ZnO nanospheres was utilised to analyse the impact of particle sizes. By manipulating the number of layers and diameters of the nanospheres, the film thicknesses and particle sizes are well controlled within the simulated thin film$_{NP}$. Herein, ZnO nanospheres with particle diameters from 20 to 40 nm were investigated individually by stacking their densely packed ZnO layers right above the previous layer (Fig. 1), forming a simple unit cell structure by the closest ZnO nanoparticles. This architecture resulted in a film porosity of ~47.7% where the film thicknesses altered between 80 and 320 nm via manipulating the number of the stacking layers.

Similar to its bulk counterpart, the ZnO film$_{NP}$ (Fig. 3a) exhibits significantly higher absorption at UV region compared to the visible light, with a clear absorbance cut-off near ~370 nm. This demonstrates the intrinsic "solar blind" nature of ZnO films regardless of their morphology and film characteristics (thickness, porosity, etc.), which is well in line with our previously reported works [9, 11, 40]. Meanwhile, despite the differences of nanoparticle size, increasing the film$_{NP}$ thickness enhances the absorbance in both UV and visible ranges (Fig. 3a). Comparing ZnO film$_{NP}$ of the same thickness, negligible changes were observed in the film absorbance with 69.8% and 69.2% absorbance for 20 nm and 30 nm sized ZnO film$_{NP}$ at the film thickness of 300 nm, respectively. Similarly, a slight increase of absorbance from 61.5% to 62.7% was achieved at the film thickness of 240 nm by increasing the particle size from 30 to 40 nm (Table 1).

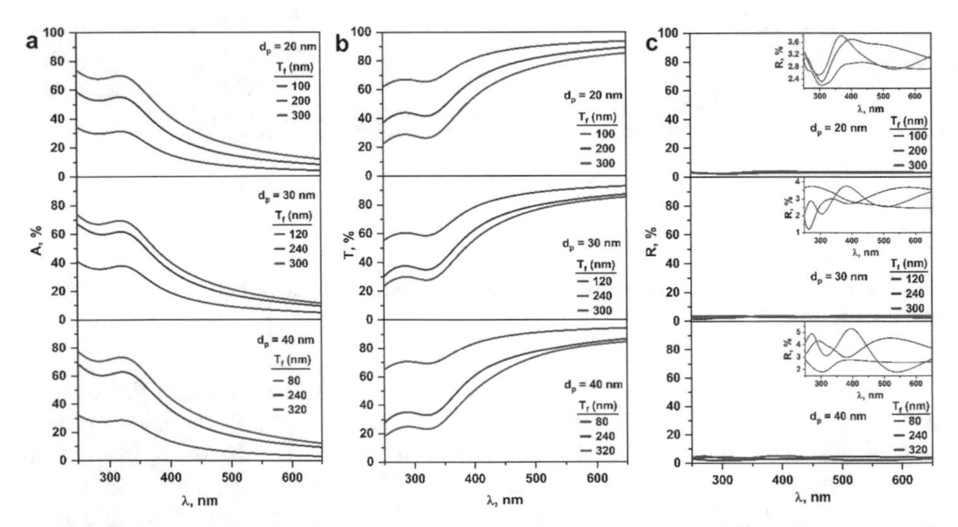

Fig. 3. The absorbance (a), transmittance (b) and the reflectance (c) of ZnO film$_{NP}$ with different nanoparticle sizes and film thicknesses, insets in (c) are the corresponding magnifications accordingly.

Table 1. The absorbance of ZnO film$_{NP}$ with different particle sizes in the UV ($\lambda = 325$ nm) and Visible ($\lambda = 600$ nm) ranges.

d_p (nm)	Tf (nm)	A (%)	
		$\lambda = 325$ nm	$\lambda = 600$ nm
20	100	31.2	4.4
	200	54.8	9.4
	300	69.8	14.0
30	120	37.8	5.6
	240	61.5	11.0
	300	69.2	13.4
40	80	28.4	3.8
	240	62.7	11.3
	320	73.1	15.0

The transmittance of the ZnO film$_{NP}$ changes in an opposite direction as present in Fig. 3b. Similar to the bulk ZnO nanofilms (Fig. 2b), the ZnO film$_{NP}$ offers high transmittance for visible light but "blocks" more UV ranged photons. Only 20% - 30% of UV light can transmit the ~300 nm thick film$_{NP}$, whereas over 75% of visible light ($\lambda \geq 500$ nm) pass through the films, indicating high UV absorbance and high transparency of porous ZnO towards visible light. Interestingly, the reflectance of ZnO film$_{NP}$ suggests an even weaker reflection property compared to its bulk counterpart with significantly low reflectance values (Fig. 3c) and unnoticeable variations in their corresponding curves (Fig. 3c inset), it indicates that the nanostructured ZnO thin films are posing a promising anti-reflection property. As a result, the ZnO nanofilm was utilised as the coverage of thin-film based solar cells to enhance its photovoltaic performance by means of significantly reducing the surface light reflectance from 14.7% to 7% [42]. The correspondingly magnified reflectance curves indicate their reflectance are all below 5.5% (Fig. 3c, insets). In addition, the smaller size of ZnO film$_{NP}$ ($d_p = 20$ nm) exhibits weaker reflectance of under 3.8% comparing to the values of under 4% and 5.5% of ZnO film$_{NP}$ with larger d_p of 30 and 40 nm, accordingly.

Table 2. The absorbance and transmittance ($\lambda = 325$ nm) of ZnO thin films with the same T_f of 300 nm.

	A (%)	T (%)
Bulk ZnO	82.7	6.2
ZnO film$_{NP}$, $d_p = 20$ nm	69.7	26.4
ZnO film$_{NP}$, $d_p = 30$ nm	69.2	27.8

Comparison of Bulk ZnO Nanofilms and Porous ZnO Film$_{NP}$.
Considering the same film thickness of 300 nm, the absorbance, transmittance and reflectance of these FDTD simulated ZnO nanofilms are present in Fig. 4a–c. The bulk ZnO thin film exhibited stronger UV absorption capacity compared to the two ZnO film$_{NP}$ with almost equal UV absorption capacity (Fig. 4a). Herein, the bulk ZnO thin film achieved a high absorbance of 82.7% at the wavelength of 325 nm (Table 2), which is 13% and 13.5% higher than the ZnO film$_{NP}$ with the nanoparticle size of 20 nm and 30 nm, respectively. This higher UV absorbance is attributed to the high packing density of bulk ZnO film in comparison with 53% packing density of ZnO film$_{NP}$.

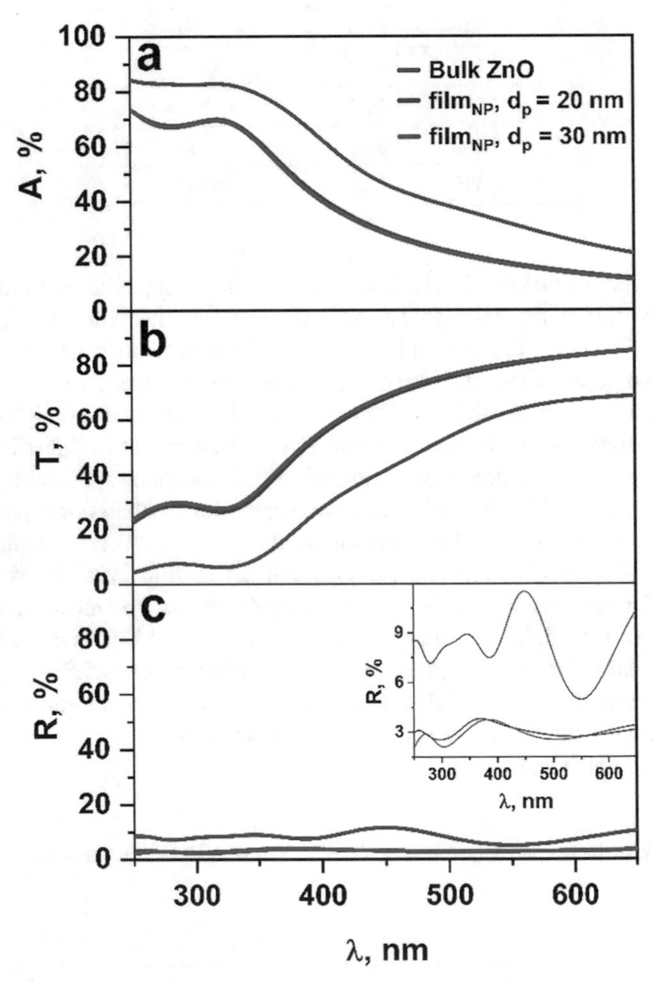

Fig. 4. The absorbance (a), transmittance (b) and the reflectance (c) of both bulk ZnO thin film and ZnO film$_{NP}$ with the same film thickness of 300 nm.

Figure 4b presents the transmittance spectra of the bulk in comparison with porous ZnO film$_{NP}$. The ZnO film$_{NP}$ demonstrated a 26%–28% transparency (26.4% for d_p of

20 nm and 27.8% for d_p of 30 nm) at the wavelength of 325 nm. However, a limited transmittance of only 6.2% was observed for bulk ZnO thin film, indicating an enhancing factor of 4–4.5 for porous films. Such higher light transmittance could be attributed to the porous nature of the thin films, allowing more photons to transmit through the gaps. The bulk ZnO thin film displays relatively higher reflectance at both UV and visible region compared to porous $film_{NP}$ (Fig. 4c). This could be related to the ideally flat and smooth surface of the bulk film in the modelling setup. It is interesting to note that both ZnO $film_{NP}$ demonstrated analogical reflectance curve along the UV-Vis range, with a reflectance peek near 385 nm. This is in stark contrast to the randomly patterned reflective spectra of bulk ZnO thin films with different thicknesses (Fig. 2c and Fig. 3c) due to the interface effects.

The Flame-Made UNN ZnO Thin Film.
The UNN ZnO thin films (Fig. 5a–b) with a porosity of 98% were successfully duplicated using FSP methods. Further details of the UNN films fabrication can be found elsewhere [11]. The film is made of orthogonally impinged ZnO nanoparticles (with an average particle diameter of 19 nm) deposited on a water-cooled glass substrate. The film thickness was ~5.3 μm (Fig. 5a) due to the shortened spraying time of 50 s, suggesting a slightly faster film growth rate of 6.4 μm/min compared to the previously reported 5.5 μm/min [11]. This could be attributed to a shorter distance between the fabricated thin film and the water-cooled substrate holder. Therefore, the heat accumulated during the deposition process can be well dissipated and consequently promotes a faster film thickness growth rate at the same time[43]. Converting the actual thickness of the UNN ZnO thin film into its bulk structure (zero porosity), the assumed "bulk" thickness (T_f') is equivalent to ~106 nm, which is very close to the simulated bulk ZnO with the film thickness of 100 nm ($FDTD_{bulk}$). This allowed us to comparatively investigate the optical properties of the FDTD simulated ZnO thin films and the experimentally fabricated UNN sample. Considering the 47.7% porosity of $film_{NP}$, an estimated film "bulk" thickness (T_f') of 105 nm was calculated from 200 nm thick $film_{NP}$ with the particle size, d_p, of 20 nm, which matches well with the thickness of bulk ZnO (100 nm) and flame-made UNN (106 nm) thinfilm. Meanwhile, the selected d_p of 20 nm for $film_{NP}$ is very close to the particle size of 19 nm for the UNN ZnO thin film measured by the transmission electron microscopy (Fig. 5b, inset).

Figure 6a–b presents the transmittance and reflectance spectra of flame-made UNN in comparison with the FDTD simulated ZnO thin films. The absorbance spectra of these nanofilms are not discussed here, as the data are quasi-chiral to their transmittance accordingly. As depicted in Fig. 6a, an extremely high transmittance of 96.6% at the wavelength of 500 nm was achieved for UNN ZnO thin film which is about 20% and 15% higher compared to the transmittance of 76.7% and 81.6% for bulk ZnO thin film and ZnO $film_{NP}$, respectively. This higher transmission towards visible region is attributed to the higher film porosity of UNN thin films (98%) and the pores between the building blocks, resulting in a better penetration of visible light through the thin film. Ignoring the slight differences in the T_f', an approximate expression of its porosity (%p) and the transmittance ($\lambda = 500$ nm) can be described as below:

$$\%p = 103.4(1 - e^{-\frac{\%T - 76.7}{6.8}})$$

Fig. 5. The microscopic images of the flame-made ZnO UNN thin film. (a) SEM image of the tilted thin film. Inset: the photographic image of the ZnO UNN thin film deposited on clean glass. (b) The higher magnification of the side view. Inset: TEM image of the ZnO nanoparticles.

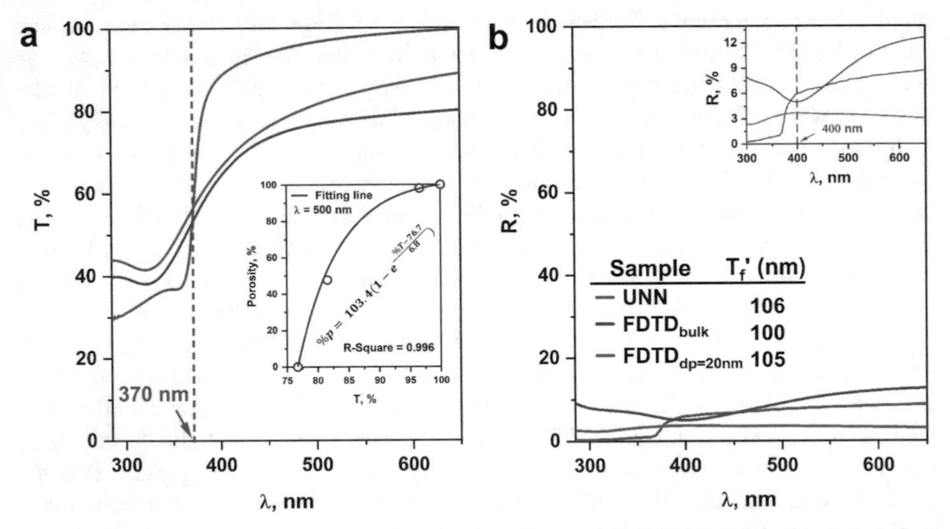

Fig. 6. The absorbance (a) and reflectance (b) of the flame-made UNN ZnO thin film (black line) and the FDTD simulated ZnO bulk film (red line) and ZnO $film_{NP}$ (blue line) with film thickness (T'_f) of ca. 100 nm. Inset of (a) depicts the relationship between film porosity and transmittance (λ = 500 nm). Inset of (b) is the magnified reflective spectra in the wavelength range of 300–600 nm.

This suggests a non-linear monotonic increasing function of the visible light transmittance to its porosity on the premise of a close T'_f around 100 nm (Fig. 6a, inset), indicating an interesting approach to improve the visible light transmission selectively by increasing the film porosity.

In regards to film transmittance in the UV region, the UNN, $FDTD_{bulk}$ and $FDTD_{dp=20nm}$ films illustrated a 35%, 38% and 41% film transmission, respectively, at the wavelength of 325 nm, showing a 'relatively small standard deviation of 3.3%. This could be attributed to the slight difference in their film thicknesses, suggesting that the conversion of the actual film thickness (T_f) into its virtual "bulk" thickness (T'_f) should offer the UV transmittance at an equivalent level. In addition, both UNN and

FDTD simulated ZnO films demonstrated a very similar cut-off wavelength of ~370 nm, which is in line with the ZnO bandgap of 3.37 eV [9–11]. As a result, it is suggested that FDTD could be a promising and powerful tool to model and validate the optical properties of porous nanostructured films fabricated by flame spray pyrolysis technique.

Figure 6b presents the reflectance spectra of flame-made UNN ZnO thin film in comparison with FDTD simulated ZnO films. A very low reflectance of <2% and <9% was measured for flame-made UNN at UV ($\lambda \leq 365$ nm) and visible ($\lambda \geq 400$ nm) regions, respectively. This is in good agreement with our previous report [10], suggesting an anti-UV reflection feature of these ultraporous UNN photodetectors [44]. The ZnO $film_{NP}$ composed of 20 nm size particles depicted similar reflectance behaviour to the UNN thin film with a comparatively low reflectance at the UV range (<3%) and a slightly lower reflectance towards visible light. The bulk ZnO, however, demonstrated a different reflectance spectrum with the lowest reflectance of 5% at UV and visible light dividing line (~400 nm) (Fig. 6b, inset), and the reflectance values increase towards both shorter and longer wavelength directions. An inflection point at ~400–405 nm in reflectance spectra has been widely observed for ZnO and ZnO-based nanomaterials [10, 44–46], however, the reason so far is uncertain. In-depth modelling and experimental investigations are required to offer instructive insights about this unknown optical behaviour of ZnO-based nanofilms.

5 Conclusion

The optical properties of both bulk ZnO (0% porosity) and porous ZnO nanofilms (47.7% porosity) with film thicknesses of 60–320 nm were investigated using FDTD simulation. The results were then validated using an experimentally fabricated flame-made UNN ZnO thin film with the equivalently similar particle size and "bulk" thickness. The ZnO thin films, in particular the porous samples, depicted a promising anti-reflection property. The light absorption of the ZnO nanofilm was enhanced by increasing the film thickness and/or decreasing the particle size as the nanofilm's building blocks. However, the film transmittance decreased significantly by increasing the film thickness. In addition, the visible light transmittance was selectively enhanced by increasing the film porosity from zero to 47.7%. The comparison study on the optical properties of simulated ZnO nanofilms with different morphologies, and the experimentally fabricated ZnO thin films demonstrates the capacity of FDTD modelling in investigating and validating the optical behaviours of nanostructured ZnO thin films.

Acknowledgment. N.N. and D.P acknowledge the financial support received from Macquarie University (MQRAS_21, ID: 174639164). Z.S and X.C acknowledge the financial support received from Macquarie University Research Training Program (RTP) scholarship.

References

1. Nasiri, N., Jin, D., Tricoli, A.: Nanoarchitechtonics of visible-blind ultraviolet photodetector materials: critical features and nano-microfabrication. Adv. Opt. Mater., 1800580 (2018)

2. Lucas, R.M., et al.: Ultraviolet radiation, vitamin D and multiple sclerosis. Neurodegener. Dis. Manag. **5**(5), 413–424 (2015)
3. Holick, M.F.: Biological effects of sunlight, ultraviolet radiation, visible light, infrared radiation and vitamin D for health. Anticancer Res. **36**(3), 1345–1356 (2016)
4. Gonzaga, E.R.: Role of UV light in photodamage, skin aging, and skin cancer. Am. J. Clin. Dermatol. **10**(1), 19–24 (2009)
5. Cadet, J., Douki, T.: Formation of UV-induced DNA damage contributing to skin cancer development. Photochem. Photobiol. Sci. **17**(12), 1816–1841 (2018). https://doi.org/10.1039/c7pp00395a
6. Nasiri, N., Tricoli, A.: Chapter 5 - Nanomaterials-based UV photodetectors. In: Thomas, S., Grohens, Y., Pottathara, Y.B. (eds.) Industrial Applications of Nanomaterials, pp. 123–149. Elsevier (2019)
7. Qu, Y., et al.: Enhanced Ga_2O_3/SiC ultraviolet photodetector with graphene top electrodes. J. Alloys Compd. **680**, 247–251 (2016)
8. Xing, J., et al.: Highly sensitive fast-response UV photodetectors based on epitaxial TiO_2 films. J. Phys. D Appl. Phys. **44**(37), 375104 (2011)
9. Nasiri, N., et al.: Structural engineering of nano-grain boundaries for low-voltage UV-photodetectors with gigantic photo- to dark-current ratios. Adv. Opt. Mater. **4**(11), 1787–1795 (2016)
10. Nasiri, N., et al.: Three-dimensional nano-heterojunction networks: a highly performing structure for fast visible-blind UV photodetectors. Nanoscale **9**(5), 2059–2067 (2017)
11. Nasiri, N., et al.: Ultraporous electron-depleted ZnO nanoparticle networks for highly sensitive portable visible-blind UV photodetectors. Adv. Mater. **27**(29), 4336–4343 (2015)
12. Xu, L., et al.: Catalyst-free, selective growth of ZnO nanowires on SiO_2 by chemical vapor deposition for transfer-free fabrication of UV photodetectors. ACS Appl. Mater. Interfaces **7**(36), 20264–20271 (2015)
13. Zhan, Z., et al.: Direct catalyst-free chemical vapor deposition of ZnO nanowire array UV photodetectors with enhanced photoresponse speed. Adv. Eng. Mater. **19**(8), 1700101 (2017)
14. Nasiri, N., et al.: Tunable band-selective UV-photodetectors by 3D self-assembly of heterogeneous nanoparticle networks. Adv. Funct. Mater. **26**(40), 7359–7366 (2016)
15. Huang, Y., et al.: Enhanced photoresponse of n-ZnO/p-GaN heterojunction ultraviolet photodetector with high-quality $CsPbBr_3$ films grown by pulse laser deposition. J. Alloys Compd. **802**, 70–75 (2019)
16. Shewale, P., Yu, Y.: Structural, surface morphological and UV photodetection properties of pulsed laser deposited Mg-doped ZnO nanorods: effect of growth time. J. Alloys Compd. **654**, 79–86 (2016)
17. Chatzigiannakis, G., et al.: Laser-microstructured ZnO/p-Si photodetector with enhanced and broadband responsivity across the ultraviolet–visible–near-infrared range. ACS Appl. Electron. Mater. **2**(9), 2819–2828 (2020)
18. Orak, I., Kocyigit, A., Turut, A.: The surface morphology properties and respond illumination impact of ZnO/n-Si photodiode by prepared atomic layer deposition technique. J. Alloys Compd. **691**, 873–879 (2017)
19. Kim, D., Leem, J.-Y.: Crystallization of ZnO thin films via thermal dissipation annealing method for high-performance UV photodetector with ultrahigh response speed. Sci. Rep. **11**(1), 1–10 (2021)
20. Pickett, A., et al.: UV–ozone modified sol-gel processed ZnO for Improved diketopyrrolopyrrole-based hybrid photodetectors. ACS Appl. Electron. Mater. **1**(11), 2455–2462 (2019)
21. Bo, R., et al.: One-step synthesis of porous transparent conductive oxides by hierarchical self-assembly of aluminum-doped ZnO nanoparticles. ACS Appl. Mater. Interfaces **12**(8), 9589–9599 (2020)

22. Chen, H., et al.: NiO–ZnO nanoheterojunction networks for room-temperature volatile organic compounds sensing. Adv. Opt. Mater. **6**(22), 1800677 (2018)
23. Mayon, Y.O., et al.: Flame-made ultra-porous TiO_2 layers for perovskite solar cells. Nanotechnology **27**(50), 505403 (2016)
24. Tricoli, A., et al.: Ultra-rapid synthesis of highly porous and robust hierarchical ZnO films for dye sensitized solar cells. Sol. Energy **136**, 553–559 (2016)
25. Liu, G., et al.: Ultraporous superhydrophobic gas-permeable nano-layers by scalable solvent-free one-step self-assembly. Nanoscale **8**(11), 6085–6093 (2016)
26. Wong, W.S.Y., et al.: Omnidirectional self-assembly of transparent superoleophobic nanotextures. ACS Nano **11**(1), 587–596 (2017)
27. Nasiri, N., et al.: Ultra-porous nanoparticle networks: a biomimetic coating morphology for enhanced cellular response and infiltration. Sci. Rep. **6**, 24305 (2016)
28. Nasiri, N., et al.: Optimally hierarchical nanostructured hydroxyapatite coatings for superior prosthesis biointegration. ACS Appl. Mater. Interfaces **10**(29), 24840–24849 (2018)
29. Teoh, W.Y., Amal, R., Mädler, L.: Flame spray pyrolysis: an enabling technology for nanoparticles design and fabrication. Nanoscale **2**(8), 1324–1347 (2010)
30. Roy, D., Samajdar, D.P., Biswas, A.: Photovoltaic performance improvement of $GaAs_{1-x}Bix$ nanowire solar cells in terms of light trapping capability and efficiency. Sol. Energy **221**, 468–475 (2021)
31. Cherniak, G., et al.: Study of the absorption of electromagnetic radiation by 3D, vacuum-packaged, nano-machined CMOS transistors for uncooled IR sensing. Micromachines **12**(5), 563 (2021)
32. Esopi, M.R., Yu, Q.: Plasmonic aluminum nanohole arrays as transparent conducting electrodes for organic ultraviolet photodetectors with bias-dependent photoresponse. ACS Appl. Nano Mater. **2**(8), 4942–4953 (2019)
33. Nasiri, N., et al.: Self-assembly dynamics and accumulation mechanisms of ultra-fine nanoparticles. Nanoscale **7**(21), 9859–9867 (2015)
34. Lu, P., Kosmas, P.: Three-dimensional microwave head imaging with GPU-based FDTD and the DBIM method. Sensors **22**(7), 2691 (2022)
35. Vallone, M., et al.: FDTD simulation of compositionally graded HgCdTe photodetectors. Infrared Phys. Technol. **97**, 203–209 (2019)
36. Wu, Z., et al.: Generation of Bessel beam sources in FDTD. Opt. Express **26**(22), 28727–28737 (2018)
37. Mohsin, A.S.M., Ahmed, F.: Study the optical property of gold nanoparticle and apply them to design bowtie nanoantenna using FDTD simulation. J. Opt. (2022)
38. Materials, S.: Refractive Index of Zinc Oxide (ZnO): Data Extracted from the Landolt-Börnstein Book Series and Associated Databases. Springer, Heidelberg (2017)
39. Bouzourâa, M.-B., et al.: Comparative study of ZnO optical dispersion laws. Superlattices Microstruct. **104**, 24–36 (2017)
40. Bo, R., et al.: Low-voltage high-performance UV photodetectors: an interplay between grain boundaries and debye length. ACS Appl. Mater. Interfaces **9**(3), 2606–2615 (2017)
41. Shinde, S.S., et al.: Photoelectrocatalytic degradation of oxalic acid by spray deposited nanocrystalline zinc oxide thin films. J. Alloys Compd. **538**, 237–243 (2012)
42. Wang, Y.-C., et al.: Photovoltaic electrical properties of aqueous grown ZnO antireflective nanostructure on Cu(In, Ga)Se_2 thin film solar cells. Opt. Express **22**(S1), A13 (2014)
43. Kim, D., Leem, J.-Y.: Crystallization of ZnO thin films via thermal dissipation annealing method for high-performance UV photodetector with ultrahigh response speed. Sci. Rep. **11**(1) (2021)
44. Cole, C., Shyr, T., Ou-Yang, H.: Metal oxide sunscreens protect skin by absorption, not by reflection or scattering. Photodermatol. Photoimmunol. Photomed. **32**(1), 5–10 (2016)

45. Alshammari, A., Bagabas, A., Assulami, M.: Photodegradation of rhodamine B over semiconductor supported gold nanoparticles: the effect of semiconductor support identity. Arab. J. Chem. **12**(7), 1406–1412 (2019)
46. Chen, H., et al.: Full solar-spectral reflectance of ZnO QDs/SiO_2 composite pigment for thermal control coating. Mater. Res. Bull. **146**, 111572 (2022)

RFID Enabled Humidity Sensing and Traceability

Hafsa Anam(✉), Syed Muzahir Abbas, Iain Collings, and Subhas Mukhopadhyay

School of Engineering, Faculty of Science and Engineering, Macquarie University, Sydney, NSW 2109, Australia
hafsa.anam@hdr.mq.edu.au, {syed.abbas,iain.collings, subhas.mukhopadhyay}@mq.edu.au

Abstract. The research proposes a semi-octagonal shaped passive chipless RFID moisture sensing tag. The tag is capable not only for traceability, but also, to sense the moisture in surroundings, hence referred as a humidity sensor chipless RFID tag. The tag has been designed and its performance has been analyzed for various substrates such as, Taconic TLX-0, FR4, Kapton HN®, Rogers RT/Duroid 5870 and Rogers RT/Duroid 5880. Moreover, Rogers RT/Duroid 5870 is used to evaluate the tag's ability to sense and measure the humidity and moisture levels. To acquire sensing behavior, Kapton tape is deployed on largest slot of the tag. As the humidity level varies, there is a change in the electrical properties of Kapton, which in return outputs sensing behavior of the proposed tag. The proposed tag is ideally capable to tag 2^{10} number of unique objects while maintaining compact dimensions of 13 mm × 7 mm. The tag proclaims an astounding sensing behavior towards varying atmospheric humidity levels. The repositioning of frequency notches is evaluated as moisture sensing. The tag anticipates a steppingstone for humidity sensing, to be deployed in agriculture industry.

Keywords: Chipless RFID · Tag · IoT · Radar Crossection (RCS) · Backscattering · Code density · Tracking · Ultra High Frequency (UHF)

1 Introduction

Radio-frequency identification (RFID) is one of the enabling technologies for Internet of Things (IoT) based systems. The system of multiple heterogeneous entities and devices interconnected via wired/wireless technologies models the IoT. The basic practice involves sharing data over the network without any human involvement. In the infrastructure of IoT, each entity is categorized using unique identifiers (UIDs) [1]. Tremendously stretching recognition and identification techniques are executed not only to label any device or object, but also holds the potential to be tagged over a human or animal [1]. Thus, a praise to radio frequency identification system as it is yielding pathways towards future IoT, and artificial intelligence (AI) based systems.

A typical radio frequency system comprises of a tag and a reader, embedded with minimum of a single antenna [2]. Transponder and interrogator are two major functional entities of RFID system, along with middle-ware supportive system. Chipless tag

N. K. Suryadevara et al. (Eds.): ICST 2022, LNEE 1035, pp. 223–237, 2023.
https://doi.org/10.1007/978-3-031-29871-4_23

holds the capability to embed the stored information in the form of binary code in the interrogator signal. The signal is backscattered towards the reader and measured as RCS response. Tags hold the capability to undergo back-scattering phenomena for wireless battery free oriented identification technique. The experimental set-up comprises of two horn antennas connected to vector network analyser (VNA) mounted at far-field distance from the tag [3].

RFID tags are categorized into active and passive, subject to tag's operation and power supply. Active tags hold privilege because of long read range as they incorporate power source. Whereas passive tags not using any power supply are inexpensive solutions towards tracking with limited read range [4]. Regardless of this tradeoff between cost and read range, passive RF identification tags are widely deployed around the world as replacement for barcodes.

Depending upon backscattered linear communication methods, the technique has been classified into; (a). Time-variant based modulated backscattered signal (b). Time-invariant based modulated backscattered signal. The former has been further distributed into non-linear time variant and linear time-variant. Whereas linear time-invariant and non-linear time-invariant are categories of later. All the classified methods have been used in motion modulated chipless RF identification as an emerging technique [5].

Chipless RF tags outperform compared to battery assisted tags because of cost efficient identification technique. The system doesn't need any encoding or decoding protocol for operation. Instead, the identification information in the form of unique ID is encoded using electromagnetic signature (EMS). Apart from low-cost outcomes, the technology is super suitable because of flexible designed chipless RFID tags. These transponders can be deployed directly on the surface of the item to be tagged. So, printability and flexibility are the latest thought-provoking areas of interest for researchers [4]. Also, the chip-free tags hold tremendous benefits compared to traditional barcode technique. Moreover, compared to QR-codes, the proposed tags hold privilege because of non-line of sight communication. Because of less-complexity and minimal power utilization/consumption, the technology finds strong roots for green IoT based applications [6]. With the passage of time, technology is incorporating in IoT system roots with rapid increase in applications, majorly sensors. Along with data capturing, today's era need is also fulfilled for environmental sensors ranging from moisture to temperature and edibles tracking [7]. [8] reports various types of sensors depending upon their classification and operation. Also, the book highlights outputs from both chip-based and chip-free tracking outcomes. Chipless RFID sensors are the technologies that meet the list of requirements for a sensing system to be perfectly able to be deployed with minimal power consumption, contact-less link, and multi-functional features [10].

Extensive research is done and still in progress in the last decade to unfold upcoming aspects/benefits of chipless RFID system [10]. In [11], multi-resonator tag is built holding 8-bit data ID. A 6-bit chipless RFID tag with embedded humidity and temperature sensing is proposed in [12]. Chen Su et al. [13] have come up with a tremendous signal processing algorithm for noise-free detection of tags in multi-tag environment. In [14, 15], recently the researcher has used a novel design technique to yield CRFID tag. Moreover, researchers of [16] and [17] recently have designed frequency domain-based chipless tags with high coding capacity. A multi-band resonator comb-shaped design [18] yields high data capacity of 40-bit word in reasonable dimensions. Most recently, Usman A. Haider et al. [19] provides future product tracking solutions via 40-bit semi-octagonal shaped chipless RF identification tag. In [20], very recent research has been conducted using four different chipless RFID tags for enhanced RCS response using conductive screen printing. Moreover, [21] proposes reconfigurable chipless RFID embedded with gas detection sensor. Flexible low cost chipless RFID tag with novelty of resonance bits constrained in a narrow frequency band is proposed by Giovanni et al. [22].

Resonators are extensively deployed for encoding data from a passive chipless RFID tag [23]. Different lengths of resonators are deployed by Mumtaz et al. [24] and Polivka et al. [25] to capture several data bits using On-Off Keying (OOK) encoding. The output is a combination of zero's and one's. Zero is embedded in the captured identification code via alteration of resonator structure in a way that the prongs of desired resonator are shorted. The tag outlined by Khan et al. [26] provides a glimpse of loop resonators, having identical properties as ring resonators. Square ring resonators deployed as chipless RFID tag with embedded flexibility is proposed for biocompatible and unobtrusive wireless application [27]. [28] illustrates a novel hair-pin array based chipless RFID tag with minimal inter-symbol interference. M. Noman et al. have acquired use of square open loop resonators for one byte data analysis in chipless RFID tag [29].

This paper presents efforts to provide cost-effective solutions for identification embedded with sensing capabilities to cope with the need of era. Kapton tape is deployed on largest slot as a sensing material to introduce a moisture sensor tag under varying humidity levels. The presented tag uses ultra-high frequency radio waves as a communication medium. Integration of stimulus-sensitive Kapton tape makes it capable of sensing the humidity variations around. Varying humidity levels results in alteration of polarizability of polymer, ultimately permittivity modification of Kapton. Because of humid sensitive nature, Kapton polyimide is widely deployed for sensing attributes. The tag holds 10-bit data words in ultra-high frequency (UHF) band.

2 Working Principle

RFID is manifested as the most sound and definitive data-capturing technology to encounter the growing communication needs of the era. A typical RFID holds two major entities. One is transponder, also referred to as RFID tag. The other is reader, also known as interrogator. The reader showers electromagnetic waves over the chipless sensor tag. As a result, the tag is excited, and it decodes the stored binary identification code backwards towards the reader. The signal is transmitted through transmitter antenna of reader and is picked-up by receiver antenna mounted at tag. The waves pass through the chipless tag and eyes for; (a). Resonators (b). Slots. The presence of resonator/metal is decoded as a bit-word '1', whereas '0' bit is acquired when the signal goes through any slot in the tag. Once all the tag's information is fetched by the bombarded signal, the signal leaves the tag and finds its way towards reader's receiver antenna [11]. The coded signal that is showered back towards the reader is known as backscattered signal and the phenomena is referred to as backscattering. Proposed identification is operated via generation of frequency signature, measured as 'RCS' on the desired frequency spectrum [30, 31]. The tag is designed and optimized using CST STUDIO SUIT®. The tag is kept at a far-field distance from the reader. Far-field is measured by Fraunhoffer distance formula [32] Eq. (1).

$$R = \frac{2D^2}{\lambda} \tag{1}$$

where, 'D' is the largest dimension/measurement of radiator and 'λ' is radio wave wavelength. The incident plane wave used to excite the chipless backscattered based tags is given by Eq. (2) [4].

$$\boldsymbol{E}(z;\ \boldsymbol{t}) = \widehat{a_x}\boldsymbol{R}\mathrm{e}\left[E_{x\circ}e^{j(\omega t+kz+\varphi_x)}\right] + \widehat{a_y}\boldsymbol{R}\mathrm{e}\left[E_{y\circ}e^{j(\omega t+kz+\varphi_y)}\right] \tag{2}$$

where '$\boldsymbol{E}$' refers to electric field, angular frequency is labelled by 'ω' and 't' and 'k' denotes time and wave vector respectively. Figure 1 illuminates the incident plane wave along with far-field illustration of proposed tag.

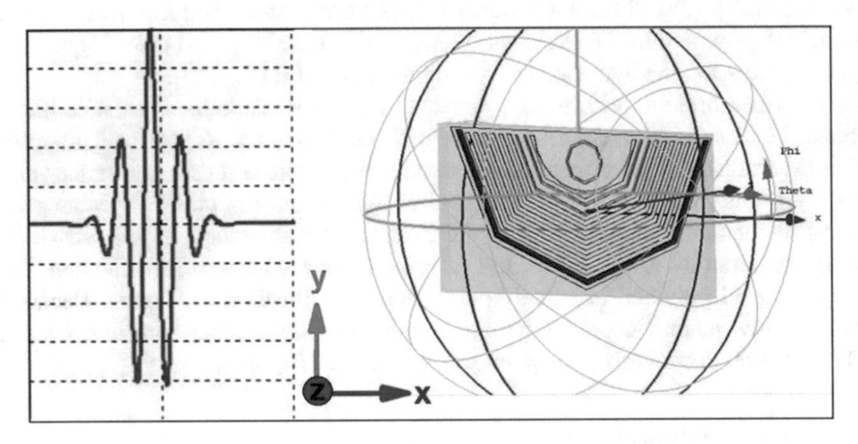

Fig. 1. Incident Plane Wave and Far-Field of Proposed Sensor Tag.

The read range of passive chipless RFID tags may vary depending upon the type of antenna used and the transmitted power. Maximum read range for proposed RF identification system is given by Eq. (3) [4].

$$R^{max} = \sqrt[4]{P_{TX}G_{TX}G_{RX}\frac{\lambda^2}{(4\pi)^3 P_{RX}}\sigma_{min}} \tag{3}$$

where 'G_{TX}' and 'G_{RX}' are the transmitter and receiver antennas gains, whereas, powers being transmitted and received are denoted by 'P_{TX}' and 'P_{RX}' respectively, plane wave wavelength and least possible detectable RCS value by reader are represented by 'λ' and 'σ_{min}' respectively. Equation (4) represents the power received at receiving horn antenna, being sent by transmitting antenna [33].

$$\frac{P_{RX}}{P_{TX}} = \left(\frac{\lambda}{4r\pi}\right)^2 G_{TX}G_{RX} \tag{4}$$

where transmitter and receiver antennas gain and powers are labelled as 'G_{TX}', 'G_{RX}', 'P_{TX}' and 'P_{RX}' respectively. Moreover, 'λ' is wavelength and antennas separation are referred as 'r'.

The proposed tag operates in a way that each metal/resonator corresponds to one dip in RCS/Frequency curve. Each dip further corresponds to 1-bit of unique identification code. Since the tag is combination of alternative slots and metal portions, each metal resonator yields a 0-bit. Ultimately the tag generates a binary combination of 1's and 0's. So, the ten resonators etched via copper/Silver/Aluminum over a substrate, output 10-bit data capacity with a unique tag ID: 1111111111. Whereas 0's are embedded in the combination via shortening of slots. Each resonator has a different length, hence corresponds to different frequency over the RCS curve. The smallest slot refers to dip at largest frequency i.e., least significant bit (LSB). Whereas largest slot resonates at smallest frequency yielding most significant bit (MSB). The strip length of each resonator is inversely proportional to resonant frequency. Consequently, decreasing the resonator length increased its resonant frequency over frequency axis, and vice versa. The resonance frequency calculation can be done using Eq. (5) [4].

$$f_r = \frac{c}{2L}\sqrt{\frac{2}{(\varepsilon_r + 1)}} \tag{5}$$

where 'c' refers to speed of light, 'L' is the length of particular slot and dielectric permittivity is represented by 'ε_r'.

3 Proposed CRFID Tag

3.1 Tag Design

The proposed 10-bit semi-octagonal shaped chipless RFID tag is shown in Fig. 2. The tag is 13 mm stretched across x-axis and 7 mm across y-axis. The tag has been designed and analyzed for various substrates along with varying conductors etched on top surface. The radiator is designed in a way that it has multiple semi-octagonal resonators. The tag exhibits $2^{10} = 1024$ data capacity.

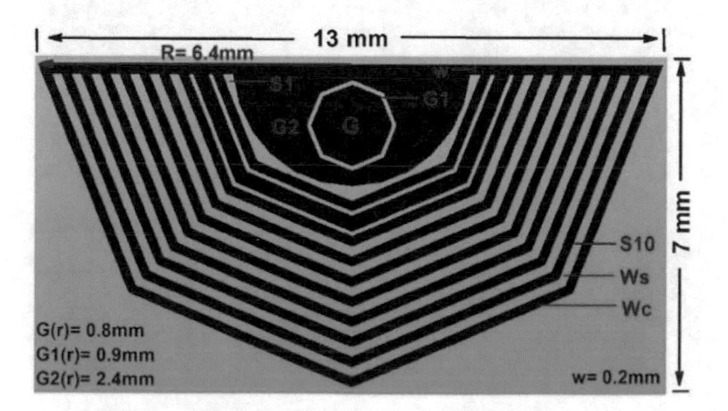

Fig. 2. Layout of proposed Tag.

3.2 Layout Dimensions

The dimensions of the proposed RF identification tag are expressed in Table 1. 'Ws' and 'Wc' are widths of slots and radiators whereas number of slots are labelled as S1-S10 respectively. R is the outer radius of the largest semi-octagon. '$G_{(r)}$' is radius of inner most semi-octagon, '$G1_{(r)}$' is the corresponding gap radius of the inner most octagon, and '$G2_{(r)}$' is the radius of only semi-circle of the tag. G and G1 are added as they fine-tuned the slots, so are part of tag optimization. Table 1 provides slots and metals widths along with their radius.

Table 1. Characterized Tag Dimensions.

No	Slot Radius (mm)	Slot Width (mm)	Resonator Width (mm)
1	2.7	0.3	0.2
2	3	0.1	0.3
3	3.4	0.1	0.2
4	3.8	0.2	0.2
5	4.2	0.2	0.2
6	4.6	0.2	0.2
7	5	0.2	0.2
8	5.4	0.2	0.2
9	5.8	0.2	0.2
10	6.2	0.2	0.2

The current distribution of proposed passive tag at 12 GHz, 16.5 GHz, 21.5 GHz and 31 Ghz is illustrated in Fig. 3.

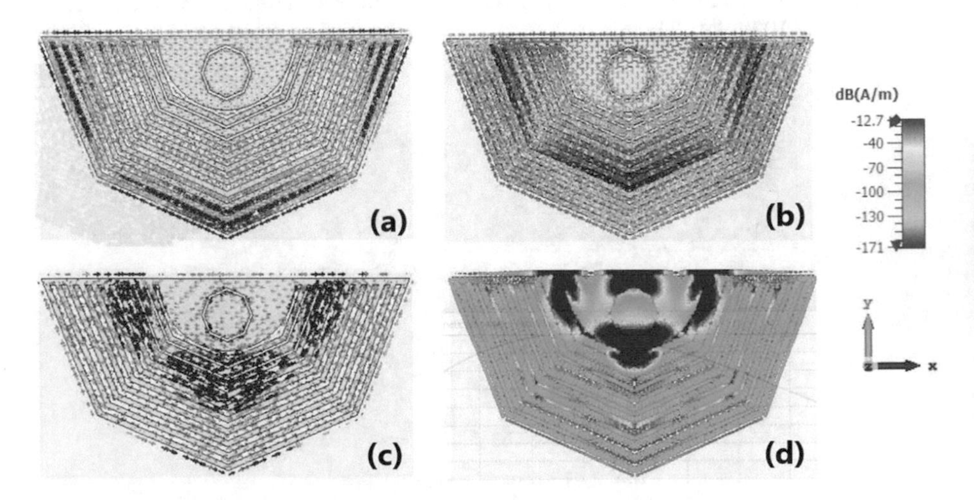

Fig. 3. Surface Current Distribution at: **(a)** 12GHz; **(b)** 16.5Ghz; **(c)** 21.5GHz; **(d)** 31GHz.

4 Results and Discussions

This section presents simulated results of proposed tag designed and optimized for various substrates. Graphical representation of the tag accompanied by various substrates carrying tag ID: 1111111111 are demonstrated. It has been analyzed that by opting different substrates along with compatible radiators, there is dissimilarity in electrical properties and as a result the graphical shift/variation is noted in frequency signature RCS response. Characteristic comparative analysis of all the tags designed is depicted in Table 2.

4.1 Taconic TLX-0 Substrate

The tag designed by incorporating copper on the Taconic TLX-0 substrate is referred to as 'Tag-2'. Electrical permittivity of utilized substrate is 2.45 while thickness of radiator is 0.035 mm. Figure 4 exhibits the RCS vs. frequency response of the mentioned tag. 12 Ghz–30 GHz is the frequency band utilization of compact 10-bit tag.

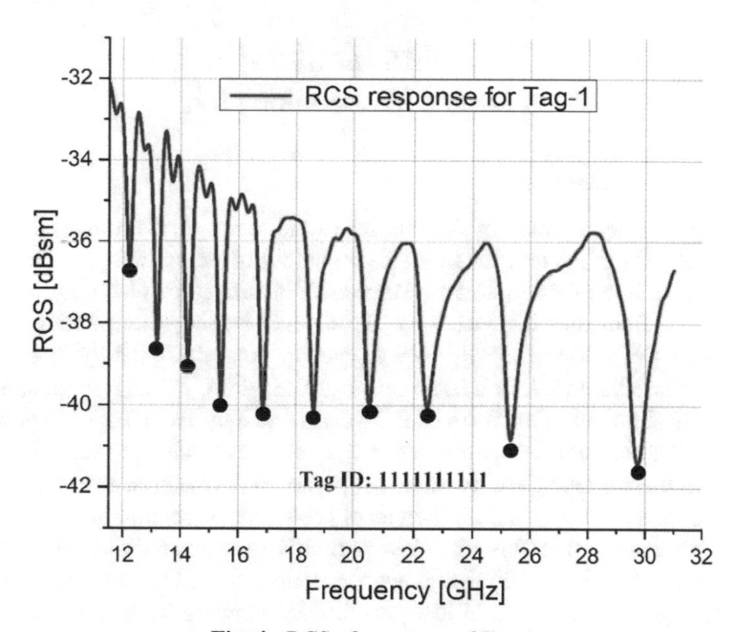

Fig. 4. RCS of response of Tag-1.

4.2 FR-4 Substrate

The design opted via utilizing FR-4 substrate along with copper conductor is labelled as 'Tag-2'. The RCS response of the corresponding tag is depicted in Fig. 5. Electrical permittivity of FR-4 is 4.3 while 0.035 mm is the thickness of etched copper radiator on the surface of substrate. 10-bit compact tag has acquired RCS response in frequency band of 9.5 GHz–25 GHz. Compared to Tag-1, it is very clear that as we opt for high permittivity substrate, the RCS curve drifts towards left over the frequency axis.

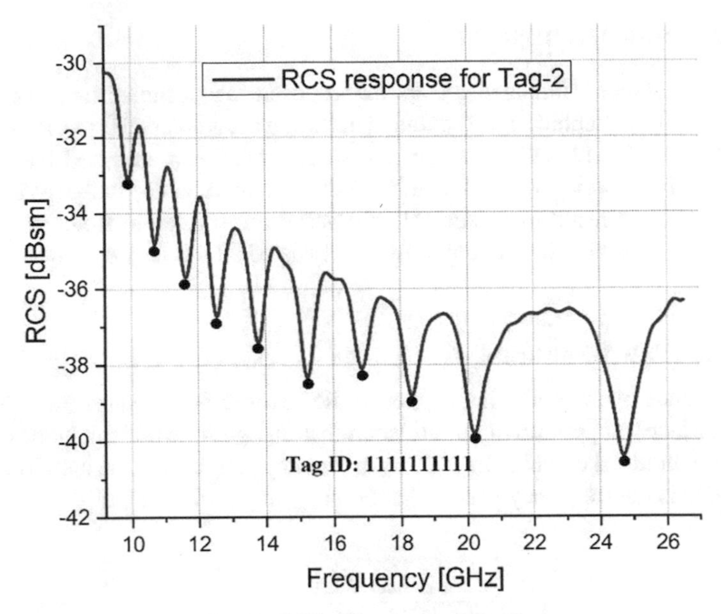

Fig. 5. RCS of response of Tag-2.

4.3 Kapton® HN Substrate

A tag acquired via deployment of Aluminum as radiator over Kapton® HN substrate is referred to as 'Tag-3'. Whereas 'Tag-4' is an output of silver nanoparticle-based ink printed on the surface of Kapton®HN. Kapton®HN defines its electrical permittivity value as 3.5. The frequency band used by Tag-3 while etching Aluminum of thickness 0.007 mm is 11 GHz–30 GHz. While the frequency band occupied by Tag-4 availing silver nano particle-based ink of 0.015 mm is 11 GHz–30 GHz. Figure 6a and Fig. 6b refers to graphical output in the form of RCS curves of Tag-3 and Tag-4 respectively.

A trend can be analyzed compared to Tag-2, as we are switching towards low permittivity substrate, there is change in electrical properties of the material and as a result there is right shift on the frequency axis. Whereas opposite is the scenario is when compared with Tag-1. We can see that Tag-1 has low permittivity compared to Kapton deployed tags. So, we can reveal assuredly that as we are opting for a high permittivity substrate compared to Tag-1 substrate i.e., of low permittivity comparatively, there is a left shift of ID response over the frequency axis. This analysis is very clearly manifested via frequency band allocation of the tags.

4.4 Rogers RT/Duriod® 5870 Substrate

Tag ID comprising of all 1's generated via impression of copper as radiator at surface of Rogers RT/Duriod® 5870 is named as 'Tag-5'. The substrate used has an electrical permittivity of 2.2 whereas thickness of radiator pasted at its surface is 0.035 mm. Tag measured graphical response is revealed in Fig. 7a. Compared to all the previous tags, this one has low permittivity. So, it is eyed that the RCS curve has drifted towards right while acquiring frequency band of 12 GHz–30.5 GHz comparatively.

Table 2. Comparison of proposed chipless tags.

Parameters	Tag 1	Tag 2	Tag 3	Tag 4	Tag 5	Tag 6	Tag 7
Substrate	Taconic TLX-0	FR-4	Kapton® HN	Kapton® HN	RT/Duroid® 5870	RT/Duroid® 5880	RT/Duroid®5870
Thickness (mm)	0.5	0.5	0.125	0.125	0.16	0.78	0.16
Permitivity	2.45	4.3	3.5	3.5	2.2	2.2	2.2
Loss Tangent	0.0019	0.025	0.0026	0.0026	0.0009	0.0009	0.0009
Conductor (mm)	Cu (0.035)	Cu (0.035)	Aluminum (0.007)	Silver ink (0.015)	Cu (0.035)	Cu (0.035)	Cu (0.035)
Data bits	10	10	10	10	10	10	9
Sensing bits	0	0	0	0	0	0	1
Flexibility	x	X	√	√	x	√	x
Frequency band (GHz)	12–30	9.5–25	11–30	11–30	12–30.5	12–31	10–31

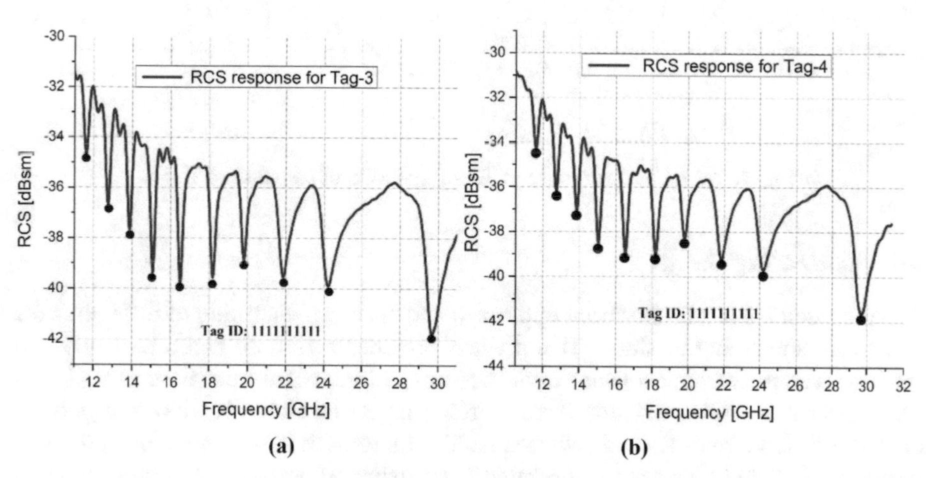

Fig. 6. (**a**) RCS of response of Tag-3; (**b**) RCS of response of Tag-4.

4.5 Rogers RT-Duriod 5880 Substrate

The chipless RF identification tag utilizing Rogers RT/Duriod-5880 as flexible substrate along with copper radiator is cited as 'Tag-6'. Permittivity of flexible substrate used here is 2.2 along with copper as radiator of 0.035 mm thickness. The RF range for tag-6 is 12 GHz–31 GHz, with an overall bandwidth of 19 GHz. This depicts that as we switch from rigid to flexible substrate, the frequency band shifts towards right with some increment in the operational bandwidth. The RCS vs. frequency response is expressed in Fig. 7b.

Table 3. Multiple Tag-ID's.

No.	Tag ID	No. of Bits	Freq. band (GHz)	Possible combinations
Ref. ID	1111111111	10	12–30.5	210
Tag-A	1110101011	10	12–30.5	2^{10}
Tag-B	1101010101	10	12–30.5	2^{10}

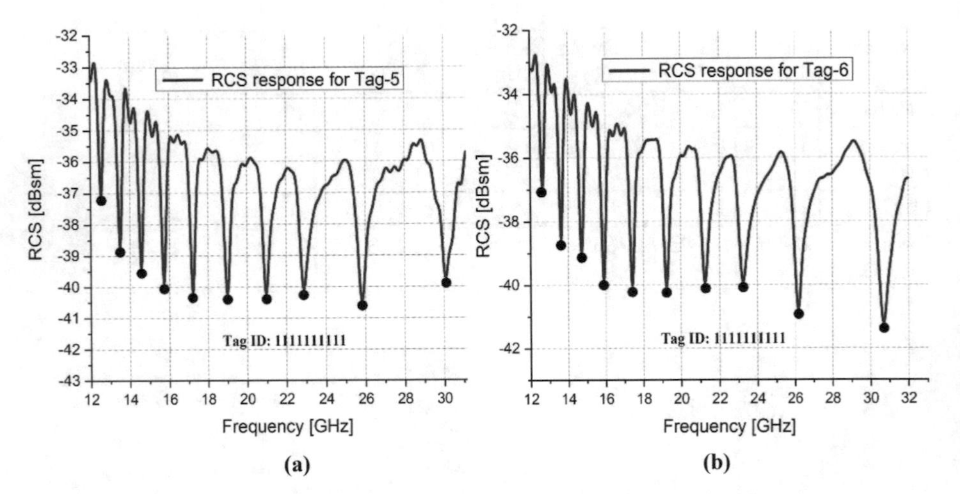

Fig. 7. **(a)** RCS of response of Tag-5; **(b)** RCS of response of Tag-6.

4.6 Shorted Tag Design

The proposed tag has further been optimized and analyzed via filling of different slots, known as 'shortening of slots'. This process provides a glimpse that a particular slot resonates at a particular resonance frequency only. Moreover, the behavior of tag is also examined while varying radiator. If we shorten a particular slot of tag, its corresponding dip at RCS curve vanishes and switches from 1-bit to 0-bit in identification code. As a result, multiple tag ID's can be generated comprising of various binary combinations. Here, the tag structure is intentionally changed to depict its binary response. Sometimes, while embedding shortening aspect, the neighboring dips of corresponding slots experience some minor fluctuations. This oscillation is just because of the current distribution at radiator surface out-turn of abolition of 0-bit slots. Figure 8 shows graphical illustration of various shorted tag ID's acquired via shortening approach.

All one's combination (1111111111) as reference tag ID, and two random ID combinations (1110101011) and (1101010101) are multiple tag-IDs generated and illustrated in Table 3.

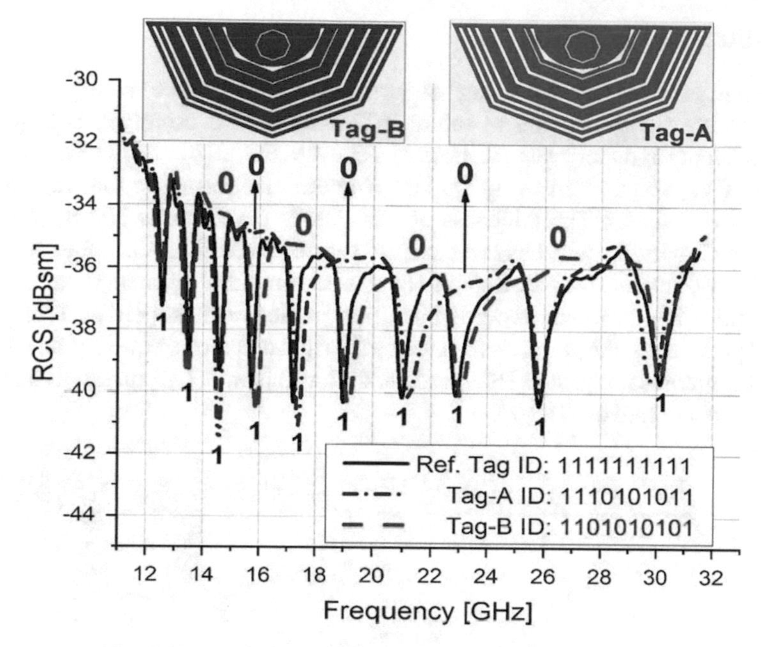

Fig. 8. Shorted Tag-ID's (Inset: shorted Tag-A and Tag-B).

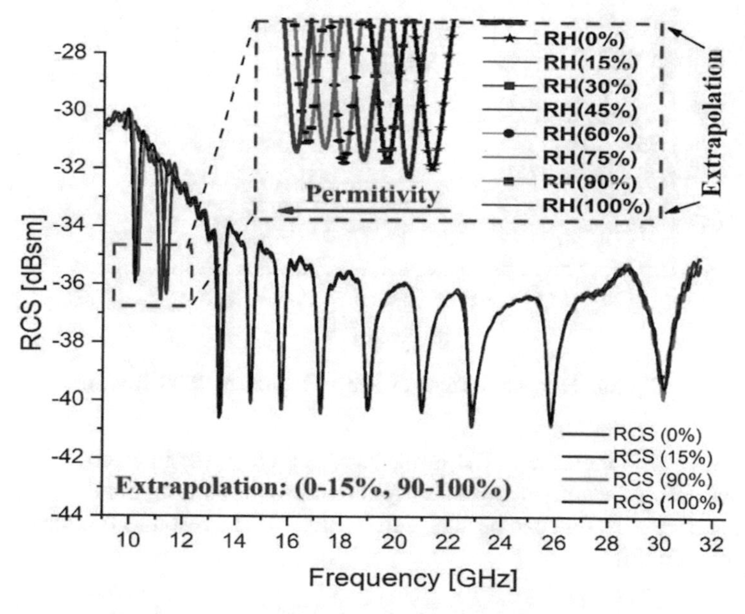

Fig. 9. Humidity sensing response of proposed CRFID.

5 Humidity Sensor Tag

The 10-bit passive chipless RFID tag has further been optimized to incorporate sensing behavior to cope with demand of future IoT systems. The prototype is acquired by etching a copper radiator over the base of RT/Duriod® 5870. The humidity sensing behavior of tag is analyzed using Dupont Kapton® HN heat resistant tape over the largest slot of radiator. The thickness of substrate is 0.125 mm while it exhibits 3.5 relative permittivity. Kapton-HN is a moisture sensitive material. When the sensor tag is surrounded by humidity clouds, Kapton-HN absorbs humidity and undergoes hydrolysis process, ultimately proceeds towards alteration in electrical properties. With varying electrical parameters, the graph shifts towards the right or left on frequency axis. Figure 9 conveys the graphical layout as RCS feedback with Kapton®HN expressed as humidity sensing curve referred as 'Tag-7'.

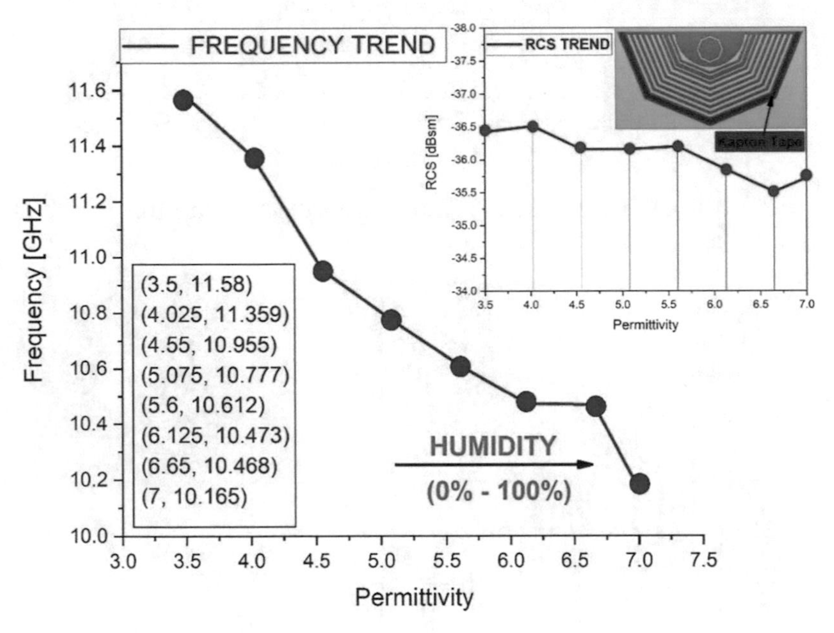

Fig. 10. Frequency Trend of Sensor Tag (Inset: RCS Trend).

With every 15% increase in RH level, there is an increase in electrical permittivity for Kapton®HN. As electrical permittivity increases, the sensing dip drifts towards the left on frequency axis, that can be analyzed in Fig. 10. The resonance frequency shift is illustrated by Eq. (6).

$$\Delta f = f_{res} - f_{(shifted)} \tag{6}$$

where 'f_{res}' is original resonating frequency of sensing slot and '$f_{(shifted)}$' shows shifted resonance dip. Alternatively, if we know the amount of shift, we can calculate shifted resonance frequency given by Eq. (7).

$$f_{(shifted)} = f_{res} - \Delta f \tag{7}$$

Table 4. Characteristic Sensing trends.

No	Humidity (%)	Electrical permittivity	Freq. (res) (GHz)	Freq. (shifted) (GHz)	Δf (GHz)	RCS (dBsm)
1	0	3.5	11.58	–	–	−36.428
2	15	4.025	11.58	11.359	0.221	−36.514
3	30	4.55	11.359	10.955	0.221	−36.163
4	45	5.075	10.955	10.777	0.178	−36.161
5	60	5.6	10.777	10.612	0.165	−36.202
6	75	6.125	10.612	10.473	0.139	−35.851
7	90	6.65	10.473	10.468	0.005	−35.51
8	100	7	10.468	10.165	0.303	−35.721

The tags performance not only depends upon its high encoding capacity, but also on compact size. Moreover, flexibility is also a major milestone for deployment of tag over flexible materials. Also, sensing behavior of tag makes it ranked among top to map with future IoT sensor demands. Increasing permittivity and corresponding resonance frequency shift values are shown in Table 4. Graphical illustration of humidity sensing trend with varying permittivity, frequency and RCS response is exhibited in Fig. 11.

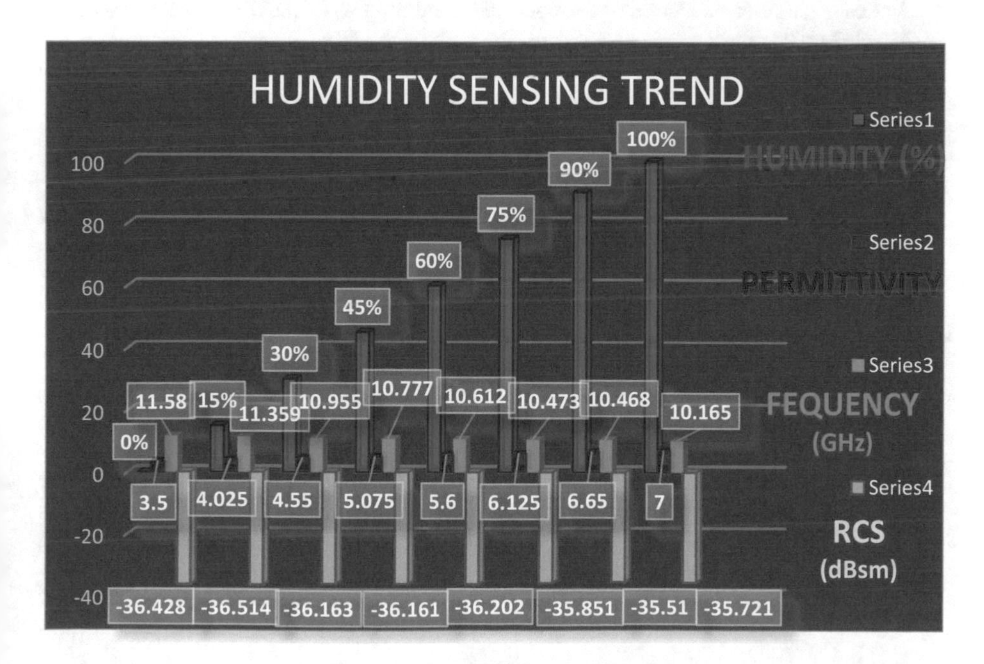

Fig. 11. Humidity Sensing Trend.

6 Conclusion

In this work, a novel and flexible chipless RFID tag loaded with alternative semi-octagonal resonators and slots is proposed. Sensing features are incorporated in tag via deployment of Kapton moisture sensitive tape. The sensor tag operates by variation of electrical properties. This chipless RF identification tag can further be optimized to load more slots until the saturation limit. The work can be extended for multi-sensing utilizing environment friendly biodegradable materials with minimal environmental footprint. Further, novel techniques are required that can address multitude of necessities/demands i.e., green electronic based, widely applicable, and low-cost wireless microwave sensors.

References

1. Mulloni, F., Donelli, M.: Chipless RFID sensors for the Internet of Things: challenges and opportunities. Sensors **20**(7), 2135 (2020)
2. Haider, U.A., et al.: A fully passive 10-digit numeric keypad sensor using chipless RFID technology. IEEE Sens. J. (2023)
3. Anam, H., Habib, A., Jafri, I., Amin, Y., Tenhunen, H.: Directly printable frequency signatured chipless RFID tag for IoT applications. Radioengineering **26**(1) (2017)
4. Habib, A., Mirza, A., Umair, M.Y., Salimi, M.N., Ahmed, S., Amin, A.: Data dense chipless RFID tag with efficient band utilization. AEU Int. J. Electron. Commun. **152**(3), 154220 (2022)
5. Azarfar, A., Barbot, N., Perret, E.: Motion-modulated chipless RFID. IEEE J. Microwav. **3**(1), 256–267 (2023)
6. Zhao, Y., Genovesi, S., Manara, G., Costa. F.: Frequency-coded mm-wave RFID tags using cross polarization. In: 3rd URSI Atlantic and Asia Pacific Radio Science Meeting (AT-AP-RASC), pp. 1–4 (2022)
7. Khalid, N., Mirzavand, R., Iyer, A.K.: A Survey on battery-less RFID-based wireless sensors. Micromachines **12**(7), 819 (2021)
8. Martin, F., Velez, P., Munoz-Enano, J., Su, L.: RFID sensors for IoT applications. Planar Microwav. Sens., 385–442 (2023)
9. Brinker, K., Zoughi, R.: Tunable chipless RFID pressure sensor utilizing additive manufacturing - model, simulation, and measurement. IEEE Trans. Instrum. Meas. (2023)
10. Gee, K.M., Anandarajah, P., Collins, D.: Use of chipless RFID as a passive, printable sensor technology for aerospace strain and temperature monitoring. Sensors **22**(22), 8681 (2022)
11. Sajeer, A., Menon, S.K., Donelli, M.: Development of enhanced range, high Q, passive, chipless RFID tags for continuous monitoring and sensing applications. Electronics **11**(1), 127 (2021)
12. Liu, L., Chen, L.: Characteristic analysis of a chipless RFID sensor based on multi-parameter sensing and an intelligent detection method. Sensors **22**(16), 6027 (2022)
13. Su, C., Zou, C., Jiao, L., Zhang, Q.: A MIMO radar signal processing algorithm for identifying chipless RFID tags. Sensors. **21**(24), 8314 (2021)
14. Noman, M., Haider, U.A., Ullah, H., Hashmi, A.M., Tahir, F.A.: Realization of chipless RFID tags via systematic loading of square split ring with circular slots. IEEE J. Radio Freq. Identif. **6**, 671–679 (2022)
15. Noman, M., Haider, U.A., Hashmi, A.M., Ullah, H., Najam, A.I.: A novel design methodology to realize a single byte by loading a square open-loop resonator with micro-metallic cells. IEEE J. Microwav. **3**(1), 43–51 (2023)

16. Noman, M., et al.: 12-bit chip-less RFID tag with high coding capacity per unit area. In: Proceedings of the 2022 IEEE International Symposium on Antennas and Propagation and USNC-URSI Radio Science Meeting, pp. 127–128 (2022)
17. Noman, M., Haider, U.A., Ullah, H., Tahir, F.A., Rmili, H., Najam, A.I.: A 32-bit single quadrant angle-controlled chipless tag for radio frequency identification applications. Sensors. **22**(7), 2492 (2022)
18. Babaeian, F., Karmakar, N.C.: Compact multi-band chipless RFID resonators for identification and authentication applications. Electron. Lett. **56**(14), 724–727 (2020)
19. Haider, U.A., Noman, M., Rashid, A., Rmili, H., Ullah, H., Tahir, F.A.: A semi-octagonal 40-bit high capacity chipless RFID tag for future product identification. Electronics **12**(2), 349 (2023)
20. Svanda, M., Machac, J., Polivka, M.: Constraints of using conductive screen-printing for chipless RFID tags with enhanced RCS response. Appl. Sci. **13**(1), 148 (2022)
21. Ayadi, H., Machac, J., Svanda, M., Boulejfen, N., Latrach, L.: Proof of concept of reconfigurable solvent vapor sensor tag with wireless power transfer for IoT applications. Appl. Sci. **12**(20), 10266 (2022)
22. Casula, G.A., Montisci, G.: A flexible narrowband multiresonator for UHF RFID chipless tag. In: Microwave Mediterranean Symposium (MMS), Pizzo Calabro, Italy, pp. 1–4 (2022)
23. Finkenzeller, K.: RFID HANDBOOK: Fundamental sand Applications in Contact-less Smart Cards and Identification, 2nd edn. Wiley, Hoboken, USA (2003)
24. Polivka, M., Havlicek, J., Svanda, M., MacHac, J.: Improvement in robustness and recognizability of RCS response of U-shaped strip-based chipless RFID tags. IEEE Antennas Wirel. Propag. Lett. **15**, 2000–2003 (2016)
25. Islam, M.A., Yap, Y., Karmakar, N., Azad, A.K.M.: Orientation independent compact chipless RFID tag. In: Proceedings of the IEEE International Conference on RFID-Technologies and Applications, RFID-TA 2012, France, pp. 137–141. IEEE (2012)
26. Vena, A., Sydänheimo, L., Tentzeris, M., Ukkonen, L.: A novel inkjet printed carbon nanotube-based chipless RFID sensor for gas detection. In: 2013 European Microwave Conference, pp. 9–12 (2013)
27. Rather, N., Simorangkir, R.B.V.B., Buckley, J., O'Flynn, B.: Flexible and semi-transparent chipless RFID tag based on PDMS-conductive fabric composite. In: 2022 International Workshop on Antenna Technology (iWAT), Dublin, Ireland, pp. 33–36 (2022)
28. Helmy, A.M., El-Khobby, H.A., Hussein, A.H., Nasr, M.E.: A new design of high capacity chipless RFID tag using size scaled arrays of hairpin resonators. In: 2022 International Telecommunications Conference (ITC-Egypt), Alexandria, Egypt, pp. 1–4 (2022)
29. Noman, M., Haider, U.A., Hashmi, A.M., Ullah, H., Najam, A.I., Tahir, F.A.: a novel design methodology to realize a single byte chipless RFID tag by loading a square open-loop resonator with micro-metallic cells. IEEE J. Microwav. **3**(1), 43–51 (2023)
30. Zafar, S., Rubab, S., Ullah, H., Tahir, F.A.: A 12-bit hexagonal chip-less tag for radio frequency identification applications. In: IEEE International Symposium on Antennas and Propagation and North American Radio Science Meeting, pp. 1503–1504 (2020)
31. Salemi, F., Hassani, H.R., Mohammad-Ali-Nezhad, S.: Linearly polarized compact extended U-shaped chipless RFID tag. AEU – Int. J. Electron. Commun. **117**, 153129 (2020)
32. Ali, A., Williams, O., Lester, E., Greedy, S.: High code density and humidity sensor chipless RFID tag. In: 7th International Conference on Smart and Sustainable Technologies (2022)
33. Habib, A., Akram, S., Ali, M.R., Muhammad, T., Zainab, S., Jehangir, S.: Radio frequency identification temperature/CO2 sensor using carbon nanotubes. Nanomaterials **13**(2), 273 (2023)

Wearable Ultraviolet Photodetector for Real Time UV Index Monitoring

Adnin Tazrih Natasha[1](✉), Xiaohu Chen[1], Binesh Puthen Veettil[2], and Noushin Nasiri[1]

[1] NanoTech Laboratory, School of Engineering, Macquarie University, Sydney, Australia
{adnin-tazrih.natasha,jayden.chen}@hdr.mq.edu.au,
noushin.nasiri@mq.edu.au

[2] School of Engineering, Macquarie University, Sydney, Australia
binesh.puthenveettil@mq.edu.au

Abstract. We present the successful development of an Ultraviolet (UV) index monitoring device for determining the daily exposure of an individual to sunlight. The core of the monitor is the calibrated measurement of the UV ray's incident on the UV nano sensor that was fabricated by self-assembling the ZnO nanoparticle aerosols on a glass substrate featuring platinum interdigitated electrodes. Using Bluetooth low energy 5.0 communication system, the device was connected to the companion mobile application named "UV Index Monitor". The UV index level varies from 1 to 10, where 1 stand for safe UV radiation and 10 stands for extremely harmful UV exposure. The alert messages will be displayed on the companion application to communicate with the end-users to manage their exposure to UV rays. The device is designed in a wearable watch frame with the dimensions of 5.0 × 3.0 × 2.0 cm and is powered by a 3.7 V rechargeable battery. The UV index data stored in the mobile application can be shared through emails or other platforms for storage.

Keywords: UV index monitoring · Wearable Technologies · Nano Sensors · Bluetooth low energy · Zinc Oxide · Wireless

1 Introduction

UV radiation is a type of electromagnetic radiation that is invisible to the human eye [1]. It falls in the range of the electromagnetic spectrum between X-rays and visible light [2, 3]. There are three types of UV radiation: UVA, UVB, and UVC. UVA radiation has the longest wavelength and is the least energetic of the three types [4]. It can penetrate deep into the skin and cause damage to the skin's connective tissue, leading to wrinkles, age spots, and other signs of aging [5, 6]. UVA radiation can also contribute to the development of skin cancer [7, 8]. UVB radiation has a shorter wavelength and more energy than UVA radiation. It is responsible for sunburn and is a major cause of skin cancer [9]. UVB radiation can also damage the eyes, leading to cataracts and other eye problems [10, 11]. UVC radiation has the shortest wavelength and the most energy of

N. K. Suryadevara et al. (Eds.): ICST 2022, LNEE 1035, pp. 238–250, 2023.
https://doi.org/10.1007/978-3-031-29871-4_24

the three types. It is mostly absorbed by the ozone layer in the Earth's atmosphere and does not reach the surface.

Generally, UV radiation is required for the body in small amounts, as it helps the body produce vitamin D. Vitamin D is essential for maintaining healthy bones and teeth, as it helps the body absorb calcium. Vitamin D deficiency can lead to a number of health problems, including rickets in children and osteoporosis in adults [12]. The body produces vitamin D when the skin is exposed to UVB radiation from sunlight. However, it is important to balance the need for vitamin D with the risks associated with excessive UV exposure. Getting a moderate amount of sun exposure, such as spending 15–20 min in the sun a day (recommended by World Health Organisation) [13], is generally enough for the body to produce the vitamin D it needs.

It's important to note that excessive exposure to UV radiation can cause various health problems, including sunburn, skin cancer, cataracts, and other eye problems [14]. Wearable UV detection devices, such as wristbands or clothing, [15–19] can help individuals monitor their exposure to UV radiation and take steps to protect themselves. These devices can also provide real-time data on UV exposure, allowing people to make more informed decisions about when to seek shade or apply sunscreen [20, 21]. Additionally, wearable UV detectors can be particularly useful for people who spend a lot of time outdoors, such as athletes, outdoor workers, and beachgoers.

Wide bandgap semiconductors are a class of materials that have a large energy gap between the valence band and the conduction band. Because of this large energy gap, they are able to absorb and detect UV radiation more efficiently than traditional semiconductors [22, 23]. This is because the energy gap of wide bandgap semiconductors is greater than the energy of UV photons. As a result, wide bandgap semiconductors can be used as UV photodetectors, which are devices that convert UV radiation into electrical signals. Materials such as Gallium Nitride (GaN), Titanium dioxide (TiO_2), Silicon carbide (SiC), and Zinc Oxide (ZnO) are popular wide bandgap semiconductors used as UV photodetectors [20, 22, 24–27]. They are used in a variety of applications including UV spectroscopy, UV sensing, and UV imaging. Some of the advantages of using wide bandgap semiconductors as UV photodetectors include high sensitivity, high speed, and high radiation resistance.

2 Materials, Devices and Methods

This section describes the materials and fabrication techniques used to develop the UV nanosensor which is the core of the UV index monitoring device.

2.1 Materials

Xylene (Sigma Aldrich) was used to dilute Zinc Naphthenate (10% Zn, Strem Chemicals) with a total Zinc atom content of 0.3 mol L^{-1} for preparing the Zinc precursor solution.

2.2 Fabrication Technique

Figure 1 shows the flame spray pyrolysis technique, which is used herein to fabricate ZnO-based UV sensors. More details about this nanofabrication technique is reported in our previous works [4, 24, 28, 29]. A syringe pump was used to deliver the prepared solution at a rate of 5 mL min^{-1} while dispersing it into a thin spray by 5 L min^{-1} of oxygen at a continuous pressure drop of 2 bars. The pilot flame was ignited by supporting premixed methane/oxygen flames (CH_4 = 1.25 L·min^{-1}, O_2 = 1.3 L·min^{-1}). A water-cooled substrate holder placed at 12 cm height above the burner (HAB) was utilised to keep the substrate temperature below 220 °C. The ZnO nanoparticle aerosols were orthogonally impinged on the water-cooled glass substrates resulting in the rapid self-assembly of ultraporous nanoparticle films. The sensor substrates were made of glass with interdigitated platinum lines with 5 μm width and spacing and an active circular area of 3.5 mm in diameter (ED-IDE3-Pt, Micrux Technologies). The prepared thin films were annealed (Compact muffle/tube 2-in-1 furnace, Zhengzhou TCH) at 300 °C for 12 h at ambient pressure.

2.3 Sensor Testing Apparatus

DC photocurrents were measured using a source meter unit (Model 2450, Keithley). Mounted LED (M375L4, ThorLabs) were used as the UV light source to provide 375 nm of UV illumination at different power density. The power density of the light illumination was controlled by LED driver (DC2200, ThorLabs). The current of the LED was varied along 20 mA, 40 mA, 80 mA and 150 mA to generate a light density (LD) of 1.22 mW cm^{-2}, 2.18 mW cm^{-2}, 4.09 mW cm^{-2} and 7.43 mW cm^{-2}, receptively (Table 1). All measurements were carried out at room temperature under atmospheric condition. The source meter provides a voltage of 3.3 V across the sensor and the sensitivity of the sensor is tested by determining the resistance change during the LED OFF and ON cycles.

Table 1. LED current and the corresponding power and irradiation light density at room temperature (25 °C).

I_{LED} (mA)	P_{LED} (mW)	LD (mW cm^{-2})
20	0.96	1.22
40	1.71	2.18
80	3.21	4.09
150	5.83	7.43

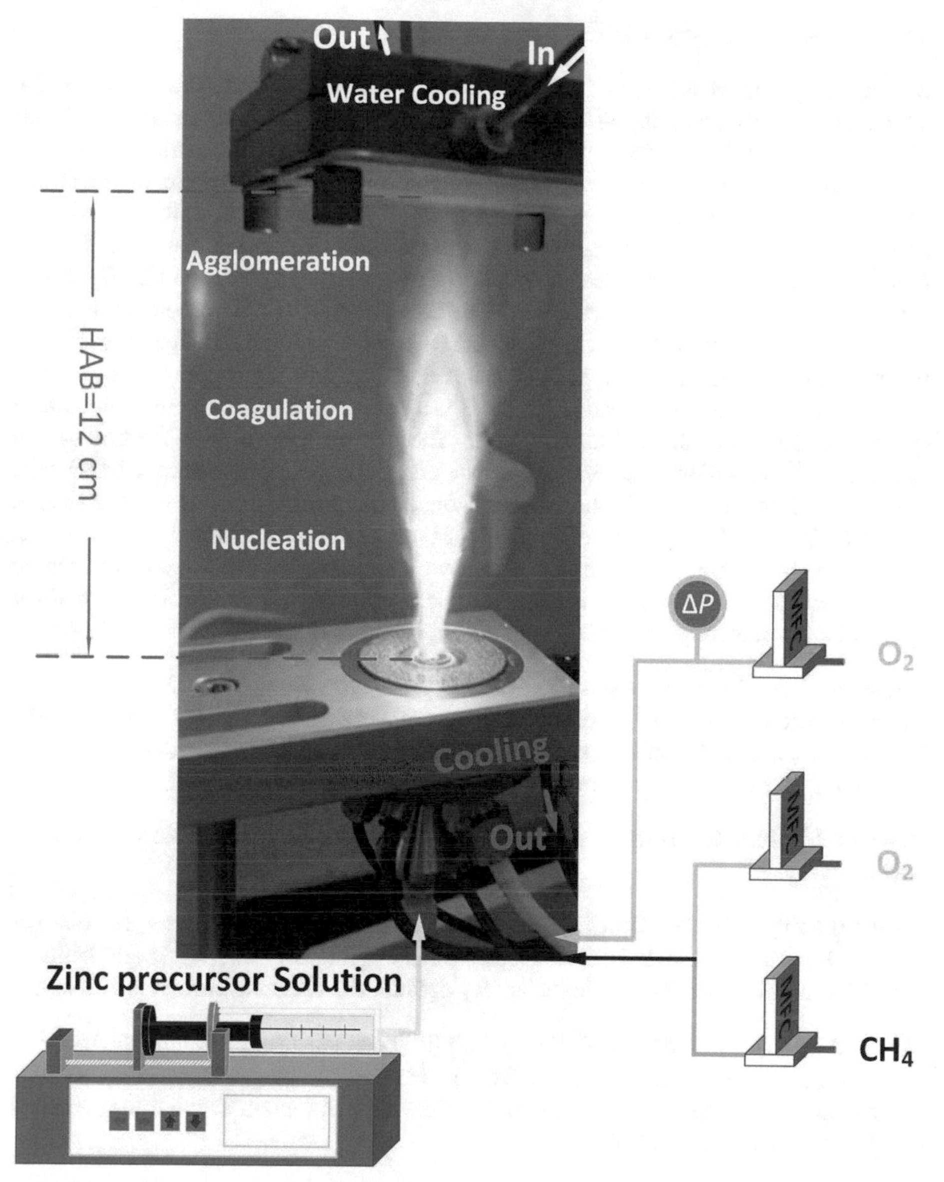

Fig. 1. Nanosensor fabrication using flame spray pyrolysis (FSP) technique

3 UV Sensor System

The UV index monitoring device has been designed in the basis of ohm meter circuit with Arduino microcontroller. The component has been selected on the basis of design structure and desired functionality of the device. In this section, the design electrical components, connections and mobile application will be discussed in detail.

3.1 Electronic Components

Arduino Nano 33 BLE Sense. The Nordic Semiconductors nRF52840, a 32-bit ARM® Cortex®-M4 CPU operating at 64 MHz, is the processor used in the Arduino Nano 33 BLE Sense, a development of the original Arduino Nano. The Arduino Nano 33 BLE sense not only is a very well form microcontroller board with splendid technical specifications but also number of embedded sensors are included in this board. The ability to run Edge Computing applications (AI) utilising TinyML is the key feature of this board, in addition to the remarkable array of sensors. The Nano 33 BLE Sense's communications chipset supports Bluetooth® Low Energy and Bluetooth® client and host devices. The purpose of selecting Nano 33 BLE sense is the primary feature of having Bluetooth low energy in the same microcontroller chip and the Bluetooth low energy makes the device more power sufficient. BLE is a type of wireless communication made specifically for short-range communication. The current consumption for this device under no UV Radiation is around 22.22 mA and during sleep mode it is 18.89 mA. Under UV illumination, the highest current consumption for the device reaches around 37 mA to 45 mA. The selection of nano 33 BLE sense not only makes the device smart in communication but also makes it possible weight for wearable device which can be wear or mount on the body and the user can carry it everywhere. BLE is designed for applications where long battery life is more important than fast data transfer rates.

Resistors. The ohm meter circuit constructed with Arduino measure the voltage drop and resistance of the unknown component by applying voltage divider formula in the circuit. Here, a 220-Ω resistor has been used to pass the current in the circuit and hence to obtain appropriate values from the sensor.

Polymer Lithium Ion Battery (LiPo). A 3.7 V 120 mAh rechargeable LiPo battery powers the system.

Adafruit Micro LiPo w/MicroUSB Jack – USB LiIon/LiPoly Charger. The charger has 5 V input via Micro-B USB connector. It is eligible for charging single Lithium Ion/Lithium Polymer 3.7/4.2 V batteries. By default, the charging current is 100 mA.

SPDT Slide Switch. The SPDT slide switch has been used for switching the mode of the device from operation to charging. The device will be charger if the switch is slide to the connection towards the charger and the device will remain active if the switch is being slide to the power mode.

Vivo Android 12 Smartphone. VIVO Y11s Dual Sim smartphone has been used for this project. The android smartphone makes it easier for mobile application for any projects easier through android operating system. It is easier for primary mobile application development in android platform other than iOS operating system.

3.2 Device Design

Figure 2a shows the device connection diagram where R_1 is the known resistor of value 220 Ω, R_2 is the UV nano sensor (resistivity across the sensor is measured for determining

UV radiation), V_{out} is the Analog (A_0) pin of the Arduino, V_{in} in the input of the Arduino and GND is the common ground. The UV nano sensor (R_2) is connected to the GND pin of the Arduino in one end and the other end is connected in series with the R_1—a known value resistor placed at the A_0 pin of the Arduino. The Arduino is powered by a 3.7 V battery connected to the SPDT switch 1 and the negative wire of the battery is connected to the GND pin of the board. The Battery charger positive wire is connected to the SPDT switch 2 and the negative with the GND pin. The SPDT switch 1 toggle between battery and the charger. The switch 1 connects with switch 2 to power the device through battery and the switch 1 connects with switch 3 to connect the charger with the battery for charging mode. The device does not operate in charging mode and vice versa. The device has been mounted on the wrist similar to a hand watch (Fig. 2b, c).

Mobile Application. The app was created using Flutter SDK and the C++ -like Dart language. Android Studio has been used as the IDE, and Dart and Plugin have been installed. Data storing software for the mobile application has been implemented using Google Firebase. User registrations and application data are stored in Firebase's Cloud Fire Store Database. As shown in Fig. 2d, e, the mobile application has a multi- page user interface. The user interface indicates the real time date, time, and day. The application indicates the data history page and store the data within the application. The data history appears date wise with a calendar date format. The application has data sharing option which allows the application to share the data with other platforms. The data can be refreshed from time to time. The user would be able to terminate the connection from disconnecting the application from the device anytime. It allows the users to create their individual profile and hence the data collection of the application is basically stored with specific user profile.

4 Results and Discussions

This section presents the test procedures and the performance of the UV nanosensor and the UV index monitor. These tests were performed both indoors and outdoors. The indoor tests were conducted under Thorlabs Laboratory modular optical tweezer setup, varying different LED current and consequently, different light density (Table 1). The outdoor tests were conducted under direct sunlight at different times and conditions of the day.

4.1 Sensor Testing Results

The sensor has been tested at four different light density of 1.22, 2.18, 4.09 and 7.43 $mW.cm^{-2}$ (Table 1) under the UV LED with the wavelength of 375 nm. The effect of temperature (25 and 40 °C) on the UV detection performance of the device has been evaluated (Fig. 3). For changing the temperature in the laboratory setup, mcte1-19913l-s Peltier module, 200 W, 50 × 50 × 3.5 mm has been used to obtain different testing temperatures. The dark current test was conducted by confining the device in the optical enclosure (XE25C11D/M, ThorLabs). Figure 3 depicts the dynamic resistance curves of

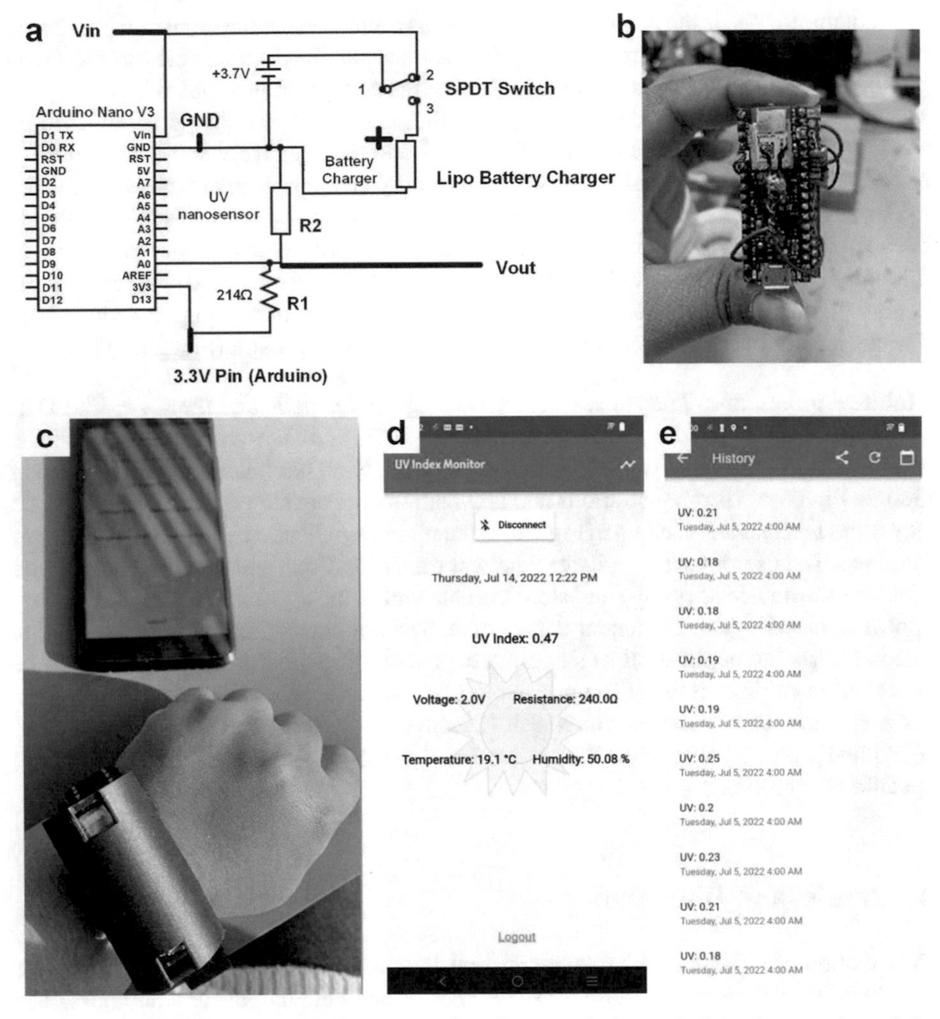

Fig. 2. a) UV index monitor device connection diagram, b,c) practical implementation of the circuit in a compact device structure., d) UV Index Monitor mobile application and e) data storage history.

the UV nanosensor exposed at different light density at 25 °C (Fig. 3a) and 40 °C (Fig. 3b). When the UV light was on, the resistance of the nanosensor decreased dramatically at first and gradually approached its minimum value, however, the resistance would quickly increase soon after the UV light was off and then slowly recover to the baseline (Fig. 3a, b).

Meanwhile, the higher light density of UV radiation would result in lower resistance in the photodetector, which could be attributed to the higher level of photogenerated charge carriers resulting in higher conductivity and lower resistance in the device. In fact, by increasing the light density from 1.22 mW.cm^{-2} to 7.43 mW.cm^{-2}, more charge carriers (photogenerated electrons and holes) were excited and consequently enhanced

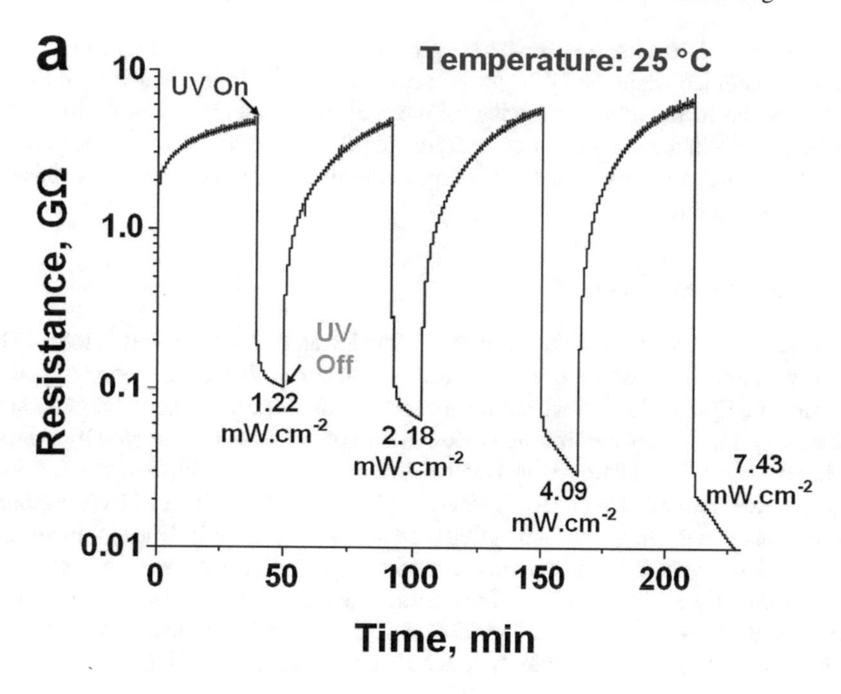

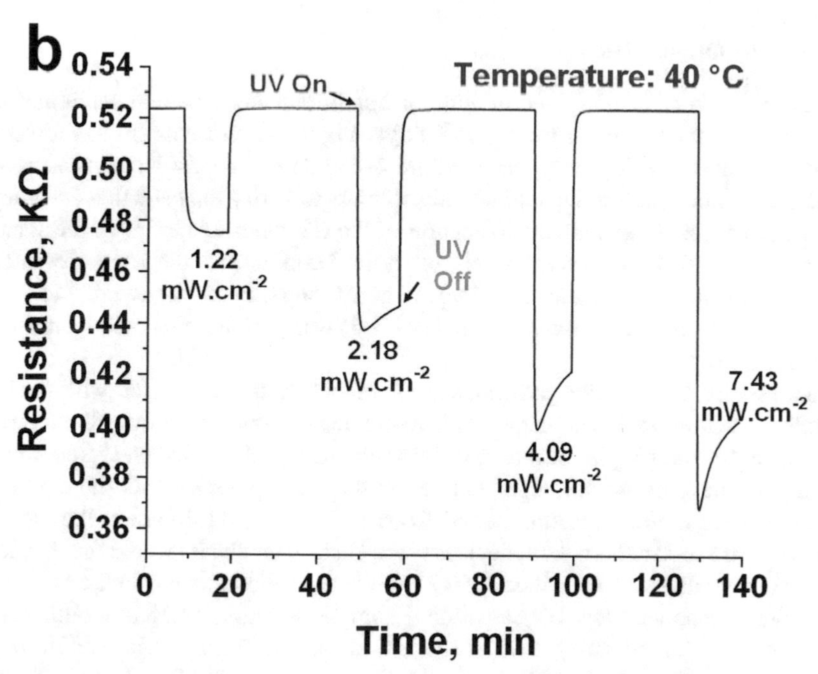

Fig. 3. Resistance versus Time graphs at different temperature of a) 25 °C, and b) 40 °C, at the light density (LD) of 1.22, 2.18, 4.09 and 7.43 mW.cm^{-2}, respectively.

the conductivity of the nanosensor. These photogenerated electron-hole pairs were ultimately recombined when the UV light was off, resulting in rapid reduction in the conductivity of the device, and showing the recovery of the resistance as a result. In addition, the nanosensor demonstrated a high stability at 40 °C with a faster response kinetics, indicating higher working temperature favors the interaction between the nanosensor and the environment.

4.2 Device Indoor Testing

The nanosensor embedded in the 3D printed watch frame was evaluated indoors (Thorlabs light setup) to verify its UV detection by comparing it with the commercial UV index meter (Digital Ultraviolet Radiometer model 6.5). Figure 4a shows the accuracy of the developed UV index monitoring device in comparison with commercially available UV index meter. The Digital Ultraviolet Radiometer device calibrates the UV index into numerical range between 0–11, where 1–2, 3–5, 6–7, 8–10 and 11 correspond to low, moderate, high, very high and extreme radiation, accordingly. The UV index monitoring device developed herein demonstrated a very similar performance compared to the commercially available device. This satisfactory performance was archived under different light density of 1.22, 2.18, 4.09 and 7.43 mW.cm^{-2}, indicating the capability of the fabricated system in accurately measuring the real-time UV index.

4.3 Device Outdoor Testing

The device has been also verified in outdoor conditions under natural sunlight during different times the day for a period of 7 days. Figure 4b presents the UV detection performance of the device during the daytime over 7 days. The UV index measurement was also carried out at evening and late-night hours to verify the detection accuracy of the device. The results explain the indication of the UV index according to the standard UV index range from 0–10 respectively of those 7 consecutive days. As presented in Fig. 4b, the device displayed an UV index of near 0 on rainy or mostly rainy days (day 1 and day 4). In contrast, a higher UV index (0–5) was measured on cloudy and sunny days (days 5–7).

Furthermore, the outdoor performance of the device in comparison with the commercially available UV index meter was investigated to evaluate the overall effective of the developed wearable platform in real-time monitoring of UV index. Figure 4c compares the accuracy of the developed UV index monitoring device with the digital UV radiometer, over a 30 min testing period from 10:50 am to 11:20 am in the morning. From 10:50 am to 11:02 am, both devices were kept under shade, where the developed UV index meter detected a UV index of 0.23 while the radiometer displayed a UV index of 0, which means very low UV radiation. From 11:04 am to 11:16 am, both devices were exposed to natural sunlight where the UV index monitoring device and the digital UV radiometer measured UV index of 4.34 and 5, respectively. The UV index monitoring device is designed calculate data for 10 min operational time and switch to sleep mode for 1 min. The upside-down peaks in the Fig. 4c depicts the switching of the device to sleep mode for 1 min for every 10 min of operation. The results suggested the

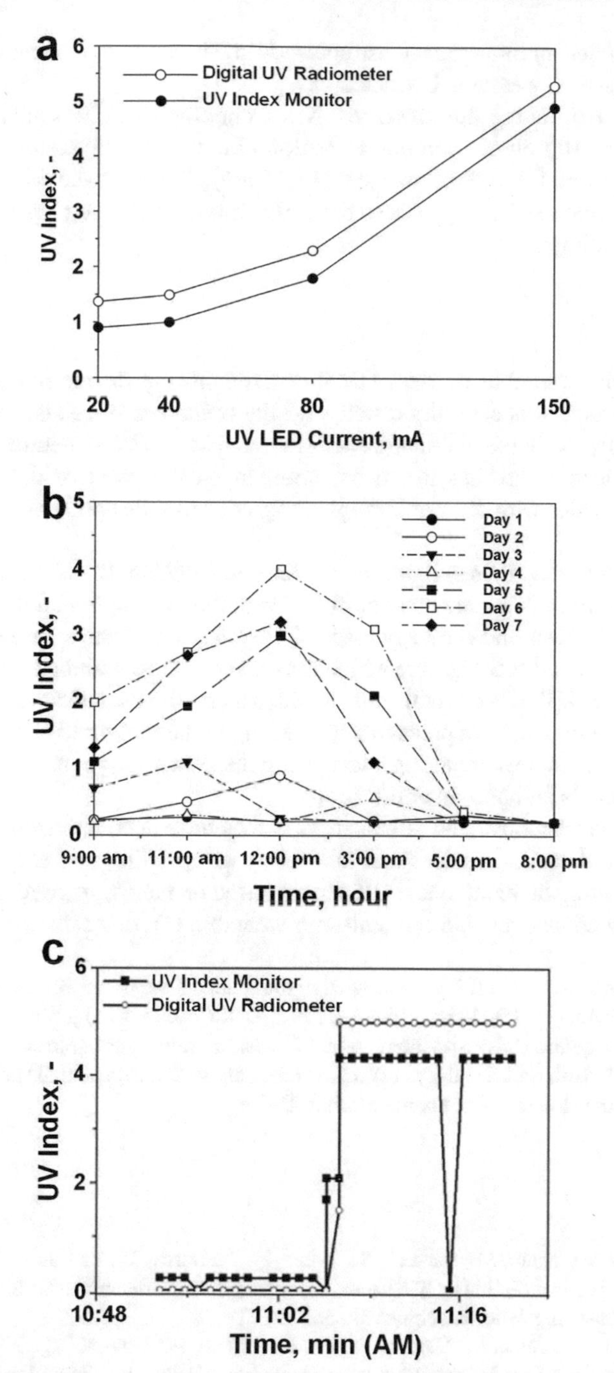

Fig. 4. a) Comparison of UV index results from this device and Digital Ultraviolet Radiometer under Thorlabs light setup in the laboratory. B) 7 days outdoor UV index by this device and c) Outdoor Comparison with Commercial UV Radiometer.

flame-made UV sensor based on nanostructured ZnO is capable of being integrated as a reliable and real-time wearable UV detector.

It is worth mentioning that the developed technology can be switched into sleep mode after every 10 min of operation to prolong the battery life resulting in a sudden drop in the measured UV index down to 0 (11:15 am). However, the device displays the same UV index upon switching back to the active mode, indicating high stability of the developed technology.

5 Conclusion

In this paper, the wearable wireless UV index monitoring device has been designed and its performance was carefully tested, and the outcomes were also compared with the commercially available UV index Radiometer. The device structure is designed in the form of a watch and thus it can be mounted on the wrist of the end-user. The device communicates with the user through designated mobile application via Bluetooth connectivity.

By exposing to the UV/sun light, the device can calculate the UV index from 0–10 numerical range to indicate the amount of UV radiation throughout a day. Meanwhile, it consumes low power and can be operated 24 h a day and 7 days a week. The device is battery operated and rechargeable which makes it easier to maintain. Such device not only measures the UV index in real-time, but also alerts the user about the UV radiation via delivering the messages of precaution according to the measured UV index. Furthermore, it offers the features of storing and sharing the data through other communication methods like emails and pop-up notifications.

The device has demonstrated satisfactory performance in comparison with the established UV photodetection devices and laboratory setup with good accuracy and high stability, suggesting the flame-made UV sensor based on nanostructured ZnO is capable of being integrated as a reliable and real-time wearable UV detector.

Acknowledgment. N.N. and B.P.V acknowledge the financial support received from Macquarie University (MQRAS_21, ID: 174639164). A.T.N and X.C acknowledge the financial support received from Macquarie University international Research Training Program scholarship (iRTP). The authors thank Amir Safari and Ryan Wreyford for help with design and 3D printing the frame structure for the developed UV index monitoring device.

References

1. Cherniak, G., Avraham, M., Bar-Lev, S., Golan, G., Nemirovsky, Y.: Study of the absorption of electromagnetic radiation by 3D, vacuum-packaged, nano-machined CMOS transistors for uncooled IR sensing. Micromachines **12**, 563 (2021)
2. Bo, R., Nasiri, N., Chen, H., Caputo, D., Fu, L., Tricoli, A.: Low-voltage high-performance UV photodetectors: an interplay between grain boundaries and debye length. ACS Appl. Mater. Interfaces **9**, 2606–2615 (2017). https://doi.org/10.1021/acsami.6b12321
3. Bo, R., et al.: One-step synthesis of porous transparent conductive oxides by hierarchical self-assembly of aluminum-doped ZnO nanoparticles. ACS Appl. Mater. Interfaces **12**, 9589–9599 (2020). https://doi.org/10.1021/acsami.9b19423

4. Nasiri, N., Bo, R., Wang, F., Fu, L., Tricoli, A.: Ultraporous electron-depleted ZnO nanoparticle networks for highly sensitive portable visible-blind UV photodetectors. Adv. Mater. **27**, 4336–4343 (2015). https://doi.org/10.1002/adma.201501517
5. Cadet, J., Douki, T.: Formation of UV-induced DNA damage contributing to skin cancer development. Photochem. Photobiol. Sci. **17**(12), 1816–1841 (2018). https://doi.org/10.1039/c7pp00395a
6. de Gruijl, F.R.: UV adaptation: Pigmentation and protection against overexposure. Exp. Dermatol. **26**, 557–562 (2017)
7. Wu, S., et al.: Cumulative ultraviolet radiation flux in adulthood and risk of incident skin cancers in women. Br J. Cancer **110**, 1855–1861 (2014). https://doi.org/10.1038/bjc.2014.43
8. Pfeifer, G.P., Besaratinia, A.: UV wavelength-dependent DNA damage and human non-melanoma and melanoma skin cancer. Photochem. Photobiol. Sci. **1**(11), 90–97 (2012). https://doi.org/10.1039/c1pp05144j
9. Gonzaga, E.R.: Role of UV light in photodamage, skin aging, and skin cancer. Am. J. Clin. Dermatol. **10**, 19–24 (2009)
10. Hoffmann, R.T., Schmelz, M.: Time course of UVA- and UVB-induced inflammation and hyperalgesia in human skin. Eur. J. Pain **3**, 131–139 (1999). https://doi.org/10.1053/eujp.1998.0106
11. Furusawa, Y., Quintern, E.L., Holtschmidt, H., Koepke, P., Saito, M.: Determination of erythema-effective solar radiation in Japan and Germany with a spore monolayer film optimized for the detection of UVB and UVA – results of a field campaign. Appl. Microbiol. Biotechnol. **50**, 597–603 (1998). https://doi.org/10.1007/s002530051341
12. Holick, M.F.: Sunlight, ultraviolet radiation, vitamin D and skin cancer. Sunlight Vitam. D skin Cancer, 1–16 (**2014**)
13. Suman, G., Suman, S.: Ultraviolet radiation-induced immunomodulation: skin ageing and cancer. In: Dwivedi, A., Agarwal, N., Ray, L., Tripathi, A. (eds.) Skin Aging & Cancer, pp. 47–58. Springer, Singapore (2019). https://doi.org/10.1007/978-981-13-2541-0_5
14. Sánchez-Pérez, J., Vicente-Agullo, D., Barberá, M., Castro-Rodríguez, E., Cánovas, M.: Relationship between ultraviolet index (UVI) and first-, second-and third-degree sunburn using the Probit methodology. Sci. Rep. **9**, 1–13 (2019)
15. Shiv, Y., Kumar, R., Kumar, V., Wairiya, M.: A lightweight authentication scheme for wearable medical sensors. In: Proceedings of 2018 8th International Conference on Cloud Computing, Data Science & Engineering (Confluence), pp. 366–370 (2018)
16. Nasiri, N., Tricoli, A.: Advances in wearable sensing technologies and their impact for personalized and preventive medicine. In: Wearable Technologies, IntechOpen (2018)
17. Nasiri, N.: Wearable devices: the big wave of innovation; BoD–books on demand (2019)
18. Nakata, S., Arie, T., Akita, S., Takei, K.: Wearable, flexible, and multifunctional healthcare device with an ISFET chemical sensor for simultaneous sweat pH and skin temperature monitoring. ACS Sens. **2**, 443–448 (2017)
19. Lymberis, A., De Rossi, D.E.: Wearable Ehealth Systems for Personalised Health Management: State of the Art and Future Challenges, vol. 108. IOS press (2004)
20. Qiu, M., et al.: Visualized UV photodetectors based on Prussian blue/TiO_2 for smart irradiation monitoring application. Adv. Mater. Technol. **3** (2018)
21. Xie, Y., et al.: A self-powered UV photodetector based on TiO_2 nanorod arrays. Nanoscale Res. Lett. **8**, 1–6 (2013)
22. Nasiri, N., Jin, D., Tricoli, A.: Nanoarchitechtonics of visible-blind ultraviolet photodetector materials: critical features and nano-microfabrication. Adv. Opt. Mater. **7**, 1800580 (2019). https://doi.org/10.1002/adom.201800580
23. Nasiri, N., Tricoli, A.: Chapter 5 - Nanomaterials-based UV photodetectors. In: Thomas, S., Grohens, Y., Pottathara, Y.B. (eds.) Industrial Applications of Nanomaterials, pp. 123–149. Elsevier (2019). https://doi.org/10.1016/B978-0-12-815749-7.00005-0pp

24. Nasiri, N., Bo, R., Hung, T.F., Roy, V.A.L., Fu, L., Tricoli, A.: Tunable band-selective UV-photodetectors by 3D self-assembly of heterogeneous nanoparticle networks. Adv. Funct. Mater. **26**, 7359–7366 (2016). https://doi.org/10.1002/adfm.201602195
25. Li, D., et al.: Realization of a high-performance GaN UV detector by nanoplasmonic enhancement. Adv. Mater. **24**, 845–849 (2012)
26. Monroy, E., et al.: AlGaN-based UV photodetectors. J. Cryst. Growth **230**, 537–543 (2001)
27. Zhang, X., et al.: 3D-branched hierarchical 3C-SiC/ZnO heterostructures for high-performance photodetectors. Nanoscale **8**, 17573–17580 (2016)
28. Nasiri, N., Bo, R., Chen, H., White, T.P., Fu, L., Tricoli, A.: Structural engineering of nano-grain boundaries for low-voltage UV-photodetectors with gigantic photo- to dark-current ratios. Adv. Opt. Mater. **4**, 1787–1795 (2016). https://doi.org/10.1002/adom.201600273
29. Nasiri, N., Bo, R., Fu, L., Tricoli, A.: Three-dimensional nano-heterojunction networks: a highly performing structure for fast visible-blind UV photodetectors. Nanoscale **9**, 2059–2067 (2017). https://doi.org/10.1039/C6NR08425G

A Wearable Microwave Technique for Early Detection of Acute Respiratory Distress Syndrome (ARDS)

Adarsh Singh[1], Sreetama Gayen[1], Debasis Mitra[1](✉), Partha Basuchowdhuri[2], Bappaditya Mandal[3], and Robin Augustine[3]

[1] Department of Electronics and Telecommunication Engineering, IIEST, Shibpur, India
debasis.mitra.telecom@faculty.iiests.ac.in
[2] School of Mathematical and Computational Science, IACS, Kolkata, India
[3] Division of Solid State Electronics, Department of Electrical Engineering, Ångström Laboratory, Uppsala University, 75121 Uppsala, Sweden

Abstract. This work targets the early detection of Pulmonary edema and acute respiratory distress syndrome (ARDS). This Microwave technique utilizes a simple, planar, non-invasive, low-cost, non-ionizing, portable wearable system which is capable to diagnose the abnormal buildup of fluid inside the lungs and predict the severity of disease using the transmission coefficient between antennas. Early prediction of disease or diagnosis at home provides better preliminary inputs for physicians or doctors that enables the opportunity for better treatment or care of the patient.

Keywords: ARDS · Pulmonary edema · Disease Detection · RBR · Bow-Tie Antenna

1 Introduction

Humans' urge to understand diseases and develop techniques to diagnose them, can be traced back to prehistoric times, and it has been an ever-evolving element of science for decades. The lungs are the most important part of the respiratory system of the human body. Developments and accumulations of watery fluid in the lungs are well-known in the medical field as symptoms of medical or surgical problems, such as heart failure, and accumulation of fluid is a fatal medical condition itself. Pulmonary edema affects roughly one in every 15 people aged 75 to 84, and just over one in every seven people aged 85 and older [1]. Shortness of breath develops, if this illness is not addressed promptly, and leads to acute respiratory distress syndrome (ARDS) which is a kind of lung failure. The chest radiograph is likely the most extensively used tool for the clinical assessment of pulmonary edema among established methods. Although radiography (x-ray scan/CT scan) is widely used, its ionizing property and lack of sensitivity prevent its application for a long time or continuous monitoring [2]. As a result, microwave-based biomedical diagnostic devices started being developed as safe, inexpensive, and portable

N. K. Suryadevara et al. (Eds.): ICST 2022, LNEE 1035, pp. 251–258, 2023.
https://doi.org/10.1007/978-3-031-29871-4_25

diagnostic and monitoring techniques. The use of microwaves for lung fluid detection was originally examined in 1973, [3] when it was discovered that water content within the lungs changes the position of the node of the standing wave due to alteration in the lungs' effective dielectric constant. This principle of microwave detection technique is broadly utilized in the detection of cancerous anomalies or pulmonary edema because of higher water content.

This work targets the diagnosis of the abnormal build-up of fluid in case of pulmonary edema or acute respiratory distress syndrome (ARDS). The presented wearable system is composed of two wearable antennas with resonance-based reflector (RBR) having a unidirectional pattern and wideband nature operating in the frequency range of 1.5–3.13 GHz (70.5%) with a provided gain of 5.7 dBi.

The viability of a simple wearable system that may be used at home to detect or monitor fluid buildup in the lungs is investigated using a Preliminary diagnostic system, it has a set of two antennas with a portable Vector network analyzer (VNA) only, which makes it low in cost and feasible for diagnosis at home also. This research is founded on the premise that the transmission coefficient of antennas varies when the dielectric properties of the medium between the transmitting and receiving antennas change [4]. Both antennas can be worn on the body either in front and back of the chest or sideways of the chest and transmission coefficients (both magnitude and phase) can be examined, any sudden change in magnitude or phase of transmission coefficient would be the indication of the abnormal buildup of fluid inside the lungs.

Previously reported works [5, 4] for fluid accumulation detection had some inadequacy so this presented work fulfills that research gap and a comparison has been shown in Table 1.

Table 1. Comparison with previously reported works

Ref No.	Transmitting and receiving antennas	Antenna bandwidth	Voxel model used	Feasibility for real-time application
[5]	Different	Narrowband	Full Human body	No (Dimension of narrowband antennas will change concerning every patient)
[4]	Same	Narrowband	Only Lungs	No (Only lungs were taken into consideration)
This work	Same	Wideband	Full Human Body	Yes (Antennas are wideband and a full human body model has been taken into consideration for design and simulations)

2 Antenna Design

The reported work had a specific requirement of a low-profile antenna with a unidirectional radiation pattern and wideband in nature. Microstrip patch antennas have the characteristics of a unidirectional radiation pattern and a low profile generally. However, the high-quality factor (Q) of MSA restricts the bandwidth and makes it a narrowband antenna [6]. Planar antennas exhibit very wideband and low-profile attributes due to the low-quality factor (Q) but their radiation patterns are omnidirectional [7]. To overcome the limitations mentioned above, a ring shape reflector has been designed [8] which works on the resonant effect of a microwave resonator.

2.1 Design of Resonance Based Reflector

A ring-based RBR was reported, [8] and similar has been used in this reported work as shown in Fig. 1(b), where r_1 is the outer radius of the ring and the width is w. The RBR has been designed on the RT duroid 6002 $\varepsilon_r = 2.94$ ubstrate of 0.8 mm thickness. This design has been designed and simulated using CST microwave studio [9].

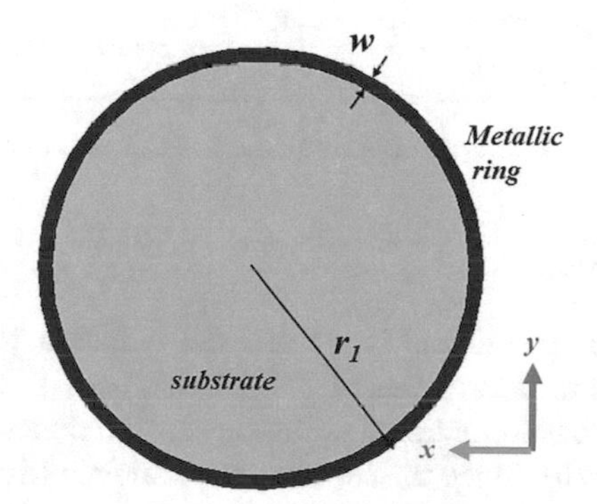

Fig. 1. Design of Resonance-based Reflector (RBR)

2.2 Design of Bow-Tie Antenna

RBR-based antenna [8] was designed for unidirectional and wideband operation for various applications such as indoor communication, remote sensing, and wireless sensor systems. Since it has the above-mentioned attributes thus in this reported work, it has been used for the detection of lung abnormalities.

In the Proposed design, a bow tie antenna surrounded by the metallic ring structure has been taken [10] and optimized using parametric analysis for wideband application (1.5–3.13 GHz) covering the 2.45 GHz ISM band. The design and dimensions of the antenna have been presented in Fig. 2 and Table 2; respectively.

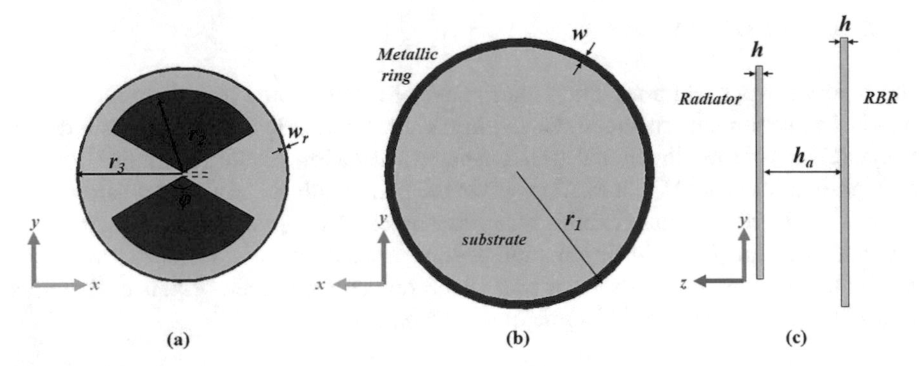

Fig. 2. Design of RBR-based bowtie antenna (a) Top view, (b) Bottom view, and (c) Side view

Table 2. Antenna dimensions.

Parameter	Dimension	Parameter	Dimension
r_1	33 mm	w_r	0.1 mm
r_2	18.75 mm	h	0.8 mm
r_3	26 mm	h_a	13.4 mm
w	2 mm	φ	120°

2.3 Antenna Parameters

The simulated reflection coefficient (S_{11}) and the radiation pattern are presented in Fig. 3, wideband nature of the antenna (1.5–3.13 GHz) is visible in Fig. 3(a) and from the radiation pattern plot, unidirectional radiation characteristics are evident with a peak gain of 5.7 dBi in Fig. 3(b). Since the Reflection coefficient (S_{11}) for the antenna has a minimum value of –20 dB at the resonant frequency it provides a low chance of noise interference in the operating frequency range.

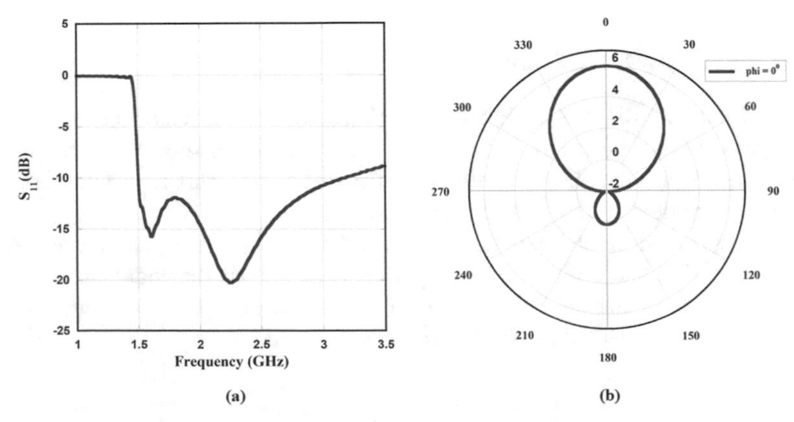

Fig. 3. Antenna parameters (a) Reflection coefficient (b) Radiation pattern of proposed RBR Bowtie antenna at f = 1.6 GHz.

3 Methodology

Two RBR-based bow-tie antennas are placed as wearable in the middle of the chest (one on the front side of the chest and another on the back of the chest) as shown in Fig. 4. One of them works as a transmitting antenna (Ant. 1) and other operates as receiving antenna (Ant. 2). The antenna is applied on the human chest and back areas, with the expectation that it will adhere to the body surface. As it could be placed in slightly different positions on the chest/back areas in practical terms (taking into account individuals with different body shapes and sizes), the effective permittivity of the environment exposed to the near field of the antenna can also vary, and so the resonant frequency [11]. A wideband antenna was chosen to overcome the same.

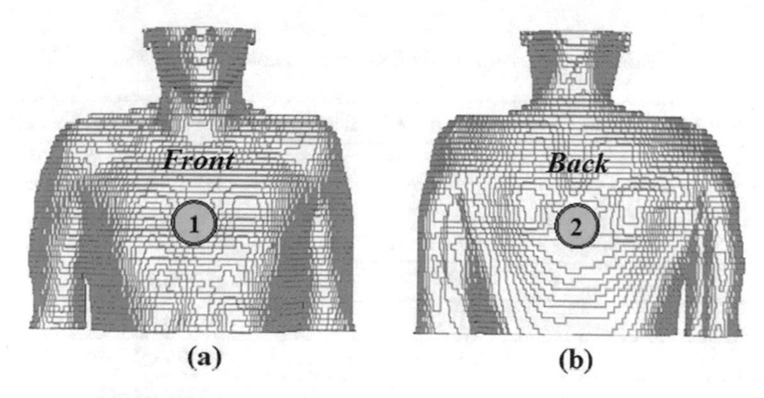

Fig. 4. Two-sensor system (a) Front view and (b) Back view

In the case of pulmonary edema or ARDS, the Alveoli of the lungs get filled with a watery fluid, which doesn't accumulate at only one position but spreads inside the lungs, unlike lung cancer. A single alveolus is estimated to be roughly 4.2×10^{-3} mm^3 in volume, and there are on average 480 million alveoli present in the human lungs. To achieve an analytical model of fluid spread, small water spheres have been introduced in the human body model inside the lung area which can be seen in Fig. 5.

To increase the severity or extent of illness, the radius of watery fluid spheres has been varied. The possible range of radius is 0 mm (for healthy lungs) to 7.5 mm (most severely ill) where the lung volume is considered 90% filled with watery fluid. It is assumed that diagnosis in the lower range of fluid accumulation (30–40% of total lung volume) is more crucial, the present work is focused on early detection or diagnosis at home. Beyond the above-mentioned limit, patients are supposed to get admitted into the hospital for proper treatment and ventilator support if required. The radius of fluid spheres vs parentage volume of lung occupied is listed in Table 3.

Using the simulation platform CST Microwave studio [7], design and simulation have been performed, and using its parametric sweep analysis, variation in the radius of fluid spheres could be done and Transmission coefficients S_{21} (magnitude and phase) have been recorded. Analysis and discussion of the changes in fluid accumulation level and their effect on transmission coefficients have been presented in the next section.

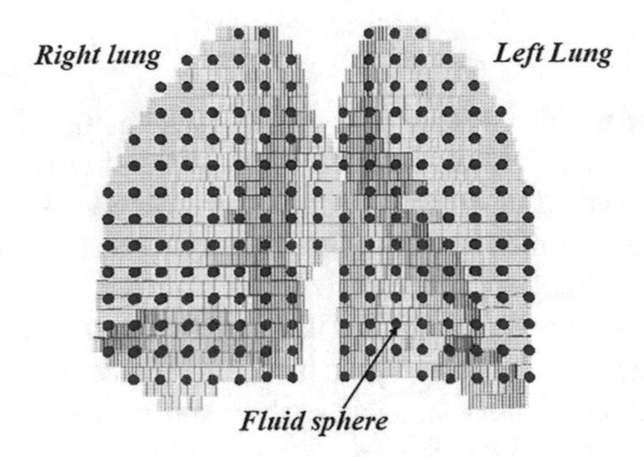

Fig. 5. Analytical model of edema fluid inside the lungs

Table 3. The radius of fluid spheres vs volume occupied

Volume occupied of the lung (%)	Radius of a sphere (mm)	Volume of a sphere (mm^3)
10	3.16	132.17
20	4.54	391.97
30	5.2	588.97
40	5.75	796.32
50	6.17	983.88
60	6.55	1177.09
70	6.9	1376.05
80	7.21	1569.98
90	7.5	1767.14

4 Results and Discussion

Since this work is focused on the detection of pulmonary edema or ARDS in the early stage so radius of the spheres has been varied for lung volume occupied from 0% (healthy) to 60% (severe) and variation in transmission coefficient S_{21} (mag) and S_{21} (phase) have been plotted in Fig. 6(a) and (b), respectively.

Variation in the magnitude of S_{21} is evident as it decreases with increasing fluid percentage. Since its value is very small 0.0001 to 0.0006 (in case of absolute value) or –100 to –55 dB (in logarithmic or dB scale), it could be a problem for measurement, and chances for error are also possible. So, the phase of Transmission Coefficient S_{21} (phase) is also plotted below, and it is observed that an apparent variation occurs with respect to fluid percentage.

Since phase variation of S_{21} is very evident and decreases gradually from 115° (healthy) to –143° (60% fluid accumulation) at 1.76 GHz, this parameter is more reliable to predict the existence of the disease.

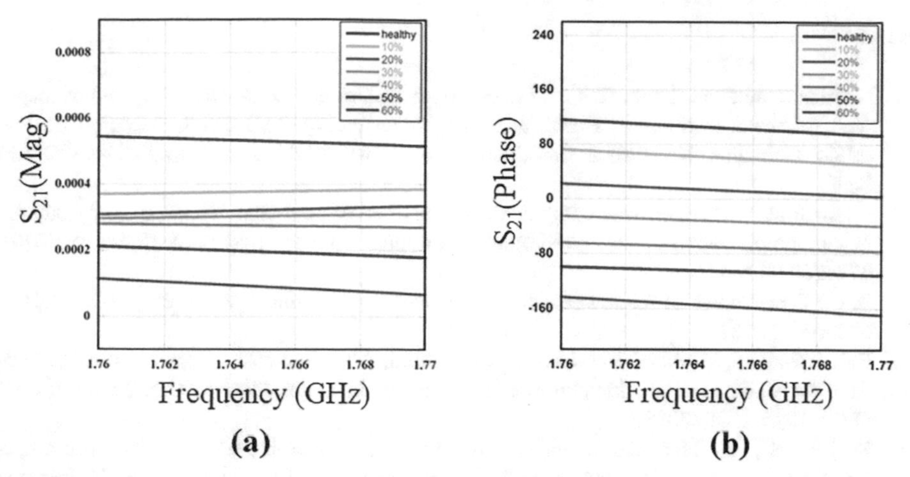

Fig. 6. Transmission Coefficient (a) S_{21}(mag) and S_{21}(phase) variation with an increment of fluid percentage.

The proposed system can predict the range of severity of the disease based on the variation of the Transmission Coefficient (S_{21}). With the combined observation of both parameters S_{21} (mag) and S_{21} (phase), a more accurate diagnosis can be done.

5 Conclusion and Future Scope

Pulmonary edema is a symptom of many diseases and it is a fatal condition in itself. Acute respiratory distress syndrome (ARDS) drew the attention of Medicos and researchers when Coronavirus hit the world in 2019, Covid-19 ARDS usually develops within a few hours to a few days which makes early detection more crucial and important. The severity of mentioned diseases has been successfully diagnosed using the proposed microwave wearable system. The outcomes in this research section describe that a simple wearable two-antenna system for the detection of lung abnormality is feasible despite the complex structure of the lung area. It is a simple, low-cost, convenient, and accurate system consisting of two low-profile antennas having a unidirectional radiation pattern (5.7 dBi peak gain) and wideband nature (1.5–3.13 GHz). In the future, Machine learning and Artificial intelligence will be utilized to predict more accurately different complications that can occur from respiratory distress, such as blood clots or deep vein thrombosis, muscle weakness, etc. Simulations and measurements will be performed for different body models available in CST and a human phantom will be developed to test the proposed system in different use cases and scenarios. The premise of this research work could serve as a stepping stone for future research on wearable systems that can detect abnormalities in different organs of the human body.

Acknowledgement. This work acknowledges "AI-based Detection of Acute Respiratory Distress Syndrome (AI-DARDS)", Dr. no. 2020–03612, Indo-Swedish DBT - Vinnova project.

References

1. National Guideline Centre (UK): Chronic Heart Failure in Adults: Diagnosis and Management. National Institute for Health and Care Excellence (UK), London, September 2018. (NICE Guideline, No. 106) 2, Introduction. https://www.ncbi.nlm.nih.gov/books/NBK536089/
2. Snashall, P.D., et al.: The radiographic detection of acute pulmonary oedema. A comparison of radiographic appearances, densitometry and lung water in dogs. Br. J. Radiol. **54**(640), 277–288 (1981)
3. S, C.: Possible use of microwaves in the management of lung disease. Proc. IEEE **61**(5), 673–674 (1973)
4. Bait-Suwailam, M.M., Al-Busaidi, O., Al-Shahimi, A.: A low-cost microwave sensing platform for water accumulation abnormality detection in lungs. Microw. Opt. Technol. Lett. **60**(5), 1295–1300 (2018)
5. Katjana, K., Sayrafian, K., Bengi, U., Dumanli, S.: A wearable wireless monitoring system for the detection of pulmonary edema. In: IEEE Global Communications Conference (GLOBECOM), pp. 1–5 (2021)
6. Peng, L., Xie, J.Y., Jiang, X., Li, S.M.: Wideband microstrip antenna loaded by elliptical rings. J. Electromagn. Waves Appl. **30**(2), 154–216 (2016)
7. Novak, M.H., Volakis, J.L.: Ultrawideband antennas for multiband satellite communications at UHF–Ku frequencies. IEEE Trans. Antennas Propag. **63**(4), 1334–1341 (2015)
8. Peng, L., Xie, J.Y., Sun, K., Jiang, X., Li, S.M.: Resonance-based reflector and its application in unidirectional antenna with low-profile and broadband characteristics for wireless applications. Sensors **16**(4), 2092 (2016)
9. CST Microwave Studio. https://www.cst.com/
10. peng, L., Xie, J.Y., Jiang, X., Li, S.M.: Investigation on ring/split-ring loaded bow-tie antenna for compactness and notched-band. Frequenz **70**, 89–99 (2016)
11. Dumanli, S.: Challenges of wearable antenna design. In: 46th European Microwave Conference (EuMC), pp. 1350–1352 (2016)

Acoustic Sensing in Microfluidic Cell

Valentin Mateev(✉), Georgi Ivanov, and Iliana Marinova

Technical University of Sofia, Sofia 1000, Bulgaria
vmateev@tu-sofia.bg

Abstract. In this work are presented early experiments for fluid velocity and pressure sensing in microfluidic cell, based on low frequency acoustic spectrum. Many of microfluidic cells and microfluidic chips internal properties could be observed only by numerical modeling. Some remote methods are applicable for contactless measurements inside the microfluidic cells channels. With advanced numerical procedures it can be reconstructed internal fluid flow properties. This paper considers measurements of acoustic signals, acquired over the fluid cell mixing chamber. Acoustic spectrum is correlated with fluid flow parameters as internal velocity and pressure. Results are verified by computational Fluid Dynamics (CFD) modeling of a microfluidic cell with known velocities and pressures.

Keywords: microfluidic cell · acoustic sensing · sound sensing · fluid velocity measurements · sonification

1 Introduction

Microfluidic cells and chips are gaining significant attention for many applications in biology, biomedicine, MEMS, micro- and nano- chemistry and many more [1, 2, 3, 4]. Fast emerging trend is integration of sensing electrodes inside the cells architecture to detect microchannel flow properties and internal content, especially from biological origin. The electrode integration meets many difficulties, where metal electrodes must be united with non-metal substrates under very strict requirements to avoid micro stress [1], fluid friction with main flow or droplet formation [3, 4], corrosion, polarization blocking and channel jamming [5]. Different technologies are considered for that purpose e.g. 3D printing methods, lithographic induced etching, surface deposition, etc.

Many of microfluidic cells and microfluidic chips internal properties could be observed only by numerical modeling [2, 6, 7, 8]. Significant work on the subject of microscale measurements and modeling have been done in the recent years, taking into account fluid properties control [5], phase imaging [6], fluid profiling [8], electrokinetics [7, 9]. This paper considers Computational Fluid Dynamics (CFD) modeling of a microfluidic cell and corresponding sound spectrums on cell upper surface. As it may be seen from the results, there is a correlation between fluid pressure and velocity, related by sound spectrum analysis of measured acoustic data on microfluidic cell surface. These results can be used for precise contactless estimations of fluid flow in inaccessible for direct measurements region as internal cell micro channels.

N. K. Suryadevara et al. (Eds.): ICST 2022, LNEE 1035, pp. 259–266, 2023.
https://doi.org/10.1007/978-3-031-29871-4_26

Microfluidic cell under consideration has two input flow channels, mixing micro chamber and one output (Fig. 1). Input channels cross-section is 50 × 50 μm, which are leading to a 50 × 100 μm output.

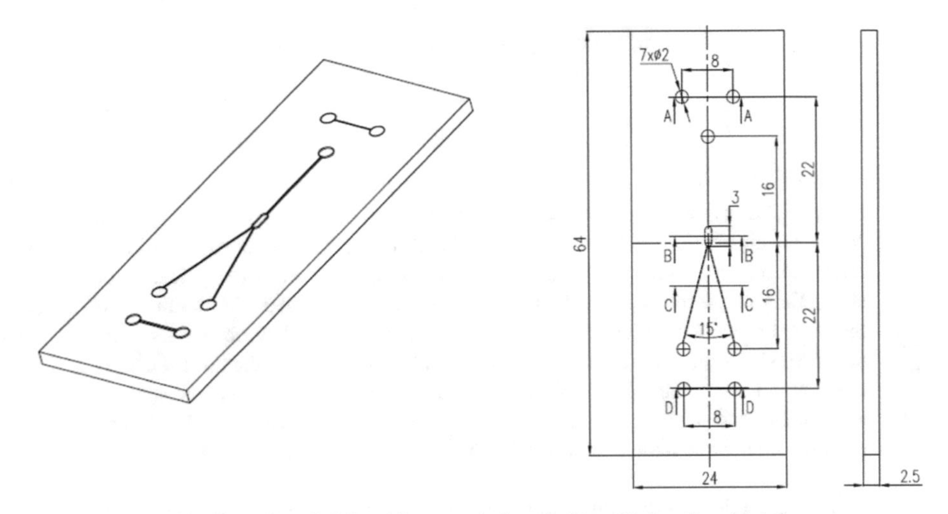

Fig. 1. Microfluidic cell general view (left), cell drawing (right).

The cell prototype is manufactured over a glass substrate by laser lithography and two pieces are assembled by ultrasound glass ablation. Therefore, glass material properties and channel wall roughness are used in the CFD modeling for wall friction estimation. Modeling is performed via Comsol software.

2 CFD Model and Governing Equations

A 3D model of the proposed microfluidic cell was created in order to study the, velocity and pressure distributions inside the cell channels. On Fig. 2 is shown drawing of created model and its sizes. Fluid-structure interaction is carried out combining fluid-flow and structural mechanics in to calculate the equivalent velocity and pressure values. The cell channel fluid flow is calculated by Navier–Stokes equations [11].

$$\rho\frac{\partial \boldsymbol{v}}{dt} + \rho(\boldsymbol{v}\nabla)\boldsymbol{v} = \nabla[-p\boldsymbol{I} + \boldsymbol{K}] + \boldsymbol{F}, \tag{1}$$

where ρ is fluid density, **v** is fluid velocity vector, **I** is indentity tensor, p is the pressure and **K** is kinematic viscosity term, μ is dynamic viscosity coefficient,

$$\boldsymbol{K} = \mu\Big(\nabla\boldsymbol{v} + (\nabla\boldsymbol{v})^T\Big) \tag{2}$$

Fluid domain is considered as incompressible,

$$\rho\nabla\boldsymbol{v} = 0. \tag{3}$$

For inlets a velocity boundary condition is set to 0.02 m/s.

The level set method is used to track the fluid interface of two phase fluid solution, normalized as $\phi = \phi_1 + \phi_2 = 1$, it uses the following equation.

$$\frac{\partial \phi}{\partial t} = -\mathbf{V}\nabla\phi\left(\varepsilon\nabla\phi - \phi(1-\phi)\frac{\nabla\phi}{|\nabla\phi|}\right), \tag{4}$$

where ε is the interface thickness controlling parameter around the interface boundary, set to 1/2 from the maximum element size in the domain. Model mesh statistics is shown in Table 2.

The density is a function of the level set function. Let ρ_1 and ρ_2 be the densities of two mixing fluids. Here, the first fluid corresponds to the domain where $\phi_1 < 0.5$, and second fluid corresponds to the domain where $\phi_2 > 0.5$. When density averaging is set to volume average, the density is defined as.

$$\rho = \rho_1 + (\rho_2 - \rho_1)\phi. \tag{5}$$

Similarly, the dynamic viscosity can be defined by setting viscosity average μ to volume average.

$$\mu = \mu_1 + (\mu_2 - \mu_1)\phi, \tag{6}$$

where μ_1 and μ_2 are the dynamic viscosities of mixing fluids, material properties are presented in Table1 [11].

More details on actual microfluidic cell could be found in [12].

Glass architecture is manufactured by laser lithography, channel wall roughness is dependent from optical power and speed of glass melting. Channel wall roughness, used for wall friction drag force **F** in Eq. (1) estimation, is set to 0.2 μm. [13].

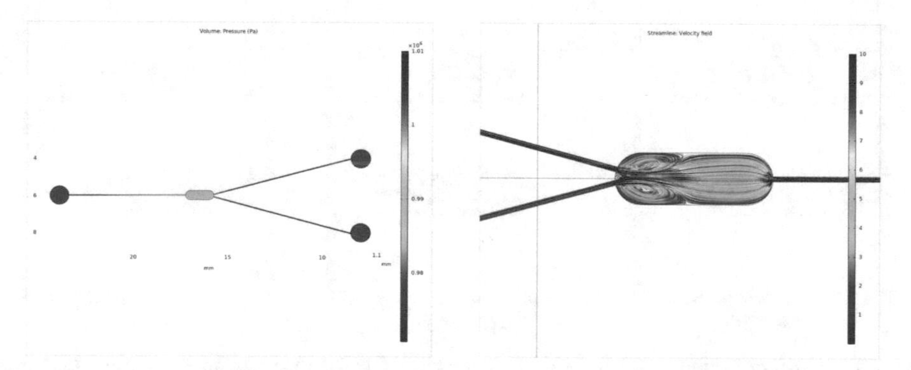

Fig. 2. Calculated pressure drop distribution in the microfluidic cell channels (left) and water only fluid flow velocity distribution (right).

3 Modeling and Measurement Results

Calculated results velocity field stream lines in the mixing chamber domain are presented in Fig. 2.

Micro-convective effects are observed in the modeled results, which are related with channel cross-section changes. Pressure drop is observed, which is dependent from achieved channel fluid velocity. Calculated pressure in the microfluidic cell mixing chamber cross-section is presented in Fig. 2.

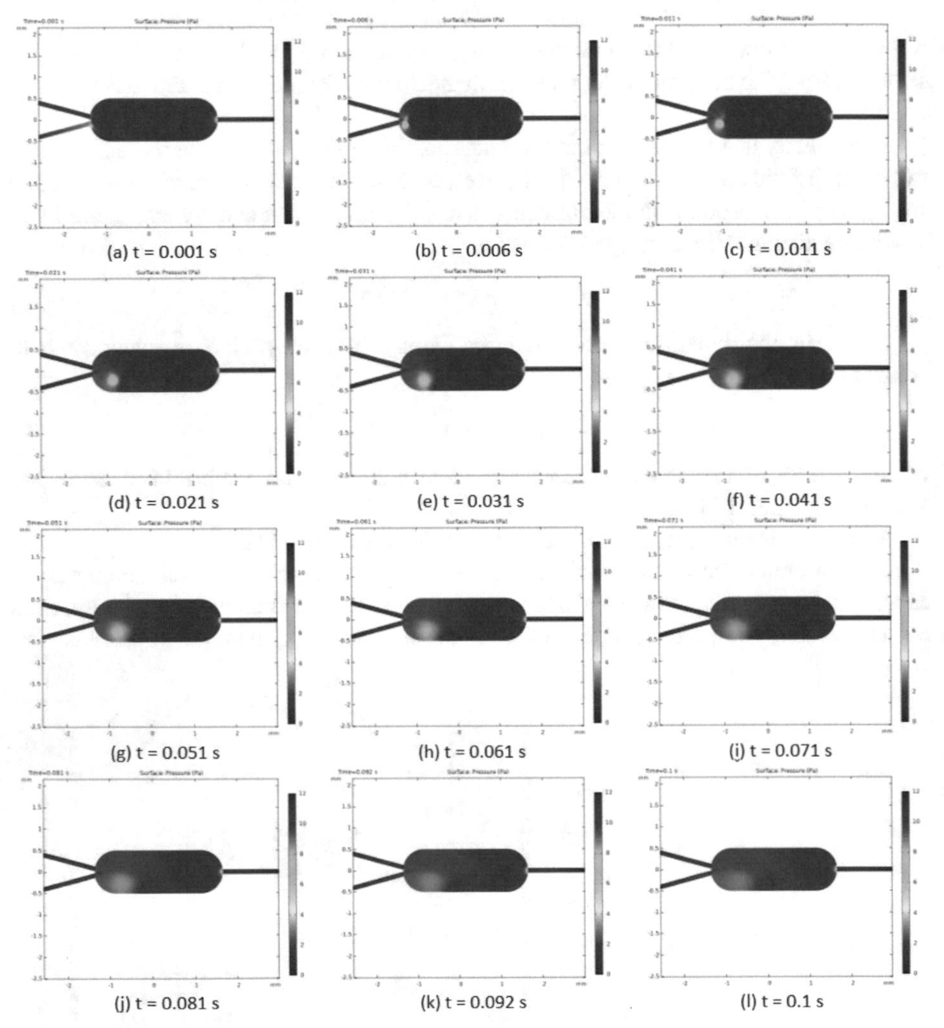

Fig. 3. Transient pressure distribution of moving oil droplet insertion in the mixing chamber domain filled with water. Simulation end time is 0.1 s.

Material properties used for CFD two phase modeling are presented in Table 1. Base fluid is water with density and dynamic viscosity as shown. Oil like fluid is injected in upper cell channel into the water. Transient pressure distribution of moving oil droplet insertion in the mixing chamber domain filled with water is shown in Fig. 3.

Modeling results for CFD problem solving are confirmed by single cell acoustic spectrum measurements. Measuring ceramic microphone is directly located over the

Table 1. Material properties.

	Unit	Oil	Water
Density ρ	kg/m^3	850	1000
Dynamic viscosity μ	Pa·s	1.85e–5	8.9e–4

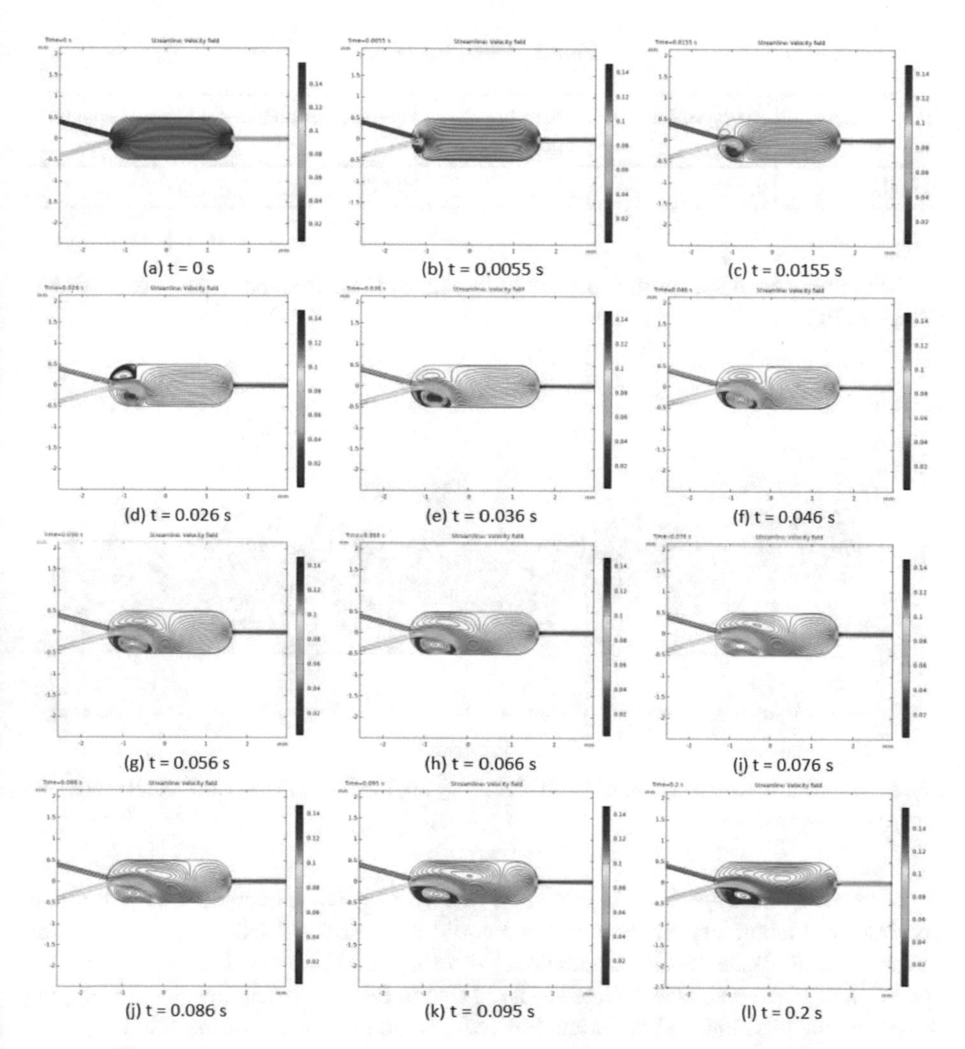

Fig. 4. Calculated velocity field stream lines in the mixing chamber domain.

cell mixing chamber with frequency bandwidth of 44 kHz. Fluids are pumped with precise Cytosurge FluidFM MFCS v2 fluid controlled with single 0.2 Pa pulses during testing.

Calculated velocity field stream lines in the mixing chamber domain, after oil droplet insertion are shown in Fig. 4. They are covering 200 ms time range. A main fluid vortex around low density droplet is visible in all frames of the simulation. Time dependent fluid pressure variations are considered as a sound source with same frequency content determined by pressure transients [14]. These pressure variations are transmitted to microfluidic cell surface and acquired by mounted microphone crystal.

Table 2. Mesh statistics.

Min element quality	Average element quality	Number of triangles	Number of quads	Edge element
0.09938	0.7439	7784	2342	540

Measured sound spectrum is presented in Fig. 5. Time-frequency data processing is shown in Fig. 6.

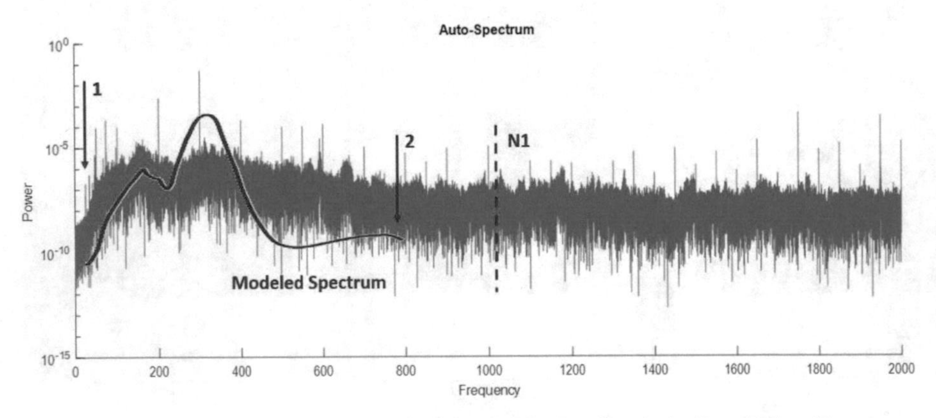

Fig. 5. Acoustic spectrum, measured (with blue noisy line) and analytically calculated by pressure time variations (red line).

Measured sound spectrum is presented in Fig. 5. Spectrum content above 1 kHz is coming from laboratory environment and could not be related with pressure variation inside the cell. These results are marked with threshold line N1. Modeled frequency spectrum of Fig. 3 results is depicted as a red line in Fig. 5. Visually results overlapping is not satisfactory, but it shows common two maximum pattern in the low frequency range below 200 Hz.

After many observations, it was noted, that most of the sound recorded is coming from outer environment and sound pulses from the micro pump unit. Other source of sound recorded is the movements of the cell with each pump pulse. At the current stage of experiments, it could not be pointed exact correlation of sound spectrum and pressure or velocity inside the mixing chamber of the cell. The noise reduction must be further improved and more accurate experimental analysis must be made.

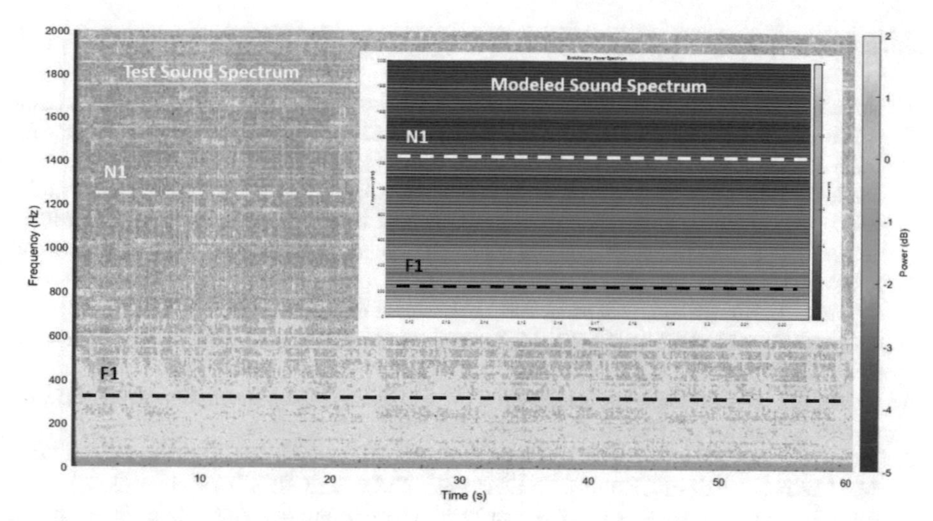

Fig. 6. Time-to-frequency spectrogram visualization of acoustic spectrum.

Time-frequency data processing, of measured so far data, is shown in Fig. 6. Both spectrums of modeled and measured plots are shown. Common frequency lines are visible in low frequency range. The comparison of these results must be further improved and numerical estimation of difference and pressure change correlation with sound spectrum must be provided.

4 Conclusion

It is presented acoustic sensing in microfluidic cell. Many of microfluidic cells and chips internal properties could be observed only by numerical modeling. Therefore, this paper considers measurements of acoustic signals, acquired over the fluid cell mixing chamber. Results are verified by computational Fluid Dynamics (CFD) modeling of a microfluidic cell with known velocities and pressures. Result analysis shows a not clear relation between fluidic and acoustic signals. Significant noise levels are observed, so further noise reduction must be provided for accurate measurements. Measurement setup vibrations are important signal noise generating source. In order to reduce them we are working on second measurement channel with 3-axis accelerometer, which will monitor the vibrations on microfluid chip surface. Bothe spectrums, acoustic and vibration, will be intersect to eliminate vibration influence and denoise acoustic signal. This will be a subject for future work. If a correlation between acoustic spectrum and actual fluid velocity in mixing chamber will provide powerful tool for contactless volumetric measurements. Acoustic tomography could extend the spatial accuracy of sub-surface reconstruction of detected properties. This way a better control and visualization of intra cell processes will be possible.

Acknowledgment. Some measuring and data processing techniques, used in this research, are developed with the support of the National Science Fund of the Ministry of Education and Science of Republic of Bulgaria within the KP-06-N47/2 contract.

References

1. Kovács, R., Borók, A., Bonyár, A.: Numerical simulations of shear stress in microfluidic channel models. In: 2022 45th International Spring Seminar on Electronics Technology (ISSE), pp. 1–5 (2022)
2. Sauli, Z., Taniselass, S., Lim, W., Rahman, N.A., Norhaimi, W.M.W., Retnasamy, V.: 2D velocity streamline visualization of microfluid flow in backward facing step microchannel. In: 2012 Fourth International Conference on Computational Intelligence, Modelling and Simulation, pp. 138–140 (2012)
3. Agarwal, A., Khushalani, D., Shrivastava,D.: Effect of illumination from laser on droplet formation in microchannel. In: 2021 IEEE India Council International Subsections Conference (INDISCON), pp. 1–5 (2021)
4. Bishnoi, P., et al.: Computational study on the dynamics of drop generation under different ambient conditions. AIP Conf. Proc. **2341**, 030021 (2021)
5. Moise, V., Marii, L., Svasta, P.M.: Pump control system in microfluidic systems. In: 2021 44th International Spring Seminar on Electronics Technology (ISSE), pp. 1– 5 (2021)
6. Park, G., Kim, G., Kim, K., Park,Y.: Quantitative phase imaging of fluid mixing in microfluid chips. In: 2016 Asia Communications and Photonics Conference (ACP), pp. 1–2 (2016)
7. Chen, C., Ran, B., Liu, B., Liu, X., Jin, J., Zhu, Y.: Numerical study on a bio-inspired micropillar array electrode in a microfluidic device. Biosensors **12**(10), 878 (2022)
8. Chen, C., Li, P., Guo, T., Chen, S., Xu, D., Chen, H.: Generation of dynamic concentration profile using a microfluidic device integrating pneumatic microvalves. Biosensors **12**(10), 868 (2022)
9. Niu, B., Wang, Y., Lv, S., et al.: Insight the role of electrolyte in electrokinetic stacking of targets by ion concentration polarization on a paper fluidic device. Microfluid Nanofluid **26**, 90 (2022)
10. Xiao, S., Wollman, Z., Xie, Q., et al.: Current monitoring in nanochannels. Microfluid Nanofluid **26**, 86 (2022)
11. Comsol Multiphysics user's guide. COMSOL AB (2018)
12. Mateev, V., Ivanov, G., Marinova, I.: Modeling of microfluidic cell with electrodes for impedance measurements. In: XXXI International Scientific Conference Electronics - ET2022, pp. 1–4 (2022)
13. Butkutė, A., et al.: Optimization of selective laser etching (SLE) for glass micromechanical structure fabrication. Opt. Express **29**, 23487–23499 (2021)
14. Ralchev, M., Mateev, V., Marinova, I.: Measurement of AC electric arc discharge acoustic spectrum. In: XXXI International Scientific Symposium Metrology and Metrology Assurance (MMA), pp. 1–4 (2021)

Development of Compact Actuator for Direct Measurement of Skin Displacement in Ultrasonic Haptics

Daisuke Mizushima(✉) and Daiki Sato

Aichi Institute of Technology, 1247, Yachigusa, Yakusa, Toyota, Aichi, Japan
d-mizushima@aitech.ac.jp

Abstract. Haptics is a research area concerning tactile sensation. Haptics are applied in meta-verse and tactile transmission, such as in telemedicine. Ultrasonic haptic actuators apply converged ultrasonic waves to the skin to produce vibratory stimulations. Ultrasonic haptic actuators can provide tactile sensations without the skin contact. To directly measure skin displacement, small actuators that can sufficiently stimulate tactile sensation have been developed, and some characteristics have been measured. Experimental results indicate that although the sound pressure generated by the small actuator sufficiently provide tactile stimulation, the sound pressure generated was approximately 70% of that of a commercial haptics development kit, thereby narrowing the frequency band over which tactile sensations could be presented as compared with the commercial product. The circular actuator also has a slightly lower threshold than the arched actuator, and the half-width of the convergent ultra-sound may influence the magnitude of the tactile stimulus.

Keywords: Haptics · Haptic Actuator · Ultrasonic Actuator · Amplitude Modulation · Tactile Stimulation

1 Introduction

Haptics is a research area in tactile sensations. They contain physiologies, sensors, and actuators. Although the primary application of haptics is in the meta-verse, it has potential applications in tactile transmission, such as in telemedicine.

Haptic actuators present tactile stimuli, such as vibration, pressure, and temperature to cause tactile sensation and are essential devices for practical applications [1]. The most widely used haptic actuator is the eccentric motor. Linear resonance motors [2] and dielectric elastomers [3] were used as haptic actuators. Moreover, a composite tactile presentation device that combines motors, a belt, and a Peltier element has been reported [4].

The ultrasonic haptic actuator used in this study applied converged ultrasonic waves to the skin to produce vibratory stimulation [5]. Ultrasonic haptic actuators are unique in that they are capable of tactile presentation without the skin contact. This feature was the focus of this study, wherein the displacement of the skin caused by ultrasonic haptic

N. K. Suryadevara et al. (Eds.): ICST 2022, LNEE 1035, pp. 267–273, 2023.
https://doi.org/10.1007/978-3-031-29871-4_27

actuators was observed. Specifically, this study aimed to quantify skin stimulation using ultrasound and to associate physical and sensory quantities.

Ultrasonic haptic actuators have already been commercialized; however, they are large and must be stationary. Additionally, the generated sound pressure is insufficient. To examine the skin displacement due to ultrasonic haptic actuators, it is desirable to obtain sufficient sound pressure from a small, easy-to-handle actuator. Therefore, in this study, a small and close focal length actuator was developed and its characteristics were compared with those of commercial products. Furthermore, a sensory evaluation was performed to investigate the threshold of the sound pressure necessary for tactile sensation and to identify the characteristics related to the intensity of tactile stimulation.

2 Theory of Tactile Stimulation

2.1 Relationship Between Skin Displacement and Tactile Stimulation

The skin has a layered structure with internal mechanoreceptors that transmit tactile stimulations. Mechanical stimulation causes skin displacement, and mechanoreceptors convert this displacement into tactile stimulation. Mechanoreceptors can be classified into four types, namely SA I, SA II, FA I, and FA II, based on differences in response time to skin deformation and the sizes of their receptive fields. The sensation of a steady stimulus is called a pressure sensation, and that of a vibrating stimulus is called a vibration sensation. The frequency responses of the displacement thresholds of the four types of mechanoreceptors to the vibration sensation are shown in Fig. 1 [6].

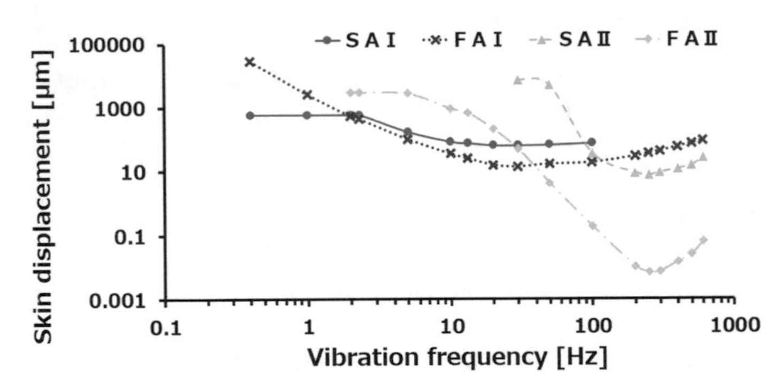

Fig. 1. Frequency response of the displacement thresholds of four types mechanoreceptors [6].

As shown in Fig. 1, mechanoreceptors detect low-frequency vibrations ranging from 1 Hz to 600 Hz. Typically, ultrasonic speakers have a resonant frequency of 40 kHz; therefore, they are amplitude-modulated at an arbitrary frequency with 40 kHz as the carrier when used as haptic actuators. Although the frequency characteristics of sensitivity differ for each receptor, the threshold of FA II is remarkably low, particularly after 100 Hz. Ultrasonic haptic actuators with a low generating force are considered to primarily stimulate this receptor.

2.2 Generation of Tactile Stimulation by Convergent Ultrasonic Wave

For simplicity, we assumed contact between the convergent ultrasonic and index fingers. To quantify the effect of convergent ultrasonic vibration on tactile sensation, the Hertzian contact theory, which is used for contact between elastic surfaces, is used. R_1, ν_1, andE_1 represent the curvature radius, Poisson's ratio, and Young's modulus of skin, respectively. R_2 indicates the curvature radius of the converged ultrasonic waves, and ν_2, E_2 indicate the Poisson's ratio and Young's modulus of air, respectively. When p is the central sound pressure of the converging ultrasonic waves, displacement δ is expressed as follows:

$$\delta = \sqrt[3]{\frac{9}{16}\frac{R_1 R_2}{R_1 + R_2}\left(\frac{1-\nu_1{}^2}{E_1} + \frac{1-\nu_2{}^2}{E_2}\right)^2 p^2} \tag{1}$$

The radius of this contact surface a is expressed as follows:

$$a = \sqrt[3]{\frac{3}{4}\frac{R_1 R_2}{R_1 + R_2}\left(\frac{1-\nu_1{}^2}{E_1} + \frac{1-\nu_2{}^2}{E_2}\right)p^2} \tag{2}$$

Although the Poisson's ratio and Young's modulus of air vary with atmospheric pressure and temperature, the change in air pressure due to sound pressure is negligible and can be approximated under the same temperature and ambient pressure conditions. The curvature radius of the convergent ultrasonic beam was determined via measurement. However, these calculations did not consider the viscosity of the skin, sound absorption, or reflection at the surface.

3 Experimental Setup

3.1 The Miniature Ultrasonic Haptic Actuator

An overview of the arched jig used to converge ultrasonic waves is shown in Fig. 2.

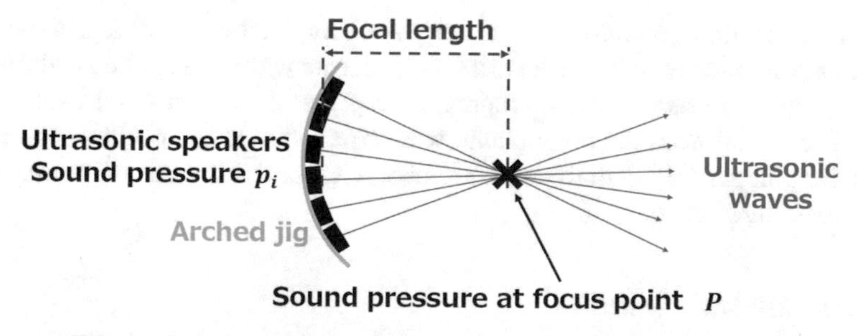

Fig. 2. Overview of the arched jig and converged ultrasonic waves

Assuming that the sound pressure applied from one ultrasonic speaker to the focal point is p, the total maximum sound pressure P is six times p. if the ultrasonic waves emitted by the six ultrasonic speakers converge. However, this is valid only when the distance from each source to the focal point is exactly the same and the sound waves are perfectly in phase. In reality, the sound pressure and the phase of sound waves generated will have discrepancies owing to misalignments in mounting the ultrasonic speaker, oscillation timing, and individual differences. If the sound pressure at the focus of each loudspeaker is P_i, and the phase shift is θ_i, the following Eq. (3) is obtained.

$$P = \sum_{i=1}^{n=6} \sqrt{2} p_i sin(2\pi f + \theta_i) \tag{3}$$

In this study, six ultrasonic speakers were arranged in an arched or concentric circle shape with focal lengths of 50 mm and 100 mm for each shape, making a total of four types. In contrast, the actuator in the commercial haptics development kit (Ultra Haptics STRATOS Explore) has 256 ultrasonic speakers, and the focal length can be adjusted from 100–700 mm. As the objective of this study was to obtain the highest possible sound pressure, the focal length was fixed at 100 mm.

3.2 Experimental Setup

A schematic of the sound pressure measurement is shown in Fig. 3.

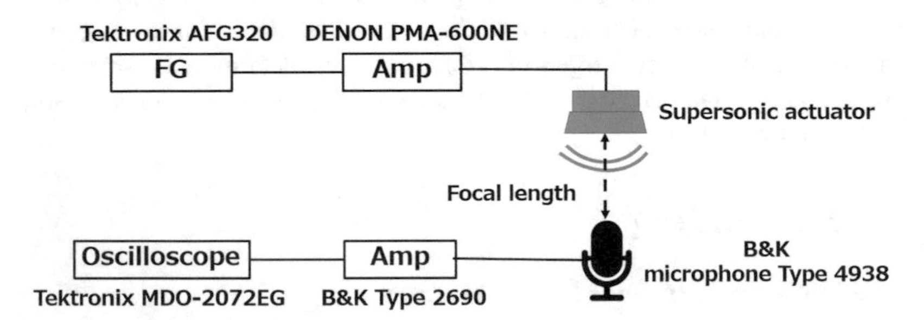

Fig. 3. Schematic of sound pressure measurement

From the function generator (FG), an amplitude modulation (AM) signal modulated at an arbitrary frequency with respect to a 40 kHz carrier wave was applied to the actuator via an amplifier. A measurement microphone was placed at the focus of the actuator, and the detection signal from the microphone was observed using an oscilloscope through an amplifier. The sensitivity of the measurement microphone is constant throughout the entire measurement band.

4 Experimental Results

4.1 Sound Pressure Characteristics

Figure 4 shows the variation in sound pressure when the voltage applied to the four types of small actuators was varied. The modulation frequency was 200 Hz and the modulation factor was 100%.

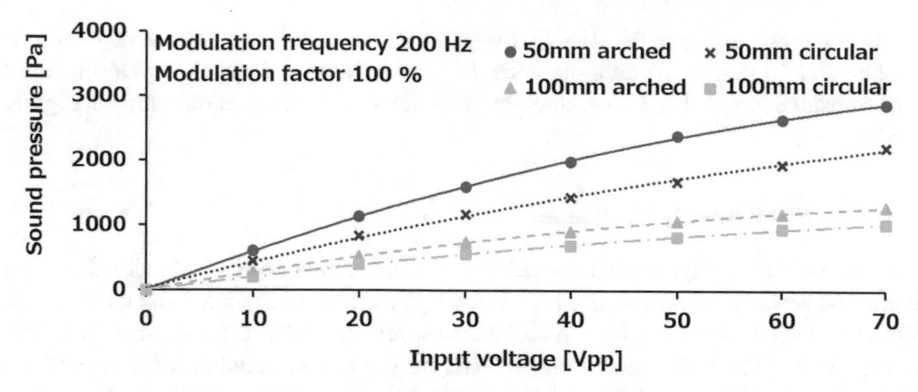

Fig. 4. Sound pressure vs. Input voltage at focal point

As shown in Fig. 4, the highest sound pressure is approximately 2800 Pa. The closer the focal distance, the higher is the sound pressure. Circular jigs have lower sound pressures than arched jigs. This is thought that the ultrasonic waves do not converge sufficiently because of the jig modeling accuracy (about 0.5 mm). The maximum sound pressure obtained from STRATOS Explore was ap-proximately 4000 Pa, and the sound pressure obtained from the small actuator was approximately 70% of this value.

4.2 Sound Pressure Distribution

Figures 5(a) and (b) show the sound pressure distribution in the focal plane of an arched and small circular actuator with a focal length of 50 mm. Figure 5(a) shows the arched image, and Fig. 5(b) shows the circular. The input voltage was 60 V, the modulation frequency was 200 Hz, and the modulation factor was 100%.

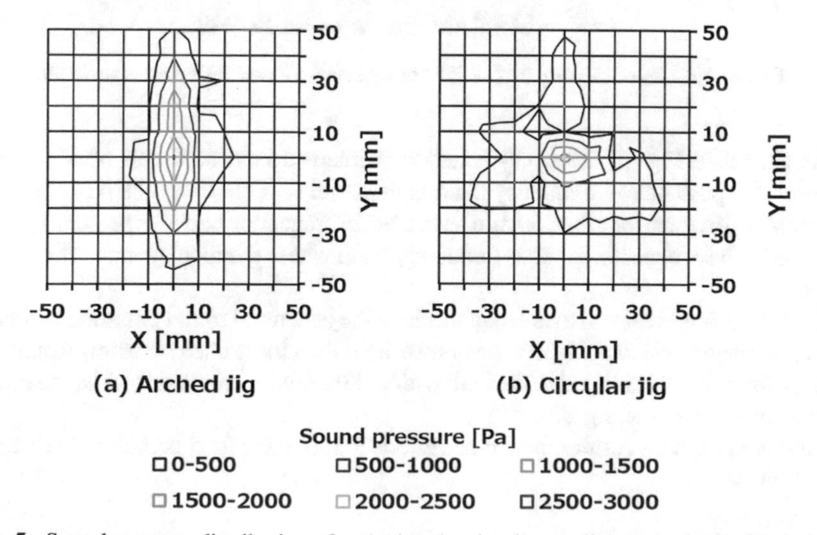

Fig. 5. Sound pressure distribution of arched and a circular small actuator in the focal plane

From Figs. 5(a) and (b), the half-widths of the sound pressure were 4 mm × 20 mm and 10 mm × 10 mm, respectively. The STRATOS Explore development kit was measured at 6 mm × 6 mm, implying that the half-widths was the same or slightly larger.

4.3 Results of Sensory Evaluation

The microphone in Fig. 3 was replaced with the index finger of the subject's dominant hand, and sensory evaluation was performed by touching convergent ultrasonic waves. The subjects included 10 males in their 20 s. At several modulation frequencies, the input voltage to the four small actuators and STRATOS Explore was gradually increased from 0 V to investigate the lower limit of the sound pressure in the tactile stimuli. The results of the sensory evaluation are shown in Fig. 6.

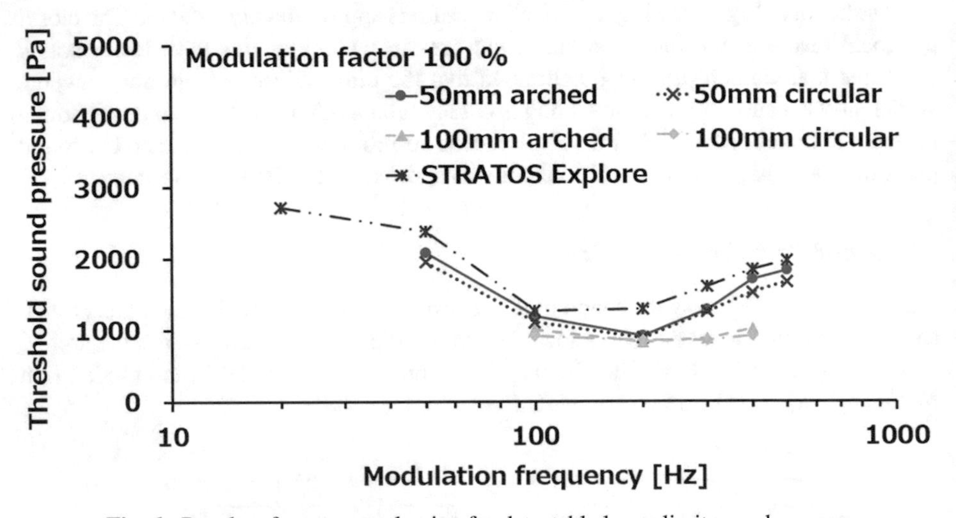

Fig. 6. Results of sensory evaluation for detectable lower limit sound pressure

As shown in Fig. 6, the small actuators perceived tactile stimuli of 50–500 Hz. However, the presentable frequency band is narrower than that for STRATOS Explore. Actuators with a focal length of 50 mm have a wider frequency band for presenting tactile sensations. These occur in the low-frequency band where the displacement threshold is higher.

Therefore, it was related to the magnitude of the generated sound pressure. Moreover, the circular actuator had a slightly lower threshold than the arched actuator. Apparently, the detectable signal is related to the half-width of the sound pressure, that is, the number of mechanoreceptors being stimulated.

The measurement results for the commercial and developed actuators are summarized in Table 1.

Table 1. Measurement results of miniature and commercial actuator

	Max. Sound pressure [Pa]	Half-width [mm × mm]	Detectable frequency [Hz]
STRATOS Explore	4300	6 × 6	20–500
5 mm arched	2464	4 × 20	50–500
5 mm circular	2053	10 × 10	50–500
10 mm arched	1595	12 × 25	100–400
10 mm circular	911	–	100–400

5 Conclusion

Small actuators that can sufficiently stimulate tactile sensation with small devices are developed to directly measure skin displacement using an ultrasonic haptic actuator, and some characteristics are measured.

From Table 1, although the sound pressure generated by the small actuator sufficiently provided tactile stimulation. It was found that the sound pressure generated was approximately 70% of that of the commercial product, therefore the detectable frequency band is narrower than the commercial product.

The circular actuator also exhibited a slightly lower threshold than the arched actuator. As the amount of displacement increases as the radius of curvature de-creases in the Hertzian contact theory, the half-width of the convergent ultrasound may influence the magnitude of the tactile stimulus.

References

1. Srinovasan, M.A., Basdogan, C.: Haptics in virtual environments: taxonomy, research status, and challenges. Comput. Graph. **21**(4), 393–404 (1998)
2. Tanabe, K., Takei, S., Kajimoto, H.: The whole hand haptics clove using numerous liner resonant actuators (III) Improvement of haptic feedback, EC2015, pp. 314–316, IPSJ, Sapporo Japan (2015). (in Japanese)
3. Hau, G., Rizzello, G., Seelecke, S.: A novel dielectric elastomer membrane actuator concept for high-force applications. Extreme Mech. Letters **23**, 24–38 (2018)
4. Inoue, y., Naka, I., Kato, F., Tachi, S.: Haptic presentation system based on haptic primary colors. TVRSJ **25**(1), 86–94 (2020). (in Japanese)
5. Freeman, E.: Enhancing ultrasound haptics with parametric audio effects.In: ICMI 2021, Montreal, Canada, pp. 692–696 (2021)
6. Bolanowski, Jr., S.J., Gescheider, G.A., Verrillo, R.T., Checkosky, C.M.: Four channels mediate the mechanical aspects of touch. J. Acoust. Soc. Am. **84**(5), 1680–1694 (1988)

An Efficient Inductance-to-Digital Converter Insensitive to Coil Resistance of Differential Type Inductive Sensors

P. P. Narayanan[1], Sreenath Vijaykumar[1(✉)], R. Rahul Raja[2], V. Sowbaranic Raj[2], and Karthi Balasubramanian[2]

[1] Department of Electrical Engineering, Indian Institute of Technology Palakkad, Palakkad, India
sreenath@iitpkd.ac.in

[2] Department of Electronics and Communication Engineering, Amrita School of Engineering, Amrita Vishwa Vidyapeetham, Coimbatore, India
b_karthi@cb.amrita.edu

Abstract. This paper proposes an Inductance-to-Digital Converter (LDC) that measures the change in inductance ΔL of a Differential-type Inductive Sensor (DIS) independent of its coil resistance. The ΔL is measured using a ratio metric approach that eliminates error due to the tolerances of the components used in the circuit. The proposed LDC is realized using a simple dual-slope conversion technique and possesses advantages such as noise immunity and high accuracy. Existing schemes for measuring differential inductance assume the coil resistance is negligible or equally matched. This assumption may not be practical always, and hence a noticeable error in inductance measurement will be introduced in the existing architectures, while the proposed design does not suffer from this limitation. In addition, the final output provided by the proposed LDC will always be proportional to the measurand, irrespective of whether the measurand has a linear or inverse relationship with the sensor inductance. The prototype developed showed superior performance with a worst-case error of 0.72% when the series coil resistance was varied from 0 to 1000 Ω in steps of 100 Ω. Moreover, the developed architecture showed a good resolution of 3.5 μH and high linearity, with a worst-case error of 0.55% when ΔL was varied from –6 mH to +6mH.

Keywords: Inductance to digital converter (LDC) · Differential-type Inductive Sensors (DIS) · Instrumentation amplifier (INA) · Differential Variable Reluctance Transducer (DVRT)

1 Introduction

Differential-type inductive sensors (DIS) have a wide range of applications in various technologies. They are used to measure different physical quantities including position [1], displacement [2], speed [3], angle [4], pressure [5] and acceleration [6] and is also used in micro-positioning of tools [7]. It is very popular in industrial, as well as in high

N. K. Suryadevara et al. (Eds.): ICST 2022, LNEE 1035, pp. 274–285, 2023.
https://doi.org/10.1007/978-3-031-29871-4_28

radiation environments due to its precision, ruggedness, reliability, and simplicity. DIS is a simple three terminal component which is equivalent to two inductors $L_{X1}$and L_{X2} connected in series. The differential pair is designed in such a way that if one of the inductance value increases due to variation in the measurand, the other inductance falls by the same amount and vice-versa. The inductance for the DIS can be represented as $L_{X1} = L_0 \pm \Delta L$ and $L_{X2} = L_0 \mp \Delta L$. Where L_0 is the offset inductance and ΔL is the change in inductance due to the variation in the measurand. In practical scenarios, coil resistances $R_{C1}$and R_{C2}, albeit generally a small value, will always be present with the differential pair L_{X1} and L_{X2} respectively.

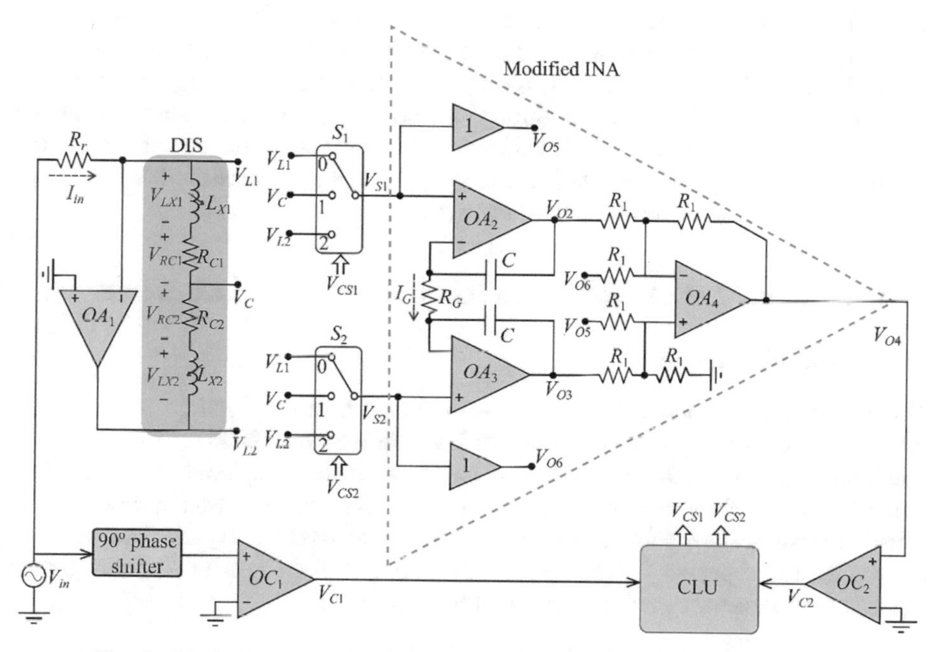

Fig. 1. Block diagram of the proposed Inductance-to-Digital Converter (LDC)

There are various architectures currently available for measuring the inductance of a DIS. Discharge model based inductance calculation is exploited in [8] and [9], where the inductance is measured by directly interfacing with a microcontroller. These schemes assume that the coil resistance is negligible, however this may not be the case in most practical scenarios. In [10], due to a mismatch of 0.5 Ω between R_{C1} and R_{C2}, an error of 0.125% is introduced. The scheme proposed in [11] is insensitive to coil resistance, but it can be interfaced only with a Linear Variable Differential Transformer (LVDT) and the system may fail if interfaced with a Differential Variable Reluctance Transducer (DVRT). A switched capacitor-based direct digital converter for DIS that accepts non-linear sensor inductances and provides a linear digital output, proportional to the measurand is proposed in [12], but the method fails in the presence of series coil resistances. A timer-based technique that performs demodulation and digitization concurrently without the use of a rectifier, mixer, or an analog to digital converter is provided in [13]. This methodology achieves high accuracy with evenly matched resistors, i.e. $R_{C1} = R_{C2}$, but in practical situations, as ΔL changes, a mismatch in the coil resistances is bound to occur that will affect the accuracy of measurement. The architecture developed in [14] efficiently measures the inductance, eliminating the effect of coupling capacitances between the moving wiper and the inductor, but this architecture also suffers from the sensitive dependence on the coil resistances. Since most of the existing schemes used for measuring ΔL are sensitive to the presence of the sensor coil resistances (R_{C1} and R_{C2}), a readout scheme that can measure ΔL independently of L_0, R_{C1} and R_{C2}would be extremely useful.

An inductance to digital converter that can provide a digital output proportional to ΔL, independent of the series coil resistances and the nominal inductance of a differential type inductive sensor is proposed in this paper. The proposed converter employs a modified dual slope technique and an instrumentation amplifier (INA) to achieve the desired output. The ΔL is measured through a ratio metric approach, thereby achieving a highly accurate measurement independent of the tolerances of the components in the circuit. The following sections detail the functionality of the proposed method, prototype developed, experimental results, and discussion.

2 Operation of the Proposed Inductance-to-Digital Converter

2.1 Circuit Description

The proposed Inductance-to-Digital Converter (LDC) is depicted in Fig. 1. The circuit makes use of the dual slope method along with a modified INA. The INA's basic architecture is preserved except for the fact that the feedback resistor is replaced with a capacitor in the gain stage of the INA, which performs the integration operation during the dual slope conversion. The modified INA is made up of op-amps OA_2, OA_3 and OA_4, resistors R_G and R_1, and capacitor C. The op-amp OA_4 along with the resistor R_1 forms a four input adder-subtractor circuit whose output is given as (1).

$$V_{O4} = (V_{S1} + V_{O3}) - (V_{S2} + V_{O2}) \tag{1}$$

The DIS is constituted by the inductors L_{X1} and L_{X2} along with their respective coil resistances $R_{C1}$and R_{C2}. The DIS is kept in the feedback of op-amp OA_1, thus effectively ensuring that the current flowing through the DIS: $I_{in} = V_{in}/R_r$. The circuit makes use of two Single Pole Triple Throw (SPTT) switches, S_1 and S_2 which are controlled by the signals V_{CS1} and V_{CS2}, respectively, generated by a Control and Logic Unit (CLU). Op-amp OA_1 is fed by a sinusoidal input V_{in} while a 90° phase shifted version of the same is given to the comparator OC_1 using a phase-shifter module. The CLU continuously monitors the phase shifted signal and the output of the INA to generate the control signals V_{CS1} and V_{CS2}. The following section explains the operation of the converter.

2.2 Operation

The proposed LDC works with a sinusoidal input signal $V_{in} = V_R \sin(2\pi ft)$ with a time period T. As mentioned in Sect. 2.1, Current $I_{in} = V_{in}/R_r$ continuously flows through the

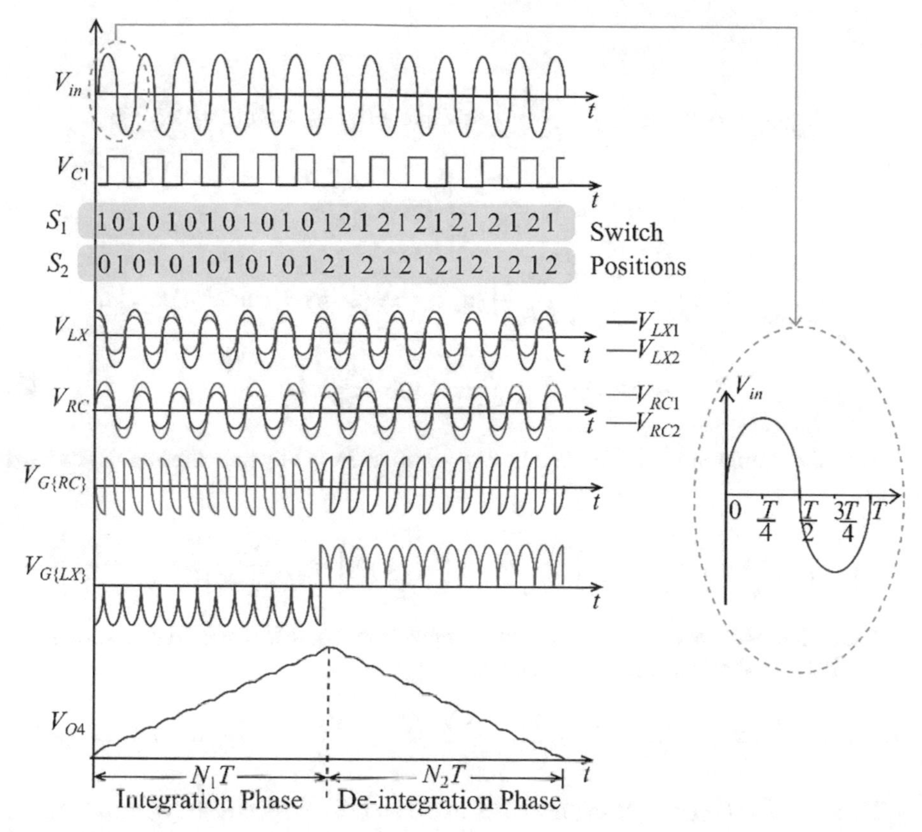

Fig. 2. Voltage waveforms depicting voltage signal V_{in}, V_{LX1}, V_{RC1}, V_{LX2} and V_{RC2}, V_{C1} and V_{O4}, and switching positions of S_1 and S_2. Note: $V_{G\{RC\}}$ is the voltage across R_G assuming $L_{X1} = L_{X2} = 0$ and $V_{G\{LX\}}$ is the voltage across R_G assuming $R_{C1} = R_{C2} = 0$. Thus the net voltage $V_G = V_{G\{RC\}} + V_{G\{LX\}}$

inductive sensor, generating voltages V_{LX1}, V_{RC1}, V_{LX2} and V_{RC2} across the components L_{X1}, R_{C1}, L_{X2} and R_{C2} respectively. The proposed LDC operates in two phases: (a) ~~the~~ integration phase for a time of N_1T sec and (b) the de-integration phase for N_2T sec, where N_1 is a predefined count and N_2 is the count to be measured. The operation of both the phases are detailed below and the corresponding waveforms are depicted in Fig. 2

Integration Phase: The integration phase starts at $t = 0$, as shown in Fig. 2. Switches $S1$ and $S2$ Are maintained at positions 1 and 0, respectively, during $0 \le t \le T/4$. Then the current IG is given as,

$$I_G = \frac{V_{S1} - V_{S2}}{R_G} = \frac{V_C - V_{L1}}{R_G} = -\frac{V_{LX1} + V_{RC1}}{R_G} = \frac{-V_R}{R_r R_G}\sin(2\pi ft)\left[j\omega L_{X1} + R_{C1}\right].$$

As this current I_G flows through the feedback capacitors C of op-amps OA_2 and OA_3, the voltages at V_{O2} and V_{O3} can be expressed as in (2) and (3) respectively during $0 \le$

$t \leq T/4$.

$$
\begin{aligned}
V_{O2\left[0,\frac{T}{4}\right]} &= V_C + \frac{1}{C}\int_0^{\frac{T}{4}} \frac{-V_R}{R_r R_G}\left[2\pi f L_{X1}\cos(2\pi ft) + \sin(2\pi ft)R_{C1}\right]dt \\
&= V_C + \frac{-V_R}{R_r R_G C}\left[L_{X1} + \frac{R_{C1}}{2\pi f}\right]
\end{aligned} \tag{2}
$$

$$
\begin{aligned}
V_{O3\left[0,\frac{T}{4}\right]} &= V_{L1} - \frac{1}{C}\int_0^{\frac{T}{4}} \frac{-V_R}{R_r R_G}\left[2\pi f L_{X1}\cos(2\pi ft) + \sin(2\pi ft)R_{C1}\right]dt \\
&= V_{L1} + \frac{V_R}{R_r R_G C}\left[L_{X1} + \frac{R_{C1}}{2\pi f}\right]
\end{aligned} \tag{3}
$$

Thus the output voltage V_{O4} can be expressed as in (4) by substituting (2) and (3) in (1).

$$
V_{O4\left[0,\frac{T}{4}\right]} = \left(V_C + V_{O3\left[0,\frac{T}{4}\right]}\right) - \left(V_{L1} + V_{O2\left[0,\frac{T}{4}\right]}\right) = \frac{2V_R}{R_r R_G C}\left[L_{X1} + \frac{R_{C1}}{2\pi f}\right] \tag{4}
$$

The switches $S1$ and $S2$ are maintained at position 0 and 1, respectively, during $T/4 \leq t \leq 3T/4$, Generating current IG given as,

$$
I_G = \frac{V_{S1} - V_{S2}}{R_G} = \frac{V_{L1} - V_C}{R_G} = \frac{V_{LX1} + V_{RC1}}{R_G} = \frac{V_R}{R_r R_G}\sin(2\pi ft)\left[j\omega L_{X1} + R_{C1}\right].
$$

ThusV_{O2} and V_{O3} can be expressed as in (5) and (6), respectively. Substituting (5) and (6) in (1) results in (7).

$$
\begin{aligned}
V_{O2\left[\frac{T}{4},\frac{3T}{4}\right]} &= V_{L1} + \frac{1}{C}\int_{\frac{T}{4}}^{\frac{3T}{4}} \frac{V_R}{R_r R_G}\left[2\pi f L_{X1}\cos(2\pi ft) + \sin(2\pi ft)R_{C1}\right]dt \\
&= V_{L1} - \frac{2V_R L_{X1}}{R_r R_G C}
\end{aligned} \tag{5}
$$

$$
\begin{aligned}
V_{O3\left[\frac{T}{4},\frac{3T}{4}\right]} &= V_C - \frac{1}{C}\int_{\frac{T}{4}}^{\frac{3T}{4}} \frac{V_R}{R_r R_G}\left[2\pi f L_{X1}\cos(2\pi ft) + \sin(2\pi ft)R_{C1}\right]dt \\
&= V_C + \frac{2V_R L_{X1}}{R_r R_G C}
\end{aligned} \tag{6}
$$

$$
V_{O4\left[\frac{T}{4},\frac{3T}{4}\right]} = \left(V_{L1} + V_{O3\left[\frac{T}{4},\frac{3T}{4}\right]}\right) - \left(V_C + V_{O2\left[\frac{T}{4},\frac{3T}{4}\right]}\right) = \frac{4V_R L_{X1}}{R_r R_G C} \tag{7}
$$

During $3T/4 \leq t \leq T$, the switches S_1 and S_2 are maintained at position 1 and 0, similar to that during $0 \leq t \leq T/4$. Thus the expressions for the voltages V_{O2}, V_{O3} and V_{O4} are obtained as in (8), (9) and (10).

$$
V_{O2\left[\frac{3T}{4},T\right]} = V_C + \frac{1}{C}\int_{\frac{3T}{4}}^{T} \frac{-V_R}{R_r R_G}\left[2\pi f L_{X1}\cos(2\pi ft) + \sin(2\pi ft)R_{C1}\right]dt
$$

$$= V_C - \frac{V_R}{R_r R_G C}\left[L_{X1} - \frac{R_{C1}}{2\pi f}\right] \tag{8}$$

$$V_{O3\left[\frac{3T}{4},T\right]} = V_{L1} - \frac{1}{C}\int_{\frac{3T}{4}}^{T} \frac{-V_R}{R_r R_G}\left[2\pi f L_{X1}\cos(2\pi f t) + \sin(2\pi f t) R_{C1}\right]dt$$

$$= V_{L1} + \frac{V_R}{R_r R_G C}\left[L_{X1} - \frac{R_{C1}}{2\pi f}\right] \tag{9}$$

$$V_{O4\left[\frac{3T}{4},T\right]} = \left(V_C + V_{O3\left[\frac{3T}{4},T\right]}\right) - \left(V_{L1} + V_{O2\left[\frac{3T}{4},T\right]}\right) = \frac{2V_R}{R_r R_G C}\left[L_{X1} - \frac{R_{C1}}{2\pi f}\right] \tag{10}$$

As a result, the net voltage V_{O4} after one complete cycle can be expressed as,

$$V_{O4} = V_{O4\left[0,\frac{T}{4}\right]} + V_{O4\left[\frac{T}{4},\frac{3T}{4}\right]} + V_{O4\left[\frac{3T}{4},T\right]}$$

$$= \frac{2V_R}{R_r R_G C}\left[L_{X1} + \frac{R_{C1}}{2\pi f}\right] + \frac{4V_R L_{X1}}{R_r R_G C} + \frac{2V_R}{R_r R_G C}\left[L_{X1} - \frac{R_{C1}}{2\pi f}\right]$$

$$V_{O4} = \frac{8V_R L_{X1}}{R_r R_G C} \tag{11}$$

Thus from (11) it can be observed that the output voltage V_{O4} is dependent only on L_{X1} and is made independent of coil resistance R_{C1}. As the operation is repeated for N_1 cycles of the input sinusoid, the resultant voltage $V_{O4[\text{int}]}$ can be expressed as given in (12).

$$V_{O4[int]} - \frac{8V_R L_{X1}}{R_r R_G C} N_1 \tag{12}$$

De-integration Phase: During this phase, the CLU produces control signals in such a way that the switches S_1 and S_2 are switched between the positions 1 and 2 as follows:

During $0 \leq t \leq T/4$: S_1 at 1and S_2 at 2.

During $T/4 \leq t \leq 3T/4$: S_1 at 2 and S_2 at 1.

During $3T/4 \leq t \leq T$: S_1 at 1 and S_2 at 2.

Thus the input to the dual slope during the de-integration phase will be V_{L2} and V_C. When V_{L1} and V_C was taken, during the integration phase, output voltage V_{O4} was obtained as given in (11). Similar to this when V_{L2} and V_C is given during the de-integration phase, the output V_{O4} can be obtained as,

$$V_{O4} = \frac{-8V_R L_{X2}}{R_r R_G C}.$$

The negative polarity of the voltage during the de-integration phase is due to the fact that the effective current I_G is in the opposite direction to that of during the integration phase. As such, the voltage V_{O4} starts decreasing linearly from $V_{O4[int]}$. This will continue during the de-integration until the voltage V_{O4} crosses zero as illustrated in Fig. 2. This zero-crossing marks the end of the de-integration phase, which is detected by the

CLU by continuously monitoring the output of OC_2. The CLU counts the number of time periods (T) required till the zero crossing of voltage V_{O4}. As mentioned in Sect. 2.1, if the total count required during the de-integration period for zero crossing of V_{O4} be N_2, then the total change in voltage V_{O4} during the de-integration period is given as (13).

$$V_{O4[de-int]} = \frac{-8V_R L_{X2}}{R_r R_G C} N_2 \tag{13}$$

Thus (14) can be obtained from (12) and (13) since the magnitude of the voltage change is same during the integration and deintegration phases: $|V_{O4[int]}|=|V_{O4[\text{de-}int]}|$.

$$\frac{L_{X1}}{L_{X2}} = \frac{N_2}{N_1} \tag{14}$$

Applying component-dividendo rule on (14), we get (15).

$$\frac{L_{X1} - L_{X2}}{L_{X1} + L_{X2}} = \frac{\pm\Delta L}{L_0} = \frac{N_2 - N_1}{N_2 + N_1} \tag{15}$$

$$\pm\Delta L = \left(\frac{N_2 - N_1}{N_2 + N_1}\right) L_0 \tag{16}$$

Thus the change in the sensor inductance ΔL for the differential pair can be expressed as (16). In cases, where the measurand (x) varies linearly with the change in inductance (ΔL), the sensor inductance can be represented as.

$$L_{X1} = L_0(1 \pm kx) \text{ and } L_{X2} = L_0(1 \mp kx) \tag{17}$$

where $kx = \Delta L/L_0$, and k represents the transformation constant. And in cases where the measurand varies inversely with respect to ΔL, the inductance can be expressed as.

$$L_{X1} = \frac{L_0}{(1 \pm kx)} \text{ and } L_{X1} = \frac{L_0}{(1 \mp kx)} \tag{18}$$

Thus substituting (17) or (18) in (15), the final output provided by the proposed LDC is obtained, which can be seen to be always proportional to the measurand. This further enhances the effectiveness of the converter.

Auto-zero Phase: The auto-zero phase is initiated initially to ensure that the output voltage V_{O4} is zero before the operation. During the auto-zero phase, the CLU monitors the polarity of voltage V_{O4}. when $V_{O4} > 0$, V_{OC2} will be high and hence the switching operations of S_1 and S_2 will be executed similar to that of the de-integration phase to bring back the voltage at V_{O4} to zero. Similarly, the integration operation will be executed by CLU if $V_{O4} < 0$ to bring the voltage at V_{O4} to zero. auto-zero phase will be terminated immediately when the comparator level changes. Hence $V_{O4} = 0$ at the end of auto-zero phase.

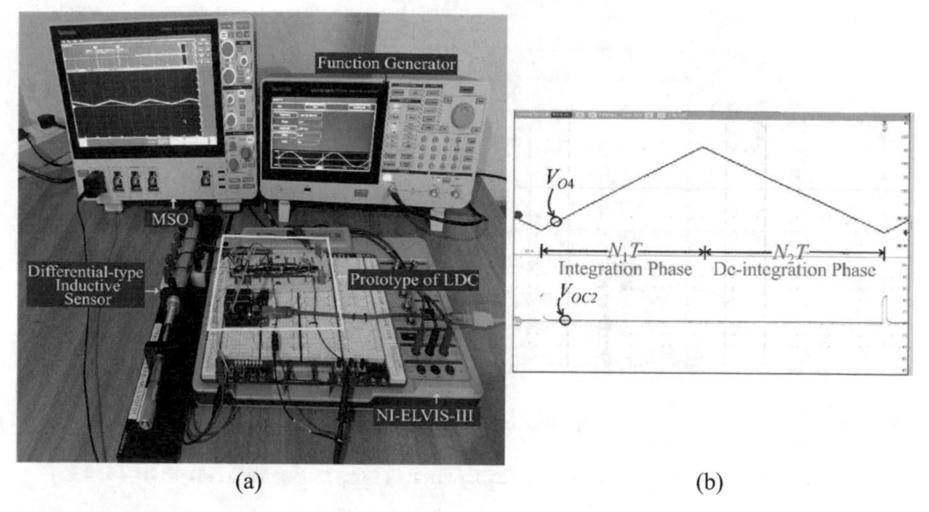

Fig. 3. (a) Developed prototype of the proposed LDC, (b) Waveform V_{O4} and V_{OC2} observed in the Mixed Signal Oscilloscope (MSO) MSO44 4-BW-200 from TEKTRONIX when $\Delta L = 3$ mH.

3 Experimental Setup and Results

3.1 Experimental Setup

A prototype of the proposed LDC was developed in the laboratory, as shown in Fig. 3(a). The op-amps OA_1, OA_2, OA_3 and OA_4 were implemented using the LF347 IC from Texas Instruments (TI). The Comparator IC- LM311 from TI was used to implement OC_1 and OC_2. The 4 channel analog multiplexer IC MAX309 from MAXIM Integrated was used to implement the SPTT switches. Both the resistors R_1 and R_G were chosen to have resistance values of 100 kΩ, and the capacitor C was set to 0.1 μF. The differential type inductive sensor was emulated using a variable inductance box with $L_0 = 14$ mH, where the inductance values were measured using 6-½ digit LCR meter E4980AL provided by KEYSIGHT. Input signal of 5 V peak to peak amplitude and a frequency of 10 kHz was generated using an arbitrary function generator AFG31051 from TEKTRONIX. The integration period was set to have $N_1 = 2000$ cycles of the input signal. The Control and logic unit was realised using Atmega328 microcontroller from Atmel Corporation, which operates with a 16 MHz clock frequency. Figure 3(b) shows the voltage V_{O4} observed when $\Delta L = 3$ mH. The details of various tests performed using the developed prototype and the results observed are detailed in the following section.

3.2 Results

Linearity Test: The linearity test of the developed prototype was performed by varying ΔL from -6 mH to 6 mH at steps of 0.5 mH with $L_0 = 14$ mH. The CLU'S serial monitor was used to record the de-integration count N_2. A worst-case linearity error of 0.55% was observed throughout the test. Figure 4 shows the linearity plot between $\frac{N_2-N_1}{N_2+N_1}$ and $\frac{\Delta L}{L_0}$.

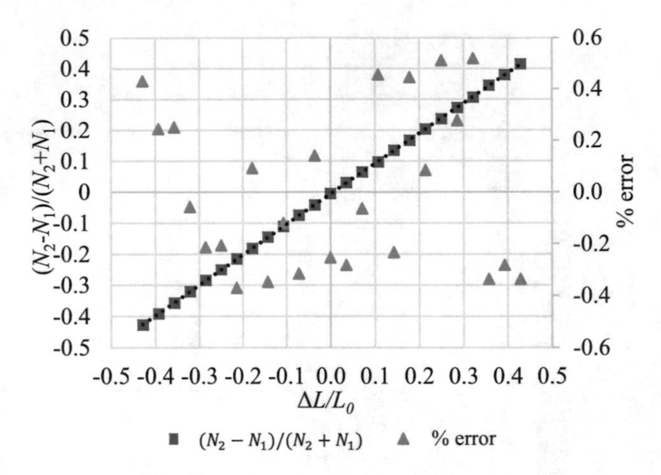

Fig. 4. Plot showing $\frac{N_2-N_1}{N_2+N_1}$ and % Error when $\Delta L/L_0$ varies from –0.43 to +0.43

Cross-Sensitivity Test: The sensitivity of the measurement towards the change in coil resistance was tested. During the test, two variable resistors were connected in series with the DIS and ΔL was set to –3 mH during the test. The insensitivity towards the change in resistance was evaluated using two methods. In method-1, both the series variable resistors were varied equally from 40 Ω to 1 kΩ in steps of 100 Ω and N_2 was noted for each step. This test observed a worst case error of 0.43%. This test demonstrates the Circuit's capability to measure inductance accurately even when differential sensors with differing series resistance are interfaced. In the second method, ΔL and the series resistor R_{C1} were kept constant at 3 mH and 40 Ω respectively, and the resistance R_{C2} was varied from 40 Ω to 1040 Ω in steps of 100 Ω. The final output was noted for each step change in R_{C2} and it was found out that the worst case error was 0.72%. This test shows the effectiveness of the circuit in measuring the inductance, even if there is a mismatch between R_{C1} and R_{C2}. The measurement error incurred at each step of the resistance variation during both the methods is tabulated in Table 1.

Table 1. Cross sensitivity test

Method-1	ΔL= - 3 mH										
R_{C1} **(in Ω)**	40	140	240	340	440	540	640	740	840	940	1040
R_{C2} **(in Ω)**	40	140	240	340	440	540	640	740	840	940	1040
FSE (%)	0	0	0.17	0.25	0.34	0.34	0.08	0.08	0.43	0.25	0.43
Method-2	ΔL=3 mH										
R_{C1} **(in Ω)**	40	40	40	40	40	40	40	40	40	40	40
R_{C2} **(in Ω)**	40	140	240	340	440	540	640	740	840	940	1040
FSE (%)	0	0.11	0.33	0.51	0.65	0.65	0.72	0.72	0.65	0.39	0.10

Table 2. Repeatability test

Parameter	Equation	Observed performance during measurement	
		ΔL= 3 mH	ΔL= -3 mH
SD (σ)	$\sqrt{\frac{1}{M-1}\sum_{i=1}^{M}(S(i)-\bar{S})^2}$	2.02 μH	4.34 μH
SNR	$10\log\frac{\sum_{i=1}^{M}(S(i))^2}{\sum_{i=1}^{M}(S(i)-\bar{S})^2}$	63.17 dB	56.77 dB
RE	$\Delta_{Smax}/(S_u - S_l)$	0.048 %	0.057 %
ENOB	$(SNR - 1.76)/6.02$	10.2 bits	9.14 bits

$S(i)$: i^{th} inductance measured value
$\bar{S}$: Average value of the measurements
Δ_{Smax} : Maximum difference between multiple measurements
S_u : Upper limit of the measurement range
S_l : Lower limit of the measurement range

Repeatability Test: The repeatability test was carried out for the developed LDC when ΔL was set at –3 mH and + mH and 500 consecutive measurements were recorded. The standard deviation (SD), signal-to-noise ratio (SNR), repeatability error (RE) and Effective Number of Bits (ENOB) for both the cases were computed using the approach mentioned in [15], and are presented in Table 2.

4 Discussion

The detailed comparison of the proposed method with the existing architectures is given in Table 3. The methods mentioned in [8, 9, 15] and [14] are sensitive to the coil resistance. The architecture in [9] reported an error of 0.125% when a mismatch of 0.5 Ω is experienced between R_{C1} and R_{C2}. The experimental results of the developed prototype show better results in terms of immunity towards the effects of the variations in the coil resistance. The proposed architecture provides accurate results with a maximum error of 0.72% when the mismatch between the resistances is 1 kΩ and an error of 0.43% when both resistances are varied equally from 40 Ω to 1040 Ω. For the proposed LDC, one count change in N_2 corresponds to a change in the sensor inductance by 3.5 μH. This resolution can be further enhanced by increasing the count of N_1.

Table 3. Comparison of the proposed LDC with existing architectures for differential-type inductive sensors

Reference	Inductance range (mH)	Sensitivity towards coil resistance	Coil resistance considered	Resolution (μH)	Linearity Error (%)
[8]	5 to 40	Sensitive	141.9 Ω	78	1
[9]	NA	Sensitive	68.5 Ω	NA	0.8
[12]	-5 to 60	Sensitive	15 Ω	NA	0.2
[13]	NA	Sensitive	NA	NA	0.5
[14]	-18 to 18	Sensitive	NA	NA	0.47
[16]	1.04 to 1.56	Sensitive	12 Ω	NA	0.4
This work	-6 to 6	Insensitive	0 to 1000 Ω	3.5	0.55

NA- Not Available

5 Conclusion

An efficient direct digital converter for differential type inductive sensors whose measurement is insensitive to the coil resistance was proposed and developed. A modified dual slope conversion technique was used to obtain the digital output proportional to ΔL. The errors due to the tolerance of the components can be eliminated in the final output by the proposed ratio metric approach, which further enhances the effectiveness of this scheme. The developed architecture efficiently eliminates the effect of the series coil resistance with a worst case error of 0.72% being observed when the mismatch between the two coil resistance is varied from 0 to 1 kΩ. This shows the efficacy of the proposed method in eliminating the mismatch in the coil resistance occurring due to the presence of the measurand or due to the defects arising from the sensor structure while manufacturing. The prototype also showed a worst case error of 0.42% when both the resistors (R_{C1} and R_{C2}) were varied equally from 40 Ω to 1 kΩ, indicating the effectiveness of the proposed easy-to-use interfacing scheme for DIS having various inductance and resistance values. The developed prototype exhibited a worst-case linearity error of 0.55% when ΔL was varied from -6 mH to + 6 mH in steps of 0.5 mH. The proposed readout circuit can be easily interfaced with a differential-type inductive sensor for various applications for efficient estimation of speed, angle, distance, and proximity.

Acknowledgement. Authors would like to thank the Department of Science and Technology-Science and Engineering Research Board (DST-SERB) for financial assistance received through Start-up Research Grant (SRG) FILE NO. SRG/2020/001021.

References

1. Sandra, K.R., Kumar, A.S.A., George, B., Kumar, V.J.: A linear differential inductive displacement sensor with dual planar coils. IEEE Sens. J. **19**(2), 457–464 (2019)
2. Drumea, A., Svasta, P., Blejan, M.: Modelling and simulation of an inductive displacement sensor for mechatronic systems. In: 33rd International Spring Seminar on Electronics Technology, ISSE. IEEE (2010)
3. Quattrone, F.,, Ponick, B.: Evaluation of a permanent magnet synchronous machine with a rotor coil for improved self-sensing performance at low speed. In: XXII International Conference on Electrical Machines (ICEM) (2016)
4. Chilan, C., Xiaogang, W., Yuewei, B., Yanchun, X., Kai, L.: The design of a new differential inductance type level. In: Third International Conference on Measuring Technology and Mechatronics Automation (2011)
5. Bakhoum, E.G., Cheng, M.H.M.: High-sensitivity inductive pressure sensor. IEEE Trans. Instrum. Meas. **60**(8), 2960–2966 (2011)
6. Zmood, R.B., Zhang, Y.C., Yu, P.L.: A non-contact inductive position sensor for use in micro machines. In: IEEE Proceedings of the International Solid-State Sensors and Actuators Conference-TRANSDUCERS 1995, vol. 1 (1995)
7. Fouse, T.M., et al.: Transducer for measuring dynamic translation by differential variable reluctance. U.S. Patent Application No. 11/805, 264.Z
8. Kokolanski, M.G., Reverter, F.: Differential inductive sensor-to-microcontroller interface circuit. In: IEEE International Instrumentation and Measurement Technology Conference (I2MTC) (2019)
9. Ramadoss, N., George, B.: A simple microcontroller-based digitizer for differential inductive sensors. In: IEEE International Instrumentation and Measurement Technology Conference (I2MTC) Proceedings (2015)
10. Ganesan, H., George, B., Aniruddhan, S.: Design and analysis of a relaxation oscillator-based interface circuit for LVDT. IEEE Trans. Instrum. Meas. **68**(5), 1261–1270 (2019)
11. Ganesan, H., George, B., Aniruddhan, S., Haneefa, S.: A dual slope LVDT-to-digital converter. IEEE Sens. J. **19**(3), 868–876 (2019)
12. Philip, V.N.: Novel switched capacitor direct digital converters for variable reluctance sensors. In: IEEE First International Conference on Control, Measurement and Instrumentation (CMI) (2016)
13. Reverter, F., Gasulla, M.: Timer-based demodulator for AM sensor signals applied to an inductive displacement sensor. IEEE Trans. Instrum. Meas. **66**(10), 2780–2788 (2017)
14. Rana, S., Boby, G., Jagadeesh Kumar, V.: Self-balancing signal conditioning circuit for a novel noncontact inductive displacement sensor. IEEE Trans. Instrum. Measur. **66**(5), 985–991 (2017)
15. Xu, L., Sun, S., Cao, Z., Yang, W.: Performance analysis of a digital capacitance measuring circuit. Rev. Sci. Instrum. **86**(5), 054703 (2015)
16. Philip, N., George, B.: Design and analysis of a dual-slope inductance-to-digital converter for differential reluctance sensors. IEEE Trans. Instrum. Meas. **63**(5), 1364–1371 (2014)
17. Prommee, P., Angkeaw, K., Karawanich, K.: Low-cost linearity range enhancement for linear variable differential transformer. IEEE Sens. J. **22**(4), 3316–3325 (2022)
18. Mohan, A., Mohanasankar, S., Kumar, V.J.: Successive approximation type digital converter for floating-wiper inductive displacement sensor. In: Eleventh International Conference on Sensing Technology (ICST), (2017)

A Novel Relaxation Oscillator Based Single-Element Type Inductance-to-Time Converter

P. P. Narayanan[1](✉), Sreenath Vijaykumar[1], Jalakam Venu Madhava Sai[2], S. B. Kavya Preetha[2], and Karthi Balasubramanian[2]

[1] Department of Electrical Engineering, Indian Institute of Technology Palakkad, Palakkad, India
122004005@smail.iitpkd.ac.in, sreenath@iitpkd.ac.in

[2] Department of Electronics and Communication Engineering, Amrita School of Engineering, Amrita Vishwa Vidyapeetham, Coimbatore, India
b_karthi@cb.amrita.edu

Abstract. This paper presents an efficient Inductance-to-Time Converter (LTC) that provides a quasi-digital output whose pulse-on time is proportional to the inductance of a single-element type sensor. The LTC operates with a negative feedback mechanism, which ensures that the final output is independent of the linearity error present in the forward path elements of the converter. The proposed relaxation oscillator-based operation can enhance the update rate of the inductance measurement. In addition, the inductance measured using the converter is made insensitive to the series coil-resistance of the sensor. A prototype of the LTC with an update rate of 10 kHz was developed in the laboratory, with an excellent linearity performance along with a reduced settling time. It was observed that the design produced a worst-case linearity error of 0.29% and a settling time of 12.4 ms when the inductance was varied from 0 to 20 mH. Furthermore, the prototype showed negligible sensitivity towards the coil resistance of the sensor, with a worst-case error of 0.65%, when the series resistance was varied from 40 Ω to 1040 Ω. The test results indicate superior performance of the developed prototype over existing architectures.

Keywords: Inductance to Time Converter (LTC) · Pulse Width Modulation (PWM) · Single-element-type Inductive Sensor · Relaxation Oscillator

1 Introduction

Single-element type inductive sensors are passive components whose inductances vary with the measurand. The inductances of such sensors are generally denoted as $L_X = L_0 \pm \Delta L$, where L_0 represents the nominal inductance of the sensor in the absence of the measurand and ΔL represents the change in inductance due to the measurand. Single-element type inductive sensors are used in a wide range of applications including measurement of proximity [1], distance [2], speed [3], angle [4], and flow [5]. These

N. K. Suryadevara et al. (Eds.): ICST 2022, LNEE 1035, pp. 286–294, 2023.
https://doi.org/10.1007/978-3-031-29871-4_29

sensors also find use in fields like crack detection [6], oil debris sensing [7], and in different biomedical applications [8].

Various techniques [9–15] have been explored in the past for efficient measurement of inductances of single-element type inductive sensors. The works reported in [9] and [10] eliminate the dependence of the inductance value on the coil resistance by using matched components in the interfacing circuit. However, it may not be practical to exactly match the coil resistance of the sensor and hence the error due to this mismatch will affect the operation. Also, [9] and [10] require a dedicated Analog to Digital Converter (ADC) for digitization. A microcontroller-based model is proposed in [11–13] to measure the sensor inductance. However, a noticeable error will be introduced in presence of the coil resistance. A dual slope based inductance to digital converter is proposed in [14] to measure the inductance of a single-element type inductive sensor. However, due to the limitation on the update rate of this converter, it is not suitable for high-speed applications.

This paper proposes an efficient Inductance-to-Time converter (LTC) that can accurately measure the inductance of a single-element type inductive sensor independent of its coil resistance and can overcome the limitations of the existing architectures. The converter can provide a pulse width modulated (PWM) output signal whose pulse on-time is proportional to the sensor inductance. In addition, an efficient feedback mechanism employed in the proposed converter provides the final output independent of the linearity errors in the forward path blocks, and in inductance to voltage, and voltage to pulse converter blocks. This feedback mechanism can also enhance the update rate of the system. In addition, the final output is made independent of coil resistance and linearity errors in the forward paths without any error-correcting circuitry. The sections below explain the LTC's operation, the experimental setup, and the results obtained.

2 Operation of the Proposed Inductance to Time Converter

2.1 Circuit Description

The proposed relaxation oscillator based LTC circuit is shown in Fig. 1. The single-element type inductive sensor is represented by inductance L_X (=$L_0 \pm \Delta L$) and coil resistor R_C. The proposed LTC consists of various blocks such as sensor to voltage converter, integrator, 90° phase shifter, sawtooth generator and switched feedback network. The sensor-to-voltage converter is made up of op-amp OA_1, inductive sensor and resistor R_1 and is powered by an AC signal ($V_{in} = V_S\, sin(2\pi ft)$). S_1 and S_2 represent Single Pole Double Throw (SPDT) switches, and S_3 represents a Single Pole Single Throw (SPST) switch. The switches S_1 and S_3 are controlled by the comparator (OC_1) output signal V_{OC1}. The integrator comprises of op-amp OA_2, the resistor R_1, and the capacitor C_1. The sawtooth generator is made up of the Op-amp OA_2, resistor R_T, capacitor C_T, switch S_3 and reference voltage $-V_R$. The comparator OC_2 and the switch S_2 comprise the switched feedback network. The comparator OC_2 compares the sawtooth signal V_T with the voltage V_{O2} to generate the control signal for the switch S_2. E_1 represents an XOR gate, where $V_O = V_{OC2} \oplus V_{OC1}$. The operation of each block of the converter is detailed in the section below.

2.2 Operation

Sensor-to-Voltage Converter: The inductive sensor is connected in the feedback path of an inverting amplifier which allows a current of value v_{in}/R_S to flow through the sensor. Thus the resultant voltage at v_{O1} can be expressed as

$$v_{O1} = -\left(\frac{R_C}{R_S} + i\frac{2\pi f L_X}{R_S}\right) V_S sin(2\pi ft). \quad (1)$$

Thus, the voltage v_{O1} is proportional to the inductance and coil resistance of the sensor.

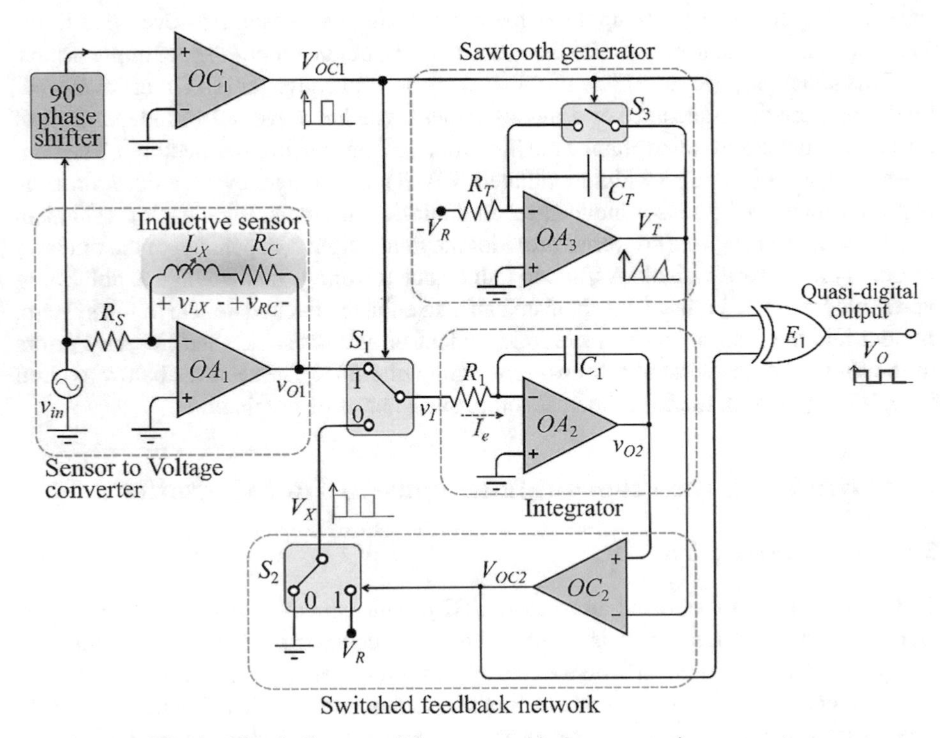

Fig. 1. Circuit diagram of the proposed inductance to time converter

Sawtooth Generator: The switch S_3 is set at the ON and OFF positions by the control signal V_{OC1}. When V_{OC1} is LOW, the switch S_3 is set at the OFF position and the voltage V_T increases linearly, generating a ramp signal with a slope given by $\frac{V_R}{R_T C_T}$. When V_{OC1} is HIGH the switch S_3 is set at ON position, which instantly discharges the capacitor C_T to zero. As V_{OC1} periodically alternates between HIGH and LOW potentials, a sawtooth wave V_T whose time period is the same as that of the input waveform is generated, as illustrated in Fig. 2.

Switched Feedback Network: The feedback network of the converter comprises of the comparator OC_2 and the switch S_2. The switch S_2 is set in position 1 or 0 when the voltage V_{OC2} is at a HIGH or LOW potential, respectively. The voltage V_{OC2} is generated by comparing the sawtooth wave V_T with the integrator voltage v_{O2}. As a result, the voltage V_{OC2} will be HIGH when $v_{O2} > V_T$ and LOW when $v_{O2} < V_T$. Hence the pulse on-time of V_{OC2} (and also V_X) increases with voltage v_{O2}. This is illustrated in Fig. 2.

Integrator: The current i_e flowing through the integrator is $\frac{v_I}{R_1}$ and the output of the integrator can be expressed as

$$v_{O2} = \frac{-1}{C_1}\int i_e dt = \frac{-1}{R_1 C_1}\int v_I dt \tag{2}$$

The voltage levels at v_I can be v_{O1} or V_X depending on the position of the switch S_1. The switch S_1 is controlled by V_{OC1}, where V_{OC1} is HIGH when $V_S\cos(2\pi\text{ft}) > 0$ and LOW when $V_S\cos(2\pi\text{ft}) < 0$. Thus the voltage at v_I during one complete cycle of v_{in} is given as,

$$v_I = \begin{cases} v_{O1} \; during \; 0 \le t \le \frac{T}{4} \\ V_X \; during \; \frac{T}{4} \le t \le \frac{3T}{4} \\ v_{O1} \; during \; \frac{3T}{4} \le t \le T \end{cases} \tag{3}$$

Substituting (3) in (2), the effective change in voltage at v_{O2} after each complete cycle can be obtained as:

$$v_{O2[0,T]} = \frac{-1}{R_1 C_1}\left\{\int_0^{\frac{T}{4}} v_{O1}dt + \int_{\frac{T}{4}}^{\frac{3T}{4}} V_X dt + \int_{\frac{3T}{4}}^{T} v_{O1}dt\right\}$$

$$= -\frac{1}{C_I R_I}\left\{\int_0^{\frac{T}{4}} -\left(\frac{R_C V_S}{R_S}\sin(2\pi ft) + \frac{2\pi f L_X V_R}{R_S}\cos(2\pi ft)\right)dt\right.$$

$$\left. + \int_{\frac{T}{4}}^{\frac{T}{4}+T_{ON}} V_S dt + \int_{\frac{3T}{4}}^{T} -\left(\frac{R_C V_S}{R_S}\sin(2\pi ft) + \frac{2\pi f L_X V_R}{R_S}\cos(2\pi ft)dt]\right)\right\}$$

$$v_{O2[0,T]} = \frac{V_S}{C_I R_I}\left(\frac{2L_X}{R_S} - T_{ON}\right);\; 0 \le t \le T. \tag{4}$$

Where T_{ON} is the pulse on-time duration of the signal v_I during $T/4 \le t \le 3T/4$. Thus, from (4), it can be clearly observed that as L_X increases, the voltage at v_{O2} also increases, which in turn increases the pulse-on time of V_X and V_{OC2}. Hence, T_{ON} increases (with L_X), settling the change in v_{O2} after each cycle to zero due to the negative feedback operation. Thus, from (4), the sensor inductance can be expressed as,

$$L_X = T_{ON}\frac{R_S}{2} \tag{5}$$

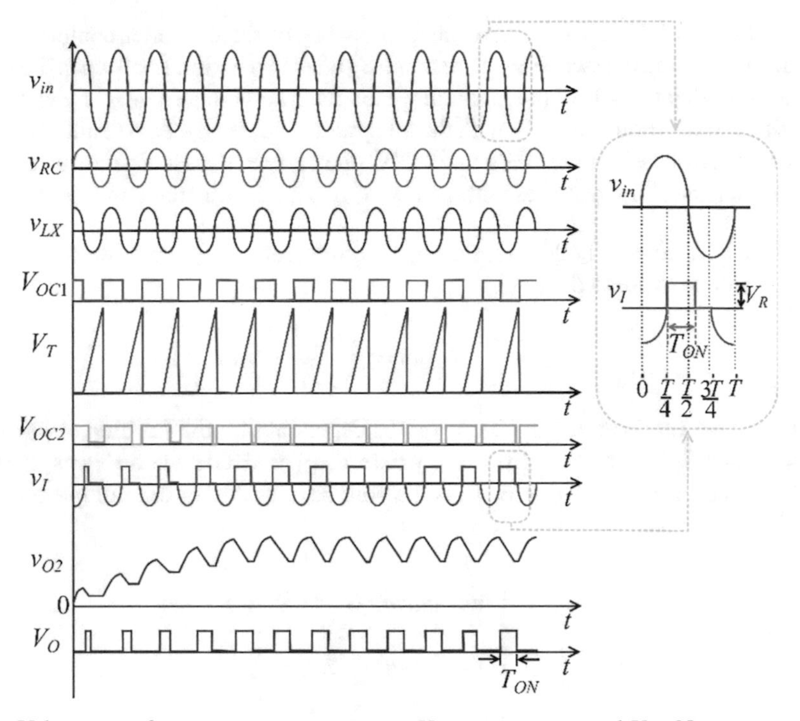

Fig. 2. Voltage waveforms v_{in}, v_{RC}, v_{LX}, v_{OC1}, V_T, v_{C2}, v_I, v_{O2} and V_O. Note: v_{RC} and v_{LX} are the voltage drops across the coil resistor and inductance of the sensor, respectively. The net voltage v_{O1} = -($v_{RC} + v_{LX}$)

Thus, with a preset value of R_S, the sensor inductance L_X can be seen to be directly proportional to the pulse on-time T_{ON}. It is to be noted that this remains true even if linearity errors are present in the forward path, thus enhancing the accuracy of the measurement. The time period T_{ON} can be measured accurately from the output V_O of the XOR gate E_1. The specifics of the experimental setup and test results are detailed in the next section.

3 Experimental Setup and Results

3.1 Experimental Setup

The prototype of the proposed LTC was developed in the laboratory to validate the functionality experimentally. OP227 IC from Analog Devices Inc. Was used to implement the op-amps OA_1, OA_2 and OA_3, while the comparators OC_1 and OC_2 were implemented using LM311 IC from Texas Instruments (TI). The switches S_1, S_2 and S_3 were implemented using MAX309 IC from MAXIM Integrated. The resistors R_S, R_1 and R_T were set to 1 kΩ, 10 kΩ and 100 kΩ respectively. The capacitors C_1 and C_T were set to 0.01 μF and 100 nF respectively. The reference voltage V_R was set to 2.5 V using a reference LM285-2.5 V IC. The input AC signal was set at 5 V peak to peak voltage and frequency of 10 kHz, using arbitrary function generator AFG31051 from TEKTRONIX. A decade inductance box, model LS-400L from IET Labs, was used to emulate the single-element inductive sensor. The developed prototype is shown in Fig. 3.

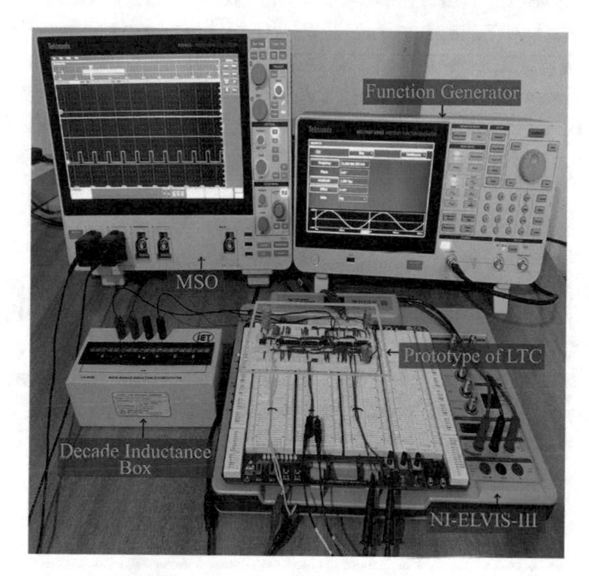

Fig. 3. Prototype of the proposed LTC along with the test equipment

3.2 Results

Linearity Test: The linearity performance of the developed prototype was tested by varying the inductance box from 0 to 20 mH in steps of 1 mH. The corresponding on-time T_{ON} was noted for each step change in L_X. The results observed are shown in

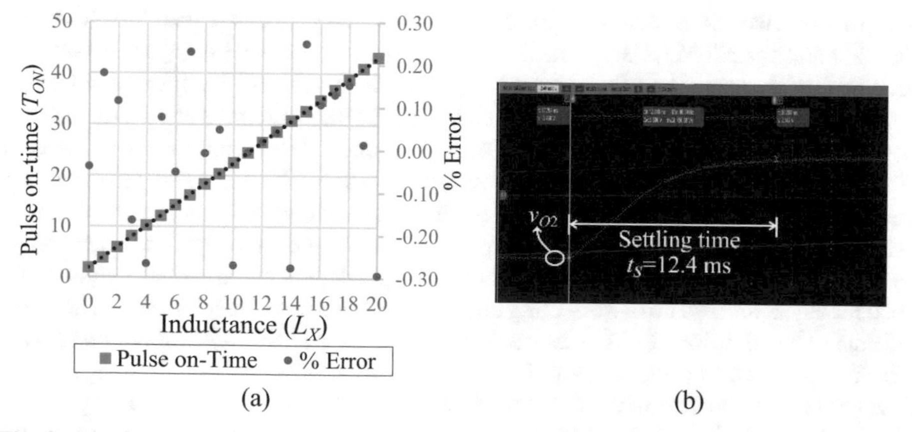

Fig. 4. **(a)** Linearity plot: pulse-on time (T_{ON}) and % Error vs change in inductance (L_X), **(b)** Settling time of v_{O2} observed in MSO when the L_X was varied instantly from 0 to 20 mH

Fig. 4(a). From the test, it was noted that the FS linearity error was 0.29% and FS error in reading was 0.49%.

Settling Time: The settling time behavior of the proposed LTC was examined by changing the inductance L_X instantly from 0 to 20 mH and observing the change in the integrator output V_{O2}. The settling time observed was 12.4 ms. This settling time was considered when the output was within the 2% limit of the final value. The change in voltage V_{O2} was observed using a Mixed Signal Oscilloscope (MSO) MSO44 4-BW-200 from TEKTRONIX. The captured output through the MSO is shown in Fig. 4(b).

Cross-Sensitivity Test: To test the dependence of the coil resistance on the inductance measurement, a variable resistance was connected in series with the inductive sensor. During the test, the inductance value was set constant at 10 mH and the series resistance was varied from 40 Ω to 1040 Ω in steps of 100 Ω and the corresponding L_X measured for each step change in resistance is noted, as tabulated in Table 1. During the test, the worst case error in reading was observed as 0.65%. This shows that the output has negligible sensitivity towards the series coil resistance of the sensor.

Table 1. Cross-sensitivity test with fixed sensor inductance $L_X = 10$ mH

R_C (in Ω)	40	140	240	340	440	540	640	740	840	940	1040
FSE (in %)	0	0.65	0.65	0.53	0.39	0.13	0.13	0	0.13	0.13	0.13

4 Discussion

The performance of the developed prototype is compared with existing architectures and the comparative analysis is shown in Table 2. The work reported in [9–13] are sensitive to series coil resistance. The LDC reported in [14] measures the ΔL independent of series coil resistance, but it is not suitable for high update rate measurement applications as the conversion time is limited to a few tens of milliseconds. The methodology outlined in [15] is unaffected by coil resistance. However, this has a very high linearity error and also the design is accomplished by mapping the inductance and coil resistance from the output using a sophisticated algorithm. In this paper, a novel inductance to time converter is developed, which can ensure that the output is independent of coil resistance without the need for any additional error-correcting circuitry or complicated mechanisms. The pulse on-time of the developed LTC was measured using an ATMEGA328 microcontroller, which was operated using a 16 MHz clock frequency. The one count change in the timer-counter in measuring the pulse-on time corresponds to a 31.25 μH change in the sensor. The resolution of this scheme can be further enhanced by using a microcontroller with a higher clock or by reducing the resistance value R_S. Furthermore, compared to existing architectures, the suggested approach provides a very low conversion time of 0.1 ms, which will be very useful for applications where the measurand varies abruptly. In addition to all these features offered by the proposed scheme, the linearity error observed is 0.29%, which is achieved using a simpler negative feedback mechanism. These results indicate that the suggested converter performs in a superior fashion as compared to the existing architectures.

Table 2. Comparison of the proposed LTC with existing architectures of single-element-type inductive sensors

Reference	Circuit complexity level	Inductance range (mH)	Sensitivity towards coil resistance	Resolution (μH)	Conversion time (ms)	Worst case linearity Error
[9]	High	0-50	Sensitive	NA	NA	2.0%
[10]	Medium	23-40	Sensitive	NA	NA	1.0%
[11]	Medium	10-100	Sensitive	97	~10	0.3%
[12]	Medium	0-8.8	Sensitive	26	NA	NA
[13]	Medium	0.1-1	Sensitive	2.2	NA	0.4%
[14]	High	69-87	Insensitive	6.25	~20	0.5%
[15]	High	0-14	Insensitive	NA	NA	4.0%
This work	Medium	0-20	Insensitive	31.25	0.1	0.29%

NA- Not Available

5 Conclusion

A novel LTC that can provide a PWM signal whose pulse-on time is proportional to the sensor inductance and is insensitive to the coil resistance is proposed and developed in this paper. The converter is also made insensitive to the linearity errors present in the inductance-to-voltage and voltage-to-pulse conversion blocks of the converter. This is achieved using an efficient negative feedback topology in the LTC. The details of various tests performed on the developed LTC are presented. The test results show that the output pulse-on time has a linear relationship with the variation in the inductance, with a worst case error of 0.29% when the inductance was varied from 0 to 20 mH. The proposed relaxation oscillator based scheme can measure the sensor inductance with an update rate of 10 kHz. A settling time of 12.4 ms was observed when the sensor inductance was varied abruptly from 0 to 20 mH. The insensitivity towards the coil resistance was validated by varying the series resistance of the sensor from 40 Ω to 1040 Ω. During this test, a worst-case error of 0.65% was noted. The proposed LTC is highly insensitive to coil resistance or non-linearity errors. As a result, any drift in these errors caused by environmental influences, measurand variations, or sensor ageing will have no effect on the LTC operation. Thus, the proposed converter is very effective in interfacing single-element type inductive sensors in various applications to measure distance, vibration, linear and angular speed.

Acknowledgement. Authors would like to thank the Department of Science and Technology-Science and Engineering Research Board (DST-SERB) for financial assistance received through Start-up Research Grant (SRG) FILE NO. SRG/2020/001021.

References

1. Kejık, P., Kluser, C., Bischofberger, R., Popovic, R.S.: A low-cost inductive proximity sensor for industrial applications. Sens. Actuators A: Phys. **110**(1–3), 93–97 (2004)
2. Fericean, S., Droxler, R.: New noncontacting inductive analog proximity and inductive linear displacement sensors for industrial automation. IEEE Sens. J. **7**(11), 1538–1545 (2007)
3. Suhas, R.C., Oberhauser, C.A.: Motor speed and position sensing using Inductive sensing (LDC) technology. In: IEEE Annual India Conference (INDICON), pp. 1–6 (2016)
4. Wogersien, A., Samson, S., Guttler, J., Beiftner, S., Biittgenbach, S.: Novel inductive eddy current sensor for angle measurement. In: IEEE SENSORS, vol. 1, pp. 236–241 (2003)
5. Lata, A., Mandal, N., Maurya, P., Roy, J.K., Mukhopadhyay, S.C.: Development of a smart rotameter with intelligent temperature compensation. In: 12th International Conference on Sensing Technology (ICST) (2018)
6. Zhang, K., et al.: Research on eddy current pulsed thermography for rolling contact fatigue crack detection and quantification in wheel tread. In: 18th International Wheelset Congress (IWC), pp. 5–11 (2016)
7. Muthuvel, P., George, B., Ramadass, G.A.: A planar inductive based oil debris sensor plug. In: 13th International Conference on Sensing Technology (ICST), pp. 1–4. IEEE (2019)
8. Yu, Y., Bhola, V., Tathireddy, P., Young, D.J., Roundy, S.: Inductive sensing technique for low power implantable hydrogel-based biochemical sensors. In: IEEE SENSORS, pp. 1–4 (2015)
9. Chattopadhyay, S., Bera, S.C.: Modification of the Maxwell–Wien bridge for accurate measurement of a process variable by an inductive transducer. IEEE Trans. Instrum. Meas. **59**(9), 2445–2449 (2010)
10. Kumar P., George, B., Jagadeesh Kumar, V.: A simple signal conditioning scheme for inductive sensors. In: Seventh International Conference on Sensing Technology (ICST), pp. 512–515. IEEE (2013)
11. Kokolanskia, Z., Jordanab, J., Gasullab, M., Dimceva, V., Reverter, F.: Direct inductive sensor-to-microcontroller interface circuit. Sens. Actuators, A **224**, 185–191 (2015)
12. Kokolanski, Z., Cubarsí, F.R., Gavrovski, C., Dimcev, V.: Improving the resolution in direct inductive sensor-to-microcontroller interface. Annu. J. Electron., 135–138 (2015)
13. Asif, A., Ali, A., Abdin, M.Z.U.: Resolution enhancement in directly interfaced system for inductive sensors. IEEE Trans. Instrum. Meas. **68**(10), 4104–4111 (2018)
14. Narayanan, P.P., Vijayakumar, S.: A novel single-element inductance-to-digital converter with automatic offset eliminator. In: IEEE International Instrumentation and Measurement Technology Conference (I2MTC), pp. 1–6 (2021)
15. Guo, Y.-X., Shao, Z.-B., Li, T.: An analog-digital mixed measurement method of inductive proximity sensor. MDPI Sens. **16**(1), 30 (2016)

Space and Time Data Exploration of Air Quality Based on PM10 Sensor Data in Greater Sydney 2015–2021

Lakmini Wijesekara(✉), Prathayne Nanthakumaran, and Liwan Liyanage

School of Computer, Data and Mathematical Sciences, Western Sydney University, Penrith, Australia
lakminikw@gmail.com, prathananthan@gmail.com, l.liyanage@westernsydney.edu.au

Abstract. Exposure to polluted air is associated with numerous adverse health effects for the general population. Therefore, it is important to monitor ambient air pollution which plays a key role in measuring the quality of the air we breathe. Particulate matter in the air with a diameter of 10 μm or less (PM10) is one of the important measurements of air quality. This paper presents a comprehensive space-time data exploration of daily PM10 measurements collected through sensors of the Greater Sydney region from 1 January 2015 to 31 December 2021 and clustering of air pollution monitoring sites based on Dynamic Time Warping (DTW) distance. According to the results, air quality was good on most days in all the places considered. The modes of the daily PM10 levels were varying spatially. Oakdale recorded the lowest mode in all the years considered. During the study period, daily PM10 levels exceeded the national air quality standards mostly in the autumn season. After 2020, the number of exceedances was reduced for all the monitoring sites except Campbelltown West and Liverpool. Further examination is needed to identify the reasons behind these exceedances. Clustering indicates four possible groups of sites according to the behaviour of the PM10 sensor data. The four clusters are Randwick-Chullora-Earlwood, Liverpool-Prospect, Bringelly and Richmond-Campbelltown West-Camden-Bargo-Oakdale.

Keywords: Clustering · Sensor data · Space-time exploration · Time series

1 Introduction

Exposure to air pollution creates numerous adverse effects on population health. Particulate matter in the air with a diameter of 10 μm or less (PM10) is one of the measurements which indicates outdoor air pollution. The common sources of PM10 include dust particles, smoke from fires, sea salt, car and truck exhausts and industry [1]. Studies report that exposure to PM10 is linked with asthma [2–4], cataract [5], cardiovascular diseases [6,7], lung cancer [8] and diabetes [9,10].

N. K. Suryadevara et al. (Eds.): ICST 2022, LNEE 1035, pp. 295–308, 2023.
https://doi.org/10.1007/978-3-031-29871-4_30

Moreover, it is reported that long-term exposure to PM10 is associated with premature death [11]. Therefore, it is essential to monitor air pollution and take necessary actions to reduce these health burdens.

National Environment Protection Council (NEPC), Australia has established air quality standards for key air pollutants including PM10 as part of the National Environment Protection Measure for Ambient Air Quality (Air NEPM). According to Air NEPM, the daily PM10 concentrations are recommended to be below 50 $\mu g/m^3$ and the goal is set to not allow for any exceedances ($>50\,\mu g/m^3$), excluding exceptional event days [12]. Exceptional events include fire or dust occurrence that is directly related to bushfire, jurisdiction-authorised hazard reduction burning or continental-scale windblown dust [13]. Therefore, analysis of exceedances is important to monitor whether the NEPC goal is achieved and to plan actions accordingly.

Sensors are used to monitor ambient air quality. Implementing cost-effective air quality monitoring systems is a growing research area [14–17]. Therefore, it is important to make the most use of the collected data. Identifying space-time characteristics of the pollutants is useful to make precautionary arrangements for the adverse effects caused by poor air quality. To the best of the authors' knowledge, no published study can be found by analyzing Greater Sydney's PM10 during the period 2015–2021 which includes Black Summer (late 2019-early 2020 bushfire season in Australia) and COVID-19 lockdown periods (first lockdown March-April 2020 and second lockdown June-August 2021 [18]). Duc et al. [18] have studied the effect of the lockdown period on air quality by considering a set of air pollutants which does not include PM10. Also, grouping locations based on the behaviour of PM10 over time will be useful in arranging similar precautionary measures in similar locations.

The objectives of this paper are to explore space-time characteristics of daily PM10 concentrations and exceedances (greater than 50 $\mu g/m^3$) in Greater Sydney from 1 January 2015 to 31 December 2021 and to identify clusters of pollution sites. The findings of this paper could be useful for the authorities to make evidence-based air policies and recommend strategies for air quality improvement.

2 Data and Methodology

New South Wales air quality monitoring network continuously measures air pollutants through a network of NATA (National Association of Testing Authorities)- accredited stations. PM10 is measured using Tapered Element Oscillating Microbalance (TEOM). This conforms to Australian and International Standards [19]. For this study, air quality data were accessed via an application programming interface (API) available on the website of the Department of Planning and Environment, NSW Government. PM10 daily measurements were downloaded from 1 January 2015 to 31 December 2021 at 18 monitoring sites in the Sydney region. A schematic representation of the methodology used in this study is given in Fig. 1.

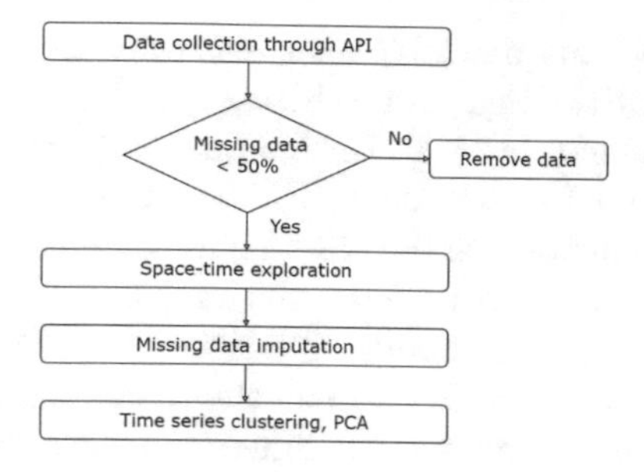

Fig. 1. Methodology

Missing value percentages, longest gap sizes and distribution of the missing values were analyzed in each of the PM10 series at different sites. Sites where the missing percentage is greater than 50% were removed from the analysis. Space-time characteristics of the rest of the sites were analysed. For the sites where the missing value percentage is less than 20%, missing values were imputed using Kalman Smoothing on Structural Time Series method in the imputeTS R package [20] as it has been proven efficient for relatively small missing values [21, 22]. Missing values with large gaps were imputed using a bi-directional method based on regularized regression models [22]. Sites with more than 30-consecutive-day gaps were removed from the analysis. Time series clustering and principal component analysis (PCA) were carried out using the remaining cleaned dataset. The final analysis includes 11 pollution monitoring sites as shown in Fig. 2.

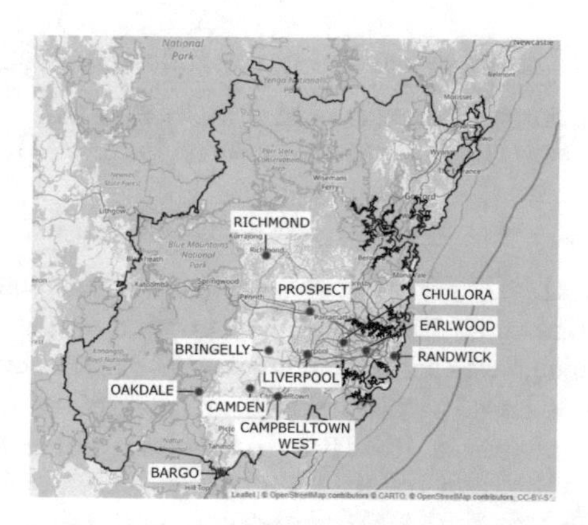

Fig. 2. Air quality monitoring sites in the Sydney Region

A variety of methods are available for time series clustering in the literature. Some of them are partitioning, hierarchical, grid-based, model-based, density based and multi-step clustering [23]. In this analysis, agglomerative hierarchical clustering [24] was applied for the sites (PM10 series). This method was chosen to understand the similarities of the sites using a dendrogram. Dynamic Time Warping (DTW) distance was used as the measure of dissimilarity. DTW is a shape-based dynamic programming algorithm that compares two series by finding the optimum warping path between them. For time series variables DTW distance make more sense than the Euclidean distance as it is less sensitive to noise, scale, and time shifts [25].

Let two series be Q and C where $Q = q_1, q_2, ..., q_n$, $C = c_1, c_2, ..., c_m$ with lengths n and m respectively. To calculate DTW distance, an n by m matrix is formed where element (i, j) contains the distance $d(q_i, c_j) =| q_i - c_j |$ between the two points q_i and c_j. Dynamic programming is employed to compute the cumulative distance matrix by applying the reiteration as follows:

$$\omega(i,j) = d(q_i, c_j) + min\{\omega(i-1, j=j-1), \omega(i-1, j), \omega(i, j-1)\} \quad (1)$$

where $\omega(i, j)$ is the cumulative distance and min represents the minimum of the cumulative distances of the adjacent elements.

The best match between two time series sequences is the one with the lowest distance path $W = \{w_1, w_2, \ldots, w_k, \ldots, w_c\}$ after aligning one time series to the other.

DTW distance is given by the Eq. 2.

$$DTW(Q, C) = \min \sum_{k=1}^{c} w_k \quad (2)$$

Warping paths are made by satisfying the following conditions.

(i) Boundary conditions: $w_1 = d(1, 1)$ and $w_k = d(n, m)$,
(ii) Continuity: $w_k = d(i_k, j_k)$ and $w_{k+1} = d(i_{k+1}, j_{k+1}), i_{k+1} - i_k \leq 1$ and $j_{k+1} - j_k \leq 1$,
(iii) Monotonicity: $i_{k+1} - i_k \geq 0$ and $j_{k+1} - j_k \geq 0$.

After calculating the DTW distances among the selected PM10 time series, agglomerative hierarchical clustering [24] was applied and the dendrograms were analysed. Average linkage was chosen to measure the inter-cluster distances. Even though DTW distance based clustering has many applications, it is rarely used to cluster pollution time series. Recently, Suris et al. [26] have used DTW-based clustering to group air pollution sites in Malaysia. They have used k-Means, partitioning, Fuzzy k-Means (FKM) algorithms and agglomerative hierarchical clustering based on complete linkage to cluster the air quality time series data. In contrast, our study used agglomerative hierarchical clustering based on average linkage and the clusters were further justified using geographical locations, elevations and principal component analysis of the time series.

Considering the sites as observations and the time points as variables, principal component analysis [24] was carried out and the data were visualized using the first two principal components.

3 Results and Discussion

The distribution of PM10 concentrations over space as well as time was explored to identify any interesting patterns. The exceedances were further analyzed to identify the locations with frequent exceedances.

3.1 Space-Time Exploration Based on Daily PM10 Concentrations

Figure 3 shows the distribution of daily PM10 concentrations at each site from 2015 to 2021 highlighting the days when PM10 concentration exceeded the Air NEPM threshold.

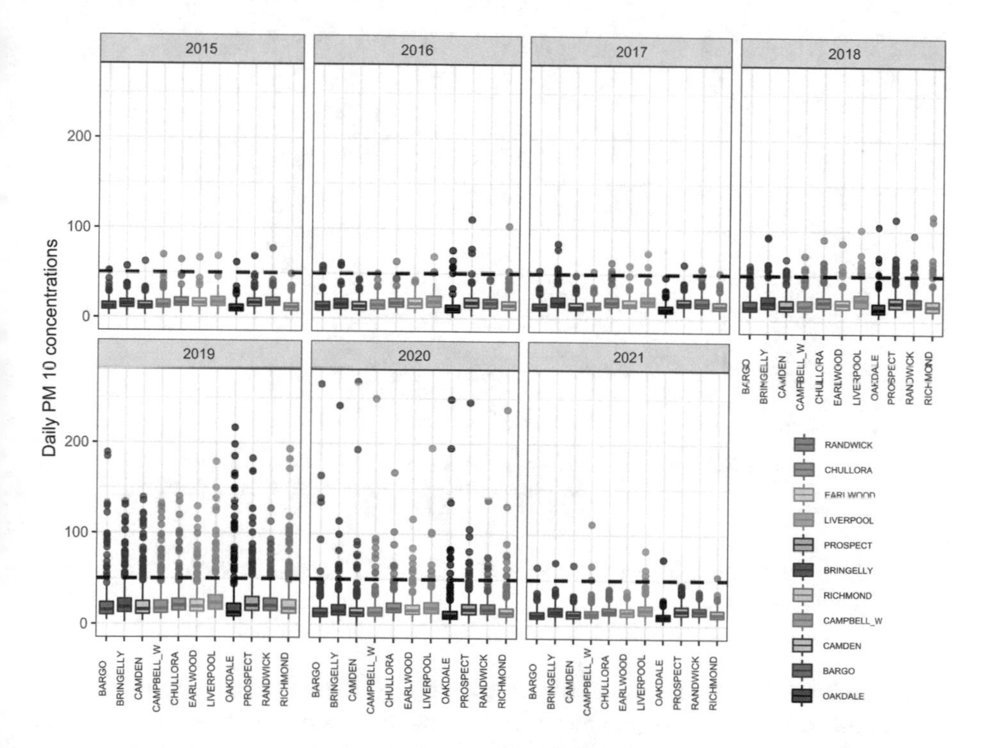

Fig. 3. Box plots showing the distributions of daily PM10 concentrations each year at different monitoring sites. The horizontal black dashed line indicates the Air NEPM threshold which is $50\,\mu g/m^3$.

It can be seen that the daily PM10 concentrations at different monitoring sites each year follow a positively skewed distribution with spatially and annually varying patterns. On most days the PM10 concentrations had reasonably good air quality. However, the number of daily PM10 exceedances has increased considerably from 2018 to 2020 followed by a drop in 2021. The increase in 2019–2020 could be due to bushfires that happened during the Black Summer period

(July 2019–March 2020). However, it is worth noting that even in 2018, there was a considerable increase in the exceedances possibly indicating any future extreme events. The drop in 2021 is possibly due to less traffic and other effects of the COVID-19 lockdown.

The spatial variation is further explored as presented in Fig. 4 which shows the mode of the daily PM10 level distributions at each monitoring site each year (i.e. the daily PM10 concentrations recorded most days at each monitoring site in each year).

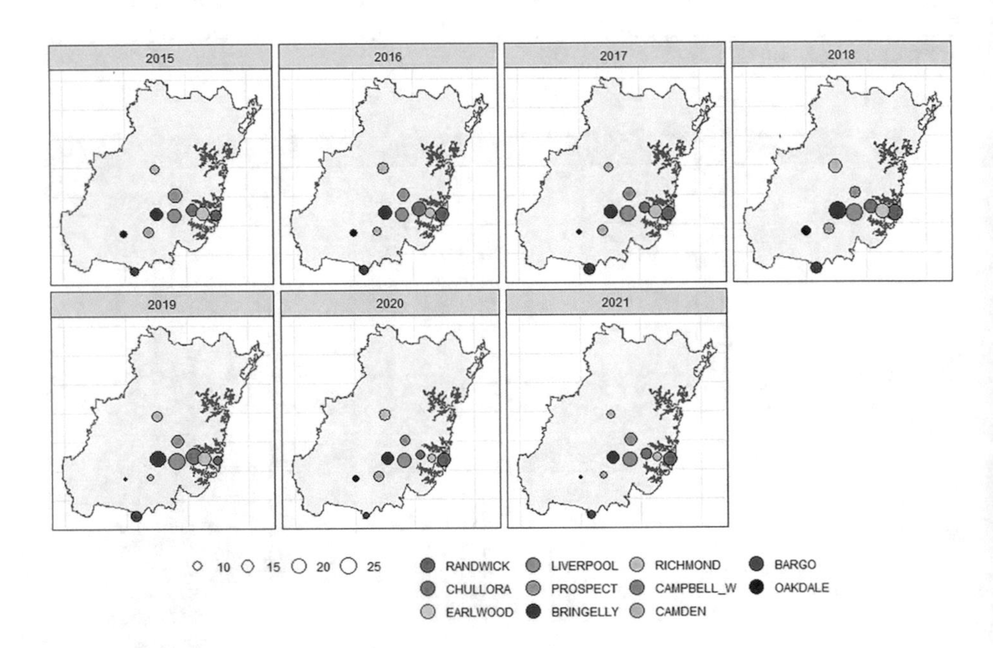

Fig. 4. Space-time variation of the mode of daily PM10 concentrations each year at different monitoring sites.

It can be seen that the modes of the daily PM10 level distributions are high in metropolitan Sydney. Oakdale recorded the lowest mode of PM10 concentration compared to other locations during the study period. However, it is worth noting that Oakdale also recorded the highest PM10 level in 2019 (refer to Fig. 3). In general, air quality in Oakdale is best (lowest PM10 level) on most days excluding extreme event days.

3.2 Space-Time Exploration Based on Daily PM10 Exceedance

Figures 5 and 6 depict the number of exceedances at each site from 2015 to 2021.

As can be seen in Fig. 5, there is a gradual increase in the number of exceedances from 2015 to 2019 followed by a sharp drop from 2019 to 2021.

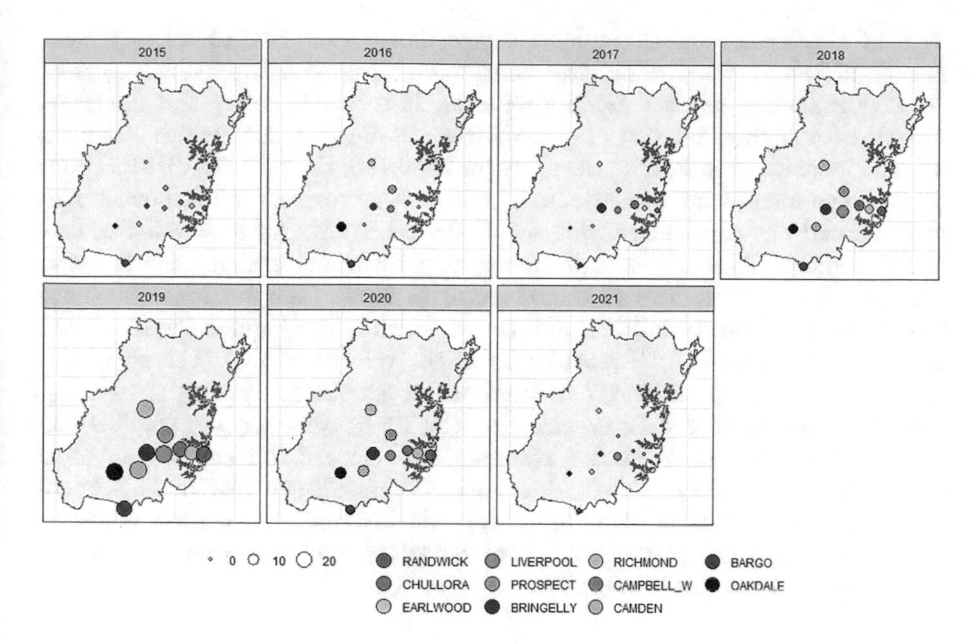

Fig. 5. Space-time variation of the daily PM10 exceedances each year during 2015–2021 at different monitoring sites.

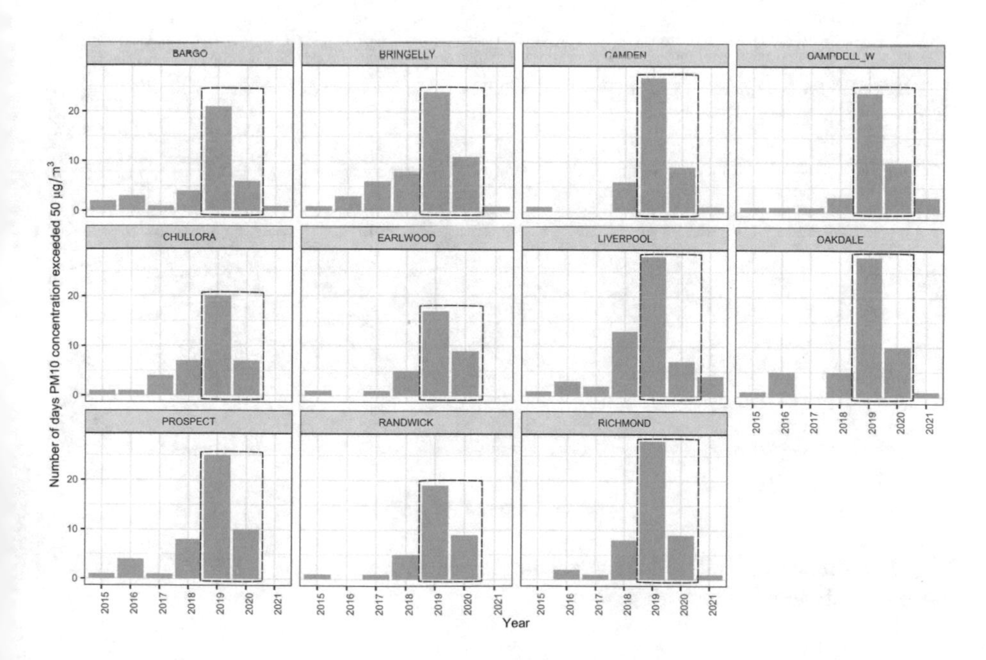

Fig. 6. The daily PM10 exceedances at different monitoring sites in each year during 2015–2021. Red dash-lined boxes highlight the period 2019–2020 which includes the Black Summer and COVID-19 first lockdown.

Figure 6 also reveals that the number of exceedances from 2015 to 2019 is increasing at all the 11 monitoring sites considered. However, the increasing trend is varying among different monitoring sites. It is worth noting that Liverpool recorded the highest number of exceedances (13 days) in 2018 which was a significant increase from 2017 (2 days). After 2020 (i.e. the first COVID lockdown period), the monitoring sites Chullora, Earlwood, Prospect and Randwick have not recorded any exceedance whereas all the other sites have recorded at least one exceedance. Again, it is worth noting that Liverpool has recorded the highest number of exceedances (4 exceedances) in 2021. Campbelltown West also recorded 3 exceedances in 2021 and one of which is the highest PM10 concentration (above 100 $\mu g/m^3$) recorded in 2021. (refer to Fig. 3). The air quality in terms of PM10 exceedance has improved at all the monitoring sites except Campbelltown West and Liverpool after the Black Summer and the COVID-19 first lockdown period. Further examination is needed to identify the reasons behind the occurrences of high exceedance at Campbelltown West and Liverpool. Additional care should be given to improve the air quality near these places.

Figure 7 further displays the seasonal variation of the exceedances from 2015 to 2021 at different monitoring sites.

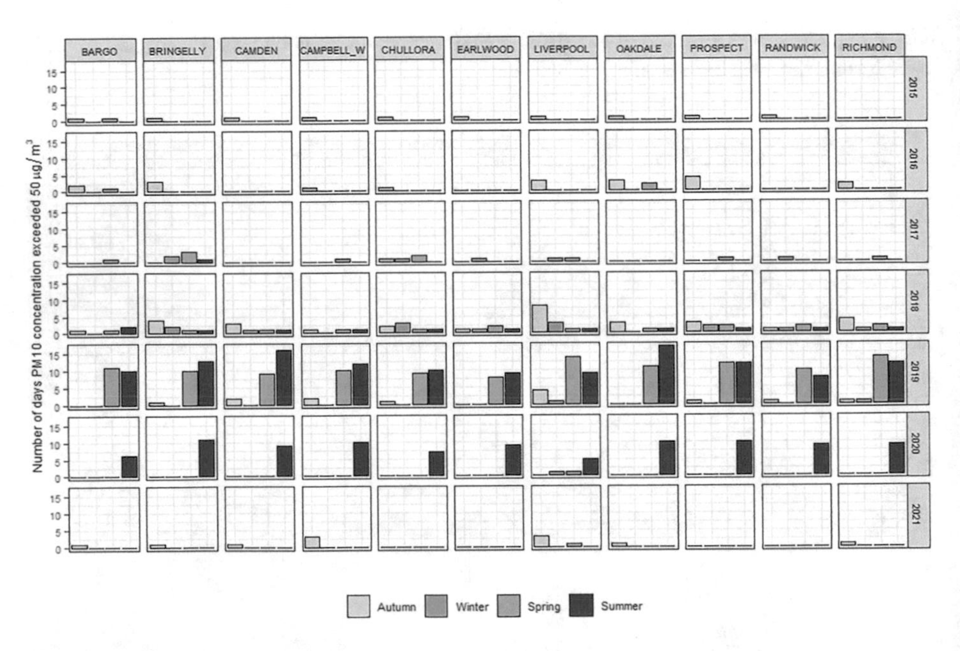

Fig. 7. Number of daily PM10 exceedances each year during 2015–2021 with four seasons at different monitoring sites

According to Figure 7, there is a slight increase in the daily PM10 exceedances during 2017–2018 compared to 2015–2016. Spring in 2019 and Summer in 2019–2020 show remarkable exceedances as can be expected due to bushfires during the Black Summer. Exceedances have rarely occurred after 2020 (after the first COVID-19 lockdown) compared to the period 2015–2018 (before the Black Summer and COVID-19 lockdown) revealing an improvement in air quality in terms of exceedances. PM10 exceedances have occurred mostly during the autumn season in all the years considered except in 2017, 2019 and 2020. In 2017, most of the monitoring sites recorded PM10 exceedance during the Spring season followed by winter.

3.3 Clustering of Monitoring Sites

To identify any groupings of sites with similar characteristics, clustering has been carried out. As an initial step, the distances among the sites have been analysed. Figure 8 shows a heat map of the distances between each pair of stations. The upper triangle of the heat map shows distances in meters while the lower triangle shows DTW distances between the PM10 series of each pair.

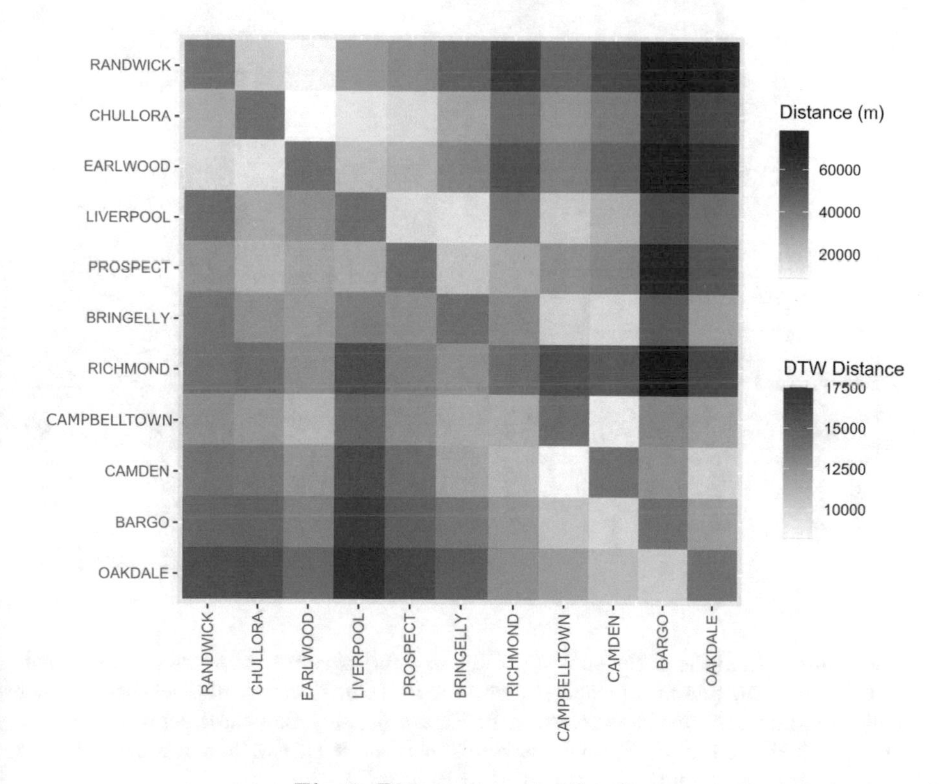

Fig. 8. Distances among the sites

Figure 8 reveals some symmetrical patterns around the diagonal of the heat map. The colour densities of the top-right and bottom-left tiles are high indicating that, when the actual distance between the two sites is high, the dissimilarity of the corresponding two PM10 series is also high.

After applying agglomerative hierarchical clustering using DTW distance as the dissimilarity measure, the dendrogram was analyzed to identify possible clusters. The cluster dendrogram is presented in Fig. 9 (a) which displays four interesting clusters. Principal component analysis was also carried out to visualize the clusters as in Fig. 9 (b). Elevations of each site are given in Fig. 9 (c) and the locations are given in Fig. 9 (d) to compare cluster locations.

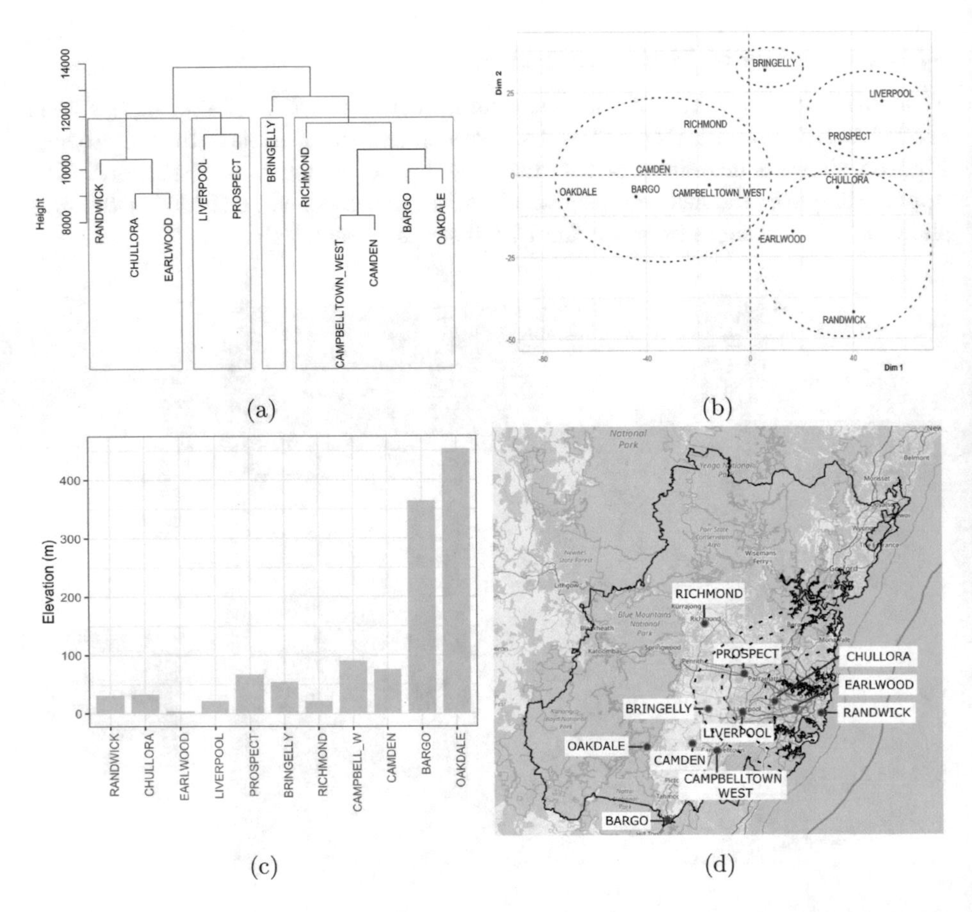

Fig. 9. Cluster Analysis of the sites. (a) Cluster dendrogram with four clusters boxed. (b) PCA plot with first two principal components (Dim1 and Dim2). Dashed ellipses roughly separate the four clusters given in Figure (a). (c) Bar chart representing the elevations of the sites. (d) Spatial distribution of each site with black dashed lines roughly separating the four clusters given in Figure (a).

There are some clusters according to the dendrogram presented in Fig. 9 (a). Camden and Campbelltown West PM10 series are closely related. Bargo and Oakdale series are also closely related even though they are far apart than Camden and Campbelltown West. All these four sites can be considered as one cluster comparatively. Richmond is closer to this cluster than all the other sites. Chullora-Earlwood and Liverpool-Prospect are two pairs which are closely related. Randwick shows the closest behaviour to the Chullora-Earlwood pair of sites. Bringelly displays a different behaviour compared to other sites. If the interested number of clusters is four, it can be represented by the four red boxes.

PCA graph of the sites shown in Fig. 9 (b) also provides further support for the identified clusters. Four ellipses represent the four clusters shown in the dendrogram. It can be seen that the sites within a cluster are located relatively closer in the 2-dimensional space of the first two principal components.

Considering the elevations presented in Fig. 9 (c), Bargo and Oakdale have relatively higher elevations which may be a reason for their PM10 similarity. Campbelltown West and Camden show approximately similar elevations and their PM10 behaviour is also closely related. However, elevation itself does not explain the other clusters.

It is also interesting that the four clusters display a pattern with the geographical locations as can be seen in Fig. 9 (d). Dashed lines approximately separate the clusters. However, note that they are only an indication to separate the clusters and do not represent any geographical boundaries or any other groupings. Geographical locations as well as the PM10 behaviour of Chullora, Earlwood and Randwick are relatively close, supporting Tobler's first law of geography that states "everything is related to everything else, but near things are more related than distant things" [27]. Liverpool and Prospect also display the same pattern (i.e. relatively closer in location as well as PM10 behaviour).

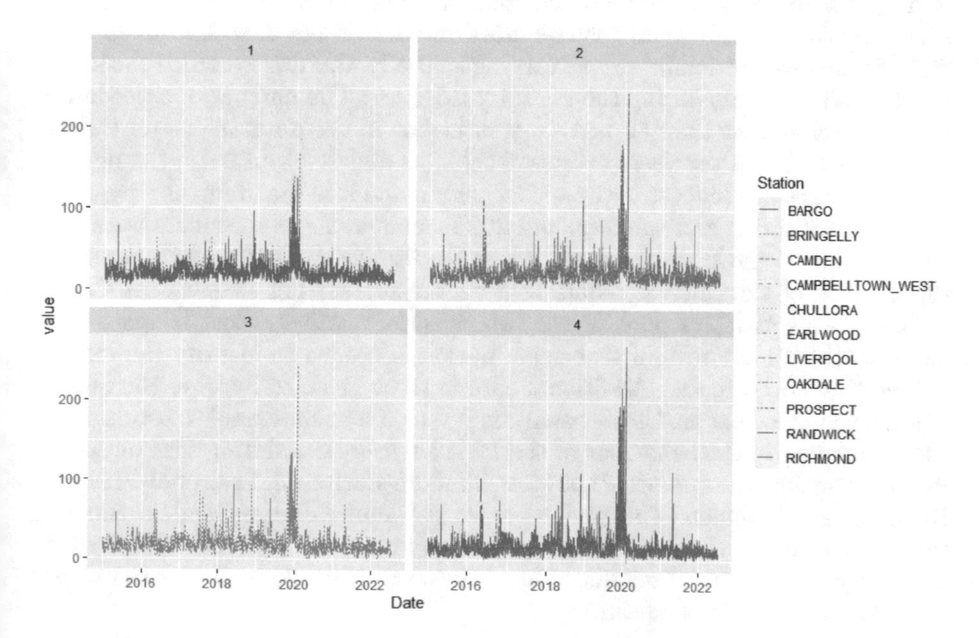

Fig. 10. PM10 time series clusters

Even though the Bringelly site is relatively closer to the Liverpool site, it is far apart considering the behaviour of the PM10 series. This is an interesting fact for further investigation. However, only the geographical location itself cannot explain the behaviour of the PM10 concentration as there may be other factors associated with the PM10 level. Richmond site is far from other sites. However, its PM10 series is related to the two pairs Camden-Campbelltown and Bargo-Oakdale.

Therefore, according to the above analysis, four reasonable clusters can be identified among the sites considered in the Greater Sydney region as below:

1. Randwick, Chullora, Earlwood
2. Liverpool, Prospect
3. Bringelly
4. Richmond, Campbelltown West, Camden, Bargo, Oakdale

Figure 10 shows the time series plots of PM10 daily concentrations for each of the four clusters providing graphical evidence of some overlapping series within clusters.

4 Conclusion

In this study, PM10 air pollution sensor data of the Greater Sydney region from 2015 to 2021 was analyzed. Exploratory analysis revealed that the distributions of daily PM10 levels at all the considered sites are positively skewed. This indicates that the air quality was good on most days near these sites. The mode of the daily PM10 levels distribution was varying spatially. Oakdale recorded the lowest mode in all the years during the period 2015–2021. The number of exceedances (PM10 level greater than $50\,\mu g/m^3$) was higher in the autumn season than in other seasons in all the years except 2017, 2019 and 2020. In 2017, the number of exceedances was highest in the spring season and it is worth further investigation. In 2019 spring and summer, and 2020 summer, the exceedances were high possibly due to the Black Summer. After the first COVID-19 lockdown period, the number of exceedances reduced at all the monitoring sites except Campbelltown West and Liverpool. These two sites are relatively close in space compared to the other sites. Further examination is needed to identify the reasons behind these occurrences. Additional care is to be given to improve the quality of the air at these places. Cluster analysis revealed four reasonable clusters of the sites according to the behaviour of the PM10 sensor data during the considered period. The four clusters are Randwick-Chullora-Earlwood, Liverpool-Prospect, Bringelly and Richmond-Campbelltown West-Camden-Bargo-Oakdale. Some of the monitoring sites in the Greater Sydney region are not included in this study due to a large number of missing values. However, it would be better to analyze those stations also in the future.

References

1. Environment Protection Authority Victoria: Pm10 particles in the air (2021). https://www.epa.vic.gov.au/for-community/environmental-information/air-quality/pm10-particles-in-the-air. Accessed 22 Sept 2022
2. Donaldson, K., Ian Gilmour, M., MacNee, W.: Asthma and pm10. Respir. Res. **1**(1), 12–15 (2000)
3. Pope III, C.A., Dockery, D.W., Spengler, J.D., Raizenne, M.E.: Respiratory health and pm10 pollution: a daily time series analysis. Am. Rev. Respir. Dis. **144**(3_pt_1), 668–674 (1991)
4. Ścibor, M., Malinowska-Cieślik, M.: The association of exposure to pm10 with the quality of life in adult asthma patients. Int. J. Occup. Med. Environ. Health **33**(3) (2020)
5. Shin, J., Lee, H., Kim, H.: Association between exposure to ambient air pollution and age-related cataract: a nationwide population-based retrospective cohort study. Int. J. Environ. Res. Public Health **17**(24), 9231 (2020)
6. Lee, B.J., Kim, B., Lee, K.: Air pollution exposure and cardiovascular disease. Toxicol. Res. **30**(2), 71–75 (2014)
7. Polichetti, G., Cocco, S., Spinali, A., Trimarco, V., Nunziata, A.: Effects of particulate matter (pm10, pm2. 5 and pm1) on the cardiovascular system. Toxicology **261**(1-2), 1–8 (2009)
8. Zhou, Y., Li, L., Hu, L.: Correlation analysis of pm10 and the incidence of lung cancer in Nanchang, China. Int. J. Environ. Res. Public Health **14**(10), 1253 (2017)
9. Orioli, R., Cremona, G., Ciancarella, L., Solimini, A.G.: Association between pm10, pm2. 5, no2, o3 and self-reported diabetes in Italy: a cross-sectional, ecological study. PLoS ONE **13**(1), e0191112 (2018)
10. Yang, J., et al.: Diabetes mortality burden attributable to short-term effect of pm10 in China. Environ. Sci. Pollut. Res. **27**(15), 18784–18792 (2020)
11. Kihal-Talantikite, W., Legendre, P., Le Nouveau, P., Deguen, S.: Premature adult death and equity impact of a reduction of no2, pm10, and pm2. 5 levels in Paris-a health impact assessment study conducted at the census block level. Int. J. Environ. Res. Public Health **16**(1), 38 (2019)
12. National Environment Protection Council (Australia): National Environment Protection (Ambient Air Quality) Measure. National Environment Protection Council (2021)
13. NSW Department of Planning and Environment: Standards and goals for measuring air pollution (2021). https://www.environment.nsw.gov.au/topics/air/understanding-air-quality-data/standards-and-goals. Accessed 22 Sept 2022
14. Abraham, S., Li, X.: A cost-effective wireless sensor network system for indoor air quality monitoring applications. Procedia Comput. Sci. **34**, 165–171 (2014)
15. Morawska, L., et al.: Applications of low-cost sensing technologies for air quality monitoring and exposure assessment: how far have they gone? Environ. Int. **116**, 286–299 (2018)
16. Liu, X., et al.: Low-cost sensors as an alternative for long-term air quality monitoring. Environ. Res. **185**, 109438 (2020)
17. Zheng, K., Zhao, S., Yang, Z., Xiong, X., Xiang, W.: Design and implementation of LPWA-based air quality monitoring system. IEEE Access **4**, 3238–3245 (2016)
18. Duc, H., et al.: The effect of lockdown period during the Covid-19 pandemic on air quality in Sydney region, Australia. Int. J. Environ. Res. Public Health **18**(7), 3528 (2021)

19. NSW Department of Planning and Environment: How and why we monitor air pollution (2022). https://www.environment.nsw.gov.au/topics/air/air-quality-basics/sampling-air-pollution. Accessed 22 Sept 2022
20. Moritz, S., Bartz-Beielstein, T.: Imputets: time series missing value imputation in r. R J. **9**(1), 207 (2017)
21. Moritz, S., Sardá, A., Bartz-Beielstein, T., Zaefferer, M., Stork, J.: Comparison of different methods for univariate time series imputation in R. arXiv preprint arXiv:1510.03924 (2015)
22. Wijesekara, L., Liyanage, L.: Air quality data pre-processing: a novel algorithm to impute missing values in univariate time series. In: 2021 IEEE 33rd International Conference on Tools with Artificial Intelligence (ICTAI), pp. 996–1001. IEEE (2021)
23. Aghabozorgi, S., Shirkhorshidi, A.S., Wah, T.Y.: Time-series clustering-a decade review. Inf. Syst. **53**, 16–38 (2015)
24. James, G., Witten, D., Hastie, T., Tibshirani, R.: An Introduction to Statistical Learning. Springer, New York (2013). https://doi.org/10.1007/978-1-4614-7138-7
25. Sardá-Espinosa, A.: Comparing time-series clustering algorithms in R using the dtwclust package. R Package Vignette **12**, 41 (2017)
26. Suris, F.N.A., Bakar, M.A.A., Ariff, N.M., Mohd Nadzir, M.S., Ibrahim, K.: Malaysia pm10 air quality time series clustering based on dynamic time warping. Atmosphere **13**(4), 503 (2022)
27. Tobler, W.R.: A computer movie simulating urban growth in the Detroit region. Econ. Geogr. **46**(Sup. 1), 234–240 (1970)

Wearable Sensors and Smart Clothing: Trends and Potentials for Research

Geethanjali Pai(✉), Subhas Mukhopadyay, and Syed Muzahir Abbas

School of Engineering, Faculty of Science and Engineering, Macquarie University, Sydney, NSW 2109, Australia

Geethanjali.pai@students.mq.edu.au, {subhas.mukhopadhyay, syed.abbas}@mq.edu.au

Abstract. This paper presents a study of the recent trends and potential of research on wearable sensors and smart clothing. Data from leading and reputable research platforms using various keywords related to wearable sensors and intelligent clothing. The data collection has been categorized based on gender (men/ women) to observe the trends and based on age classification (adult/infant). With the advancements in wearable technology, the increasing popularity, and the continuous development of sensors and technical clothing, it is vital to address the growing trends. This study and results provide deeper insights into the research community and help identify potential areas for research in this domain. Moreover, the findings will equip new researchers with a good understanding of the development and relevant research platforms.

Keywords: Wearable Sensors · Smart Clothing · Infants · Sleep · Thermoregulation · Skin Comfort

1 Introduction

The growing demand for in-hand applications, an interest in mimicking nature, and the advancements in safe wearable technologies has led researchers to explore newer possibilities in sensors and wearable technologies for a wide range of applications for adults, kids, and infants.

Infants sleep two-thirds of the day. Sleep ensures the development of the regulatory mechanisms and the complete well-being of the infant (Heraghty 2008) (Yeler 2021) (Wang 2021). Infants cannot maintain a thermal balance like adults (Bach 2002) (Bach V. 2022) (Fluhr 2010). Additionally, the hypothalamus of the brain in adults and infants controls the sleep-wake cycle (Overeem 2021), respiration, thermoregulation (Fukushi 2019), and thermal sensations from the skin (Lan 2019) (Cabanac 1981) (Harris 1957) (Tauber 2019), placing infants at high risk. As SIDS or Sudden Infant Death Syndrome is a major cause of infant fatality (Moon 2016) (Wilson 2005), infant sleep clothing or garments with sensors capable of monitoring physiological parameters like humidity, respiration, body temperature, and heart rate (Mukhopadhyay 2014.) can promote sleep and aid in the infant's well-being by keeping the adult well informed about the infants status during sleep.

N. K. Suryadevara et al. (Eds.): ICST 2022, LNEE 1035, pp. 309–315, 2023.
https://doi.org/10.1007/978-3-031-29871-4_31

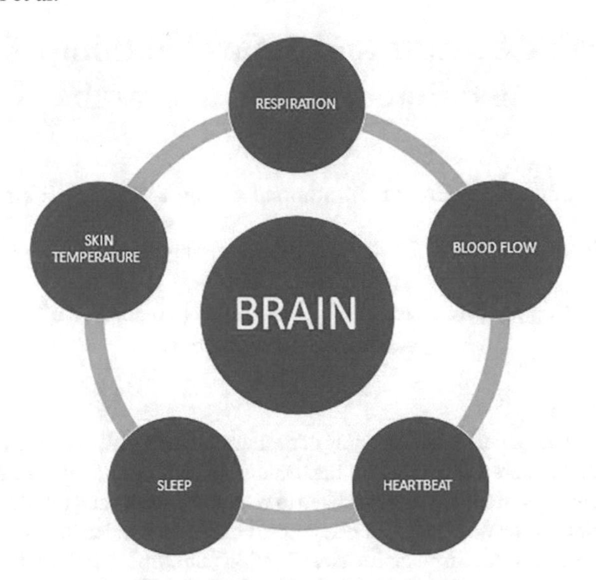

Fig. 1. Role of brain in the Infant

SIDS commonly occurs in winter (Guntheroth 1992.) (Brown 2022.). Since, room temperatures, choice of bedclothes and sleep clothing for infants play a vital role in the cause of SIDS (Kemp 1993) (Hoyniak 2022), sleep clothing with wearable sensors can mitigate the risks associated with SIDS. Hence there is a significant need to understand the current research trends in smart clothing and wearables for infant sleep.

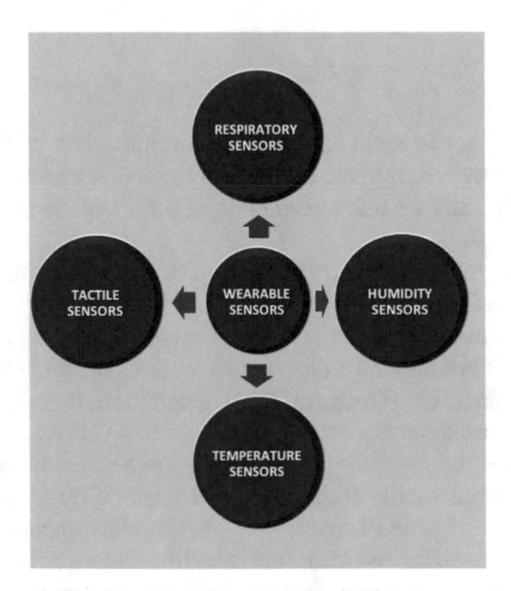

Fig. 2. Wearable sensors for infant sleep.

This paper analyses the number of research publications on sensors, wearables, smart garments, and related papers on various research platforms for upcoming researchers to understand and identify the progress or inadequacies in wearable technologies, sensors, and technical clothing. Data from leading, reputed research platforms based on gender and age were collected using keywords related to wearable sensors and functional clothing for various categories. The paradigm is to identify the current research trends and potential involved in wearable sensors and intelligent clothing. The study findings determined that the publication knowledge in sensor applications for infants was limited compared to adults. Moreover, the latest trends in sensors, the design placement, and the limitations are identified and studied further. In addition, highlighting the regulatory mechanisms and the risks involved in infant sleep ascertains the scope and prominence of addressing this domain.

2 Identification of the Number of Research Publications on Wearable Sensors for Infant Sleep

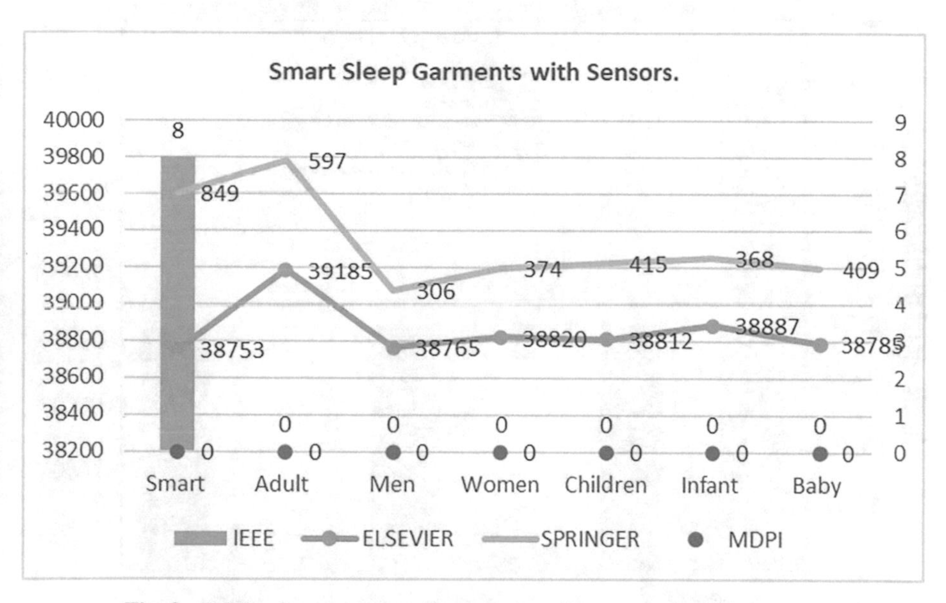

Fig. 3. Publications corresponding to smart sleep garments with sensors.

Wearable sensors are very beneficial as they can be programmed to sense physiological parameters. Given many publications, IEEE, MDPI, ELSEVIER, and SPRINGER were selected specifically for their source of literature content. A thorough search to identify the total number of publications in each journal with headings similar to the following-Wearables, Smart, Garments, Sleep, and Smart Sleep Garments with Sensors, further categorized as Adult, Men, Women, Children, Infant, and Baby, was conducted (Fig. 3).

All the popular journals presented publications on wearable sensors with generalized search headings used for sensor technology. However, streamlining the search headings to the different categories significantly decreased the number of publications. In addition, the lists on the search heading "Smart Sleep Garment with Sensors" found no accurate matches.

2.1 Comparison of Publications on Sleep Sensors for Adults and Infants

A comparison of the number of publications on sleep sensors for adults and infants, Fig. 4, showed research publications on adults remained the highest on all topics. However, the research publications on "smart sleep garments with sensors" were similar for adults, infants, and babies. Though the number of publications on smart sleep garments with sensors for adults, infants, and babies was equal, the published information was more generalized books and not related to the specific genres.

Based on the analysis, the consolidated number of research publications on infant and baby sleep sensors was low in value.

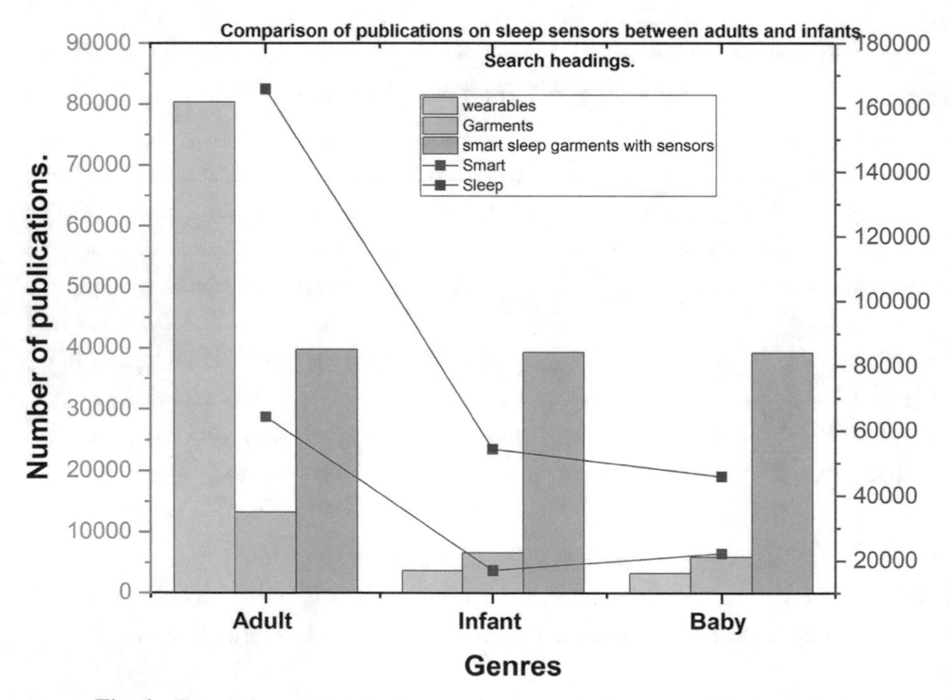

Fig. 4. Comparison of publications on sleep sensors between adults and infants.

2.2 Identification of Publications on Sleep Sensors for Infants in Google Scholar

In addition, as the search headings used in the popular journals were generalized, another search conducted on google scholar related to infant sleep and sensors developed for infant sleep identified and studied 50 research publications (2018–2022) on sensors for infant sleep. The study categorized sensors to monitor various physiological parameters of the sleeping infant like (Fig. 5).

- Position
- Heartbeat
- Respiration
- Body temperature
- Sleep-wake cycle and
- Movement.

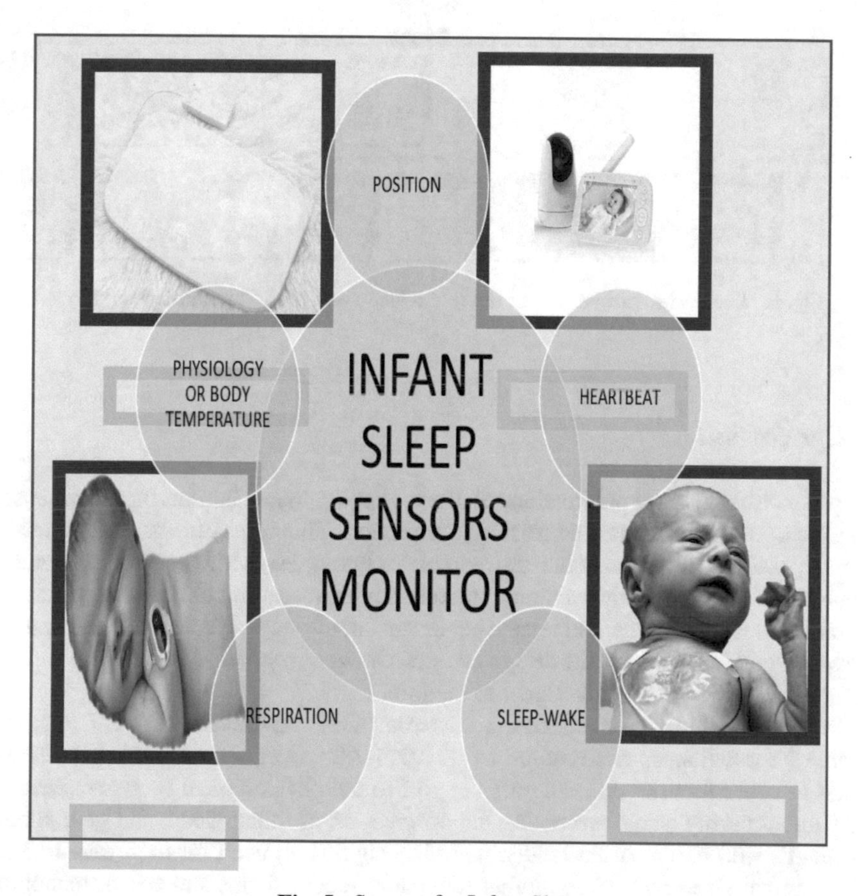

Fig. 5. Sensors for Infant Sleep

However, simple, more practical, and durable everyday sensors in infant sleep clothing need more investigation.

2.3 The Design Placements of the Infant Sleep Sensors and Possible Limitations

The developed wearable infant sleep sensors are flexible substrates or rigid robots, generally placed in the bedding material, sewn into the garment, used as patches or a wristband, and placed inside the cradle or room to monitor the infant's parameters. Although the wearable sensors are removable during washing, the sensors may not be skin-friendly, produce inaccurate results, or may not be durable to environmental factors, and can either cause injuries or be fatal (Fig. 6).

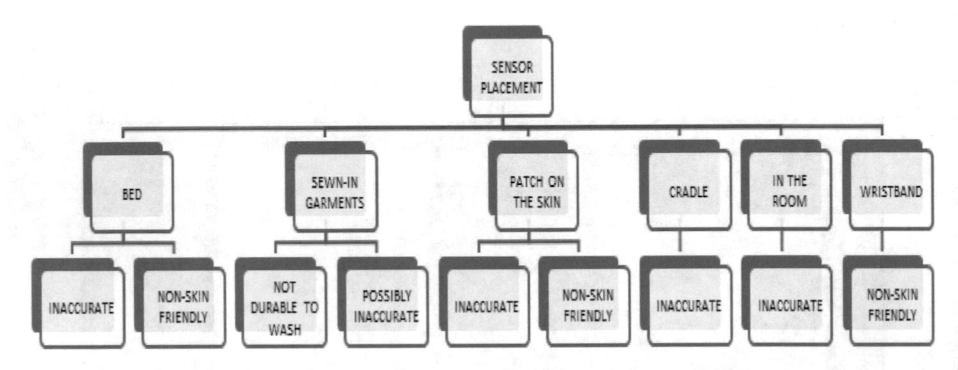

Fig. 6. The design placements of the infant sleep sensors and possible limitations.

3 Conclusion

The paper determines all publications platforms have valuable outputs based on sensors, wearables, smart garments, and related information. This report further ascertains that there has been drastic progress in safe wearable technologies, with a considerable amount of published literature. However, compared to adults, research publications on sensors, wearables, smart garments, and related papers on infant sleep lack passable publications. Moreover, the limitations in the design and placement of sensors in infant sleep clothing and bedding pose a substantial threat to the infant.

To prevent SIDS and to reduce the exacerbated risks associated with SIDS, sensors suitable for infant sleep need more investigation. Moreover, the sensor-based research should be more focused on rendering support to the physiological processes involved in the infant during sleep, promoting the progression of infant sleep, and guaranteeing the infant's well-being. Additionally, there is a significant need for more research outputs on infant sleep physiology, infant regulatory mechanisms, and sensor technology development suitable for infant sleep.

References

Adamson, K.G.: The influence of thermal factors upon oxygen consumption of the newborn human infant. J. Pediatr. **66**(3), 495–508 (1965)

Bach, V.: Hyperthermia and heat stress as risk factors for sudden infant death syndrome: a narrative review. Front. Pediatr. **10** (2022)

Bach, V.: The interaction between sleep and thermoregulation in adults and neonates. Sleep Med. Rev. **6**(6), 481–492 (2002)

Brown, R.: Sudden infant death syndrome, pulmonary edema, and sodium toxicity: a grounded theory. Diseases **10**(3), 59 (2022)

Cabanac, M.: Physiological signals for thermal comfort. In: Studies in Environmental Science, vol. 10, pp. 181–192. Elsevier (1981)

Fluhr, J.D.: Functional skin adaptation in infancy–almost complete but not fully competent. Exp. Dermatol. **19**(6), 483–492 (2010)

Fukushi, I.: The role of the hypothalamus in modulation of respiration. Respir. Physiol. Neurobiol. **265**, 172–179 (2019)

Gough, M.: Temperature changes during neonatal surgery. Arch. Disease Child. **35**(179), 66 (1960)

Guntheroth, W.R.: A seasonal association between SIDS deaths and kindergarten absences. Public Health Rep. **107**, 319+ (1992)

Harris, J.: Special pediatric problems in fluid and electrolyte therapy in surgery. Ann. N. Y. Acad. Sci. **66**(4), 966–975 (1957)

Heraghty, J.H.: The physiology of sleep in infants. Arch. Disease Child. **93**(11), 982–985 (2008)

Hoyniak, C.P.: The physical home environment and sleep: what matters most for sleep in early childhood. J. Family Psychol. **36**(5), 757 (2022)

Kemp, J.: Unintentional suffocation by rebreathing - a death scene and physiological investigation of a possible cause of sudden infant death. J. Pediatr. **122**(6). 874–880 (1993)

Kuzawa, C.: Adipose tissue in human infancy and childhood: an evolutionary perspective. Am. J. Phys. Anthropol. **107**(s 27), 177–209 (1998)

Lan, L.: Mean skin temperature estimated from 3 measuring points can predict sleeping thermal sensation. Build. Environ. **162**, 106292 (2019)

Moon, R.: SIDS and other sleep-related infant deaths: evidence base for 2016 updated recommendations for a safe infant sleeping environment. Pediatrics **138**(5), e20162940 (2016)

Mukhopadhyay, S.: Wearable sensors for human activity monitoring: a review. IEEE Sens. J. **15**(3), 1321–1330 (2014)

Overeem, S.: Sleep disorders and the hypothalamus. In: Handbook of Clinical Neurology, vol. 182, pp. 369–385 (2021)

Tauber, M.: Prader-Willi syndrome: a model for understanding the ghrelin system. J. Neuroendocrinol. **31**(7), e12728 (2019)

Wang, Z.: Energy metabolism in brown adipose tissue. FEBS J. **288**(12), 3647–3662 (2021)

Wilson, C.A.: Thermal insulation and SIDS—an investigation of selected 'Eastern' and 'Western' infant bedding combinations. Early Human Dev. **81**(8), 695–709 (2005)

Yeler, O.: Performance prediction modeling of a premature baby incubator having modular thermoelectric heat pump system. Appl. Therm. Eng. **182**, 116036 (2021)

Performance Analysis of Multiple Cavity Dielectric Modulated Tunnel FET Biosensor with High Detection Sensitivity

Sumeet Kalra(✉)

Indian Institute of Technology Delhi, New Delhi, India
sumeetkalra5@gmail.com

Abstract. In this paper, we demonstrate and analyze the sensing performance of a biosensing device based on tunnel field effect transistor (TFET). The sensing principle of the device relies upon dielectric constant modulation when biomolecules are introduced in the nanocavities embedded in the gate dielectric region. The proposed biosensor has an advantage of using the dielectric properties of analyte biomolecules to result in electrostatic doping of the silicon layer, thereby, forming source and drain regions below the respective nanocavities. Such an induced doping causes additional change in the ON-state current and hence, a higher detection sensitivity of the device. Using 2D TCAD simulations, the device is shown to exhibit a high sensitivity. Additionally, the doping less approach of realizing the device results in reduced variability, relaxed processing requirements, and elimination of many doping related intricacies, hence, resulting in an easy and low-cost device fabrication. These desirable features make the proposed device attractive for low-cost, low-power, reliable biosensing and potentially more suitable for point-of-care testing and IoT of healthcare applications.

Keywords: Biosensor · Dielectric Constant Modulation · Dielectric Modulated FET · Doping less · Field Effect Transistor · Nanocavity · Sensitivity. Tunnel FET

1 Introduction

Field Effect Transistor (FETs) of nanoscale dimensions have shown tremendous potential for use in biosensing applications, especially at the point-of care [1–3]. This is because of their advantages of high surface to volume ratios and scale match with the biomolecules that results in efficient interaction of biomolecules on the surface of such devices. The response of FET biosensing devices is assessed mainly by a parameter called detection sensitivity, which, in general, is the relative change in device current from absence to presence of immobilized biomolecules on the device sensing surface. It has been shown that the operation of a nanoscale FET biosensor in the subthreshold region, i.e., with applied gate bias close to threshold voltage (V_T) of the device, results in an enhanced detection sensitivity. This is because in sub-threshold region of operation, the device current has an exponential dependence on the applied gate bias [4, 5], implying a sharp turn ON characterized by a steep subthreshold slope (SS) of the device characteristics.

N. K. Suryadevara et al. (Eds.): ICST 2022, LNEE 1035, pp. 316–324, 2023.
https://doi.org/10.1007/978-3-031-29871-4_32

The lower the SS of the FET, the higher is its detection sensitivity during biosensing operation. However, the traditional FETs suffer from a fundamental limit on SS in the subthreshold region. To overcome this limit, several steep subthreshold devices have been proposed which be leveraged for increased detection sensitivity. One such steep subthreshold device is Tunnel Field Effect whose principle of operation is fundamentally different from that of the traditional FET, and relies on quantum mechanical tunneling. Therefore, a TFET can achieve a much lower SS [4, 5]. Consequently, biosensors based on TFETs exhibit a significantly higher sensing performance than the traditional FET based biosensors [6–8].

However, just as traditional nanoscale FETs, TFETs also suffer from the doping related intricacies. The introduction of dopants necessitates a high thermal budget diffusion or ion implantation steps followed by the annealing process. Additionally, at lower doping concentrations, the random fluctuations in the introduction and distribution of dopant atoms lead to a variability in the device characteristics [9–16]. This has been shown to result in a high variability in detection sensitivity and hence, random variations in the sensing performance of the device [10–12]. Such variations render the device unreliable for use in critical applications such as healthcare and diagnostics. Also, the random fluctuations in location and density of introduced dopant atoms result in a higher off-state current in the device and hence, deteriorated I_{on}/I_{off} ratios. Since the sensitivity of a TFET based biosensor is directly proportional to the I_{on}/I_{off} ratio, a degraded I_{on}/I_{off} implies a reduced detection sensitivity [12–14].

Further, as a TFET operates by quantum mechanical tunnelling effect, it requires sharp doping profiles at source/channel (S/C) junction in order to attain higher electric fields and hence, higher tunnelling currents across the junction. However, the high temperature requirements for dopant activation causes the dopant atoms to diffuse from the source (S) and drain (D) into the channel (C) region, making it hard to achieve a sharp profile at the junction. As a result, the tunnelling probability across the junctions is significantly reduced which lowers the tunnelling current and therefore, the sensitivity of the TFET based biosensing device. Furthermore, the dopant atoms segregate and cause defects in the material bandgap and hence, deteriorate the band-to-band generation rate and tunneling currents.

For the above-mentioned concerns, we have proposed and analyzed a doping less dielectric modulated TFET biosensor structure with multiple nanocavities to accommodate the biomolecules. Based on the charge plasma doping effect [17–19], the dielectric constant of biomolecules is utilized to result in the doping of the source and drain regions in undoped silicon layer. Using 2D TCAD simulations, we demonstrate the usefulness of the proposed device as a high sensitivity biosensor with a potential of low-cost fabrication and reduced variability, due to elimination of chemical doping.

2 Device Structure and Operating Principle

Device Structure and Parameters. The cross-sectional view of the given multiple cavity DM-TFET biosensor is illustrated in Fig. 1. As depicted, the nanoscale cavities accommodate the biomolecules for their sensing purpose. The 2D TCAD simulation parameters used are:- silicon thickness $t_{si} = 10$ nm, thickness of each nanocavity t_{cavity}

= 10 nm, HfO_2 thickness t_{HfO2} = 10 nm, gate length L_{gate} = 75 nm, the work function (WF) used for gate electrode metal layer is 4.33 eV. The length of each of the three nanocavities L_{cavity}, unless otherwise stated, is 25 nm.

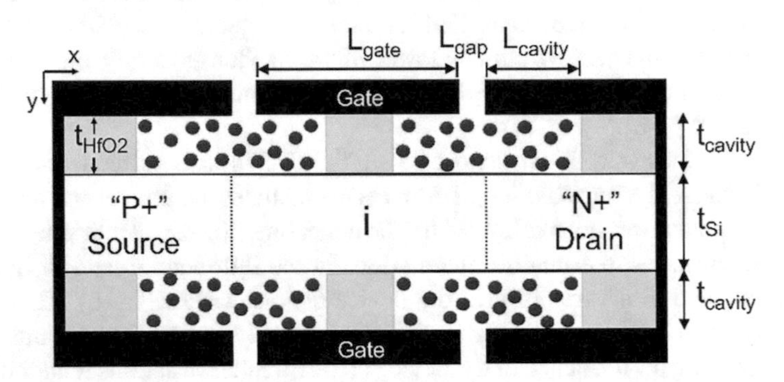

Fig. 1. The proposed dielectric modulated TEFT biosensor with multiple nanocavities underneath source, gate and drain electrodes to accommodate the biomolecules during the sensing operation.

Utilizing the charge plasma-based doping effect, p-type source and n-type drain regions are formed in the undoped silicon film [17, 18]. For this, appropriate metal electrodes are selected. For "p+" source, high WF metal such as Platinum with WF of 5.93 eV is chosen. For the formation of "n+" drain region, low WF electrode such as Hafnium having a work function = 3.9 eV is chosen. For sensitivity and process variability analysis, the gap between S-G and G-D metal electrodes L_{gap} is varied from 7–15 nm.

Operating Principle. The immobilization of biomolecules with different dielectric constants in the nanocavities changes the total dielectric constant between the metal electrodes and the silicon film in the cavity regions [3]. The value of dielectric constant $K = 1$ indicates that no biomolecules are present in the nanocavities, or that the nanocavities are filled with air. On the other hand, $K > 1$ is taken to indicate the presence of biomolecules in the nanocavities. During the sensing operation, the biomolecular conjugations in gate nanocavities modulate the cavity's dielectric constant and hence, the gate (G)-channel (C) coupling in the device. This changes the electric field at the source-channel and drain-channel junctions and hence, the tunnelling ON-current in the device when biomolecules with $K > 1$ are immobilized in the nanocavities.

In addition, when the target biomolecules having different dielectric constants undergo binding in the nanocavities, they cause change in the carrier concentration of S and D areas in the silicon film. The combined effect of these two phenomena, i.e., dielectric constant modulation in gate nanocavities and change in the doping concentration in S-D regions, together result in a significant change in current upon the conjugation of biomolecules. Consequently, a high detection sensitivity is achieved in the proposed device. The sensitivity is further enhanced due to the presence of multiple cavities which result in an increased sensing area, and hence, a greater number of biomolecules to be accommodated. This causes a greater number of biological conjugations being registered in the same device footprint, thereby resulting in a higher detection sensitivity.

3 Simulation Methodology, Results and Discussions

The 2D TCAD simulations were done using Silvaco Atlas device simulator 5.19.20.R [20]. The models enabled are: Fermi–Dirac, Shockley–Read–Hall (SRH), Lombardi mobility and bandgap narrowing. To consider the lateral point tunneling, nonlocal band-to-band tunneling model was employed.

Figure 2 shows the induced concentration of electrons and holes in the silicon regions below source and drain nanocavities in the OFF-state and ON-state for (a) $+V_{GS}$ or n-type and (b) $-V_{GS}$ or p-type like mode.

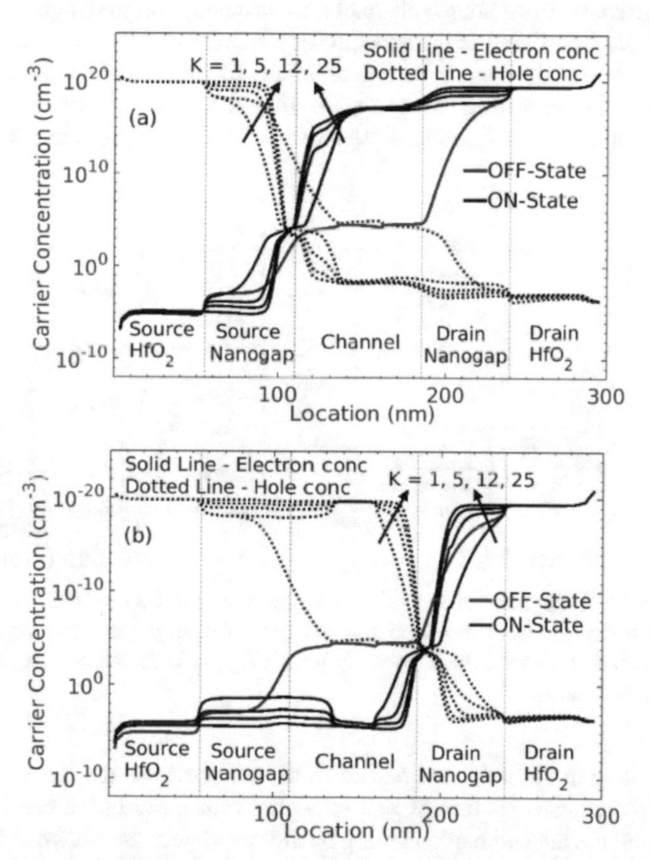

Fig. 2. Induced electron and hole concentrations below source and drain nanocavities for varying biomolecular dielectric constant in (a) OFF-state and ON-state n-type mode with $V_{GS} = 1.5$ V (b) OFF-state and ON-state p-type operation mode with $V_{GS} = -1.5$ V. $L_{cavity} = 45$ nm and $L_{gap} = 10$ nm.

As evident, in the OFF-state, the carrier concentration below source and drain nanocavities is low $\sim 10^{18}$ cm^{-3}. Also, the free carrier concentration in channel is low.

It should be noted that though in the ON-state with no biomolecules (K = 1), the doping concentration below the source and drain HfO_2 regions may be high $\sim 10^{20}$ cm^{-3}, the carrier concentration below the source and drain nanocavities is still quite low $\sim 10^{18}$ cm^{-3}. With the introduction of biomolecules, and hence, increasing K values, the carrier concentration in S/D of silicon film rises considerably ($\sim 10^{20}$ cm^{-3}), resulting in an increased ON state tunneling current. It is also worthy to note that since the tunneling occurs at silicon surface of the junction, a high induced carrier concentration ($\sim 10^{20}$ cm^{-3}) at the surface validates the proper functioning and efficacy of the device.

Figure 3 shows the band profiles with varying biomolecular dielectric constant values immobilized in the nanocavities in the (a) ON-state of n− type mode with $+V_{GS}$, and (b) ON-state of p-type operation mode with $-V_{GS}$. When the device is in OFF state (not shown here for the sake of clarity), the non-alignment of bands prevents the tunneling of carriers across source/channel and channel/drain junctions. Also, when no biomolecules are considered in the nanocavities, represented as K = 1, a wider tunneling barrier window at the source/drain to channel junctions can be seen in both Fig. 3a and 3b. This is due to a lower gate to source/drain junction coupling resulting from air-filled nanocavities. Additionally, K = 1 results in low doping concentration in both source/drain regions.

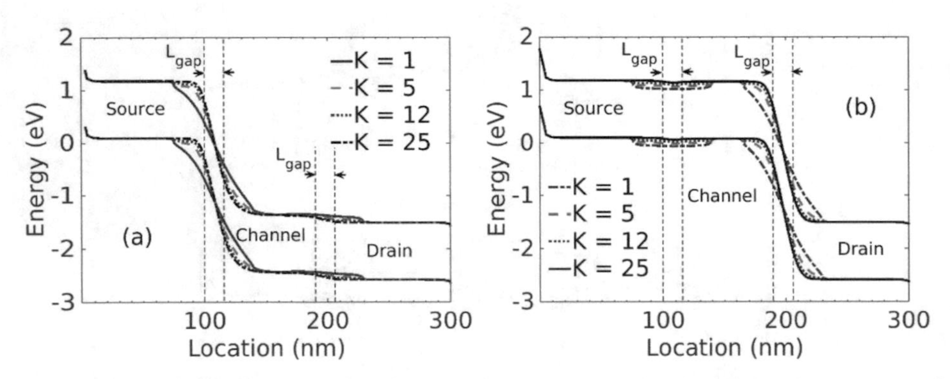

Fig. 3. Energy band diagrams of the proposed biosensor for varying dielectric constants of biomolecules in the (a) ON-state n-type and the (b) ON-state p-type like operation mode of the device. For both the cases, the nanocavity length L_{cavity} is 25 nm and gap between metal electrodes L_{gap} is 15 nm.

When binding of biomolecules occurs in the nanocavities, the increasing dielectric constant of the nanocavities from K = 1 to higher values causes the bending of energy bands. This reduces the width of tunneling barrier window at the source to channel junction for $+V_{GS}$ and the channel to drain junction region for $-V_{GS}$. As a result, the junction electric field across the source/channel and drain/channel junctions (depending upon gate voltage polarity) increases with increasing K values. Since the band-to-band generation rate (R_{BTBT}) increases exponentially with the electric field; hence, a significantly high generation rate is achieved. This also illustrated in Fig. 4 which shows increasing R_{BTBT} with increasing K values. Consequently, an increase in the tunnel current and sensitivity is seen with increasing dielectric constant in the nanocavities. Moreover, with the introduction of biomolecules, the carrier concentration in source/drain regions of silicon film below the nanocavities rises. This further increases the device current, that can be sensed and measured for detection.

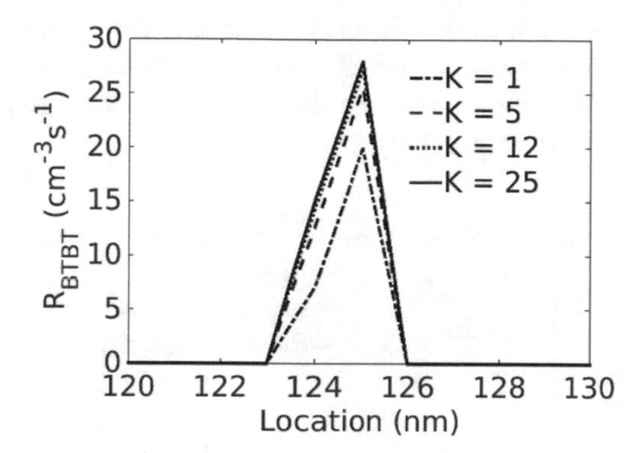

Fig. 4. The plot of band-to-band generation rate as a function of the biomolecular dielectric constant. The nanocavity length L_{cavity} is 25 nm and electrode gap L_{gap} is 15 nm.

Figure 5 shows the sensitivity (characterized as a ratio of device ON current with and without immobilized biomolecules, I_{BIO}/I_{AIR}) as a function of the biomolecular dielectric constants in the nanocavities. The sensitivity, for the reasons discussed above, increases as the biomolecular dielectric constant increases in the three (source, channel and drain) nanocavities. It must be noted that changing the length of the nanocavities L_{cavity} doesn't alter the detection sensitivity of the device. This is because the operation of a TFET is a phenomenon wherein the tunneling takes place at the source-channel or channel-drain junctions. Therefore, the tunneling current is not dependent on the length of the nanocavities.

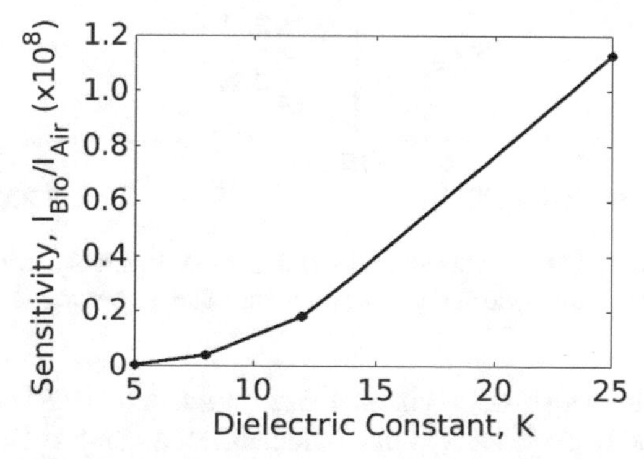

Fig. 5. Detection sensitivity (ratio of device currents with and without biomolecules) as a function of biomolecular dielectric constant for n-type operating mode with positive gate voltage V_{GS} = 1.5 V. The nanocavity length L_{cavity} is 25 nm and electrode gap L_{gap} is 15 nm.

Unlike the cavity lengths, the ON current and device sensitivity are considerably affected by the gap lengths between source to gate and gate to drain electrodes (L_{gap}). As the gap region is devoid of metal electrode to cause the charge-plasma doping effect, hence, these gaps act as depletion regions between the source-to-channel and the channel-to-drain junctions. This is also evident from Fig. 6 which shows that with increasing L_{gap}, tunneling window at the junction widens which reduces the effect of gate electric field on the tunnel junctions. This, in turn, reduces the tunneling current and device sensitivity, as shown in Fig. 7a and 7b, respectively.

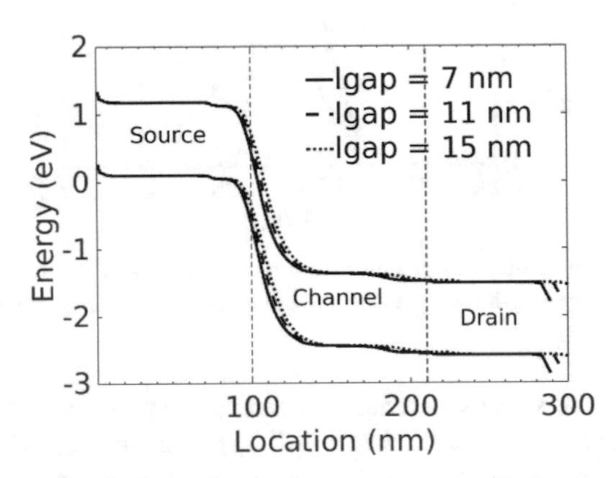

Fig. 6. The net effective shift in threshold voltage $\Delta V_{T_{net}}$ considering the combined impact of charge effect and dielectric constant modulation effect in n-channel DMFET biosensor.

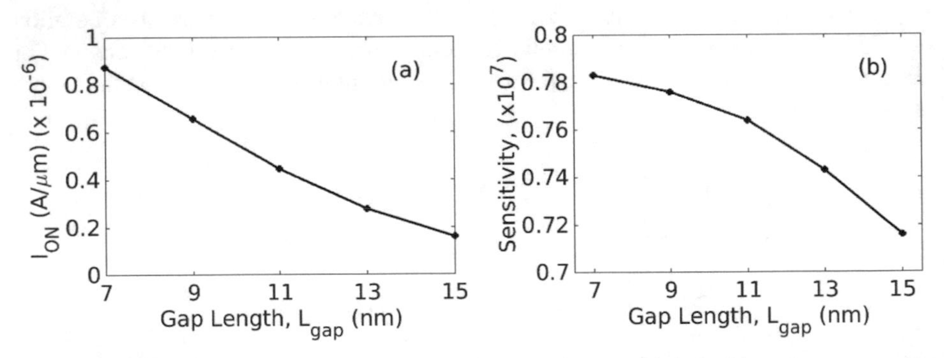

Fig. 7. The plot showing (a) ON current and (b) detection sensitivity, with respect to varying electrode gaps between source/gate and gate/drain metal. The nanocavity length L_{cavity} is 25 nm.

Since the detection sensitivity is characterized as the ratio of ON current of the device with and without the biomolecules in the nanocavities, a reduced-ON current implies a lower detection sensitivity. Hence, to achieve higher detection sensitivity, a low value of L_{gap} is desirable, as smaller gap length causes more band bending and hence, reduced tunneling window or a higher probability of tunneling of carriers. Therefore, enhanced current and sensitivity is achieved. However, small gap lengths may hinder the transport of biomolecules into the respective nanocavities. Therefore, an optimal value of the gap length must be chosen to maximize the device sensitivity without affecting the flow of biomolecules. In this work, we have chosen L_{gap} value of 15 nm, as the gap size between 10–15 nm has been shown to permit an easy passage and immobilization of biomolecules [3, 7, 21].

4 Conclusions

We have presented and analysed the detection performance of a dielectrically modulated TFET biosensing device in which the electrostatic doping of source/drain in the silicon film underneath the respective nanocavities is achieved by using the dielectric constant of the analyte biomolecules. In addition to the modulation of source/drain doping concentration by charge plasma effect, the dielectric constant modulation effect in the nanocavities above gate area is used for carrying out the sensing operation in the proposed device. Due to the absence of chemical-doping, the device does not suffer from doping related intricacies and variability in sensing characteristics. This makes the device a reliable and robust biosensing platform suitable for low power point of care and IoT healthcare applications. Due to a similarity in the process of the given doping less TFET and the traditional DMFET devices, we believe that work will provide a motivation for further investigation of the device.

References

1. Bergveld, P.: The development and application of FET-based biosensors. Biosensors **2**(1), 15–33 (1986)
2. Cui, Y., Wei, Q., Park, H., Lieber, C.M.: Nanowire nanosensors for highly sensitive and selective detection of biological and chemical species. Science **293**, 1289–1292 (2001)
3. Im, H., Huang, X.J., Gu, B., Choi, Y. K.: A dielectric-modulated field-effect transistor for biosensing. Nat. Nanotechnol. **2**(7), 430–434 (2007)
4. Seabaugh, A.C., Zhang, Q.: Low-voltage tunnel transistors for beyond CMOS logic. Proc. IEEE **98**(12), 2095–2110 (2010)
5. Ionescu, A.M., Riel, H.: Tunnel field-effect transistors as energy efficient electronic switches. Nature **479**(7373), 329–337 (2011)
6. Sarkar, D., Banerjee, K.: Proposal for tunnel-field-effect-transistor as ultra-sensitive and label-free biosensors. Appl. Phys. Lett **100**(14), 143108 (2012)
7. Abdi, D.B., Kumar, M.J.: Dielectric modulated overlapping gate-on-drain tunnel-FET as a label-free biosensor. Superlattices Microstruct. **86**, 198–202 (2015)
8. Gao, A., Lu, N., Wang, Y., Li, T.: Robust ultrasensitive tunneling-FET biosensor for point-of-care diagnostics. Sci. Rep. **6**, 22554 (2016)
9. Leonelli, D., Vandooren, A., Rooyackers, R., Gendt, S. D., Heyns, M. M., Groeseneken, G.: Optimization of tunnel FETs: impact of gate oxide thickness, implantation and annealing conditions. In: Proceedings of the European Solid State Device Research Conference, pp. 170–173, Spain (2010)
10. Nair, P.R., Alam, M.A.: Design considerations of silicon nanowire biosensors. IEEE Trans. Electron Devices **54**(12), 3400–3408 (2007)
11. Yang, X., Zang, P., Frensley, W., Zhou, D., Hu, W.: Doping fluctuation induced performance variation in SiNW biosensors. In: 13th IEEE International Conference on Nanotechnology (IEEE-NANO 2013), pp. 285–288, Beijing (2013)

12. Leung, G., Chui, C.O.: Stochastic variability in silicon double gate lateral tunnel field-effect transistors. IEEE Trans. Electron Devices **60**(1), 84–91 (2013)
13. Damrongplasit, N., Kim, S.H., Liu, T.-J.K.: Study of random dopant fluctuation induced variability in the raised-Ge-source TFET. IEEE Electron Device Lett. **34**(2), 184–186 (2013)
14. Kalra,S., Kumar, M. J., Dhawan, A.: Induced dielectric modulated tunnel field effect transistor biosensor (I-DMTFET): proposal and investigation. In: Proceedings of 17th International Conference on Nanotechnology (IEEE-NANO), pp. 97–100, USA (2017)
15. Mikolajick, T., Weber, W.M.: Silicon nanowires: fabrication and applications. In: Li, Q. (ed.) Anisotropic Nanomaterials. NT, pp. 1–25. Springer, Cham (2015). https://doi.org/10.1007/978-3-319-18293-3_1
16. Bjork, M.T., Schmid, H., Knoch, J., Riel, H., Riess, W.: Donor deactivation in silicon nanostructures. Nat. Nanotechnol. **4**(2), 103–107 (2009)
17. Rajasekharan, B., Hueting, R.J.E., Salm, C., Hemert, T.V., Wolters, R.A.M., Schmitz, J.: Fabrication and characterization of the charge plasma diode. IEEE Electron Device Lett. **31**(6), 528–530 (2010)
18. Hueting, R.J.E., Rajasekharan, B., Salm, C., Schmitz, J.: The charge plasma P-N diode. IEEE Electron Device Lett. **29**(12), 1367–1368 (2008)
19. Kalra, S., Kumar, M.J., Dhawan, A.: Dielectric modulated field effect transistors for DNA detection: impact of DNA orientation. IEEE Electron Device Lett. **37**(11), 1485–1488 (2016)
20. ATLAS Device Simulation Software, Silvaco, Santa Clara, CA, USA (2012)
21. Jang, D.-Y., Kim, Y.-P., Kim, H.-S., Park, S.-H.K., Choi, S.-Y., Choi, Y.-K.: Sublithographic vertical gold nanogap for label-free electrical detection of protein ligand binding. J. Vac. Sci. Technol. **25**(2), 443–447 (2007)

Integrated Platform for Microfluidics Based Cell Culture Applications

Sohan Dudala[1,2], Satish K. Dubey[1,3], Arshad Javed[1,3], Aritz Ozcariz[4], Ignacio R. Matias[4,5], and Sanket Goel[1,2](✉)

[1] MEMS, Microfluidics and Nanoelectronics Lab, Birla Institute of Technology and Science Pilani, Hyderabad Campus, Hyderabad 500078, Telangana, India
sgoel@hyderabad.bits-pilani.ac.in

[2] Department of Electrical and Electronics Engineering, Birla Institute of Technology and Science Pilani, Hyderabad Campus, Hyderabad 500078, Telangana, India

[3] Department of Mechanical Engineering, Birla Institute of Technology and Science Pilani, Hyderabad Campus, Hyderabad, Telangana 500078, India

[4] Department of Electrical, Electronic and Communications Engineering, Universidad Publica de Navarra, 31006 Pamplona, Spain

[5] Institute of Smart Cities (ISC), Universidad Publica de Navarra, 31006 Pamplona, Spain

Abstract. The ongoing developments in the field of microfluidics-based cell culture have led to a significant interest in systems for their visualization and control. The existing cell culture systems are limiting due to cost, complexity, and the need for regular user interventions. This work delves into developing a cost-effective, portable, and automated cell culture and monitoring platform with fluid management and integrated lens free imaging capabilities. The developed system, with a small footprint, weighing approximately 1.7 kg, is highly portable and can be used with existing CO_2 incubator systems. With the component cost of the developed system being approximately USD 300, the platform is highly affordable for an integrated cell imaging and automated media replacement system. The system can be used with conventional culture dishes as well as microfluidic systems.

Keywords: Automated Cell Culture · Microfluidics · Lens Free Microscopy

1 Introduction

Cell culture is one of the most vital molecular and cellular biology tools. Cell culture is pivotal in pharmaceutical development, disease modelling, virology, tissue regeneration studies, etc. [1]. In the existing framework for cell and tissue culture, certain critical peripheral equipment are required, including biosafety cabinet, incubator, fluid handling systems (such as pipettes or pumps) and microscopes. Commonly, the culture dishes are worked upon in a biosafety cabinet, observed in microscopic systems and fluid management is done using pipettes or pumps. Post the observation or follow-up treatment, the culture dishes are placed in an incubator to maintain conducive growth conditions. The primary issue in the existing framework is the requirement for the culture dishes to be

N. K. Suryadevara et al. (Eds.): ICST 2022, LNEE 1035, pp. 325–333, 2023.
https://doi.org/10.1007/978-3-031-29871-4_33

moved out of their optimal growth conditions for observation or biochemical treatments. The process of transporting the culture dishes back and forth from the incubator to the biosafety cabinets or microscopic system increases the possibility of contamination and undue mechanical stresses on the culture,owing to improper handling and disruption of the cell adhesion process, amongst other issues. The solution to these can be addressed in two parts. Firstly, by enabling fluid management for media and other chemical treatments within the incubator; and secondly, optical monitoring within the incubator.

Few attempts have been made towards the development of media and fluid management systems for cell culture applications. Takasago fluidic systems, Japan offers two systems for culture medium replacement – CME-0200 Portable Medium Replacement System and CEIM-010x Live Cell Imaging Fluidic System [2, 3]. The CME-0200 portable medium replacement system offers automatic medium replacement capabilities. The systems, costing approximately US$ 1800, claim to have a minimum flow rate of few microliters per minute. The MultiFlo FX multi-mode dispenser from Agilent technologies [4] is a multi-mode reagent dispenser for multitude of applications such as cell seeding, washing, media exchange, etc., but is applicable for use in high throughput situations such as for use with 6 to 348 well plates. The discussed systems can primarily be employed for static culture conditions wherein media is replaced periodically. Pitingilo et al. developed an automatic cell culture platform where dynamic perfusion of cell culture media and other constituents was demonstrated, and a successful demonstration of human-induced pluripotent stem cell differentiation was established [5]. In all the previously described systems, no optical characterization and monitoring has been integrated.

The limiting factor preventing the use of microscopy-based observation techniques from being integrated with automated media and fluid management systems is the size of the microscopy systems. The existing optical characterization equipment are bulky, and it is nearly impossible to accommodate them inside tabletop incubators making real time imaging difficult. Lens-free microscopy has been identified as a potent alternative to conventional microscopy, specifically for cell and tissue culture applications owing to the reduced footprint, cost and complexity involved [6]. These systems enable continuous and real-time monitoring of cells, which is nearly impossible with conventional systems. Lens-free microscopy uses the principle of digital incline holography where coherent or semi-coherent light sources are used. The sample is placed close to the imaging sensor, where the phase and amplitude data is recorded, and mathematical/computational reconstruction is done to obtain a representative image. Digital inline holography can be implemented with cost-effective components such as LEDs, common CMOS sensors, single board computers and 3D printed components. Significant work has been done in the development of lens-free microscopy systems and few notable ones being the works done by Tesng et al. [7], Kesavan et al. [8], Amann et al. [9], Berdeu et al. [10] and the commercial system - CytoSMART Lux2 [11]. All these systems were developed to improve portability for real-time visualization. However, none of these systems integrates fluid management, thereby requiring the user to perform additional steps for media replacement or drug treatments.

To address the discussed issues with simultaneous fluid management and imaging systems, this work reports on the development of an integrated lens-free imaging and fluid management platform for cell culture applications. The standalone system is amenable to be used with existing culture dishes as well as microfluidic devices. The functionalities of the developed system have been compared with existing relevant technologies and comparison of primary features is presented as Table 1.

Table 1. Benchmarking Table

#	System Name/ Reference	Automated Fluid Management	Integrated Microscopy
1	CME-0200 Portable Medium Replacement System, Takasago Fluidic Systems [2]	Yes	No
2	CEIM-010x Live Cell Imaging Fluidic System, Takasago fluidic systems [3]	Yes	No
3	MultiFlo FX, Agilent Technologies [4]	Yes	No
4	Pitingolo et al., 2020 [5]	Yes	No
5	Tseng et al., 2010 [7]	No	Yes
6	Kesavan et al., 2014 [8]	No	Yes
7	Amann et al., 2019 [9]	No	Yes
8	Berdeu et al., 2018 [10]	No	Yes
9	Lux 2 [11]	No	Yes
10	This System	Yes	Yes

2 Experimental

2.1 3D Printed Housing and Fluid Management System

The enclosures were designed, followed by printing using the Sigma D25 (BCN3D, Spain) 3D printer. A 2.85 mm polylactic acid (PLA) filament (BCN3D and Mitsubishi Chemical Performance Polymers, Inc.) was used for printing the components. The fluid management subsystem comprised of a set of 3D printed peristaltic pumps) (Fig. 1–1) and a set of 25 ml reagent bottles (Borosil, India) (Fig. 1–3). A set of stepper motors (NEMA 17) were used in peristaltic pumps. The fluid flow was achieved using autoclavable silicone tubing. The first reagent bottle along with the first peristaltic pump, acts as an introduction fluid circuit for new media or reagents. The reagent bottle can be replaced with prefilled reagents within the incubator as per need, without disturbing the culture system. The second fluid circuit is used for discarding the waste.

2.2 Electrical Control and Lens Free Microscopy System Integration

The overall system integration and control were achieved using a single board computer (Raspberry Pi 4 B). The control for stepper motors was established by using its controlling

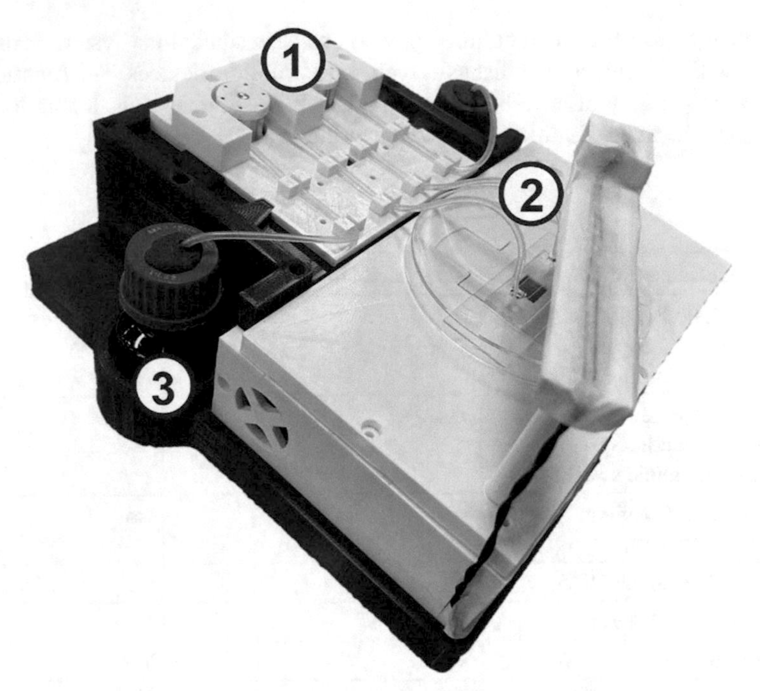

Fig. 1. Assembled prototype with subsystems – 1) dual 3D printed peristaltic pumps for exchange of media and reagents; 2) stage for culture dish or microfluidic device with CMOS sensor exposed to the top and the support structure light source; 3) reagent bottle for new media or reagents, identical component on the opposite side for collection of discards.

driver (DRV8825 driver from Polulu, USA) in 1/32 micro-stepping configuration via the GPIO ports on board the Raspberry Pi. A high-quality camera (RPI-HQ-CAMERA from Raspberry Pi) was used as the sensor for the lens-free microscopy subsystem. The camera comprises of a 12.3-megapixel CMOS sensor (IMX477 from Sony, Japan). The camera assembly was stripped down to the sensor fitting in the predesigned slot in the 3D printed base (Fig. 1–3). The source comprised of a 530 nm LED and a pinhole. Two temperature sensors (PT100 RTD, Element 14) along with a RTD to digital converter (MAX31865) were used for monitoring internal temperature. A mini-DC fan was added for temperature management. The system was powered using a custom-built lithium-polymer (LiPo) battery pack. All components were UV sterilized and assembled inside a class II biosafety cabinet. A GUI was developed using python to enable image capture, analysis and pump control. The system can be remotely controlled via any computing system over the same wireless network using the VNC viewer, thus empowering the user with on-the-go modification capabilities.

2.3 Microfluidic Device

Two variants of polydimethylsiloxane (PDMS) based microfluidic devices were developed for cell culture. The first variant was fabricated using a laser-cut polyimide sheet as the mould (Fig. 2A). The second variant (Fig. 2C) was developed using a silicon wafer-based mould with a negative photoresist. The device was designed with a main central chamber width of 200 μm. The auxiliary channels were intended for diffusion-based media exchange; however, they were not used in the current study. The PDMS polymer to curing agent ratio was used as 10:1 in line with the well-established protocol [12]. Post curing, inlet and outlets were punched using a biopsy punch. The thickness of the PDMS layer above the microchannel was controlled at 2 mm, to ensure gas permeability. The PDMS layer was bonded to a borosilicate glass slide using an oxygen plasma bonding system (CUTE, Femto Scientific, ROK). The microfluidic devices (Fig. 2B and 2D) were autoclaved, and UV sterilized prior to use.

2.4 Cell Culture

Experiments were carried out using H9c2 (2–1) rat cardiomyocytes cell line (CRL-1446, ATCC). The cells were cultured in high glucose, pyruvate Dulbecco's Modified Eagle's medium (DMEM) (Gibco, USA) with 10% fetal bovine serum (Gibco, USA). To prevent bacterial and fungal contamination Streptomycin, Amphotericin B, Penicillin (1x Antibiotic-Antimycotic, Gibco, USA) were used. The cells were incubated in a CO_2 incubator (Galaxy R48, Eppendorf, Germany) at 5% CO_2, 37 °C and saturated humidity conditions [13]. A 0.25% trypsin-EDTA solution (Gibco, USA) with phenol

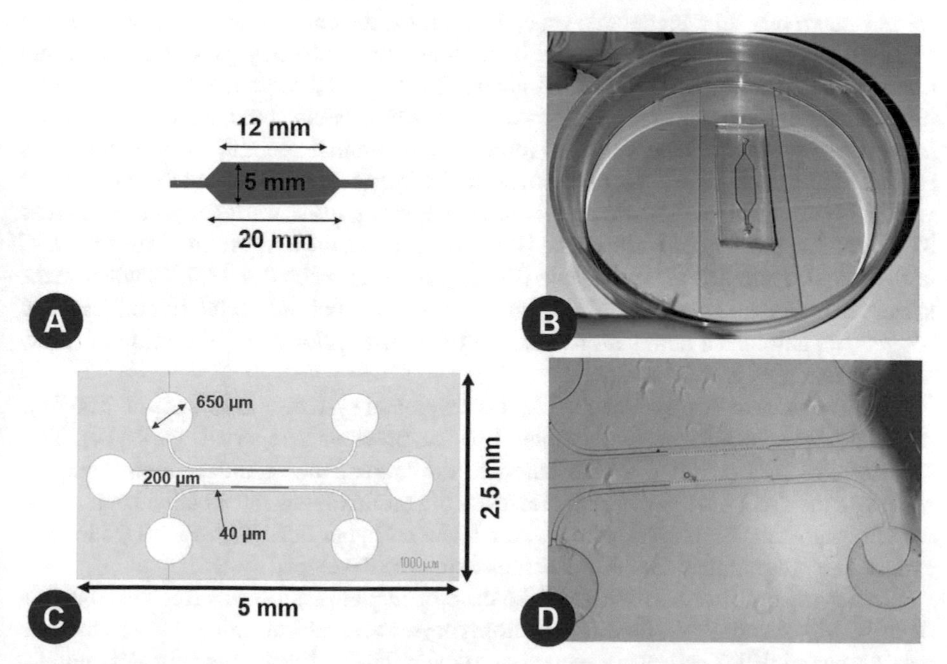

Fig. 2. A) Dimensions of the first variant microfluidic channel B) Sterilized microfluidic device developed by utilizing laser cut polyimide mould C) dimensions for the second variant - silicon wafer mould based microfluidic device D) Sterilized microfluidic device with the 200 μm width of the main cell seeding chamber. The 40 μm auxiliary channels were not used in current demonstration.

red indicator was used for cell detachment as and when required. Whenever necessary, phosphate buffered saline (PBS, pH 7.4, Gibco, USA) was used. Neubauer improved counting chamber (Blaubrand, Brand GMBH) and 0.4% Trypan Blue Solution (Gibco, USA) was used for cell counting. Standard cultures were maintained in 100 mm cell culture dishes (EasYDish Dishes, Nunc).

3 Results and Discussions

The pumping subsystem comprising a pair of stepper motors is the primary component for power management for the developed system. The micro stepping driver was used to limit the current per phase to 1.4 A. Although the maximum rated current is 2.2 A, it was realized that the operation of the motor would not involve high torque conditions, thereby eliminating the requirement for high amperage. The minimum operating voltage for the driver was 8.2 V. Keeping in view the current and voltage requirements, a 25C LiPo battery pack was used. Power management for the constituent components was one of the primary challenges in system development. A battery management system was also implemented for regulated supply to the components. A regulated supply was earmarked for Raspberry Pi to ensure continual operation. To reduce the power consumption of Raspberry Pi, the USB controller, HDMI output, and the onboard LEDs were disabled. Also, a GUI-based option was introduced wherein the system could be set in sleep mode till a predefined time. To monitor the temperature of battery pack, a PT100 temperature sensor was placed in contact with the battery pack. The maximum operating temperature was set to 42 °C, and at any moment the temperature of the battery pack reaches the set temperature, the system initiates a safety shutdown for 3 h, providing ample time for the battery pack to return to the nominal operating temperature. The second temperature sensor was placed next to the image sensor intended to be in contact with the culture dish or the microfluidic device. Further, the first trigger point was set as 37.5 °C. Also, the DC fan was switched on to bring down the temperature to under 37 °C with enhanced circulation. The second trigger point was enabled at 38.5 °C, wherein the system executed an emergency shutdown to protect the cultured cells. In both cases of triggers, an automated notification was sent to the user prior to the execution of safety shutdowns.

The lowest flow rate obtained using the developed system was 19 μl/min. The flow rate was higher than expected compared to the previously reported work [14]. This increase in flow rate could be attributed to the larger inner diameter tubing and six rollers on the spool instead of eight. However, the minimal flow rate obtained performed well for one of the developed microfluidic devices. Higher flow rates up to 1.75 ml/min can also be programmed for culture dishes using the developed system.

To demonstrate the capabilities of the developed platform for lens free visualization of cells, visualization of H9c2 rat cardiomyocytes seeded into microfluidic channels was attempted. H9c2 cells were seeded onto the microfluidic devices (Fig. 2B and 2D) using the process described for bubble-free loading for culturing mammalian cells [15]. Cells were seeded at the recommended seeding density of 1 x 10^4 cells per cm^2 [16] with adjustments for the available surface area. Post cell seeding, the microfluidic device (Fig. 2B) was connected to the inlet and outlet fluid circuits on the system. The system was programmed for a periodic media replacement every 12 h. The cells were grown in static conditions with fixed media replacement duration as it was identified that power requirements for the continuous perfusion approach could not be catered to using the battery pack. The developed peristaltic pump design lacked capabilities for ultra-low flow rates, which are an essential requirement for the microfluidic device (Fig. 2D) developed with 200 μm chamber. Thus, the follow-up media replacements were done using pipetting; however, visualization was done using the developed system.

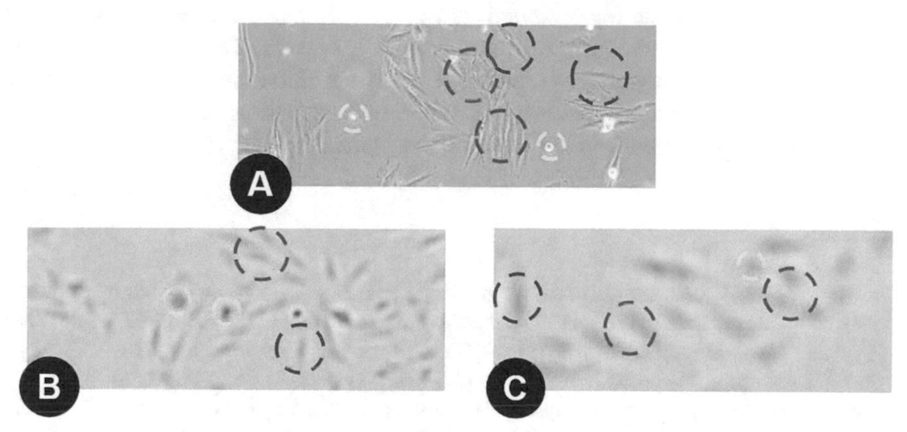

Fig. 3. A) Optical microscope image of the cells in the microfluidic device at 400x magnification, 24 h from cell seeding. B) reconstructed image segment for 5 mm wide microfluidic device (Fig. 2B) C) reconstructed image segment for 200 μm wide microfluidic device (Fig. 2C); yellow: non-viable cells, blue: isolated viable single cell, orange: clusters of viable cells.

The approach used for lens-free microscopy in the developed system relies on lens-free shadow imaging. The distance between the sensor and the substrate with cells was maintained under 1500 μm. Due to the small sample-sensor distance, the diffracting patterns impinge directly below the sample, with only a few microns away from the geometric centre. The 12.3 MP CMOS sensor captures this data. A modified version of lens-free shadow imaging described [17] was employed along with a reconstructive algorithm. Firstly, thresholding separates the diffraction patterns with the residual background, followed by the development of the binary image and obtaining filtered pixels. The binary image and filtered pixel data is used to process the image. The modifications were made keeping in view the computational capabilities of the single-board computer. The reconstructed images are shown in Fig. 3B and 3C for microfluidic devices shown in Fig. 2B and 2D, respectively. The algorithm can also estimate the viability, size, and number of cells. The reconstructed image is processed for edge detection. The circular or low aspect ratio elliptical features are treated as non-viable cells. The high aspect ratio elements are characterized as live cells. Work is currently being carried out to improve the algorithm to resolve clusters of closely packed cells. As evident from Figs. 3B and 3C, the developed algorithm provides an image capable of cell morphology evaluation. The system can be employed to detect significant changes in morphological changes due to environmental stresses or biochemical interactions.

4 Conclusions

The development and validation of standalone automated fluid management and imaging platform for cell culture applications have been demonstrated. The overall integrated platform can be used with existing standard CO_2 incubators. The component cost of the developed system is approximately USD 300, making it highly affordable for an integrated cell imaging and automated media replacement system. The integrated fluid management and visualization in an incubator friendly setup addresses the issues associated with the earlier reported systems and commercial products. The system can be used with conventional culture dishes as well as microfluidic systems. The primary limitations identified for the current work are improvement in resolution and achieving ultra-low flow rates. Work is currently underway to address these issues and also incorporate fluorescence-based imaging applications into the system.

References

1. Segeritz, C.P., Vallier, L.: Cell culture: growing cells as model systems In Vitro. Basic Sci. Methods Clinical Res. **151** (2017). doi: https://doi.org/10.1016/B978-0-12-803077-6.00009-6
2. Portable Medium Replacement System - Takasago Fluidic Systems. https://www.takasago-fluidics.com/products/portable-medium-replacement-system-1 (Accessed 01 Jul 2022)
3. Live Cell Imaging Fluidic System - Takasago Fluidic Systems. https://www.takasago-fluidics.com/products/live-cell-imaging-fluidic-system?variant=32962798256260 (Accessed 01 Jul 2022)
4. MultiFlo FX Microplate Dispenser - Overview. https://www.biotek.com/products/liquid-handling-multi-mode-washer-dispensers/multiflo-fx-multi-mode-dispenser/ (Accessed 01 Jul 2022)
5. Pitingolo, G., He, Y., Huang, B., Wang, L., Shi, J., Chen, Y.: An automatic cell culture platform for differentiation of human induced pluripotent stem cells. Microelectron. Eng. **231**, 111371 (2020)
6. Roy, M., Seo, D., Oh, S., Yang, J.W., Seo, S.: A review of recent progress in lens-free imaging and sensing. Biosens. Bioelectron. **88**, 130–143 (2017). https://doi.org/10.1016/J.BIOS.2016.07.115
7. Tseng, D., et al.: Lensfree microscopy on a cellphone. Lab Chip **10**(14), 1787–1792 (2010). https://doi.org/10.1039/C003477K
8. Kesavan, S.V., et al.: High-throughput monitoring of major cell functions by means of lensfree video microscopy Scient. Reports **4**(1), 1–11 (2014). doi: https://doi.org/10.1038/srep05942
9. Amann, S., von Witzleben, M., Breuer, S.: 3D-printable portable open-source platform for low-cost lens-less holographic cellular imaging. Sci. Rep. **9**(1), 11260 (2019). https://doi.org/10.1038/S41598-019-47689-1
10. Berdeu, A., et al.: Lens-free microscopy for 3D + time acquisitions of 3D cell culture. Scient. Reports 8(1), 1–9 (2018). doi: https://doi.org/10.1038/s41598-018-34253-6
11. CytoSMART Lux2 live-cell imager | CytoSMART. https://cytosmart.com/products/lux2 (Accessed 01 Jul 2022)
12. Dudala, S., Dubey, S.K., Goel, S.: Fully integrated, automated, and smartphone enabled point-of-source portable platform With microfluidic device for nitrite detection. IEEE Trans Biomed Circuits Syst **13**(6), 1518–1524 (2019). https://doi.org/10.1109/TBCAS.2019.2939658

13. Zhan, R., Guo, W., Gao, X., Liu, X., Xu, K., Tang, B.: Real-time in situ monitoring of Lon and Caspase-3 for assessing the state of cardiomyocytes under hypoxic conditions via a novel Au–Se fluorescent nanoprobe. Biosens. Bioelectron. **176**, 112965 (2021). https://doi.org/10.1016/J.BIOS.2021.112965
14. Davis, J.J., et al.: Utility of low-cost, miniaturized peristaltic and Venturi pumps in droplet microfluidics. Anal. Chim. Acta **1151**, 338230 (2021). https://doi.org/10.1016/J.ACA.2021.338230
15. Sahlm S., Elitas, M.: A simple, bubble-free cell loading technique for culturing mammalian cells on lab-on-a-chip devices. RSC Chips and Tips (2017). (Accessed 30 Jun 2022). https://blogs.rsc.org/chipsandtips/2017/02/28/a-simple-bubble-free-cell-loading-technique-for-culturing-mammalian-cells-on-lab-on-a-chip-devices/?doing_wp_cron=1614595663.6872100830078125000000
16. H9c2(2–1) | ATCC. https://www.atcc.org/products/crl-1446 (Accessed 30 Jun 2022)
17. Roy, M., Jin, G., Seo, D., Nam, M.H., Seo, S.: A simple and low-cost device performing blood cell counting based on lens-free shadow imaging technique. Sens. Actuators, B Chem. **201**, 321–328 (2014). https://doi.org/10.1016/J.SNB.2014.05.011

Performance Analysis of Split Gate Schottky Barrier Tunnel FET Biosensor

Sumeet Kalra(✉)

Indian Institute of Technology Delhi, Delhi, India
sumeetkalra5@gmail.com

Abstract. In this paper, the performance and sensing characteristics of a highly promising Schottky contacted bio-transistor are investigated. Compared to the conventional Schottky Barrier FET (SBFET) based biosensor, the device exhibits a boost in sensitivity due to independently biased split gate structure and device operation in the subthreshold regime when used in Tunnel FET operation mode. By controlled local gating of the Schottky junctions with split gate electrodes, the barrier characteristics can be changed to suit the requirements. Also, due to ambipolar operation, the polarity of the device can be forced electrostatically by proper biasing of split gate electrodes. Using 2D TCAD simulations, the study also carries out an analysis of the impact of liquid gate voltage and electrolyte strength on the sensing performance of the device operated in two different modes. The results of the simulation-based analysis show the device to be highly promising for biosensing applications with all the desirable characteristics of high sensitivity, increased reliability and robustness, adaptable performance, low false positives, and less fabrication complexity.

Keywords: Biosensor · Dynamic Range · Field Effect Transistor · Multifunctional · Reconfigurable · Schottky Barrier · Sensitivity · Tunable · Tunnel FET

1 Introduction

The detection, quantification and analysis of bio/chemical processes is of utmost importance in diagnostics and healthcare technology. Of the various transduction mechanisms, the FET based electrical detection of biomolecules is the most viable approach for making low-cost point-of-care analyzers and integrated bioanalytical systems [1–3]. Due to their direct, label-free detection, CMOS capability and down scalability, FETs are ideally suited for simple and portable biosensing applications. Therefore, more recently, nanostructures such as nanowires, nanosheets and nanoribbons have been utilized in FET configurations for biosensing [4–8]. The high surface area to volume ratios and scale match with target biomolecules render such nanostructures as highly sensitive biosensors. For this, FET based nanoscale biosensors qualify well as the building blocks of the future high-performance, integrated bioanalytical systems.

FET biosensors primarily rely on the field effect produced by the biomolecular charges on the sensing surface of the device, which is usually the exposed gate dielectric

N. K. Suryadevara et al. (Eds.): ICST 2022, LNEE 1035, pp. 334–343, 2023.
https://doi.org/10.1007/978-3-031-29871-4_34

region. The change in the distribution of surface charges corresponding to the binding events of biomolecules results in a change in channel conductivity and hence, the device current. Accordingly, the sensitivity of a biosensor is defined as the relative change in current $\Delta I/I_o$, where I_o is the current in the absence of biomolecules. Several methods have been reported to improve the detection sensitivity of nanoscale FET biosensors. These include, lowering the channel doping concentration, reducing the width of the nanowire [8–10], and regulating the gate voltage to operate the FET device in subthreshold regime [11, 12].

Accordingly, several studies have been reported on FET based approach of biosensing. Although most of the works employ FETs with ohmic contacts, the Schottky contacted FETs exhibit higher sensitivity due to their undoped channel, elimination of ion-implantation and dopant activation anneals for contact formations; and sharp junction profiles [13–17]. In addition, because of the undoped channel, the Schottky contacted devices offer larger screening lengths, hence, reduced concern of counter-ion screening of biomolecule charges [6–8]. Consequently, they have emerged as promising candidates for FET based electronic biosensing.

Despite their advantages, the Schottky biosensors suffer from integrability issues and a major limitation which can be understood as follows: The sensing mechanism in the Schottky Barrier (SB) FET biosensors is primarily based on two effects: (i) the Schottky Junction (SJ) effect [13–16] and/or (ii) the channel gating effect [15–18]. In the former, the adsorption of charged biomolecules modulates the SB characteristics, and hence, the device current, while in the later, the electrostatic gating of the channel by the charged biomolecules modulates the device conductivity.

In the Schottky junction effect, since the device current critically depends on the Schottky barrier characteristics, even a few adsorbed molecules in the junction area can lead to a detectable change and high detection sensitivity [15–18]. While more sensitive to the biomolecular interactions, the Schottky junction effect is less reproducible. This is because the contact areas are susceptible to even slight process variations and the residue accumulation from the wet processes. In addition, for use in electrolyte environment or physiologically relevant buffer solutions, the metal contacts are passivated to prevent leakages and electrochemical degradation. This causes the contact areas to be electrostatically isolated and inaccessible to the biomolecules [13–18]. Therefore, channel gating effect is the preferred mechanism. However, with the channel gating effect, the high Schottky junction SJ resistances dominate and reduce the device sensitivity due to a significantly reduced transconductance [18–20]. All this calls for a mechanism that enables an independent control or elimination of the Schottky barrier resistances to achieve high sensitivity.

Addressing these issues, a few recent works have proposed an SBFET biosensor with a mechanism that enables an independent regulation of the Schottky junctions [17, 18]. For this, an independent split gate (ISG) SB FET structure has been employed. The biasing applied on split gates, called as control gates CG1 and CG2, serves to modulate the two Schottky barrier resistances and control the carrier transport through SJs to achieve an enhanced detection sensitivity of the biomolecular interactions. By individually controlling the two SB characteristics, the device can be electrostatically forced to operate in multiple modes with different sensing characteristics. Additionally, with controlled Schottky junctions the electrical configuration and polarity can be adapted by adjusting the local biasing of the two control gates, thereby, allowing a considerably wider detection range than the traditional FET biosensors.

With an asymmetric biasing of control gates, the device is shown to operate as a Tunnel FET (TFET) with quantum mechanical tunneling as the dominant transport mechanism. Further by proper biasing of the two gates, both n-i-p type and p-i-n type TFET characteristics can be obtained. In both these configurations, the device exhibits lower subthreshold slope (SS) which is a figure of merit of a FET for biosensing applications [11, 12]. Since lower SS translates to a higher detection sensitivity, TFET mode operation of the device is preferred for biosensing applications.

In this paper, we carry out an in-depth analysis of the ISG SB FET device using 2D TCAD simulations and discuss further on its TFET operation mode. We also investigate the impact of reference gate voltage and strength of electrolytic solution on the working and detection sensitivity of ISG SB TFET biosensor. In addition, the work further establishes bias controlled tunability of operational mode as a potential means to achieve adaptable response of FET biosensors.

2 Device Structure and Operating Principle

The schematic view of the given biosensing device structure is shown in Fig. 1.

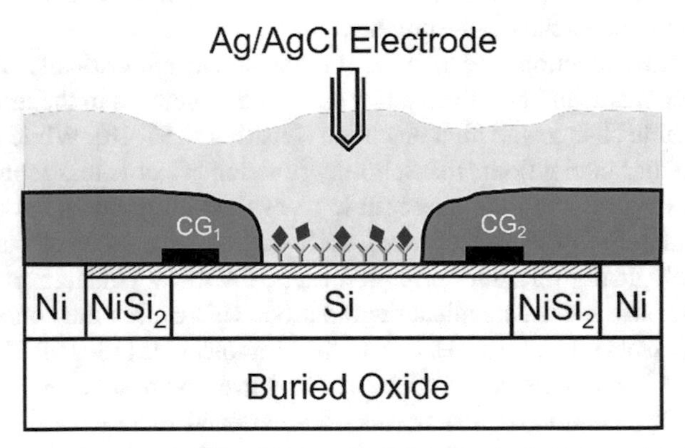

Fig. 1. The schematic illustration of independent split gate SB TFET biosensor view of dielectric modulated FET (DMFET) biosensor.

As depicted, the device is made using silicon-on-insulator (SOI) technology and consists of an ultra-thin p-type active silicon device layer serving as the channel and metallic Nickel Silicide ($NiSi_2$) source (S) and drain (D) junctions forming Schottky contacts with the channel. The active device layer is covered with a high-K dielectric, HfO_2. The split gate electrodes over the two Schottky junctions, CG1 and CG2 control the Schottky barrier heights and carrier tunneling across the Schottky junctions. The ungated portion of the channel serves as the sensing area and is functionalized to capture and immobilize the receptor biomolecules. These receptor molecules have an affinity towards the target biomolecule species and hence, bind specifically to them. To stabilize the solution potential and minimize the drift in sensing measurements, a standard Ag/AgCl reference or liquid gate electrode is incorporated through which a fixed potential (V_{LG}) is applied to the electrolytic solution. For insulation of the contact regions and metal electrodes against the wet environment of the electrolytic solution containing biomolecules, a thick passivating layer of oxide (Al_2O_3) or SU-8 resist covers the whole device but the active sensing area.

3 Working Principle and Energy Band Profiles

The working principle of ISG SB TFET biosensor is explained using the band profiles of Fig. 2. As evident, the voltages applied at CG1 and CG2 cause the bands to bend locally around the S/D Schottky junctions, allowing only one type of charge carrier to tunnel through the Schottky junction into the channel and at the same time blocking the other type of carrier. A positive (negative) voltage at CG1 and CG2 allows the electrons (holes) to tunnel through the SJs, hence, determining the device polarity. The potential barrier and transport in the channel are controlled by the biomolecular charges. Such an independent local biasing of SJs by CG1 and CG2 enables many advantages in the design of biosensors, such as, tailoring the device polarity, sensitivity, and detection limits without any amplification scheme or chemical modifications. By lowering and independently modulating two SB resistances, the control gates are also responsible to increase the detection sensitivity of ISG-SB TFET device over the existing traditional SB FET biosensors wherein the junctions are passivated for operation in electrolyte solutions and hence, are not accessible [17, 18].

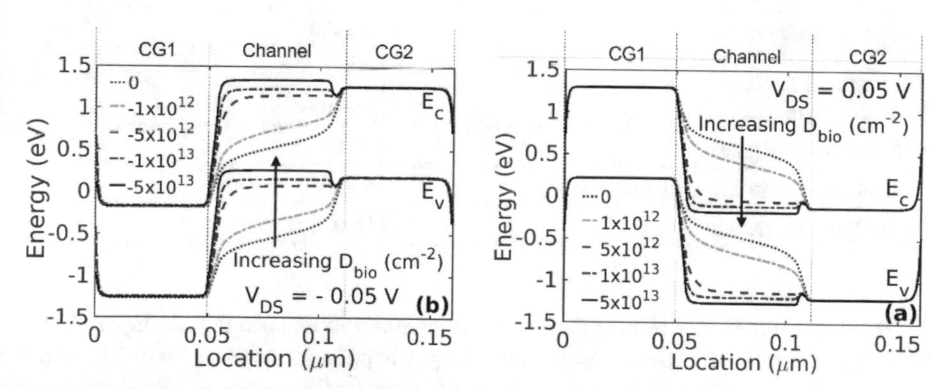

Fig. 2. The plots of energy band profiles of ISG SB TFET biosensor for varying biomolecular surface charge densities and $V_{LG} = 0V$ in (a) n-i-p TFET mode with $+V_{CG1}$ and $-V_{CG2}$ applied; and (b) p-i-n TFET modes with $-V_{CG1}$ and $+V_{CG2}$.

As illustrated in the band diagram of Fig. 2a, when an asymmetric bias is applied at CG1 and CG2, such that voltage at V_{CG1} is positive and V_{CG2} is negative, the device operates as n-i-p type TFET. Similarly, with negative V_{CG1} and positive V_{CG2} applied, the device works as a p-i-n type TFET, shown in Fig. 2b. Also as depicted for n-i-p and p-i-n TFET modes of Fig. 2, the positive (negative) charges of biomolecules immobilized on the gate sensing surface lower (raise) the energy bands in the channel region, thereby, thinning down the tunnelling barrier and allowing the band-to-band tunnelling of carriers across the respective tunnelling junctions.

4 Results and Discussions

The device dimensions and other parameters used for 2D TCAD simulations are indicated in Table 1. All the simulations were done with Silvaco Atlas device simulator [21] using the models calibrated with experimentally reported data for both SBFET device and the electrolyte solution [22, 23]. The charge transport in the channel is modeled as hydrodynamic carrier transport. In agreement to earlier reports [16]-[18], for carrier transport through SJs, the models for thermionic emission and WKB approximation of BTBT are employed.

Table 1. Parameters used for TCAD simulations of the given ISG SB TFET biosensor.

Parameter	Value
SOI film thickness	10 nm
Silicon channel doping (N_A)	1 x 10^{14} cm^{-3}
Schottky barrier height for electrons	0.66 eV
Schottky barrier height for holes	0.45 eV
Tunnelling mass of electrons and holes	$0.12*m_0$, $0.16*m_0$
HfO_2 Gate oxide thickness	5.1 nm
BOX thickness	300 nm
Channel length	160 nm
Length of control gates	50 nm
Control gate electrodes work function	4.7 eV
Electrolyte and strength	[NaCl], 1 mM

To model the effects of electrolytic environment and electric double layer (EDL) formed at the oxide-electrolyte interface, Gouy-Chapmann model is used. The other device models included are: bandgap narrowing, Lombardi mobility and SRH recombination. The charges of biomolecules are assumed to be uniformly distributed as a thin sheet on the gate sensing surface. As described in [24, 25], the electrolytic solution of strength 1mM (unless otherwise mentioned) containing the biomolecules is modeled as a semiconductor material with properties attuned to reproduce the solution behavior. Further details of the TCAD framework employed for wet environments can be found in the earlier works [24, 25].

Figure 3 shows the band-to-band tunnelling (BTBT) generation rate of carriers for both p-i-n and n-i-p TFET modes. As discussed in section III, with an increasing density of biomolecule charges, the tunnelling barrier width at the junctions decreases which results in a high band-to-band tunnelling (BTBT) generation rate of carriers.

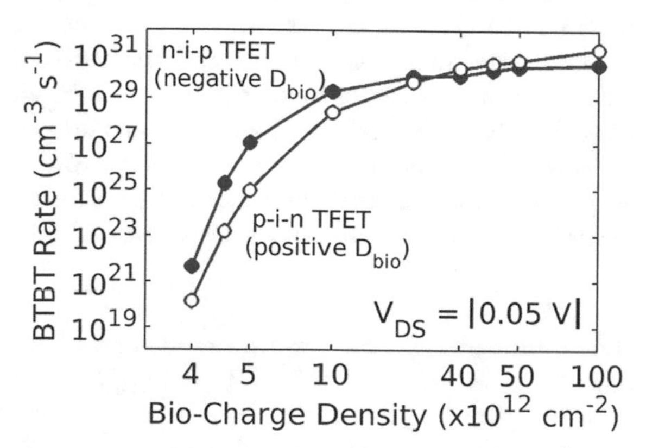

Fig. 3. Band-to-band tunneling (BTBT) generation rates of carriers as a function of biomolecular charge density (cm^{-2}) for both n-i-p and p-i-n TFET operation modes.

Figure 4 shows the plot of device current with varying liquid gate voltage V_{LG}. For n-i-p and p-i-n modes, V_{DS} is set to -0.05 V and 0.05 V, respectively. A sweep of liquid gate voltage is carried out from -2 to 2 V for both the modes. To calculate the voltage response i.e., the change in threshold voltage $\Delta V_T = V_{Tbio}$-V_{T0}, constant current method of V_T extraction is used at 1 nA/μm device current. It can be seen from Fig. 4a that for a positive liquid gate voltage in n-i-p mode of operation, the curves get closer with increasing biomolecular charge density, implying reduced voltage response ΔV_T of the biosensor. This is because + V_{LG} lowers the potential barrier in the channel by pushing the bands downwards, hence, widening the tunneling window and making the device less responsive. Similarly, for the p-i-n TFET mode shown in Fig. 4b, negative V_{LG} pulls the bands upwards, thereby reducing the response of the sensor.

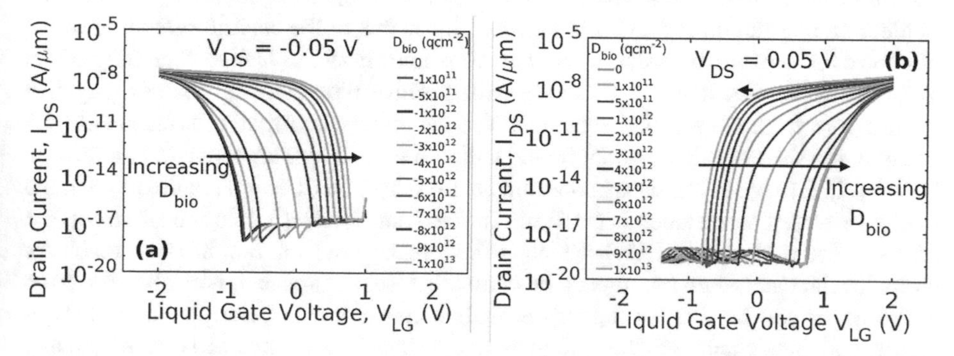

Fig. 4. The plots of transfer characteristics (I_{DS} vs. liquid gate voltage V_{LG}) of the given ISG SB TFET biosensor in (a) n-i-p and (b) p-i-n operation modes for various biomolecular charge densities (D_{bio}) in 1mM electrolyte concentration.

Additionally, as expected, for V_{LG} from -1 V to 0 V in n-i-p mode of operation, increasing negative biomolecular charges moves the curves in positive V_{LG} direction due to thinning of tunneling barrier with increasing surface charge density. Likewise, for V_{LG} swept from 0 V to 1 V for p-i-n mode shown in Fig. 4b, with increasing positive biomolecular charges, the tunnelling window is narrowed causing the device to turn ON at a reduced V_T, hence, shifting the curves in the negative V_{LG} direction.

The shift in threshold voltage V_T follows a linear trend with respect to the biomolecular charge density. This is evident from the Fig. 5 for TFET modes. As can be observed, the linearity of both p-i-n and n-i-p modes is similar (> 99%). Hence, the efficacy of both the operational modes is equally good and covers a wide span of biomolecular charge densities.

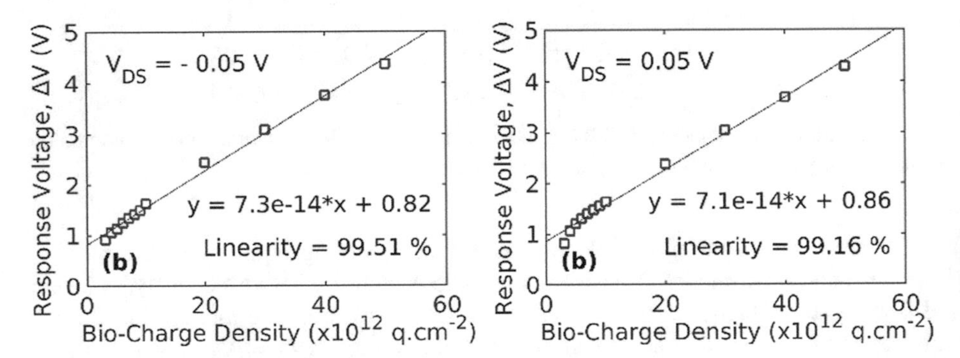

Fig. 5. The plots of sensor's responsivity ($\Delta V_T = V_{Tbio}$-V_{T0} with V_T extracted at 10 nA/μm) vs absolute values of biomolecular charge densities in 1mM NaCl electrolyte solution for (a) n-i-p TFET and (b) p-i-n TFET modes. The blue symbols show linear behavior of the voltage response of the given ISG SB TFET biosensor.

Figure 6 shows the influence of electrolytic solution concentrations on the potential across the EDL formed at the interface of oxide to electrolytic solution in n-i-p and p-i-n operation modes, respectively. The potential changes exponentially in the bulk of electrolytic solution for both the modes. This is due to the electric screening of the biomolecular charges by solution counter-ions known as the Debye screening effect discussed in section I. For p-i-n mode with positive biomolecular charges, negative counter-ions from the solution accumulate at the oxide-electrolyte interface causing a decay in the electrostatic potential on the solution side. A similar behavior can be seen for n-i-p TFET mode. The decay constant, as given by Gouy-Chapmann theory, is equal to Debye screening distance λ (<10 nm and <1 nm for NaCl solution of strength 1 mM and 150 mM, respectively), which is the length in the solution beyond which the biomolecular charges do not impact the channel. Also, as expected, it can be seen from both Fig. 6a and 6b that for higher electrolyte concentrations, the potential changes rapidly and these changes at the oxide-electrolyte interface are shielded due to a greater number of counter ions, hence, reducing the sensor's response quality.

As with any other FET biosensor, the sensor's responsivity (ΔV_T) is also influenced by the strength of the electrolytic solution containing the biomolecules. Therefore, we

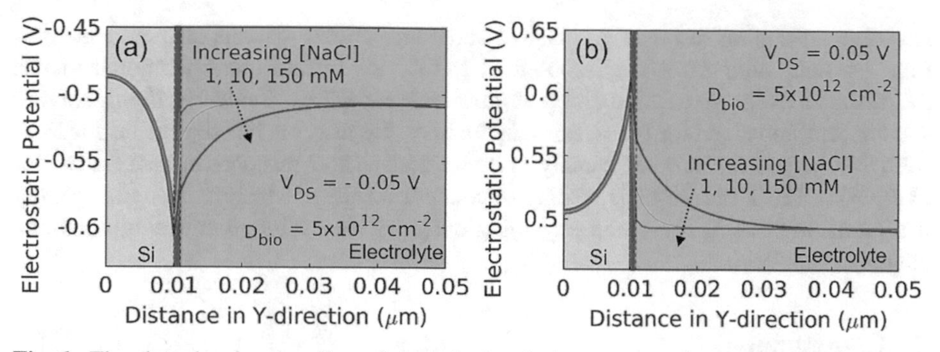

Fig. 6. The plots showing the effect of electrolytic solution concentration on the potential across the double layer formed at gate oxide to electrolyte interface in (a) n-i-p and (b) p-i-n operation mode. The regions shown correspond to a cutline taken vertically across the center of the device.

have also studied the ionic concentration dependence of the response characteristics by sweeping V_{LG} and by solving the Poisson Boltzmann equation for different salt concentrations. Figure 7 shows the corresponding results for varying biomolecular charge densities $|D_{bio}|$ from 1×10^{11} and 1×10^{13} cm^{-2}. As discussed earlier, due to the formation of double layer at oxide-electrolyte interface, the biomolecular charges are screened by the solution ions. For n-i-p (p-i-n) mode with negative (positive) bio-charge density, positive (negative) counter ions accumulate at the oxide-electrolyte interface and shield the biomolecular charges. This reduces the field gating and band bending in the channel, consequently, preventing the tunneling of carriers through a wide tunneling barrier width. As a result, the sensor response is reduced. Since at increasing salt concentrations, less effective charge is present on the surface, the response ΔV_T is lower at higher electrolyte concentrations.

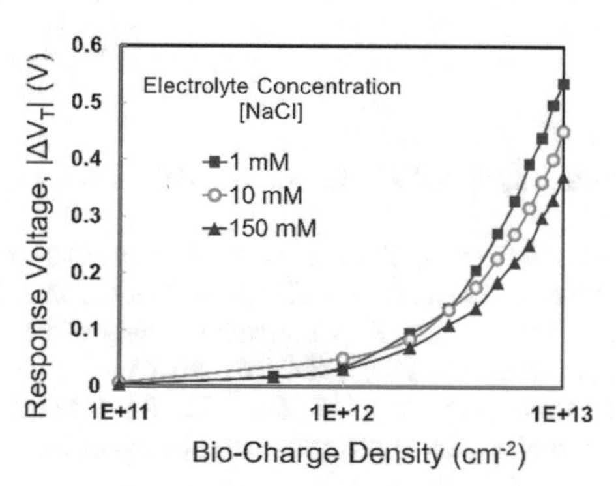

Fig. 7. The plot showing influence of salt concentrations in the electrolytic solution [NaCl] on the sensing response ΔV_T (with V_T extracted at 1 nA/μm) as a function of biomolecular charge density.

However, it is worthy to note here that the response of the given biosensor is nearly independent of the electrolyte concentration, consistently with the existing studies that low channel doping result in less Debye screening effects. This is also evident from Fig. 7 which summarizes the response of the device in both the operational modes following an exposure to three different ionic concentrations. Evidently, though higher salt concentrations reduce the sensor responsivity, the effect is not significant for ISG-SB TFET biosensor. Hence, the analysis shows that sensors response in both the modes of ISG-SB TFET biosensor is weakly dependent on the electrolyte strength, making it a promising device for use in physiologically relevant solution concentrations and diagnostic applications.

5 Conclusions

In summary, the work has investigated the performance of a recently introduced Schottky barrier bio transistor with independently biased split gate structure and band-to-band tunneling as the dominant mechanism of transport through the channel. The device performance has been analyzed in two operation modes and discussed in detail with calibrated numerical simulations. The impact of various parameters such as electrolyte strength, liquid gate voltage, biomolecular charges has been discussed. The numerical framework employed is based on Poisson–Boltzmann and Gouy-Chapmann models to simulate the biosensor characteristics. Due to controlled Schottky barrier resistances, low subthreshold slopes and high band-to-band generation rates, the device exhibits a high detection sensitivity in both the operational modes and for a wide range of biomolecular charges. It is also shown that the device operates with high sensitivity even at higher solution concentrations. The simulation results showing the features of low subthreshold slopes, controlled polarity, less Debye screening, good linearity and detection limits of charges, further substantiate the promise of the device for future multiplexed, adaptable and integrable biosensors. Therefore, we believe that the outcomes of the analysis will provide guidelines for experimental explorations of ISG-SB TFET biosensor.

References

1. Bergveld, P.: The development and application of FET-based biosensors. Biosensors **2**(1), 15–33 (1986)
2. Cui, Y., Wei, Q., Park, H., Lieber, C.M.: Nanowire nanosensors for highly sensitive and selective detection of biological and chemical species. Science **293**, 1289–1292 (2001)
3. Moon, D.I., Han, J.W., Meyyappan, M.: Comparative study of field effect transistor based biosensors. IEEE Trans. Nanotech. **15**(6), 956–961 (2016)
4. Mu, L., Chang, Y., Sawtelle, S.D., Wipf, M., Duan, X., Reed, M.A.: Silicon nanowire field-effect transistors—a versatile class of potentiometric Nanobiosensors. IEEE Access **3**, 287–302 (2015)
5. Stern, E., et al.: Label-free immunodetection with CMOS-compatible semiconducting nanowires. Nature **445**(7127), 519–523 (2007)
6. Stern, E., Wagner, R., Sigworth, F.J., Breaker, R., Fahmy, T.M., Reed, M.A.: Importance of the Debye screening length on nanowire field effect transistor sensors. Nano Lett. **7**(11), 3405–3409 (2007)
7. Nair, P.R., Alam, M.A.: Screening-limited response of NanoBiosensors. Nano Lett. **8**(5), 1281–1285 (2008)

8. Nair, P.R., Alam, M.A.: Design considerations of silicon nanowire biosensors. IEEE Trans. Electron Devices **54**(12), 3400–3408 (2007)
9. Elfström, N., Juhasz, R., Sychugov, I., Engfeldt, T., Karlström, A.E., Linnros, J.: Surface charge sensitivity of silicon nanowires: size dependence. Nano Lett. **7**(9), 2608–2612 (2007)
10. Li, J., Zhang, Y., To, S., You, L., Sun, Y.: Effect of nanowire number, diameter, and doping density on nano-FET biosensor sensitivity. ACS Nano **5**(8), 6661–6668 (2011)
11. Gao, X.P.A., Zheng, G., Lieber, C.M.: Subthreshold regime has the optimal sensitivity for nanowire FET biosensors. Nano Lett. **10**(2), 547–552 (2010)
12. Gao, A., Lu, N., Wang, Y., Li, T.: Robust ultrasensitive tunneling-FET biosensor for point-of-care diagnostics. Sci. Rep. **6**(1), 22554–22563 (2016)
13. Zörgiebel, F.M., et al.: Schottky barrier-based silicon nanowire pH sensor with live sensitivity control. Nano Res. **7**(2), 263–271 (2014). https://doi.org/10.1007/s12274-013-0393-8
14. Yoo, S.K., An, J.Y., Yang, S., Lee, J.H.: Subthreshold operation of Schottky barrier silicon nanowire FET for highly sensitive pH sensing. Electron Lett **46**(21), 1450–1452 (2010)
15. Hu, Y., Zhou, J., Yeh, P.-H., Li, Z., Wei, T.-Y., Wang, Z.L.: Supersensitive, fast-response nanowire sensors by using Schottky contacts. Adv. Mater. **22**(30), 3327–3332 (2010)
16. Heinzig, A., Slesazeck, S., Kreupl, F., Mikolajick, T., Weber, W.M.: Reconfigurable silicon nanowire transistors. Nano Lett, **12**(1), 119–124 (2011)
17. Kalra, S., Kumar, M.J., Dhawan, A.: Reconfigurable FET biosensor for a wide detection range and electrostatically tunable sensing response. IEEE Sensors J. **20**(5), 2261–2269 (2019)
18. Kalra, S., Kumar, M.J., Dhawan, A.: Schottky barrier FET biosensor for dual polarity detection: a simulation study. IEEE Electron. Device Lett. **38**(11), 1594–1597 (2017)
19. Pregl, S.: Fabrication and characterization of a silicon nanowire based Schottky-barrier field effect transistor platform for functional electronics and biosensor applications. In: Ph.D. dissertation, Inst Mater Sci, Max Bergmann Center Biomater, TU Dresden, Germany (2015)
20. Heller, I., Janssens, A.M., Männik, J., Minot, E.D., Lemay, S.G., Dekker, C.: Identifying the mechanism of biosensing with carbon nanotube transistors. Nano Lett. **8**(2), 591–595 (2008)
21. ATLAS Device Simulation Software: Silvaco. CA, USA (2012)
22. Marchi, M.D., et al.: Polarity control in double-gate, gate-all-around vertically stacked silicon nanowire FETs. In: 2012 International Electron Devices Meeting, CA, pp. 8.4.1–8.4.4, (2012)
23. Chung, I.-Y., Jang, H., Lee, J., Moon, H., Seo, S.M., Kim, D.H.: Simulation study on discrete charge effects of SiNW biosensors according to bound target position using a 3D TCAD simulator. Nanotechnology **23**(6), 065202 (2012)
24. Pittino, F., Palestri, P., Scarbolo, P., Esseni, D., Selmi, L.: Models for the use of commercial TCAD in the analysis of silicon-based integrated biosensors. Solid State Electron **98**(xx), 63–69 (2014)
25. Choi, B., et al.: TCAD-based simulation method for the electrolyte–insulator–semiconductor field-effect transistor. IEEE Trans. Electron Devices **62**(3), 1072–1075 (2015)

A Capacitive Cantilever-Based Flow Sensor

H. Harija[1(✉)], K. Sri Hari Charan[2], Boby George[2], and Arun K. Tangirala[1]

[1] Department of Chemical Engineering, Indian Institute of Technology Madras, Chennai, India
ch16d302@smail.iitm.ac.in, arunkt@iitm.ac.in

[2] Department of Electrical Engineering, Indian Institute of Technology Madras, Chennai, India
ee17b132@smail.iitm.ac.in, boby@ee.iitm.ac.in

Abstract. This paper presents a novel capacitive transduction mechanism for a cantilever-based flow sensor. The sensor has a stainless steel cantilever attached inside the pipe. The cantilever is fitted in such a way that it bends depending on the flow rate and direction of the flow. The proposed sensing mechanism senses the bending angle of the cantilever using specially shaped capacitive electrodes introduced outside the insulating water pipe. The capacitive sensor electrodes are connected to a measurement unit, but no electrical wiring is needed to the cantilever, ensuring a contactless bending angle read-out scheme. This method is new and helps to realize low-cost and easy-to-install flow sensors. The functionality of the sensing mechanism is first verified using finite element analysis (FEA). Later, a suitable sensor was developed in the laboratory and tested on a pipeline test bench. The change in capacitance due to the cantilever deflection with respect to fluid flow is obtained. A prototype sensor has been developed, and its performance for liquid flow measurement is evaluated. A suitable signal measurement circuity that obtains the difference in capacitance corresponding to the bending angle, and hence the water flow, is developed and employed for the experimental studies. The FEA and experimental test results showed that the proposed transduction and read-out mechanisms are suitable for the cantilever-based water flow sensor.

Keywords: Capacitive Sensor · Proximity Sensing · Cantilever deflection · Flow Measurement

1 Introduction

Capacitive sensors are widely used in industrial automation, military, and consumer applications, including object detection, safety, inventory management, robotics, etc. As the capacitance of a sensor is a function of the distance between the electrodes, surface areas of the electrodes, and permittivity of the medium in between, it is possible to design capacitive sensors that can track the movement of an object/target in a specified area or volume [1–3]. The object could be conductive or insulating in nature. Other methods of sensing the proximity of an object include techniques based on ultrasonic, optical, eddy current, or inductive [1–6]. An inductive or eddy current proximity sensor can only detect metal targets; on the other hand, capacitive proximity sensors are not limited to metallic target detection [6]. The possible targets for capacitive sensors include the

N. K. Suryadevara et al. (Eds.): ICST 2022, LNEE 1035, pp. 344–351, 2023.
https://doi.org/10.1007/978-3-031-29871-4_35

following but are not limited to water, glass, plastics, wood, powders, resins, etc. Some unique features of capacitive sensors are contactless sensing of an object, long life span, fast response, and low power consumption [7].

In the process industry, flow sensors are crucial to monitor and manage the fluid flow rate. Flow sensors are required to monitor water flow in closed pipes or open channels for irrigation, residential, and commercial uses. The important performance parameters of typical flow sensors are acceptable accuracy and precision, ease of manufacturing, installation and maintenance, and low power consumption. Additionally, the cost of the sensor should be considered. New flow sensors with the above characteristics are highly sought-after in the current data-driven age. Flow measurements in large diameter pipes are conventionally sensed using electromagnetic, ultrasonic, turbine, vortex, and orifice-type meters [8]. These types of meters are expensive but accurate and widely used in industrial applications. Affordable flow sensors without compromising accuracy are in high demand, especially for residential and irrigation sectors. One of the low-cost water flow sensors is a turbine flow meter. Hall Effect or inductive based-pickup coils-based schemes are used to sense the turbine's rotation [8]. Due to the wear and tear of the moving parts, these types of sensors have poor accuracy as they age. A low-cost orifice plate-based water flow meter for a household application has been presented in [9]. It causes a significant pressure drop in the fluid and is incompatible with debris flow. Cantilever-based flow sensing has been reported for microfluidics applications. Those are microelectromechanical systems (MEMS) cantilevers. The commonly used methods employed for cantilever deflection are based on strain gauge, optical, image, and capacitive sensors [8]. The strain gauge requires direct contact with the cantilever. The optical and image-based approach requires a suitable transparent opening and line-of-sight. Capacitive MEMS-based flow sensor has been reported in [10, 11]. To measure flow in domestic and agricultural applications, the MEMS cantilevers are unsuitable as the pipe diameter is a few centimeters. Implementation of a cantilever-based flow sensor using an image processing approach has been reported in [12]. The authors have adopted the image-based approach mainly for characteristic evaluation in a laboratory environment [13]. Despite the cantilever bending angle vs. flow rate being repeatable and easy to model, it is less practical to implement as the camera-based approach requires a transparent pipe window. Therefore, a non-intrusive transduction mechanism to sense the bending angle of the cantilever is essential.

This paper presents a novel capacitively coupled cantilever sensor (C^3 S) suitable for flow measurement. It uses a cantilever, and the bending angle of the cantilever due to a change in flow rate is sensed in the form of a change in capacitance. The sensing mechanism, measurement approach, and test results are presented below.

2 Capacitively Coupled Cantilever Sensor

Figure 1(a) shows the diagram of the cantilever and the location of the proposed capacitive sensing electrodes introduced outside the pipe. A simplified 3D view of the same is available in Fig. 1(b), which is the model used for the finite element analysis (FEA). A capacitive electrode, named receiver R, is introduced above the fixed end of the cantilever. The cantilever is made of stainless steel. A fixed coupling capacitance C_R is formed between the receiver electrode and the cantilever. There are two triangular-shaped capacitive electrodes, T_1 and T_2 introduced, outside the pipe, at the free end of the cantilever. This forms three capacitances, one between the cantilever and the transmitter T_1, named C_{x1} and indicated in the simplified electrical equivalent circuit in Fig. 2. The second capacitance C_{x2}, is formed between the cantilever and T_2. The third one is not shown in the figure as we are not utilizing it for sensing; it is between the T_1 and T_2. In this study, the triangular-shaped electrodes are designed and used, to realize a differential capacitive sensor configuration which is known to provide certain benefits [7]. In the proposed sensor, as the cantilever bends due to fluid flow inside the pipe, the capacitances C_{x1} and C_{x2} change in a push-pull manner for a certain range. This change in capacitance can be measured across the electrodes T_1 and R, and T_2 and R, and utilized to estimate the bending or deflection angle.

As seen in Fig. 1(a), when a cantilever is subjected to uniformly distributed force due to flow, the cantilever changes the deflection angle [13]. Or, a change in fluid flow velocity causes the cantilever to deflect. As shown in Fig. 1(b), a non-contact read-out mechanism using capacitive sensors is proposed to sense this change in deflection inside the pipe. The deflection angle, defined as the inverse tangent of deflection to the length of the cantilever [12], is given in Eq. (1). With some assumptions, it can be used to represent the flow rate Q as in Eq. (2) [13].

$$\theta = tan^{-1}\left(\frac{D}{L}\right) \tag{1}$$

$$Q \propto \sqrt{tan\theta} \tag{2}$$

Due to the geometry employed, the capacitance $C_R >> C_{x1}, C_{x2}$. Thus, the effective capacitance between T_1 and R can be approximated to C_{x1}, and that between T_2 and R can be approximated to C_{x2}. To measure the difference between C_{x1} and C_{x2}, T_1 and T_2 are excited using sinusoidal sources Vs $\angle 0°$ and Vs $\angle$ 180°, respectively. The receiver is maintained on the virtual ground. The current through C_R is passed through an i-to-v converter with feedback resistance Z_F. The output voltage V_0 can be represented as in Eq. (3), where ω is the angular frequency of the excitation signal.

$$V_o = -V_s \cdot \omega \cdot (C_{x1} - Cx_2) \cdot Z_f \tag{3}$$

2.1 Finite Element Analysis

Although capacitive sensors are used for target detection, no study is reported to determine the angular deflection of a cantilever inside pipes. Therefore, an FEA has been conducted after building a model of the cantilever inside a pipeline along with the external electrodes in COMSOL 5.5 Multiphysics® software. The capacitances C_{x1} and C_{x2} with respect to the deflection angle are obtained through the FEA. This gives the trend of change in the sensor capacitance as a function of the cantilever deflection. The 3D model was developed using Electrostatics (es) in the AC/DC module of the tool, as shown in Fig. 1b. As in the measurement scheme, the transmitting electrodes T_1 and T_2 were provided with equal and opposite voltages. The materials used in the simulation

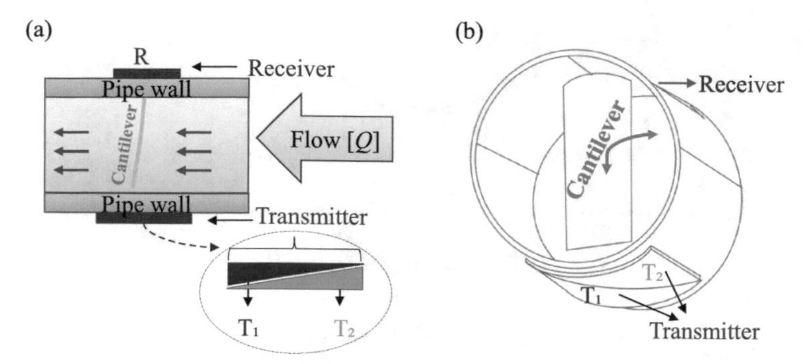

Fig. 1. Schematic diagram of the proposed sensor. **(b)** Finite element model used in COMSOL Multiphysics tool.

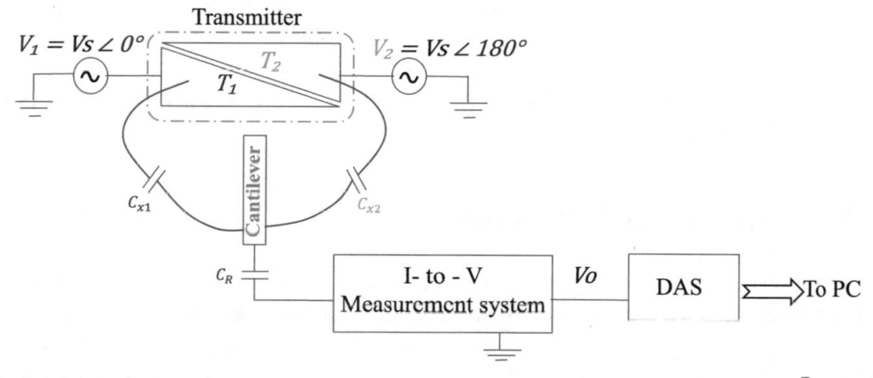

Fig. 2. An equivalent electrical circuit of the capacitive sensor with the cantilever (C^3S) in the vicinity of the electrodes. A Block diagram of the measurement system is also shown.

are stainless steel, copper, and PVC for the cantilever, electrodes, and pipe, respectively. The materials and dimensions used in the simulation match the prototype sensor built. In FEA, the cantilever deflection angle changed from $-45°$ to $45°$ in steps of $5°$, and the corresponding changes in the capacitance values were obtained. The characteristic is not-linear but it has a predictable shape. At $0°$, the ΔC was observed as 2.926 pF; hence an offset correction was performed before plotting and given in Fig. 3. From these plots, it is clear that the deflection angle can be estimated for a wide range, say, $\pm 15°$, which is sufficient for this application. A linear response is feasible if the range of deflection is limited. To operate in this region, at the design stage, the cantilever's stiffness and thickness can be selected considering the flow rate range.

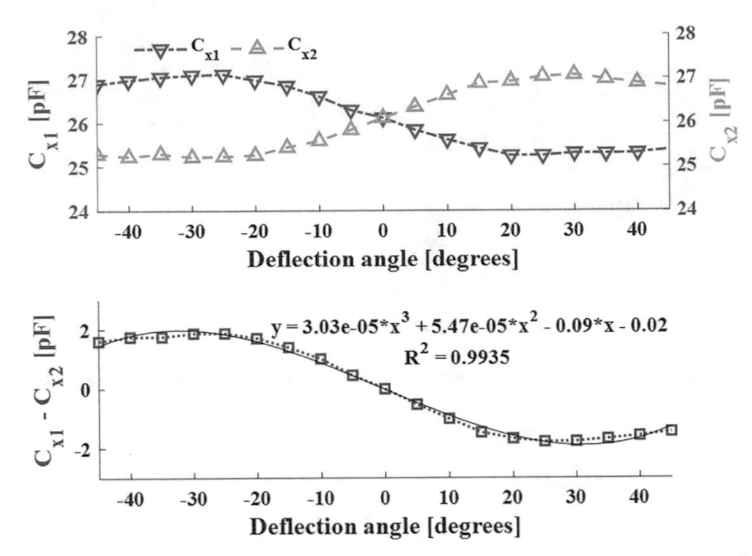

Fig. 3. Capacitance change ($\Delta C = C_{x1} - C_{x2}$) as a function of cantilever deflection (θ) from the Finite Element Analysis (FEA).

3 Experimental Set-Up and Results

The proposed proximity sensor has been developed and evaluated on a laboratory scale. The schematic diagram of the experimental set-up is illustrated in Fig. 4.

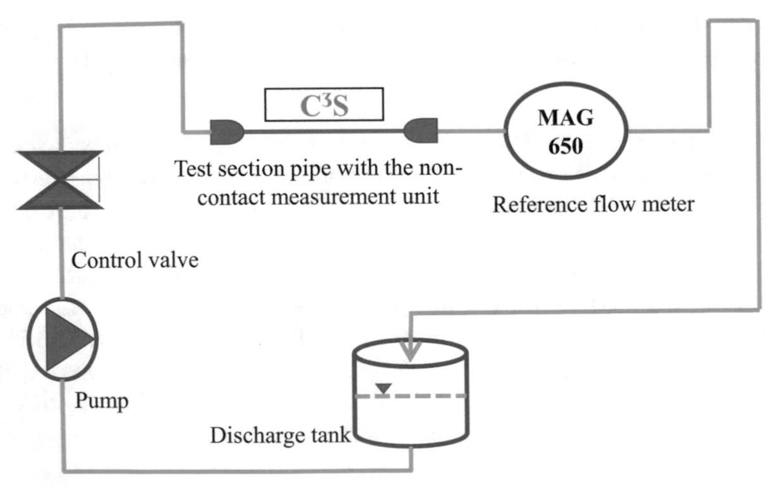

Fig. 4. Experimental set-up line diagram.

The test section pipe, along with the non-contact measurement unit, is shown in Fig. 5. The cantilever dimensions used were 50 mm × 20 mm, and it has a thickness of 0.15 mm. The main objective of this experiment is to verify the functionality of

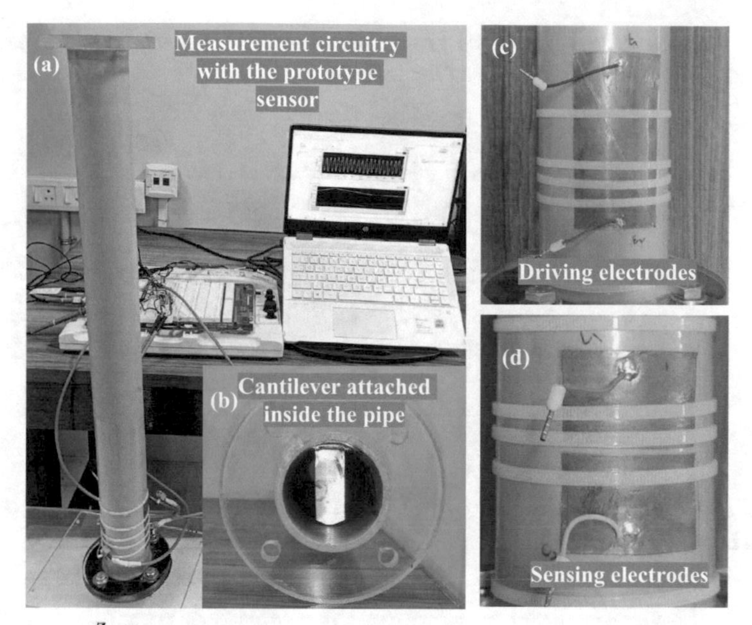

Fig. 5. (**a**) & (**b**) C^3 S prototype sensor with the measurement circuit. (**c**) &(**d**) Driving and sensing electrodes made of copper tapes.

the proposed sensor in sensing the deflection of the cantilever due to water flow. The prototype sensor is tested in a PVC pipe of 63 mm diameter. PVC pipe is chosen in this work as they are commonly used for water mains, and irrigation and are not prone to corrosion. In the prototype, the electrodes were formed using copper tapes with a few cable ties attached to the pipe; the dimensions are 77 mm $\times$ 40 mm.

A voltage of 10 V (peak-to-peak), at a frequency of 200 kHz, was provided to the transmitter from the function generator of the NI ELVIS II + board. The receiver is connected to a measurement circuit (i-to-v) using an Op-amp IC LF347 that gives the output as in (3). Output voltage data was logged and displayed with the help of a LabVIEW program. Different sets of readings were recorded, and measurements were repeatable with a maximum standard error of 0.2 mV. The result obtained from the prototype sensor known as the calibration model is shown in Fig. 6. Similar trend is observed in the simulation. The experimental data are separated into training and test data. Training data are used to build the calibration model. Validation aims to evaluate the effectiveness of the model created, and the flow rate is estimated from the sensor prototype output with the help of test data. The test data in this study are considered to be fresh data from an entirely new set of experiments. The error is the absolute difference between the estimated and actual flow rates. Figure 7 compares the reference flow meter's flow rate with the estimated flow rate from the calibration model and displays the percentage non-linearity error for each data point. The flow measurement error is expressed as a percentage of the full scale. The sensitivity of the output of the prototype sensor can be improved by either increasing the value of the feedback resistor in the i-to-v measurement circuit or by increasing the excitation signal frequency. In this current study, flow rates from 0–7 m^3/hr varied and studied.

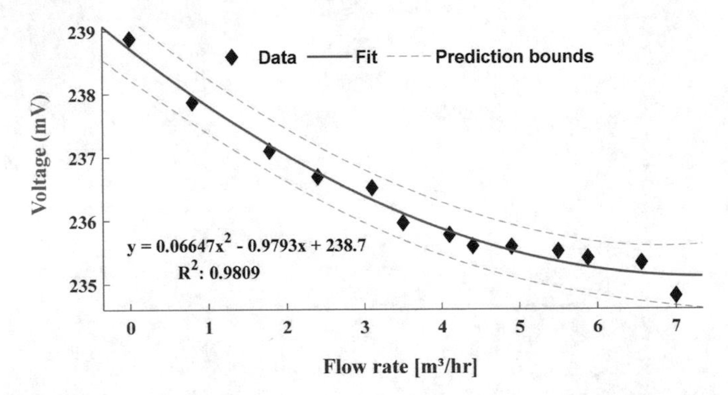

Fig. 6. Output voltage vs. flow rate characteristics plot of the capacitive sensor. This is for deflection in one direction.

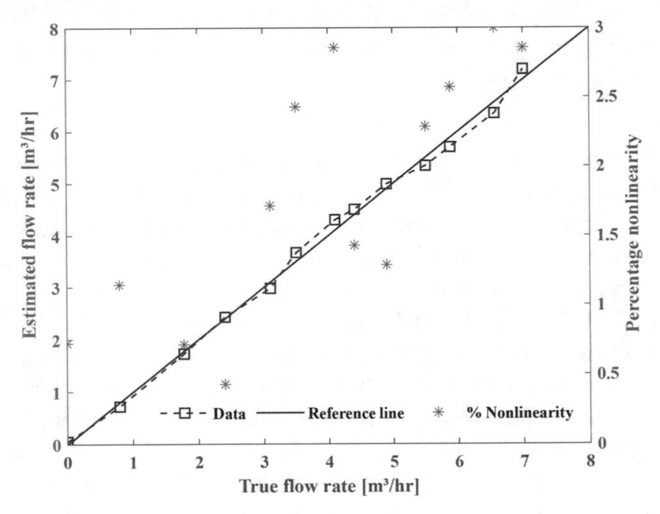

Fig. 7. Comparison plot between True flow rate and Estimated flow rate with percentage non-linearity. The sensor shows an error of 3% full scale.

4 Conclusion

A novel, low-cost, capacitively coupled cantilever-based sensor for flow measurement is presented. The transduction mechanism presented helped to realize a non-intrusive sensing scheme to estimate the cantilever deflection angle. The proposed sensor senses the cantilever deflection due to fluid flow inside a pipeline. A prototype sensor was built, and the performance was evaluated in the laboratory. It correctly measured the deflection angle of the cantilever depending on the fluid flow. The maximum error in the measurement noted was 0.3 m^3/hr. The test results indicate that the proposed read-out mechanism is well suited for the cantilever-based fluid flow sensor.

References

1. Ranjan, A., George, B., Mukhopadhyay, S.C.: Interdigital proximity sensor: electrode configuration, interfacing, and an application. Interdigital Sensors, pp. 23–34 (2021)
2. Moheimani, R., Hosseini, P., Mohammadi, S., Dalir, H.: Recent Advances on Capacitive Proximity Sensors: From Design and Materials to Creative Applications **8**(2), 26 (2022)
3. George B., Zangl H., Bretterklieber T., Brasseur G.: A novel seat occupancy detection system based on capacitive sensing. In: Proceedings of 2008 IEEE Instrumentation and Measurement Technology Conference, pp. 1515–1519 (2008)
4. Nguyen, T.D., Kim, T., Noh, J., Phung, H., Kang, G., Choi, H.R., SkinType proximity sensor by using the change of electromagnetic field. TIE, 2379–2388 (2021)
5. Ye, Y., Zhang, C., He, C., Wang, X., Huang, J., Deng, J.: A review on applications of capacitive displacement sensing for capacitive proximity sensor. IEEE Access **8**, 45325–45342 (2020)
6. George, B., Hubert, Z., Thomas, B., Georg, B.: A combined inductive-capacitive proximity sensor and its application to seat occupancy sensing. In: 2009 IEEE Instrumentation and Measurement Technology Conference, pp. 13–17 (2009)
7. Baxler, L.K.: Capacitive sensors design and application. IEEE Press, New York (1997)
8. Doebelin, E.O.: Measurement Systems—Application and Design, 5th edn. McGraw-Hill, New York (2004)
9. Muller, R.I., Booysen, M.J.: A water flow meter for smart metering applications. In: South African Conference on Computational and Applied Mechanics, 2014-SACAM (2014)
10. Ejeian, F., et al.: Design and applications of MEMS flow sensors: a review. Sens. Actuators, A **295**, 483–502 (2019)
11. Panahi, A., Ghasemi, P., Sabour, M.H., Magierowski, S., Ghafar-Zadeh, E.: Design and modeling of a new MEMS capacitive microcantilever sensor for gas flow monitoring. IEEE Canadian J. Electrical Comput. Eng. **44**(4), 467–479 (2021)
12. Naveen, H., Narasimhan, S., George, B., Tangirala, A.K.: Design and development of a low-cost cantilever-based flow sensor. In: IFAC 2020-PapersOnLine, vol. 53(1), pp. 111–116 (2020)
13. Harija, H., George, B., Tangirala, A.K.: A cantilever-based flow densor for domestic and agricultural water supply system. IEEE Sens. J. **21**(23), 27147–27156 (2021)

Interference Mitigation via NMF for Radio Astronomy Applications: A Feasibility Study

Felipe Barboza da Silva[1,2](✉), Ediz Cetin[1], Wallace Alves Martins[3], and John Tuthill[4]

[1] School of Engineering, Macquarie University, Sydney, NSW, Australia
felipe.barboza-da-silva@hdr.mq.edu.au, ediz.cetin@mq.edu.au
[2] Coppe, Federal University of Rio de Janeiro, Rio de Janeiro, Brazil
[3] Interdisciplinary Centre for Security Reliability and Trust (SnT), University of Luxembourg, Esch-sur-Alzette, Luxembourg
wallace.alvesmartins@uni.lu
[4] Commonwealth Scientific and Industrial Research Organisation (CSIRO), Space and Astronomy, Canberra, Australia
john.tuthill@csiro.au

Abstract. This work assesses the feasibility of using *nonnegative matrix factorization* (NMF) for *radio frequency interference* (RFI) mitigation in radio astronomy applications. Two NMF-based mitigation approaches are proposed, one using RFI frequency information extracted from the received signals and the other using an RFI template. The suitability and efficacy of these approaches are evaluated by targeting *automatic dependent surveillance-broadcast* (ADS-B) RFI using data collected from the Parkes radio telescope in Australia. Results show that the proposed approaches can mitigate the RFI with minimal degradation to the underlying observation of a double pulsar, and without discarding any received data, indicating the applicability of NMF-based approaches as potential RFI mitigation tools in radio astronomy applications.

Keywords: Interference mitigation · nonnegative matrix factorization · ADS-B suppression · radio astronomy

1 Introduction

The radio spectrum is becoming increasingly congested with an almost countless variety of *radio frequency* (RF) signals that are, in general, far stronger than the faint signals from celestial sources that are of interest to radio astronomers. Mitigation strategies for this *RF interference* (RFI) will be critical to the success of emerging radio astronomy instruments and this area is gaining much attention in the radio astronomy community [1–4].

There are several RFI mitigation approaches reported in the literature for radio astronomy applications. In [3], the authors propose a method based on the

N. K. Suryadevara et al. (Eds.): ICST 2022, LNEE 1035, pp. 352–364, 2023.
https://doi.org/10.1007/978-3-031-29871-4_36

mean absolute deviation (MAD) algorithm for pulsed-type RFI, whereas in [5] a framework based on spectral kurtosis is proposed for interference suppression. Other techniques such as the one in [6,7] devise subspace projections to alleviate the RFI issue. Those methods present good RFI rejection performance, however, do not concentrate on minimizing the distortion of the weak signals from outer space, which may impair the analysis of the cosmic events. Another popular mitigation technique in the radio astronomy domain is *flagging*, which involves manually selecting and removing the frequency channels corrupted by interference to alleviate the impact of RFI [8]. Nonetheless, the discarded information may contain useful astronomical data. Furthermore, flagging may disrupt the analysis of the transient properties of pulsars [8].

In order to mitigate RFI while causing minimal disruption to the celestial signals and to avoid data discarding, we propose a new framework based on the *nonnegative matrix factorization* (NMF). NMF is employed in a myriad of applications, such as remote sensing [9], *global navigation satellite system* (GNSS) [10–12], song suggestion [13], acoustic signal processing [14,15], to mention a few. In the context of radio astronomy, the main idea is to use NMF to separate the interference from the so-called signal of interest, basically constituted by the thermal noise of the receiver and the cosmic event signal, e.g., from pulsars. In this work, we employ signals captured by the Parkes radio telescope in New South Wales, Australia. Further, following on from our initial work in [16], where we proposed a *time-frequency* (TF) domain *automatic dependent surveillance-broadcast* (ADS-B) detector, we describe how to extract the RFI characteristics using NMF with focus on ADS-B signal mitigation. The ADS-B signals transmitted by aircraft broadcasting their flight ID, position, altitude, and velocity, among other parameters, are one of the main sources of interference to the Parkes radio telescope due to its close proximity to an airport. In addition, we use NMF to obtain the frequency features of the signal of interest, which aid the RFI mitigation with minimal corruption of the underlying signal of interest. Results indicate that the proposed framework achieves good suppression performance resulting in better-defined pulsar profiles than when no mitigation is employed and without discarding any of the received signal samples.

The paper is organized as follows: Sect. 2 provides an overview of pulsar processing. Section 3 describes some of the basic NMF concepts. Section 4 details the proposed NMF-based frameworks whereas the performance analysis and comparison are given in Sect. 5. Finally, concluding remarks and future work are given in Sect. 6.

2 Pulsar Processing

In order to analyze the pulsar characteristics, the received signal acquired by the radio telescopes undergoes a process called pulsar folding [1,17]. The main idea is to integrate the pulsar signals according to their rotational period so that they can be observed while compensating for the *ionized interstellar medium* (IISM). This frequency-dependent effect disperses the emitted waves along the path between the pulsar and the observer [1], with behavior similar to multipath,

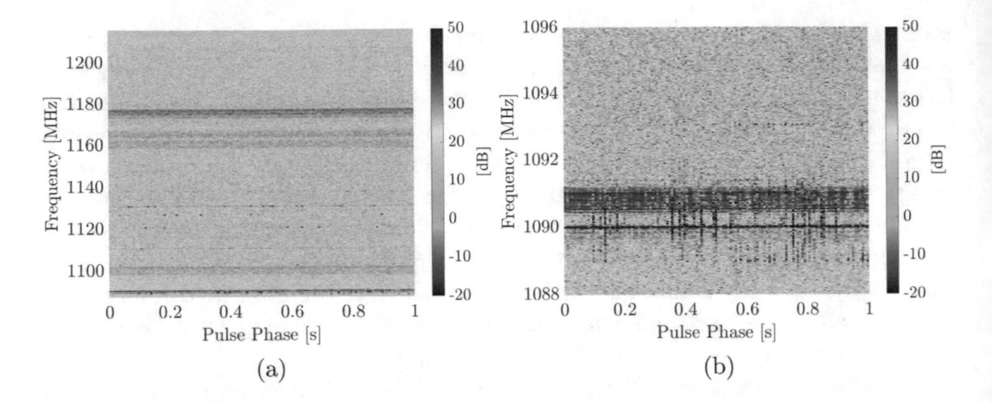

Fig. 1. DSPSR spectrogram (a) with its zoomed-in version around 1, 090 MHz (b).

altering the waves' time of arrival [18], degrading the performance of pulsar-based timing and synchronization systems. A software package broadly used by astronomers for this post-processing is the `DSPSR` [17].

Considering the Parkes radio telescope dataset used in this work, the observed pulsar is the J1939+2134, a double pulsar (two neutron stars in orbit around each other), with a rotational period around 1.56 ms. Using a 1-s long signal of the Parkes dataset as input to the `DSPSR` software, we generate a spectrogram-like plot, hereafter called the DSPSR spectrogram, shown in Fig. 1(a), with a zoomed-in version around the ADS-B center frequency (1, 090 MHz) in Fig. 1(b). These spectrograms are calculated considering the sum of both signal polarizations. The `DSPSR` software provides information about the strongest RFI within the frequency range, highlighting the most powerful RFI at 1, 090 MHz, indicating that the ADS-B signals represent a threat to the pulsar observation. Further, within the 1,120–1,140 MHz band very sparse signals can be observed; these are related to the *distance measurement equipment* (DME) signals used for aircraft guidance. It is worth mentioning that there are also other interference signals in the Parkes dataset such as the ones at 1, 100 MHz and within the 1,160–1,178 MHz band. However, the sources of such RFI are still unknown to radio astronomers. The pulsar profile is computed by integrating the frequency bins of the DSPSR spectrogram, resulting in a curve as shown in Fig. 2, which consists of a pulsar phase *vs.* electromagnetic flux plot. Ideally, the peaks around 0.1 and 0.6 would be more pronounced, with lower "noise" levels. However, due to the presence of RFI, the profile estimation becomes less accurate, which directly impacts the observation and calculation of the pulsar characteristics.

3 Nonnegative Matrix Factorization

The NMF is a low-rank decomposition method for nonnegative data. For instance, consider a dataset represented as a nonnegative matrix $\mathbf{X} \in \mathbb{R}_{+}^{N \times M}$.

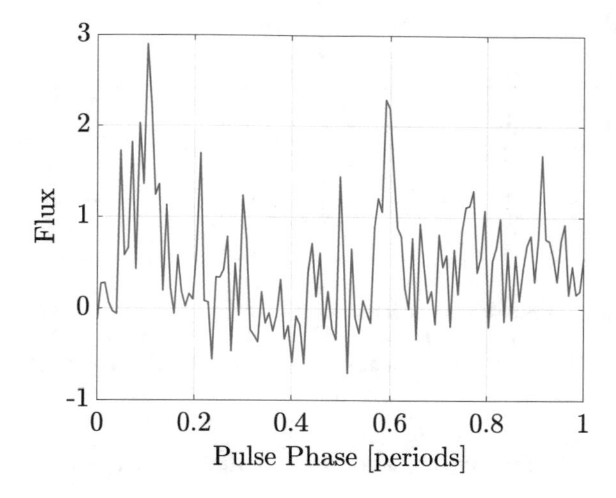

Fig. 2. Pulsar profile without RFI mitigation.

After the NMF decomposition, $\mathbf{X}$ can be expressed as the product of two lower-rank matrices as $\mathbf{X} \approx \mathbf{WH}$, $\mathbf{W} \in \mathbb{R}_+^{N \times S}$, $\mathbf{H} \in \mathbb{R}_+^{S \times M}$, with the rank of $\mathbf{W}$, $\mathbf{H}$ smaller than $\mathbf{X}$'s, i.e., $S < \min(N, M)$. NMF became a popular technique following the work in [19], where it was employed to decompose facial images, with $\mathbf{W}$ representing the face features, i.e., eyes, nose, etc., and $\mathbf{H}$ translating how the feature will be summed up to represent a face. The key difference of NMF from other decomposition methods, such as *principal* and *independent component analysis*, PCA and ICA respectively, is the physical interpretation of the decomposition. NMF is employed in a large variety of applications, such as in data science for topic modeling [20], and acoustic signal processing for source separation [14,15]. Considering the latter, $\mathbf{X}$ is usually a magnitude spectrogram calculated via the *short-time Fourier transform* (STFT) [10], where $\mathbf{W}$ and $\mathbf{H}$ represent time and frequency information respectively. These matrices can be calculated by minimizing a given similarity function $\mathcal{L}(\mathbf{X}, \mathbf{WH})$ with nonnegative constraints on $\mathbf{W}$ and $\mathbf{H}$, whose optimum point can be reached recursively [21]. Regarding the similarity function, the Kullback-Leibler (KL) divergence is commonly used in the literature [14].

4 Proposed RFI Mitigation Techniques

In this section, we detail the proposed NMF-based techniques to mitigate RFI, with a major focus on the ADS-B signals, and analyze the impact of RFI suppression on the signal of interest, i.e., the pulsar observation. The dataset employed in this work was acquired at the Parkes radio telescope, located in Parkes, New South Wales, Australia. The radio receiver uses a dual-polarized antenna, with corresponding polarizations named polA and polB respectively. The system is capable of receiving signals from 704 to 4,032 MHz, covering a bandwidth of

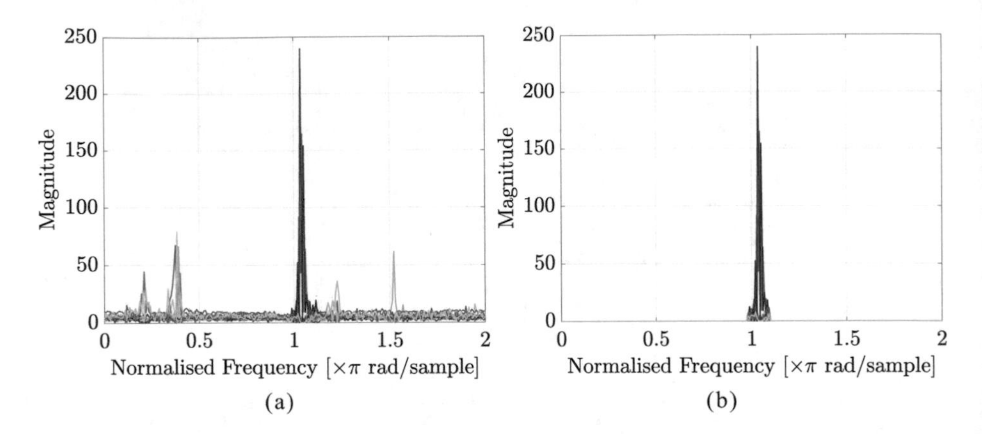

Fig. 3. Estimated $\mathbf{W}_{\mathrm{RFI}}$ before (a) and after (b) frequency selection.

3,328 MHz divided into 26 sub-bands of 128 MHz. Each sub-band signal is complex sampled using 16 bits, down-converted to DC, and down-sampled to 128 Msamples/s. Further details can be found in [1]. The ADS-B signal falls into sub-band 3, which spans [1,088,1,216] MHz. According to [1], 58% of sub-band 3's spectrum has useful data for astronomy purposes. Therefore, it is of paramount importance to detect and mitigate the ADS-B signals. In addition to the ADS-B signals, there are other interference sources that fall into sub-band 3. One such interference source is the DME signals used for aircraft guidance [11]. Parkes radio telescope data can be accessed through the Australia Telescope Online Archive (ATOA) [22].

4.1 Supervised NMF with Frequency Selection

The supervised NMF relies on the availability of clean versions of the RFI signals, and of the signal of interest. In the context of radio astronomy, the signals from cosmic events are captured at very low power levels, usually below the receiver thermal noise floor. Thus, one can assume that the signal of interest is in fact the thermal noise. In real scenarios, including the signals captured by the Parkes radio telescope, the RFI and signal of interest are mixed. Further, due to the short time duration of pulsed-type RFI, the signal frames corrupted by interference are mostly composed of the signal of interest, with only a few samples within each 10-ms signal frame corrupted by ADS-B. Thus, when estimating $\mathbf{W}_{\mathrm{RFI}}$, the majority of its components represent the signal of interest, which, according to prior simulations, induced a larger disruption of the signal of interest. In order to alleviate this issue, we manually extract the excerpts of the Parkes dataset composed mainly of ADS-B signals and use them for evaluating $\mathbf{W}_{\mathrm{RFI}}$. After estimating the RFI frequency content, we employ a post-processing scheme to eliminate spurious values in $\mathbf{W}_{\mathrm{RFI}}$, since the ADS-B bandwidth is defined by international standards [23], we can easily select the appropriate ADS-B frequency range. Using 20 signal frames corrupted by ADS-B, considering the Kullback-Leibler divergence as distance function, and $S = 10$ to compute $\mathbf{W}_{\mathrm{RFI}}$ and $\mathbf{W}_{\mathrm{SOI}}$, Figs. 3(a–b) show the estimated $\mathbf{W}_{\mathrm{RFI}}$ before, and after frequency selection respectively, where each line represents a single column of $\mathbf{W}_{\mathrm{RFI}}$.

4.2 Supervised NMF with RFI Template

One of the drawbacks of the frequency selection detailed in the previous subsection is that within the ADS-B signal bandwidth we also have, albeit a small amount, the signal of interest. Therefore, to alleviate this problem, we employ the ADS-B template depicted in Fig. 4 as $\mathbf{W}_{\mathrm{RFI}}$ instead. Consequently, the ADS-B frequency content in $\mathbf{W}_{\mathrm{RFI}}$ is represented by a single component i.e. a vector.

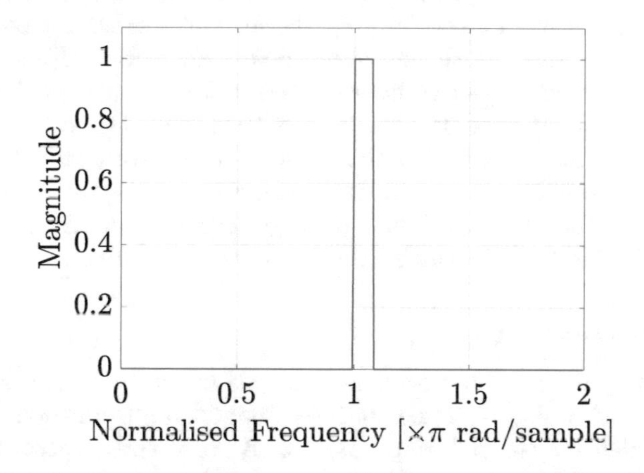

Fig. 4. ADS-B frequency template.

5 Performance Evaluation

In this section, we assess the performance of the supervised NMF with frequency selection, and with RFI template presented in Subsects. 4.1 and 4.2 respectively in mitigating RFI. Further, we also provide a discussion about the impact of the mitigation process on the signal of interest. While supervised NMF can be used to mitigate not only ADS-B signals but also other RFI types, its training would have to be modified accordingly. For instance, for DME mitigation a proper frequency selection for such interference would need to be undertaken. Hence, in this work, we only evaluate the performance of ADS-B signal suppression. The supervised NMF with frequency selection and supervised NMF with RFI template are hereafter referred to as NMF-Sel and NMF-Temp respectively.

5.1 Figures of Merit

The performance of interference mitigation techniques in RF-based communication systems is generally evaluated by means of the *signal-to-interference-plus-noise ratio* (SINR) [24]. In this case, one has a certain level of knowledge about the signal of interest and noise powers. In the area of audio processing, where the final goal is to separate different sources, such as instruments and voice, the SINR is employed along with subjective figures of merit, which measure the degradation of the separated signals from the human hearing perspective [25]. In the context of radio astronomy, the focus is to suppress RFI, while preserving the very low-level signal of interest and, at the same time, not introducing artifacts that may bias radio astronomy observations. In [26], a figure of merit to evaluate the performance of time-frequency RFI mitigation techniques in radio astronomy signals was proposed. Nonetheless, the proposed figure of merit assumes prior knowledge about the RFI power and does not take into account the degradation of the signal of interest. In [27], the authors assess the performance of the techniques by analyzing the spectrum before and after mitigation by visual inspection, without an objective figure of merit. In fact, the performance evaluation relies on the specific science case, with no one figure of merit being a consensus among astronomers. For this reason, in this work we evaluate the performance in two ways: 1) the magnitude spectrograms of the signal of interest after RFI mitigation, which we refer to as pre-folding, and 2) based on pulsar profiles described in Sect. 2, which we refer to as post-folding.

5.2 Experimental Setup

The data employed for the performance analysis in this work is from the Parkes dataset, described in Sect. 4. The first 100 signal frames from the Parkes dataset are used to calculate the pulsar profiles, with 76 of them corrupted by ADS-B and/or other RFI, and 24 without any interference. Regarding the STFT parameters, we use a rectangular window of length $L = 256$, number of DFT bins $N = 256$ and hop size of $R = L/2 = 128$ samples. As for the NMF parameters, we employ the KL divergence, random initialization of $\mathbf{W}$ and $\mathbf{H}$, and 10 components to represent both signal of interest and RFI signals, except for the case where we employ the ADS-B template where we use a single component. During the training phase, 20 signal frames are employed to compute $\mathbf{W}_{\mathrm{SOI}}$, and 20 signal frames out of the 76 are used to calculate $\mathbf{W}_{\mathrm{RFI}}$ (for NMF-Sel). It is worth mentioning that the NMF-based frameworks are employed on the signals of both polarizations i.e. polA and polB, and the matrices $\mathbf{W}_{\mathrm{RFI}}$ and $\mathbf{W}_{\mathrm{SOI}}$ are calculated considering each polarization individually. All proposed frameworks share the same $\mathbf{W}_{\mathrm{SOI}}$ (computed using polA), displayed in Fig. 5. Moreover, in the scenario considered, NMF achieves convergence within 300 iterations.

5.3 Pre-folding Results

In this subsection, we describe the mitigation results by means of reconstructed spectrograms of the signal of interest. For the sake of simplicity, we only display

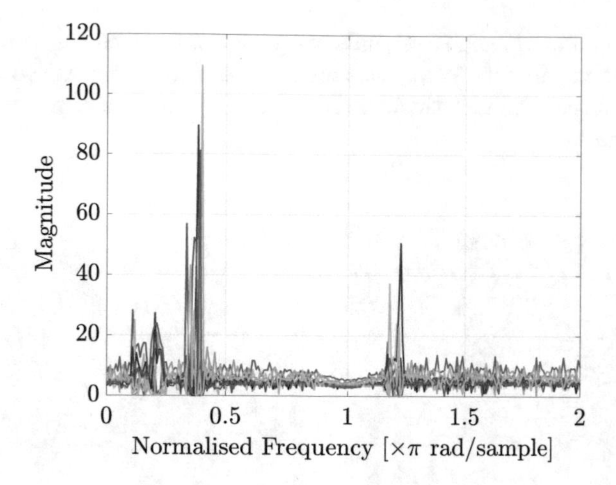

Fig. 5. Estimated $\mathbf{W}_{\mathrm{SOI}}$ matrix during the training phase.

the spectrograms corresponding to polA. Out of the 76 signal frames used for performance evaluation, we show the corresponding spectrograms of the two most representative ones. Figures 6(a–b) show the spectrograms corrupted by RFI from the Parkes dataset, where the black and blue rectangles point out the ADS-B and DME signals in the time-frequency domain respectively. In these figures it is possible to observe strong ADS-B signals, and in Fig. 6(b) two DME signals at around 1.5π rad/sample (1,120 MHz).

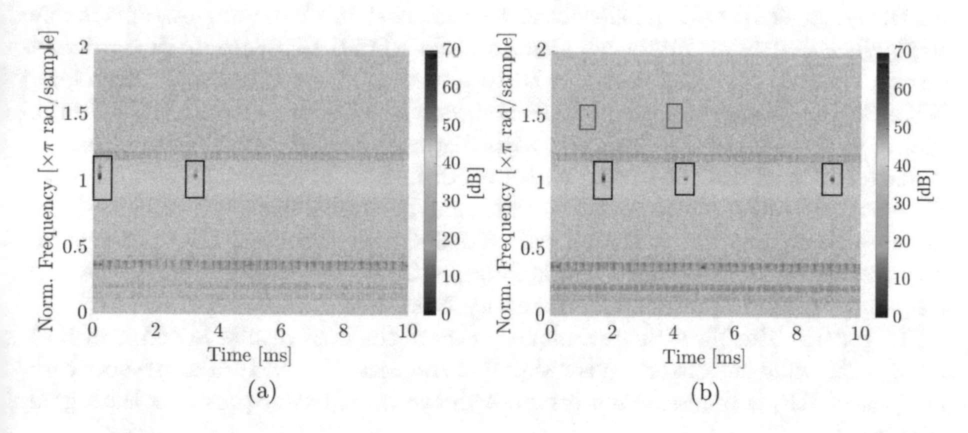

Fig. 6. Spectrograms corrupted by (a) ADS-B and (b) ADS-B & DME.

The NMF-Sel's reconstructed spectrograms are shown in Figs. 7(a–b). Overall, the ADS-B signals are drastically suppressed, with only a residual fraction of ADS-B RFI within the leftmost rectangle in Fig. 7(a). Also, the signal of interest does not seem to be attenuated, especially within the ADS-B signal frequency band. As expected, the DME signals were not mitigated, as can be noted in Fig. 7(b), since the matrix $\mathbf{W}_{\mathrm{RFI}}$ was calculated only using ADS-B signals. Further, this reinforces the fact that the NMF-Sel framework is not corrupting the signal of interest.

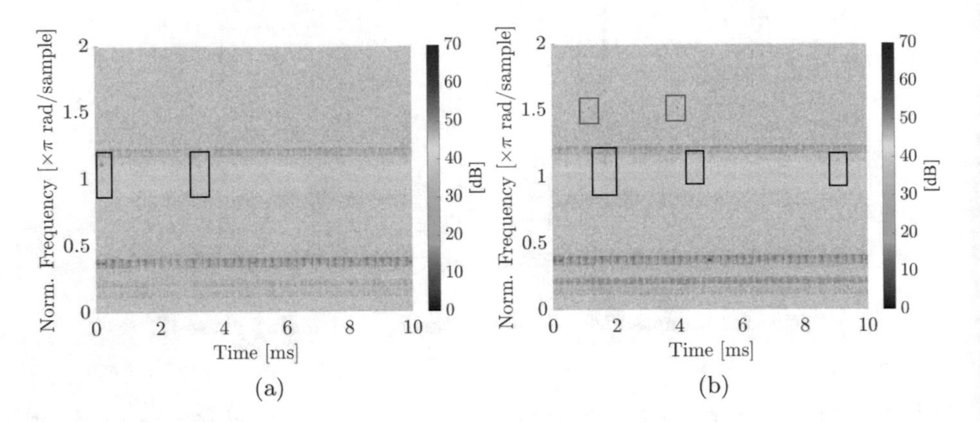

Fig. 7. Reconstructed spectrograms of the signal of interest using the NMF-Sel technique from Fig. 6(a) and Fig. 6(b) respectively.

In regard to the NMF-Temp results, Figs. 8(a–b) display the reconstructed spectrograms of the signal of interest with respect to Figs. 6(a–b). As observed, there is a significant ADS-B spectrum in the signal of interest, more specifically for the stronger ADS-B signals within the leftmost black rectangles. This is due to the fact that NMF-Temp employs a simple ADS-B template in $\mathbf{W}_{\mathrm{RFI}}$, i.e., a vector, whereas NMF-Sel uses $S = 10$ to estimate $\mathbf{W}_{\mathrm{RFI}}$ (Fig. 3(b)). Therefore, NMF-Sel is able to represent the ADS-B signals in a better way than NMF-Temp, taking into account ADS-B signals with distinct strengths and bandwidths. As a matter of fact, this spectrum varies according to numerous factors, such as distance and angle between the aircraft and the radio telescope dish, and the antenna characteristics of the aircraft. NMF-Temp, however, does not require estimating the ADS-B signal characteristics from the received signal, nor the frequency selection process undertaken by NMF-Sel.

In general, the NMF-Sel technique presents the best results in terms of RFI mitigation while not distorting the signal of interest. The NMF-Temp framework has poorer RFI suppression performance, however, it better preserves the signal of interest.

5.4 Post-folding Results

In this section, we describe the mitigation results in terms of the pulsar profile, generated using the `DSPSR` software. Figure 9 shows the pulsar profiles generated using the reconstructed signal of interest respective to each of the NMF-based

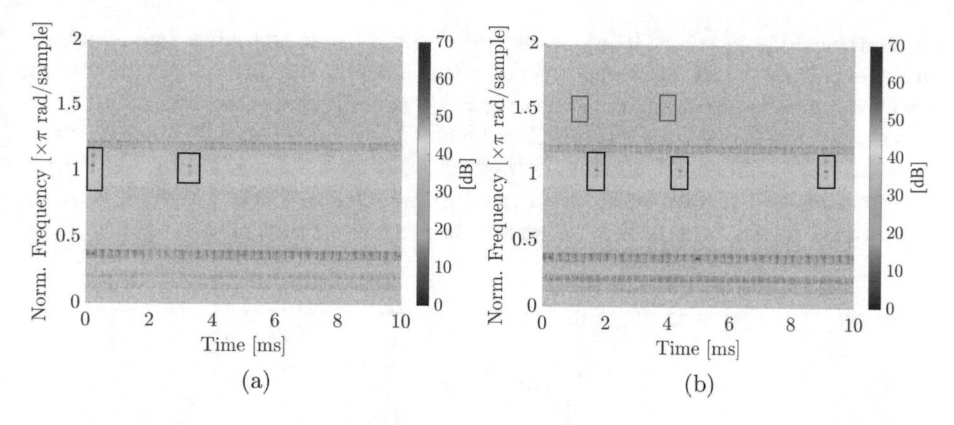

Fig. 8. Reconstructed spectrograms of the signal of interest using the NMF-Temp technique from Fig. 6(a) and Fig. 6(b) respectively.

frameworks. Surprisingly, NMF-Sel and NMF-Temp have very similar pulsar profiles, despite the worse ADS-B suppression performance of the latter. One of the hypotheses is that the residual ADS-B spectrum in the signal of interest is averaged out by the pulsar folding, hence the similar results.

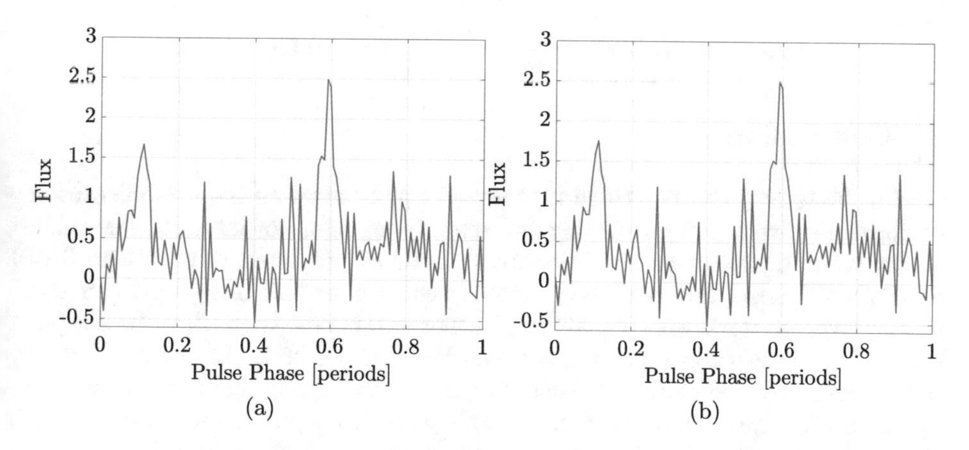

Fig. 9. Pulsar profiles corresponding to (a) NMF-Sel, and (b) NMF-Temp.

Figures 10(a–b) display the pulsar profile calculated from raw data (same as in Fig. 2), and from flagged signals, which means that some frequency channels of the DSPSR spectrogram corrupted by RFI are manually selected and have their samples zeroed out across all time instants. This way, the pulsar profile is generated neglecting the zeroed frequency channels, with the impact of RFI alleviated [8]. However, as previously mentioned, flagging may affect the analysis of the transient properties of pulsed-type cosmic events, such as pulsars [8].

Comparing Figs. 9(a) and 10(a), it can be observed that the noisy behavior outside the region of the main peak (pulse phase around 0.1 and 0.6) is alleviated. Further, the leftmost peak in Fig. 9(a) is more pronounced when compared to its neighbors, despite having a lower flux level. On the other hand, the rightmost peak has been enhanced, also with higher prominence. Considering Fig. 10(b), the spurious peaks have been attenuated along with the rightmost peak. Also, the average level of the profile in between the peaks is now larger.

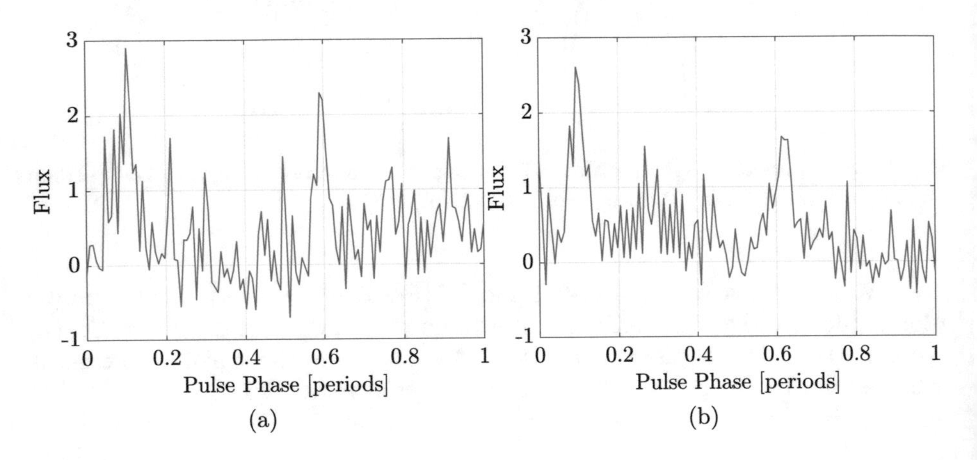

Fig. 10. Pulsar profile from (a) raw data, and (b) flagged signals.

6 Conclusions

Overall, the techniques described in this paper present promising RFI mitigation performance, while not corrupting the underlying signal of interest. The main target for mitigation was the interference sources presented in sub-band 3 of the Parkes dataset, however, the proposed schemes for RFI mitigation can also be employed to work with signals from other sub-bands, with other RFI types involved. This work evaluated the potential of using NMF in the context of radio astronomy, indicating its feasibility and pointing out the need for further research. Moreover, while the focus of this work is on pulsar radio astronomy, the NMF technique for RFI suppression presented here is likely to enhance other areas of radio astronomy, such as spectral line observations, continuum surveys and the new rapidly growing field of fast radio transients. Finally, the results described in this work call for a deeper evaluation by radio astronomy experts, which will bring more insights into the interpretation of the pulsar profiles and the establishment of agreed-upon appropriate figures of merit for performance assessment.

Acknowledgment. We would like to thank Dr. George Hobbs from the Commonwealth Scientific and Industrial Research Organisation (CSIRO), Space and Astronomy (S&A), for his valuable insights and discussions regarding pulsar profile interpretation.

References

1. Hobbs, G., et al.: An ultra-wide bandwidth (704 to 4032 MHz) receiver for the Parkes radio telescope. Publ. Astron. Soc. Aust. **37**, e012 (2020)
2. Ford, J.M., Buch, K.D.: RFI mitigation techniques in radio astronomy. In: 2014 IEEE Geoscience and Remote Sensing Symposium, Quebec City, QC, Canada, pp. 231–234 (2014)
3. Buch, K.D.: Impulsive RFI mitigation: CASPER library block. Accessed 01Mar 2022. https://casper.ssl.berkeley.edu/wiki/Impulsive_RFI_Excision:_CASPER_Library_Block
4. Roy, J., Gupta, Y., Pen, U.-L., Peterson, J.B., Kudale, S., Kodilkar, J.: A real-time software backend for the GMRT. Exp. Astron. **28**, 25–60 (2010)
5. Gary, X.D.E., Nita, G.M., Wang, H.G.: Spectral domain algorithms for RFI excision in real time. In: Proceedings of the URSI General Assembly (2008)
6. Sardarabadi, A.M., van der Veen, A., Boonstra, A.: Spatial filtering of RF interference in radio astronomy using a reference antenna array. IEEE Trans. Signal Process. **64**(2), 432–447 (2016)
7. Hellbourg, G.: RFI subspace smearing and projection for array radio telescopes (2018)
8. Cendes, Y., et al.: RFI flagging implications for short-duration transients. Astron. Comput. **23**, 103–114 (2018)
9. Lv, X., Wang, W., Liu, H.: Cluster-wise weighted NMF for hyperspectral images unmixing with imbalanced data. Remote Sens. **13**(2), 268 (2021)
10. da Silva, F.B., Cetin, E., Martins, W.A.: Radio frequency interference detection using nonnegative matrix factorization. IEEE Trans. Aeros. Electron. Syst. **58**(2), 868–878 (2022)
11. da Silva, F.B., Cetin, E., Martins, W.A.: DME interference mitigation for GNSS receivers via nonnegative matrix factorization. In: XXXIVth General Assembly and Scientific Symposium of the International Union of Radio Science (URSI GASS), Rome, Italy, pp. 1–4 (2021)
12. da Silva, F.B., Cetin, E., Martins, W. A.: Radio frequency interference mitigation via nonnegative matrix factorization for GNSS. IEEE Trans. Aeros. Electron. Syst., 1–13 (2022)
13. Benzi, K., Kalofolias, V., Bresson, X., Vandergheynst, P.: Song recommendation with non-negative matrix factorization and graph total variation. In: 2016 IEEE International Conference on Acoustics, Speech and Signal Processing (ICASSP), Shanghai, China, pp. 2439–2443 (2016)
14. Cichocki, A., Zdunek, R., Amari, S.: New algorithms for non-negative matrix factorization in applications to blind source separation. In: 2006 IEEE International Conference on Acoustics Speech and Signal Processing Proceedings, Toulouse, France, vol. 5, p. V (2006)
15. Smaragdis, P.: Convolutive speech bases and their application to supervised speech separation. IEEE Trans. Audio Speech Lang. Process. **15**(1), 1–12 (2007)
16. da Silva, F.B., Cetin, E., Martins, W.A.: ADS-B signal detection via time-frequency analysis for radio astronomy applications. In: IEEE International Symposium on Circuits and Systems (ISCAS), Daegu, South Korea, pp. 1–4 (2021)
17. van Straten, W., Bailes, M.: DSPSR: digital signal processing software for pulsar astronomy. Publ. Astron. Soc. Aust. **28**(1), 1–14 (2011)
18. Krishnakumar, M.A., et al.: High precision measurements of interstellar dispersion measure with the upgraded GMRT. A&A **651**, A5 (2021)

19. Lee, D.D., Seung, H.S.: Learning the parts of objects by non-negative matrix factorization. Nature **401**(6755), 788–791 (1999)
20. Chen, Y., Zhang, H., Wu, J., Wang, X., Liu, R., Lin, M.: Modeling emerging, evolving and fading topics using dynamic soft orthogonal nmf with sparse representation. In: IEEE International Conference on Data Mining, pp. 61–70 (2015)
21. Févotte, C., Idier, J.: Algorithms for nonnegative matrix factorization with the β-divergence. Neural Comput. **23**(9), 2421–2456 (2011)
22. Accessed 01 Mar 2022. https://www.atnf.csiro.au/observers/data/index.html
23. Reception of automatic dependent surveillance broadcast via satellite and compatibility studies with incumbent systems in the frequency band 1087.7–1092.3 MHz. M Series-Mobile, radiodetermination, amateur and related satellite services (2017)
24. Andrews, J.G., Baccelli, F., Ganti, R.K.: A tractable approach to coverage and rate in cellular networks. IEEE Trans. Commun. **59**(11), 3122–3134 (2011)
25. Hershey, J.R., Chen, Z., Le Roux, J., Watanabe, S.: Deep clustering: discriminative embeddings for segmentation and separation. In: 2016 IEEE International Conference on Acoustics, Speech and Signal Processing (ICASSP), Shanghai, China, pp. 31–35 (2016)
26. Querol, J., Onrubia, R., Alonso-Arroyo, A., Pascual, D., Park, H., Camps, A.: Performance assessment of time-frequency rfi mitigation techniques in microwave radiometry. IEEE J. Sel. Topics Appl. Earth Obs. Remote Sens. **10**(7), 3096–3106 (2017)
27. Kesteven, M., Manchester, R., Brown, A., Hampson, G.: Field trials of a RFI adaptive filter for pulsar observations. In: RFI Mitigation Workshop, p. 23 (2010)

Implementation of an Extrapolated Single-Propagation Particle Filter for Interference Source Localization Using a Sensor Network

Duc Dung Vu[1(✉)], Sanat K. Biswas[2], Alan Kan[1], and Ediz Cetin[1]

[1] School of Engineering, Macquarie University, Sydney, NSW, Australia 2109
duc-dung.vu@hdr.mq.edu.au, {alan.kan,ediz.cetin}@mq.edu.au
[2] IIIT Delhi, New Delhi, India
sanat@iiitd.ac.in

Abstract. Global Navigation Satellite System (GNSS) signals are vulnerable to intentional or unintentional Radio Frequency Interference (RFI) due to their low-received signal power levels. In order to geo-locate and track the RFI source(s), distributed network of sensors are used. However, the sensor measurements tend to be corrupted or degraded due to the environment and system noise. In addition, the non-linear nature of the position estimation based on Angle of Arrival (AOA) and Time Difference of Arrival (TDOA) measurements further complicates the problem. This paper presents the hardware design of an extrapolated single-propagation particle filter (ESP-PF). The ESP-PF was proposed to overcome the challenges in RFI source localization based on hybrid AOA and TDOA measurements from and between the sensor nodes, respectively. The performance of this hardware implementation is evaluated and compared with its fully-software counterpart in terms of position estimation accuracy. The results show that the hardware implementation performs comparably with that of the floating-point software-only implementation.

Keywords: Interference · GNSS · Particle Filter · Geo-localization

1 Introduction

The timing and positioning information provided by the Global Navigation Satellite Systems (GNSS) play a key role in enabling the functioning of modern infrastructure and numerous services in many aspects of our daily lives. The low-received signal power levels of the GNSS signals, however, makes them vulnerable to intentional and/or unintentional Radio Frequency Interference (RFI). The threat of RFI presents a problem to the integrity of the systems that rely

This work was supported in part by the Macquarie University International Research Training Program Scholarship for Master of Research (iRTP-MRES).

N. K. Suryadevara et al. (Eds.): ICST 2022, LNEE 1035, pp. 365–373, 2023.
https://doi.org/10.1007/978-3-031-29871-4_37

on GNSS to function. As such, GNSS itself has become a critical infrastructure and its vulnerability is a major concern which must be alleviated.

To alleviate this vulnerability, several approaches have been reported in the literature considering the detection, mitigation, and geo-localization [1–9] of RFI. In the context of geo-localization, the unknown nature of the RFI necessitates passive systems consisting of spatially-distributed Sensor Nodes (SNs) [10]. Typically, these systems use Received Signal Strength (RSS), source Angle of Arrival (AOA), Time Difference of Arrival (TDOA), or a combination of AOA/TDOA, or Frequency Difference of Arrival (FDOA) to estimate the RFI position [10]. The GNSS Environmental Monitoring System (GEMS) [11–13], the precursor to the GRIFFIN system used in this work, is one such system which combines AOA and TDOA measurements to geo-locate narrow– and wide–band RFI signals. The system consists of spatially-distributed SNs, where AOAs to the RFI are estimated and are wirelessly connected to a Central Node (CN) for TDOA estimation and RFI geo-localization.

Sequential estimators, such as Extended Kalman Filter (EKF), Unscented Kalman Filter (UKF) and the Single-Propagation Unscented Kalman Filter (SPUKF), can be used to combine the information on the source and AOA/TDOA observation to obtain a localisation solution. An analysis conducted in [14] showed that the SPUKF provides a better performance in terms of estimation accuracy, confidence and processing compared to the EKF and UKF, when the RFI source is static or moves at constant speed. In order to deal with highly non-linear estimation problems, a low-complexity Extrapolated Single-propagation Particle Filter (ESP-PF) for GNSS interference source localization was introduced in [15] and its performance evaluated in MATLAB. In this work, we explore the hardware implementation of this particle filter on a System-on-Chip (SOC) embedded platform which combines the software programmability of ARM-based processor(s) with the hardware programmability of a Field Programmable Gate Array (FPGA). We use High Level Synthesis (HLS) to accelerate the design and the implementation process, and evaluate the resulting hardware implementation using field test data from the GRIFFIN system.

The rest of the paper is organized as follows: Sect. 2 provides a brief summary of the mathematical model of the ESP-PF. Section 3 introduces the hardware implementation of the ESP-PF on a Xilinx SoC platform, and the experiment setup for verification. The location estimation performance of the hardware implementation is provided and compared with the MATLAB model in Sect. 4. Conclusions and future work are given in Sect. 5.

2 Extrapolated Single-Propagation Particle Filter Algorithm

The particle filter is often used to solve highly non-linear estimation problems, and its performance depends on the number of particles used in the algorithm [16,17]. Each particle is a random sample state vector, which is propagated to compute the corresponding *a priori* sample at the next observation epoch. Besides, the weight corresponding to each particle is proportional to the likelihood of the particle given the observation. The weighted average of the propagated samples is considered as the *a posteriori* mean state vector.

As mentioned earlier, measurements at sensor nodes may contain errors and impacted by noise for various reasons. Hence, the unknown AOA bias, clock bias, and drifts in TDOA measurement should be estimated in addition to the position and velocity of the RFI source. Thus, the state vector for RFI localization at time t is defined as [15]:

$$\mathbf{X}(t) = [\mathbf{r}^T \quad \mathbf{v}^T \quad \mathbf{b_T}^T \quad \mathbf{d_T}^T \quad \mathbf{b_a}^T]^T \tag{1}$$

where:

$$\begin{aligned}
\mathbf{r} &= [x \quad y]^T \\
\mathbf{v}^T &= [\dot{x} \quad \dot{y}]^T \\
\mathbf{b_T} &= [b_{T_{12}} \quad b_{T_{13}} \quad b_{T_{23}}]^T \\
\mathbf{d_T} &= [d_{T_{12}} \quad d_{T_{13}} \quad d_{T_{23}}]^T \\
\mathbf{b_a} &= [b_{a_1} \quad b_{a_2} \quad b_{a_3}]^T
\end{aligned} \tag{2}$$

$\mathbf{r}$ and $\mathbf{v}$ are denoted as position and velocity vectors of the RFI source, $d_{T_{i/j}}$ is the clock drift and $b_{T_{i/j}}$ is the clock bias for the ith and j^{th} sensor nodes, respectively, and b_{a_i} is the AOA bias at the ith sensor node. In general, the particle filter will estimate the position and velocity of the RFI source. Clock bias and drift are also included to improve the estimation. Assuming that the clock biases, drifts and the AOA biases are constants, the state space vector model can be formulated as:

$$\begin{bmatrix} \dot{\mathbf{r}} \\ \dot{\mathbf{v}} \\ \dot{\mathbf{b}}_\mathbf{T} \\ \dot{\mathbf{d}}_\mathbf{T} \\ \dot{\mathbf{b}}_\mathbf{a} \end{bmatrix} = \begin{bmatrix} \mathbf{A}_{4\times 4} & \mathbf{O}_{4\times 9} \\ \mathbf{O}_{9\times 4} & \mathbf{O}_{9\times 9} \end{bmatrix} \begin{bmatrix} \mathbf{r} \\ \mathbf{v} \\ \mathbf{b_T} \\ \mathbf{d_T} \\ \mathbf{b_a} \end{bmatrix} + \begin{bmatrix} \mathbf{O}_{2\times 1} \\ \mathbf{a}_{2\times 1} \\ \mathbf{O}_{9\times 1} \end{bmatrix} + \mathbf{v}(t) \tag{3}$$

The AOA observation θ_i from the ith SN and the TDOA measurement δt_{ij} between the ith and j^{th} SN at the kth epoch can be computed as [10]:

$$\delta t_{ij} = \frac{1}{c}(||\mathbf{r}_i - \mathbf{r}(k)|| - ||\mathbf{r}_j - \mathbf{r}(k)||) + b_{T_{ij}} + kd_{T_{ij}} + \eta_T(k) \tag{4}$$

$$\theta(t)_i = tan^{-1}\left(\frac{x(k) - x_i}{y(k) - y_i}\right) + b_{a_i} + \eta_a(K) \tag{5}$$

where η_T and η_a are the TDOA and AOA measurement noise, respectively. The position of the ith or j^{th} SN in x and y coordinates are denoted as $x_{i/j}$ and $y_{i/j}$, and the vector $r_{i/j}$ is the position vector of either the ith or the jth SN.

According to [15], the Probability Density Function (PDF) of the state vector is assumed to be Gaussian and 2,000 particles are drawn from the initial

PDF assumption. However, for hardware implementation, we decided to have the number of particles as a power of 2. Hence, 2,048 particles were used in the software-only implementation of ESP-PF, and provides similar estimation performance to 2,000 particles. Similar to the conventional particle filter, there are three phases in the filtering algorithm, namely: "sampling","importance", and "resampling" [18]. However, this paper implements a computationally efficient variant of the traditional particle filter, the ESP-PF, which propagates the mean state vector and extrapolates other state vectors through the Richardson Extrapolation method to the sample particle [19]. The propagation of the mean particle is computed through integrating Eq. (3).

3 Methods

3.1 Hardware Implementation

The SoC Platform used to implement the ESP-PF algorithm is the Xilinx Radio Frequency System on Chip (RFSoC) XCZU28DR platform. The RFSoC integrates, amongst other functionalities, Programmable Logic (PL) with $\approx$ 930k logic cells and 4,272 Digital Signal Processing (DSP) slices, and a Processing System (PS) consisting of Quad-core ARM Cortex-A53 processors, with PL and PS connected via multiple high-speed Advanced eXtensible Interface (AXI) buses. We used Xilinx Vitis software for application-based software-hardware co-design.

We first converted the MATLAB algorithm given in [15] into C/C++ code, ensuring in the process that all the dynamic function calls and dynamic memory allocations were changed to static ones in order to use HLS for implementing the ESP-PF algorithm. Further, the C/C++ code was made to use fixed-point arithmetic to speed up calculations using the DSP slices on the RFSoC [20]. During the code conversion, we conducted an analysis investigating the relationship between the fixed-point representation and the number of particles used in ESP-PF, where we ran ESP-PF using different fixed-point representations with different power of 2 numbers of particles, particularly 512, 1,024, and 2,048. The analysis results show that ESP-PF using 1,024 particles and word-length of 32-bits consisting of 11 integer and 21 fractional bits can provide comparable position estimation accuracy as the software only implementation in [15] using 2,048 particles and 64 bits floating-point representation. It was intriguing to observe that the word-length and arithmetic quantisation process resulted in a positive impact on the position estimation performance. Based on this fixed-point design space exploration process, the ESP-PF hardware design used 1,024 particles, and 32-bit fixed-point representation including 11 integer and 21 fractional bits, further reducing the hardware complexity and resource usage.

In this hardware implementation, the ESP-PF algorithm would be placed into PL, which is referred to as a kernel by Vitis. On the other side, the host application created in PS is used to feed the observation data from SNs stored in 4 $\times$ 1 GB DDR4 RAMs to the kernel for RFI source localization. Moreover, a Vitis platform of the XCZU28DR is needed to provide hardware properties to the kernel, including a 100 MHz clock, a 32-bit interrupt and 32-bit AXI connection for PS/PL communication. This platform, the host application in PS and the kernel in PL are combined together by Vitis IDE as shown in Fig. 1.

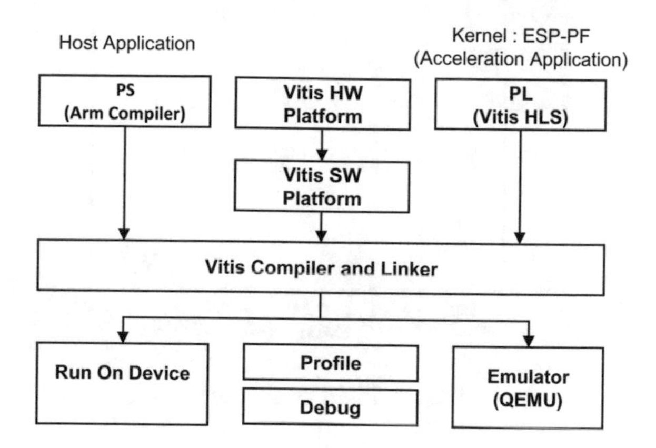

Fig. 1. Acceleration platform for acceleration application

3.2 Hardware Performance Testing

The hardware implementation is verified and its performance evaluated using real field trial data from a network of three prototype GRIFFIN SNs covering an area of 400×800 m^2 as shown in Fig. 2, with system performance evaluated using various narrow- and wide-band RFI sources positioned at different locations. The RFI source power levels were set so as not to disrupt users outside the boundary of the field trial area and GNSS monitoring stations were set up at the perimeter to monitor the signal levels. In this work, we consider a static RFI case, with the source location along with locations of some of the trees in the field test area depicted in Fig. 2. The ground truth data were obtained using a GNSS L1/L2-base station and a GPS RTK rover, and the locations of the SNs were surveyed prior to the field trial allowing for a less ambiguous evaluation of the geo-localization performance.

4 Results and Discussion

Overall, 52 observation data points were used from the field trial to evaluate the RFI location estimation performance using the hardware implementation on a Xilinx XCZU28DR emulator. The localization accuracy was compared to that of the MATLAB implementation in [15] using the same observation data points to verify and evaluate the performance of the hardware implementation. Moreover, PL resources required to implement the ESP-PF was also established to assess how well the implementation is optimized.

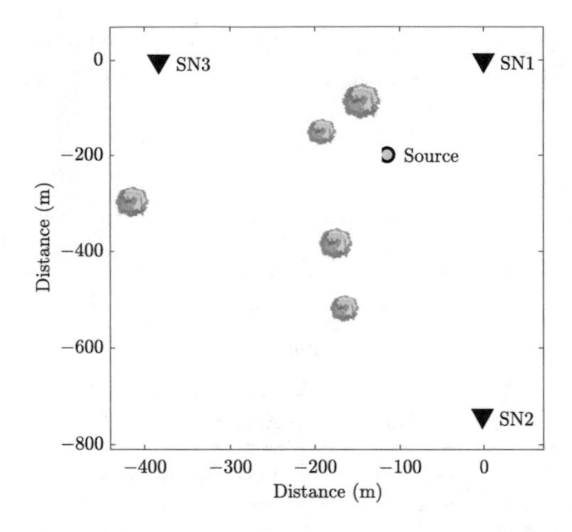

Fig. 2. Field trial setup

The mean and the standard deviation of the location estimation error in the $X-$ and Y–directions for the MATLAB model and the hardware implementation are shown in Fig. 3. As can be observed from the figure, the mean error of the hardware implementation was from 1 to 1.5 m while mean error of the MATLAB implementation was around 0.5 m in the X–direction. For the Y–direction, the mean error of the hardware implementation and MATLAB model are very close to one another, with a location estimation error of approximately 1 m. Considering the localization error standard deviations, as can be observed from the figure, the standard deviation error of the hardware implementation is smaller than that of the MATLAB model's in both the $X-$ and Y–directions, which indicates a higher confidence position estimation.

Hardware resource utilisation targeting the Xilinx XCZU28DR device is shown in Table 1. As can be observed, the current implementation uses a significant amount of the available Look-Up Tables (LUTs) and DSP resources. Hence, optimization of the hardware implementation forms part of our future research and development in this area.

Table 1. Particle filter implementation resource usage for XCZU28DR

BRAMs	DSPs	FFs	LUTs
177 (8%)	1203 (28%)	218126 (25%)	364819 (85%)

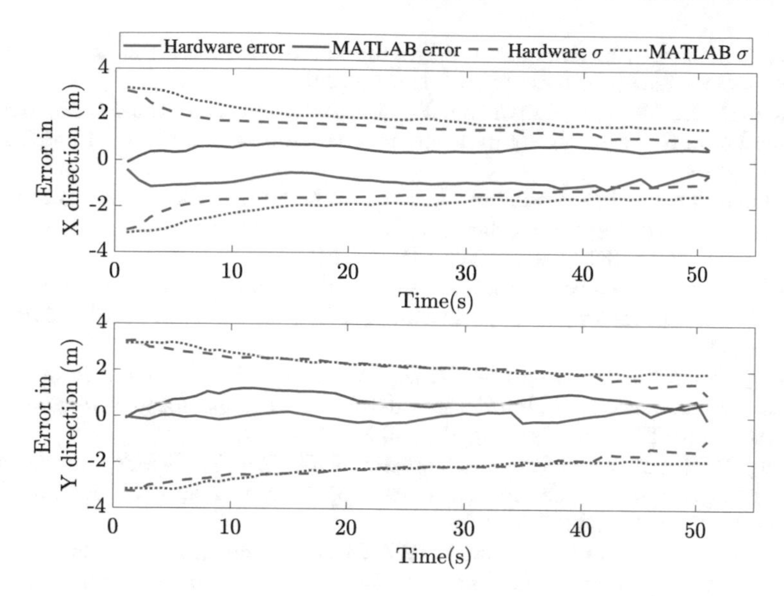

Fig. 3. Position estimation error and standard deviation for MATLAB and hardware implementation in the $X-$ and $Y-$ directions

5 Conclusion

Hardware implementation of the ESP-PF algorithm is presented targeting the Xilinx XCZU28DR SoC platform for RFI source geo-localization. The results based on real field trial data show that the fixed-point hardware implementation on the emulator performs comparably to that of the floating-point software-only implementation in MATLAB. During the hardware implementation it was also discovered that a 1,024 particle ESP-PF implemented in hardware provided same results for a 2,048 particle ESP-PF implemented in MATLAB. We think this is due to the "dithering" effect introduced by the quantization process having a positive impact on the estimation performance of the ESP-PF algorithm. Theoretical analysis of this forms part of our future work in this domain. While the hardware implementation results are promising, resource utilization is not as efficient as expected with large resource utilization observed for the PL components. Reducing this forms part of our future research in this area.

References

1. Borio, D., Camoriano, L., Savasta, S., Presti, L.L.: Time-frequency excision for GNSS applications. IEEE Syst. J. **2**, 27–37 (2008)
2. Tani, A., Fantacci, R.: Performance evaluation of a precorrelation interference detection algorithm for the GNSS based on nonparametrical spectral estimation. IEEE Syst. J. **2**, 20–26 (2008)

3. Bours, A., Cetin, E., Dempster, A.G.: Enhanced GPS interference detection and localisation. Electron. Lett. **50**, 1391–1393 (2014)
4. Axell, E., Eklöf, F.M., Johansson, P., Alexandersson, M., Akos, D.M.: Jamming detection in GNSS receivers: performance evaluation of field trials. Navigation **62**, 73–82 (2015)
5. Broumandan, A., Jafarnia-Jahromi, A., Daneshmand, S., Lachapelle, G.: Overview of spatial processing approaches for GNSS structural interference detection and mitigation. Proc. IEEE **104**, 1246–1257 (2016)
6. Borio, D., Dovis, F., Kuusniemi, H., Lo Presti, L.: Impact and detection of GNSS jammers on consumer grade satellite navigation receivers. Proc. IEEE **104**, 1233–1245 (2016)
7. Wang, P., Cetin, E., Dempster, A.G., Wang, Y., Wu, S.: Time frequency and statistical inference based interference detection technique for gnss receivers. IEEE Trans. Aeros. Electron. Syst. **53**(6), 2865–2876 (2017)
8. Wang, P., Wang, Y., Cetin, E., Dempster, A.G., Wu, S.: GNSS jamming mitigation using adaptive-partitioned subspace projection technique. IEEE Trans. Aeros. Electron. Syst. **55**, 343–355 (2019)
9. da Silva, F.B., Cetin, E., Martins, W.A.: Radio frequency interference detection using nonnegative matrix factorization. IEEE Trans. Aeros. Electron. Syst. **58**(2), 868–878 (2022)
10. Dempster, A.G., Cetin, E.: Interference localization for satellite navigation systems. Proc. IEEE **104**(6), 1318–1326 (2016)
11. Trinkle, M., Cetin, E., Thompson, R.J.R., Dempster, A.G.: Interference localisation within the GNSS environmental monitoring system (GEMS) – initial field test results. In: Proceedings 25th International Technical Meeting of the Satellite Division of the Institute of Navigation (ION GNSS 2012), vol. 4, pp. 2930–2939 (2012)
12. Cetin, E., Thompson, R., Trinkle, M., Dempster, A.: Interference detection and localization within the GNSS environmental monitoring system (GEMS)—system update and latest field test results. In: Proceedings of 27th International Technical Meeting of Satellite Division of Institute of Navigation (ION GNSS+ 2014), pp. 3449–3460 (2014)
13. Cetin, E., Thompson, R.J.R., Dempster, A.G.: Passive interference localization within the GNSS environmental monitoring system (GEMS): TDOA aspects. GPS Solut. **18**(4), 483–495 (2014). https://doi.org/10.1007/s10291-014-0393-5
14. Biswas, S.K., Cetin, E.: GNSS interference source tracking using kalman filters. In: 2020 IEEE/ION Position, Location and Navigation Symposium (PLANS), pp. 877–882 (2020)
15. Biswas, S.K., Cetin, E.: Particle filter based approach for GNSS interference source tracking: a feasibility study. In: 2020 XXXIIIrd General Assembly and Scientific Symposium of the International Union of Radio Science, pp. 1–4 (2020)
16. Le Gland, F., Oudjane, N.: Stability and uniform approximation of nonlinear filters using the hilbert metric and application to particle filters. Ann. Appl. Prob. **14**(1), 144–187 (2004)
17. Crisan, D., Doucet, A.: A survey of convergence results on particle filtering methods for practitioners. IEEE Trans. Signal Process. **50**(3), 736–746 (2002)
18. Arulampalam, M., Maskell, S., Gordon, N., Clapp, T.: A tutorial on particle filters for online nonlinear/non-Gaussian Bayesian tracking. IEEE Trans. Signal Process. **50**, 174–188 (2002)

19. Biswas, S.K., Dempster, A.G.: Approximating sample state vectors using the ESPT for computationally efficient particle filtering. IEEE Trans. Signal Process. **67**(7), 1918–1928 (2019)
20. Inacio, C., Ombres, D.: The DSP decision: fixed point or floating? IEEE Spect. **33**(9), 72–74 (1996)

Towards Wide-Band Laser Heterodyne Radiometer for Space Applications

Vu Hoang Thang Chau[1(✉)], James Bevington[2], Yiqing Lu[1], and Ediz Cetin[1]

[1] School of Engineering, Macquarie University, Sydney, NSW 2109, Australia
{Vuhoangthang.Chau,Yiqing.Lu,Ediz.Cetin}@mq.edu.au
[2] Interplanetary Exploration Institute (IXI) Ltd., Sydney, NSW, Australia
James@ixi.org.au

Abstract. Laser Heterodyne Radiometer (LHR) is state of the art device for remote sensing and has been widely used for atmospheric studies. This work looks at how the conventional narrow-band LHR sensor can be adopted for spaceflight, where orbital speed of the spacecraft necessitates fast acquisition times along with space, weight, and power constraints. We explore wide-band LHR to address these requirements and highlight challenges associated with achieving such a system.

Keywords: Laser Heterodyne Radiometer · Methane Characterization

1 Introduction

According to the report prepared by the Australian Academy of Science, global warming has contributed to many extreme weather events in terms of their frequency and severity [1]. As stated in the same report, the path forward is to reduce global greenhouse gas emissions and limit warming to well below 2 °C as envisaged under the Paris Agreement [1]. Among the greenhouse gases, methane is the second most potent gas and has a considerable impact on the climate of Earth. Away from Earth, for space exploration, such as on Mars, the presence of methane implies many theories such as serpentinisation, biological origin, etc. [2]. Hence, detecting and characterising methane emissions has wide-ranging applications, such as for use in regulating methane emissions on Earth, as well as identifying the origins of methane on Mars and beyond.

Various sensors have been developed for use in situ, ranging from an aerial platform to a spacecraft, with a view to detect and measure methane concentration in the atmosphere. Of these sensors, the Laser Heterodyne Radiometer (LHR) provides the best resolution needed for methane characterisation. The state-of-the-art LHR provides high spectral resolution ($\lambda/\delta\lambda \approx 6 \times 10^7$) in a

This work was supported in part by the Macquarie University International Research Training Program Scholarship for Master of Research (iRTP-MRES).

N. K. Suryadevara et al. (Eds.): ICST 2022, LNEE 1035, pp. 374–381, 2023.
https://doi.org/10.1007/978-3-031-29871-4_38

relatively compact size [3]. This high resolution motivates us to develop a spaceborne LHR sensor that would not only enable methane detection but also enable us to identify its source, e.g. natural gas vs coal seam vs agriculture. One of the challenges of developing such a sensor for space applications, in addition to the typical size, weight, and power constraints, is the need for fast acquisition times given the orbital speed of the satellite relative to Earth's surface. This challenge is further compounded by the fact that a nadir pointing sensor, rather than a limb profile one used in [4], is required, which increases the sensor's sensitivity requirements further. The Mini-LHR [5], which we refer to as a narrow-band sensor, operates by capturing the spectra one wavelength at a time, taking several minutes to complete. While this is suitable for ground-based applications where the device stays stationary, it is not suitable for spaceflight where fast acquisition times are required, given the orbital speeds. Recent work [3] has shown that it is possible to improve the integration time and the Signal-to-Noise Ratio (SNR) of the narrow-band sensor. Motivated by this, in this work we explore wide-band LHR sensor development.

The rest of the paper is organised as follows: Sect. 2 provides an analysis of narrow-band LHR operation and details challenges associated with the wide-band operation. In this section, we also carry out an analysis of the solution set-up to overcome these challenges with a view to future wide-band LHR realisation. Concluding remarks and future work are given in Sect. 3.

2 Laser Heterodyne Radiometer

2.1 Background, Analysis and Comparison

An LHR is a device that can measure and detect the absorption features of substances illuminated by a light source (e.g. sunlight, blackbody sources). A simplified block diagram of a narrow-band LHR operation where samples are captured by scanning a local oscillator, usually a laser, across a wavelength spectrum is shown in Fig. 1. As can be observed from the figure, the signal which contains the desired information is located at very high frequencies (tens to hundreds of THz), and it is down-converted to Radio Frequency (RF) using a photomixer after mixing with the laser – the Local Oscillator (LO) – in a 2×2 optical coupler. The photomixer is an optical-to-electrical device that produces a photoelectric current that the RF receiver can process. The RF receiver amplifies the signal with a Low Noise Amplifier (LNA) before filtering out the noise from high and near DC frequencies by a Bandpass Filter (BPF). The signal power is then determined by the Square Law Detector (SQD) and captured by a Lock-in Amplifier (LIA). The LIA demodulates the electrical signal, with the frequency reference from the Mechanical Chopper (MC), and averages the signal like a low pass filter. The output of the RF receiver provides a sample every time the LO is adjusted along the entire spectrum. The disadvantage of the narrow-band LHR is that its working mechanism is sample-by-sample capture, and an integration time to capture all samples is required. As more samples are required, the total time to visualise an entire signal spectrum increases significantly.

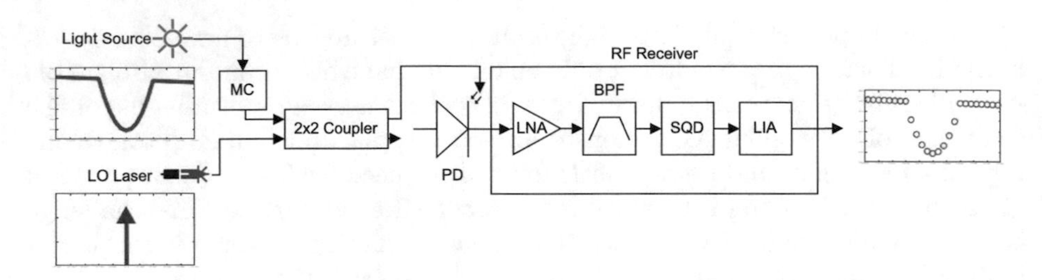

Fig. 1. Diagram of a conventional LHR capturing a Light Source spectrum. The Mechanical Chopper (MC) modulates the input signal before mixing it with the laser via a 2×2 Optical Coupler. The Photodiode (PD) downconverts the high-frequency signal to RF to be processed by the RF Receiver, which has Low Noise Amplifier (LNA), Bandpass Filter (BPF), Square-law Detector (SQD) and a Lock-in Amplifier (LIA). The output is collected at the LIA.

The following demonstrates and analyses the mechanisms of the narrow-band LHR operation via MATLAB simulations. The LHR benefits from the continuing improvements of lasers as an LO in terms of their compact size, stable, narrow-width and high power output [5]. The result of the narrow-band LHR modelling and simulation is depicted in Fig. 2, where the LHR scans across a large bandwidth of a blackbody source after absorbing a substance at 182.88 THz. The absorbing substance's profile is arbitrary and customisable with different input gases. A final rescaling was needed due to the optical power loss from the 2×2 optical coupler. The simulation result acts as a guide to understand the operation of LHR correctly and serves as a future key measurement. However, the study also raises many challenges that may influence the wideband LHR differently from the narrow-band LHR due to the mechanisms of capturing the signal. Because the wideband LHR intends to capture many samples across the entire light spectrum rather than capture samples gradually, signal excess noise and signal image presence can be problematic. These problems arise not only due to the mode of operation of the laser but also due to the operation of the photomixer as a square-law detector and the insufficient ability of the optical back-end of the LHR to pre-process the critical RF signal within the source signal.

2.2 Challenges for LHR Wideband Detection

Three main factors contribute to challenges associated with realising a wideband LHR. These are laser mode of operation, photomixer and RF receiver. While these are not of concern in narrow-band LHR systems where the laser frequency is adjusted, this is not the case for the wideband LHR solution needed for space applications. This section, therefore, concentrates on analysing the photomixer and RF Receiver. Different methods are explored with a view to addressing these challenges, and their performances are analysed via simulations. According to Deng *et al.* [6], the photoelectric current of the photomixer is proportional to

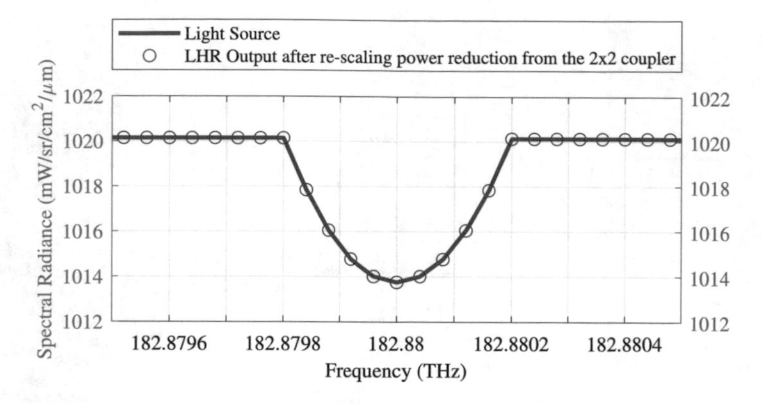

Fig. 2. Simulation result of capturing the spectrum of a Blackbody source with an arbitrary absorption feature at 182.88 THz (blue). The LHR outputs are captured in every LO scan (red) along the entire spectrum window. The absorption line is customizable.

the square of the total electromagnetic field of the source light $E_S(t)$ and the laser $E_L(t)$ as:

$$I(t) = R|E(t)|^2 = R|E_S(t) + E_L(t)|^2 \tag{1}$$

$$= R\left(cos\ \omega_S t + cos\ \omega_L t\right)^2 \tag{2}$$

where R is the responsivity of the photomixer, ω_S and ω_L are the wavelengths of the source signal and the laser, respectively. However, a wide-band source is a sum of numerous wavelengths, hence Eq. (2) becomes:

$$I(t) = R \cdot \left(\sum_{n=1}^{k} cos\ \omega_{s_n} t + cos\ \omega_L t\right)^2 \tag{3}$$

$$= R \cdot \left(\sum_{n=1}^{k} cos^2\ \omega_{S_n} t + \sum_{n=1}^{k} cos\ \omega_{S_n} . cos\ \omega_L t + cos^2\ \omega_L t\right) \tag{4}$$

Signal Excess Noise Problem. The first term in Eq. (4) is the *signal excess noise*, whilst the second term is the RF signal (useful signal), and the third term is the laser excess noise. The signal excess noise introduces excessive power to the entire bandwidth of the system if the total optical power is large enough. The noise origins from the source signal beating against itself within the photomixer. Additionally, this introduces the possibility of burying the desired dips, representing the absorption lines, at the output of the photomixer. Whether or not the important information is hindered depends on the power spectral density of the source light. Figure 3 illustrates the impact of a relatively wide-band white light source with an average power of 1 mW and a linewidth of 0.01 THz. Furthermore, the effects of signal excess noise on the SNR and the bandwidth of the RF signal are essential questions that need to be answered. A solution for this problem, adopted from De Arruda Mello *et al.* [7] is to apply optical balanced,

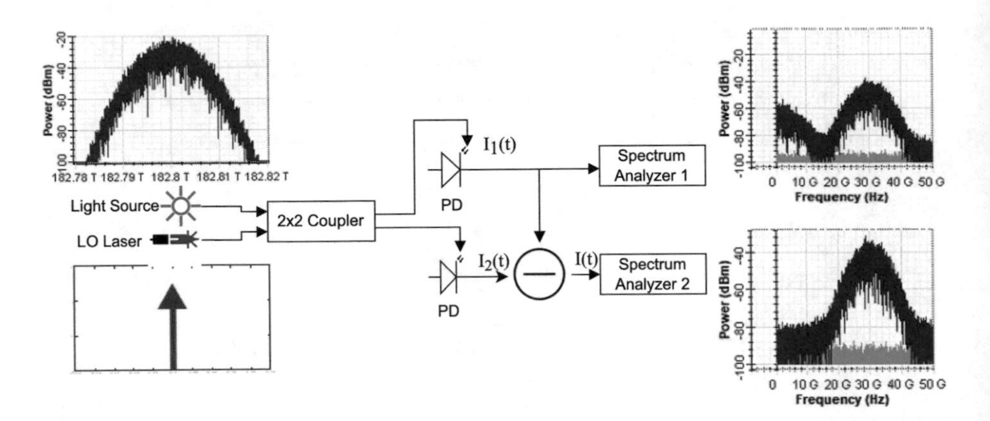

Fig. 3. Balanced Detection Circuit Architecture with Two Photodiodes to cancel out Signal Excess Noise. Spectrum Analyzer 1 shows Signal Excess Noise near DC, whilst Spectrum Analyzer 2 shows desired signal only with excess noise cancelled.

coherent detection, which utilises two photomixers to cancel out the signal excess noise, as shown in Fig. 3. Using this approach, the total photoelectric current will be two photoelectric currents flowing in opposite directions:

$$I(t) = I_1(t) - I_2(t) \tag{5}$$

According to the transfer function of a 2×2 Optical Coupler,

$$I_1(t) = R \cdot \left(\sum_{n=1}^{k} cos\ \omega_{S_n} t + cos\ \omega_L t \right)^2 \tag{6}$$

$$I_2(t) = R \cdot \left(\sum_{n=1}^{k} cos\ \omega_{S_n} t - cos\ \omega_L t \right)^2 \tag{7}$$

Expanding Eqs. (6), (7) and substitute into Eq. (5), the total photoelectric current remains only for the RF term:

$$I(t) = 4 \cdot R \cdot \left(\sum_{n=1}^{k} cos\ \omega_{S_n} t . cos\ \omega_L t \right) \tag{8}$$

As shown in Eq. (8) and can be observed from the Spectrum Analyzer 2 output in Fig. 3, in the end only the desired RF signal will remain.

Image Noise Problem. A common problem in RF signal processing when down-converting a high-frequency signal by mixing with a local oscillator without proper filtering is the inability to distinguish between the desired signal and its image, which is a signal that has the same frequency distance to the local oscillator as the frequency distance from the desired signal to the local oscillator, but on the other side of the spectrum [8]. The wide-band LHR may suffer from

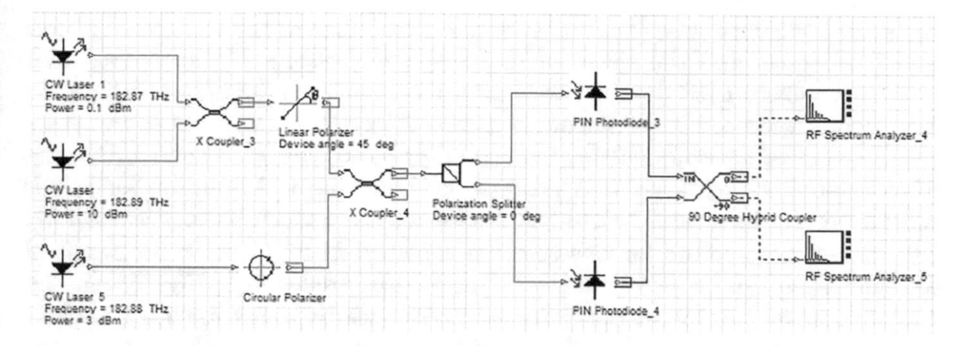

Fig. 4. Image Rejection Circuit with two Lasers (182.87 THz, 0.1 dBm and 182.89 THz, 10 dBm) and an LO Laser (182.88 THz) as referenced frequency. Outputs are obtained at the 90° hybrid coupler by the two spectrum analyzers.

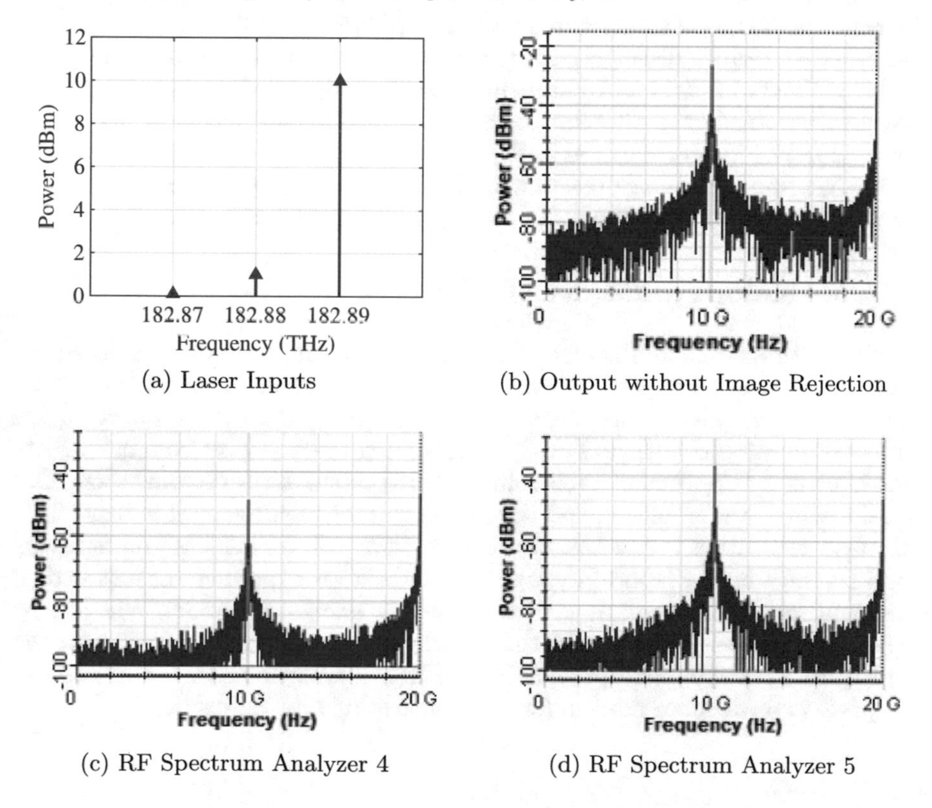

(a) Laser Inputs

(b) Output without Image Rejection

(c) RF Spectrum Analyzer 4

(d) RF Spectrum Analyzer 5

Fig. 5. Simulation result from Fig. 4. Figure 5a shows the frequency (THz) and power (dBm) of two light input Lasers and an LO Laser. Figure 5c shows the detection of CW Laser 1 at 182.87 THz (first peak at 10 GHz). Figure 5d shows the detection of CW Laser 2 at 182.89 THz (first peak at 10 GHz). Notice that the power difference between the two peaks is the same as the power difference of the two lasers.

the same issue with wide-band operation as the image of the wide-band source signal collapses onto the signal of interest at the output of the photomixer. The narrow-band LHR does not encounter the same issue because although the detection is wide-band, every sample of the detection is captured from a very narrow-band of the source light. On the other hand, the wide-band LHR aims to capture all samples within a much wider bandwidth. A solution for this is to incorporate the optical heterodyne image-rejection mixer as suggested in [9]. The circuit architecture for this solution is shown in Fig. 4, with the result of the simulation can be viewed in the RF Spectrum Analyzers in Figs. 5c and 5d. The inputs for this simulation are two narrow-width lasers assumed as light sources at 182.87 THz, and 182.89 THz with power levels of 0.1 dBm and 10 dBm, respectively. These input signals are heterodyned with an LO laser at a frequency of 182.88 THz, with 1 dBm power as shown in Fig. 5a. The two light source lasers are symmetric with respect to the LO laser in terms of their frequencies. Without the image rejection circuit, the heterodyned signals overlap on top of one other and are indistinguishable from each other in terms of their power levels, as shown in Fig. 5b. Additionally, the overlapping of the two frequencies after beating inside the photomixer results in an increase in power, as can be observed from Fig. 5b compared to Figs. 5c and 5d. Figure 5, on the other hand, shows that when the image rejection architecture is employed, one can identify the difference in power between the two signals ($\approx$ just under 10 dBm).

3 Conclusions and Future Work

Starting with modelling a narrow-band LHR operation, this work explored the challenges associated with wide-band LHR sensor development suitable for space flight which necessitates fast acquisition times given the orbital speed of the spacecraft. Several challenges have been identified to transition from narrow– to wide-band operation, and possible solutions are presented to overcome them. Our initial findings are promising and suggest that implementing a wide-band LHR with fast acquisition is feasible. Our future work will concentrate on verifying this feasibility in the laboratory environment via an experimental test bed. If successful, our next steps would be to have a bench-top model, followed by exploring approaches to minimize the size and power consumption of the system to produce a proof of concept wide-band LHR culminating in the development of a payload for a space mission for on-orbit testing and validation.

References

1. Australian Academy of Science: The Risks to Australia of a 3 °C Warmer World, March, 2021
2. Passmore, R.L., Bowles, N.E., Smith, K.M., Appleton, R.: The Development of High Sensitivity Spectroscopic Techniques for Remote Sensing of Trace Gases in the Martian Atmosphere – First Year Report (2009)
3. Zenevich, S., et al.: A concept of 2U spaceborne multichannel heterodyne spectroradiometer for greenhouse gases remote sensing. Remote Sens. **13**(12), 2235 (2021)

4. Weidmann, D., et al.: The methane isotopologues by solar occultation (MISO) nanosatellite mission: spectral channel optimization and early performance analysis. Remote Sens. **9**, 10 (2017)
5. Wilson, E.L., et al.: A portable miniaturized laser heterodyne radiometer (mini-LHR) for remote measurements of column CH4 and CO2. Appl. Phys. B Lasers Opt. **125**(11), 1–9 (2019)
6. Deng, H., et al.: Laser heterodyne spectroradiometer assisted by self-calibrated wavelength modulation spectroscopy for atmospheric CO2 column absorption measurements. Spectrochimica Acta Part A Mol. Biomol. Spectrosc. **230**, 118071 (2020)
7. de Arruda Mello, D.A., Barbosa, F.A.: Digital Coherent Optical Systems. ON, Springer, Cham (2021). https://doi.org/10.1007/978-3-030-66541-8
8. Razavi, B.: RF Microelectronics. Prentice Hall Communications Engineering and Emerging Technologies Series, 2nd edn. Prentice Hall, Upper Saddle River (2012)
9. Iwashita, K.: Optical FSK heterodyne detection using image rejection mixer. Electron. Lett. **25**, 255–256 (1989)

Design, Development, and Implementation of an IoT-Enable Sensing System for Agricultural Farms

Brady Shearen(✉), Fowzia Akhter, and S. C. Mukhopadhyay

School of Engineering, Macquarie University, Sydney, NSW 2109, Australia
brady.shearan@students.mq.edu.au

Abstract. A novel interdigital sensor has been proposed for soil carbon measurement. The sensor has been characterized using Electrochemical Impedance Spectroscopy (EIS) method for various soil samples. A calibration curve has been developed to determine carbon concentration from soil samples. Additionally, a smart sensing system has been developed to determine various soil and air quality parameters to understand the impact of carbon loss on the soil and air. Installation of this system in agricultural farms will allow the farmers to understand the current scenario and change of soil carbon to identify the ecological risks along with productive remedies to address the big issue. Farmers can also be informed about Carbon credit to get benefits from the government.

Keywords: carbon loss · global warming · agriculture · soil quality

1 Introduction

Climate change i.e., the long-term shifts in temperatures and weather patterns are mainly due to human activities such as burning fossil fuels like coal, oil and gas. In COP26, Nations have promised to end deforestation, curb methane emissions and stop public investment in coal power [1]. To achieve the targeted goal, governments, NGOs, scientists along with communities need to work together. Global warming is caused by greenhouse gases, predominantly by carbon dioxide (CO_2), almost 74% and New South Wales (NSW) and Queensland generates highest amount of CO_2 in Australia. All of the greenhouse gases need to be reduced significantly to address the problem of global warming. The predominant component CO_2 is generated due to many activities, coal fired electricity is a topic of hot discussion in recent times, the CO_2 emission from agricultural activities contribute 15%, which is the second highest source of greenhouse gases [2]. It is well understood that for our survival we need to produce crops, raise animals and continue agricultural activities in which soil carbons are disturbed and produce CO_2 [3].

The project aims to develop and implement an intelligent carbon sensing system to help to address one of the world's most critical crisis, i.e. global warming. The depletion of Carbon (C) stock in the soil increases the concentration of CO_2 in the atmosphere significantly, and Australian soils are at risk of becoming greenhouse source. One solution

N. K. Suryadevara et al. (Eds.): ICST 2022, LNEE 1035, pp. 382–393, 2023.
https://doi.org/10.1007/978-3-031-29871-4_39

to this problem is regular monitoring of the soil carbon using distributed intelligent Internet of Things (IoT) enabled smart carbon sensing devices. In addition, this project will build an intelligent sensing system that will monitor change of SOC with time along with soil temperature, moisture and a few other environmental parameters (air temperature, humidity, air pressure, CO_2, TVOC) from any remote location. Monitoring additional environmental parameters along with soil carbon will help to understand amount of carbon loss from the soil and effect of soil carbon loss on environmental parameters.

2 Necessary Parameters for Agricultural Applications

The important parameters for soil are Carbon, soil moisture, soil temperature and some essential environmental parameters that are affected by the change in soil parameters are rainfall, air temperature, humidity, CO_2 and total volatile organic component (TVOC) [4]. This section describes the essential parameters and the sensor used for measuring those parameters.

2.1 Soil Carbon

A low-cost novel planar interdigital sensor of circular configuration is used similar to our prior research for detection of soil measurement [5]. The sensing area of the proposed sensor is 6.25 mm^2. Figure 1 shows the proposed sensor soil carbon measurements. Figure 3 shows the proposed sensor. The impedance of the electromagnetic sensors is a function of the interaction of the electric field of carbon stock.

Fig. 1. Proposed MEMS based interdigital sensor for soil carbon sensing.

The sensor is tested for five soil samples using the Electrochemical Impedance Spectroscopy (EIS) method. Soil has been collected from agricultural farms and samples have been prepared with different percentage of Carbon for laboratory experiments. The sensor's responses towards those samples have been analysed for a frequency sweep of 10 Hz–10 kHz. The change in resistance (R) and reactance (X) of the sensor for various soil samples has been recorded. However, the behaviour of the sensor's resistance (R) has been included in this article (Fig. 1) because the resistance (R) changes more significantly as compared to the reactance (X) of the sensor for different soil samples. It is clear from Fig. 1 that the sensor is able to detect and differentiate various soil samples. When the carbon percentage in soil increases, the value of resistance reduces due to the increased conductivity. The resistance values at 1 kHz are taken for analysis to express the soil carbon percentage as a function of the sensor's resistance values. The frequency of 1 kHz is selected because it exists in a frequency band (10 Hz–1.2 kHz) where carbon concentration variations are noticeable. Figure 2 depicts the relationship between soil carbon percentage and resistance. This curve can be used as a standard to measure the unknown samples.

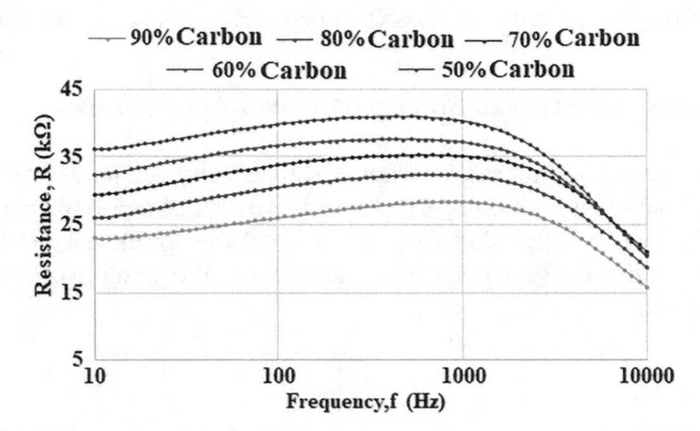

Fig. 2. Change in sensor's resistance (R) for various soil carbon percentage (50%–90%) for the frequency range from 10 Hz to 10 kHz at 25 °C and 40% humidity.

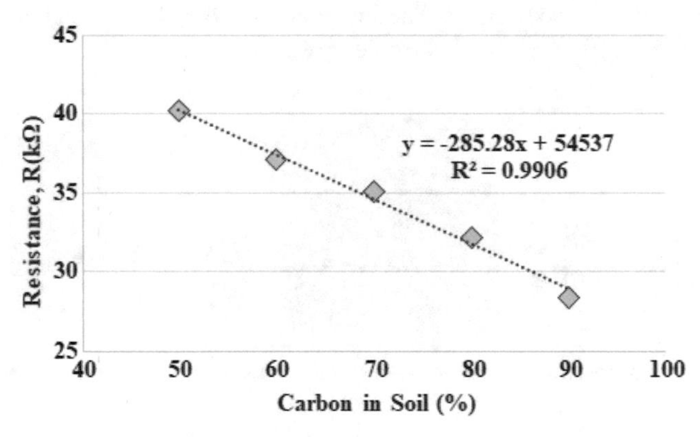

Fig. 3. The relationship between the sensor's resistances (R) with soil carbon percentage at 1 kHz frequency at 25 °C.

The standard formula for calculating soil carbon percentage is obtained from resistance values. It is found from Fig. 3 that the coefficient of determination, R^2 is 0.9906 for resistance (R). The value of R^2 is nearer to one using the curve fitting technique; the more accurate values can be obtained using the calibration curve [6].

$$Soil\ Carbon(\%) = \frac{54537 - R_{s\_soil}}{285.28} \tag{1}$$

where Soil Carbon (%) is the percentage of carbon in soil calculated using the sensor by measuring the resistance (R_{s_soil}). An impedance analyser IC will be used to interface the carbon sensor with microcontroller based system.

2.2 Rainfall

Rainfall is important to track because of the impact that rainfall patterns and distribution have on the ecosystem under observation. Australia has a climate that is both unpredictable and diverse. High temperatures, limited rainfall, and large evapotranspiration are common in Australia. Drylands are described as areas where water shortage exists as a result of evapotranspiration far surpassing precipitation. As a result, when analyzing the water-balance equation, quantifying the amount of rainfall becomes an incredibly significant component [6, 7].

The use of a tipping-bucket rain gauge has been the most popular method for point-source rainfall measurements in the past. When water is filtered into the system through funneling, it leans a mechanical fulcrum. The amount of water that has fallen over time can be linked to the capacity of each half of the fulcrum [7]. Low-cost approaches have become more prominent in research applications in recent years. These include capacitive and ultrasonic sensors, both of which use low-cost, low-power ways to determine the water level of a container after rainfall.

Due to the process of rainwater being kept and level-measured in a container before being emptied, an ultrasonic sensor was utilized inside our proposed system. This allows for the easy integration of water quality sensors into the system at some point, offering further insight on the rainwater's quality.

In terms of larger-scale applications, recent research findings have looked into the use of remote sensing technologies to track rainfall across wide areas. Estimating rainfall by detecting the change in telecommunication signal intensity between two or more radio towers is one of the most current techniques used. This technology is still in its early stages of study and development in Australia [8].

2.3 Soil Moisture and Temperature

Because of the impact of groundwater recharge, soil temperature and moisture are crucial parameters to monitor for both agricultural applications such as food harvesting [9] and for various hydrological models based on the water balance equation. Because the hydrothermal properties of soil are known to be relatively stable, changes in temperature within the soil over time can provide important information about soil moisture conditions. [10, 11].

All soil moisture detection systems have their own set of restrictions and obstacles, with suitable sensors determined by size, accuracy, cost, response, robustness, and other factors. The sensors themselves are frequently not expensive in these applications; rather, the interfacing/sampling electronics that obtain data from the sensors are. Low-cost temperature and moisture sensors will be used to measure soil moisture. Large capacitive soil moisture sensors are the most frequent low-cost soil moisture sensors because they are a very cost-effective technique of monitoring the moisture content, albeit they are prone to quick corrosion and inaccuracy over lengthy periods of usage.

As a result, a low-cost sensor such as the DFRobot soil moisture sensor is recommended. The SHT-10x is used to measure temperature since it is a surface-mounted capacitive sensor with anticorrosive qualities that is insulated with a low-cost copper sintered mesh. These sensors assess soil parameters based on a single point source. There is now research being done on using fiber-optic cabling to assess changes in soil temperature over long distances, albeit these techniques still come with high development and deployment costs.

2.4 Air Temperature and Humidity

The temperature and humidity of the environment play a big role in the intensity of evapotranspiration episodes. Temperature and humidity are particularly important to track because they have a significant impact on the availability and viability of agricultural applications such as farming. Measuring these parameters also provides an increase and contrast to spatial and temporal resolution of current temperature data [12].

There are many different types of ambient temperature and humidity sensors that may be purchased for a modest price. The low-cost device we want to use in our system is the "CCS811/BME280 Environmental Combination Sensor," which has been used in numerous scientific projects within our Macquarie Research lab. This sensor measures atmospheric humidity and temperature, as well as measuring a variety of other environmental variables such as Carbon dioxide, atmospheric pressure, and total volatile organic compounds (VOCs).

2.5 Atmospheric Pressure

Atmospheric pressure is an important parameter to monitor due to its indication of the environments weather condition. When a low-pressure system moves through an environment, it generally indicates there are clouds, high winds and precipitation present. Conversely, high-pressure systems indicate calm and clear weather conditions. Measuring the air pressure in combination with the other parameters can provide a much clearer understanding of the environmental conditions and how they are affecting soil and crop growth as well as water/rain quality.

2.6 Carbon Dioxide (CO_2)

Carbon dioxide is the main contributed towards greenhouse gases. The rapid rise of CO_2 emissions due to the industrial revolution has caused change in weather patterns, water supplies and affected crop growth [13]. Monitoring of CO_2 is an essential part of air-quality monitoring and can provide information into soil quality as well as rain and potable water quality.

2.7 Total Volatile Organic Component (TVOC)

TVOC is a term that refers to a collection of organic chemical compounds that have been grouped together to make monitoring easier when they are found in the environment or in emissions. Many compounds, including natural gas, fall under the category of volatile organic compounds (VOCs). VOCs are used to characterize pollutants in polluted air, and they often refer to gaseous vapors emitted by chemicals rather than the liquid phase. Monitoring VOCs give us a clear indication in the air quality and will allow for us to analyze and draw relationships to its effects on the other monitored parameters.

3 Sensing System Development and Implementation

The prototype system utilizes an array of three soil moisture and three soil temperature sensors [14], an Arduino Uno microcontroller [15], an environmental combination sensor [16], an AD5933 impedance analyzer [17], a rain gauge [18], a 6W solar panel, solar power conversion shield [19] and a rechargeable battery to maintain energy autonomy.

The soil moisture sensors are connected to the Arduino as analogue inputs and the soil temperature sensors are connected using the I2C communication protocol, through an I2C multiplexer as the temperature sensors all share the same local address. The AD5933 is utilized alongside the carbon sensor to provide impedance measurements. This IC requires calibration to calculate the most appropriate feedback resistor, system gain factor and phase values. After calibration, the sensors impedance value over a wide frequency range are recorded and stored via I2C and the magnitude and phase angles are converted into resistance and reactance.

Both the solar power conversion and the LoRa shields are linked with the Arduino microprocessor through Arduino compatible shields [20]. The power management shield is connected to the 6000 mAh battery pack as well as the 6 V solar cell.

Table 1 summarizes the many types of sensors and other electronics, while Fig. 4 depicts the proposed system and depicts the system's internal electronic architecture.

Table 1. Electronics used in the prototype system.

Component Name	Description
Arduino Uno	Ultrasonic
DFRobot Soil Moisture Sensor	Capacitive soil moisture sensor with protective coating to prevent corrosion and scratching
SHT-30 Temperature Sensor	Weatherproof temperature sensor with metal encasing for soil deployment
CCS811/BME280 Environmental Combo	BME280 provides humidity (%), temperature (°C), and barometric pressure (Pa), Carbon dioxide (CO_2)
I^2C Multiplexer	I^2C multiplexer board to combat device addressing issues
Rain Gauge	Tipping Bucket/Ultrasonic Rain gauge
Solar Panel	6 W, 6 V Solar panel
Seed Solar Shield	Arduino compatible Solar power conversion shield
LoRa Shield for Arduino	LoRa 915 MHz Arduino compatible transceiver shield
Battery	6000 mAh 3.7 V Polymer Lithium-Ion rechargeable battery
ANT-916-CW-HWR-SMA	External antenna

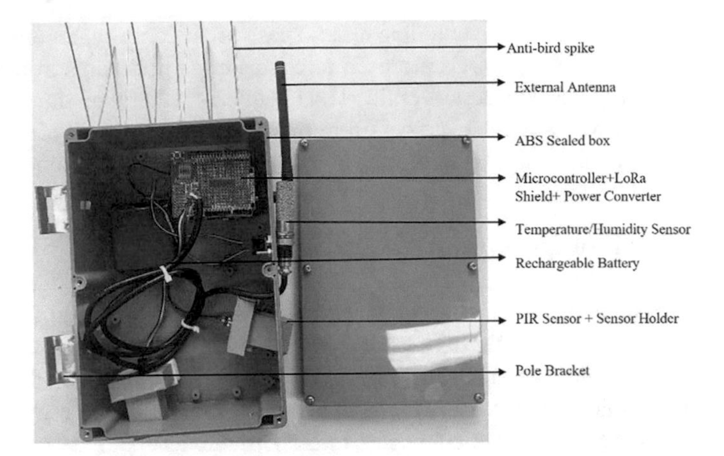

Fig. 4. Integration of the electronics in the sealed box

4 System Installation and Results

A testing site was offered by the University of Sydney to our Macquarie University research group for use for initial scoping of deployment requirements and future use of deployment of sensor nodes. The location was L'lara farm, which is part of a large commercial farm partially owned by USYD and utilized for agricultural research purposes.

There are currently over 32 professionally installed soil moisture sensors installed on the property, which operate utilizing a public LoRaWAN network. These sensors required a surrounding caging to prevent livestock from damaging the sensors, as seen in Fig. 5. These encaged locations alongside usage of the LoRaWAN network were

Fig. 5. Installation site of node

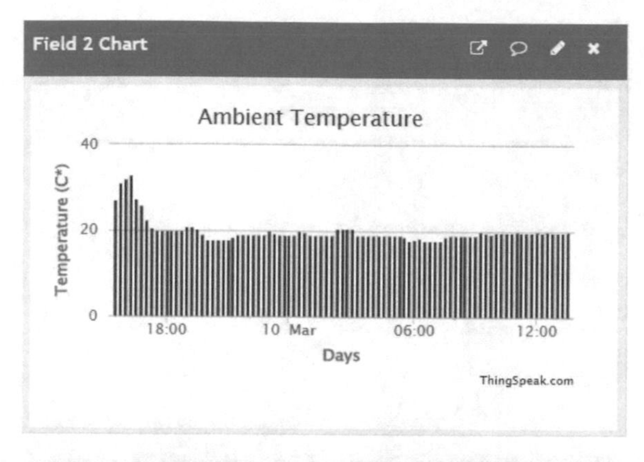

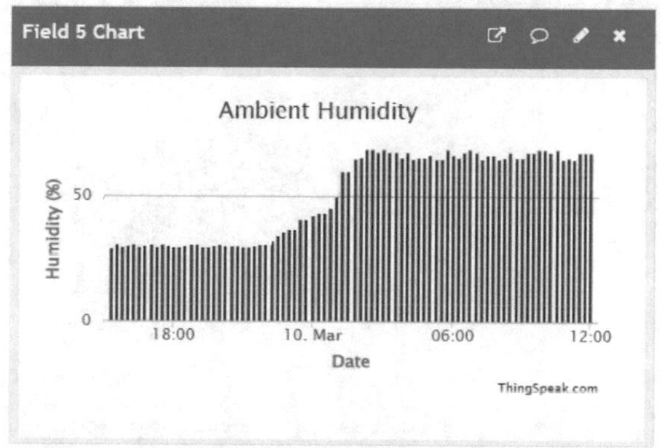

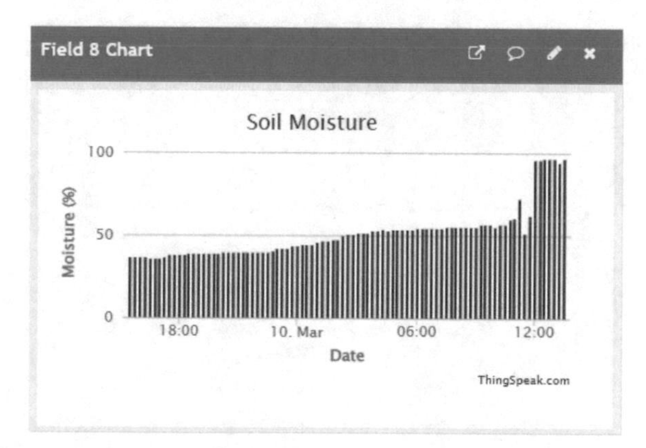

Fig. 6. Uploaded sensor data to the ThingSpeak cloud server (a) Ambi-ent Temperature (°C), (b) Ambient Humidity (%), (c) Soil Moisture (°C), (d) Top Soil Temperature (°C), (e) Rainfall (mm), (f) Carbon Dioxide CO2 (ppm), (g) Carbon (%), (h) TVOC (ppb), (g) carbon dioxide (ppm) and (f) Total Vol-atile Organic Component (ppb).

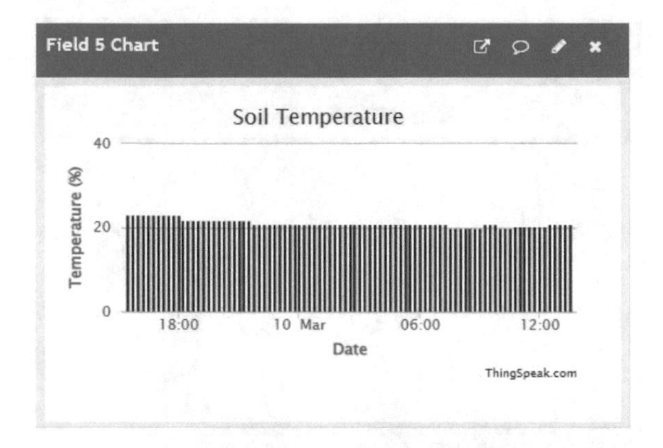

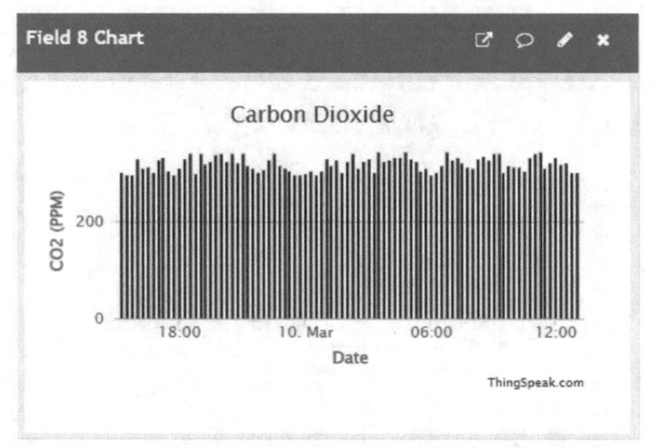

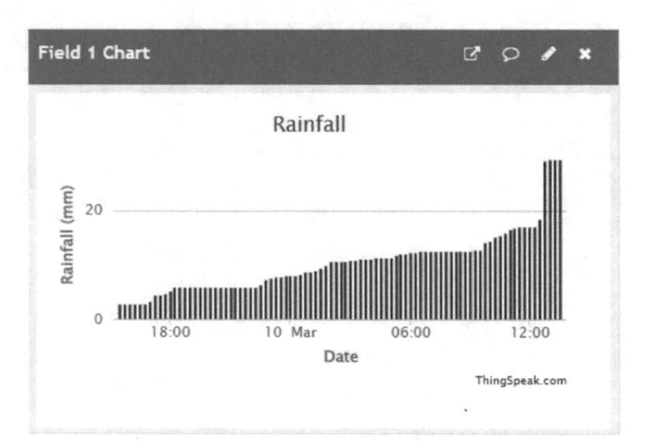

Fig. 6. (*continued*)

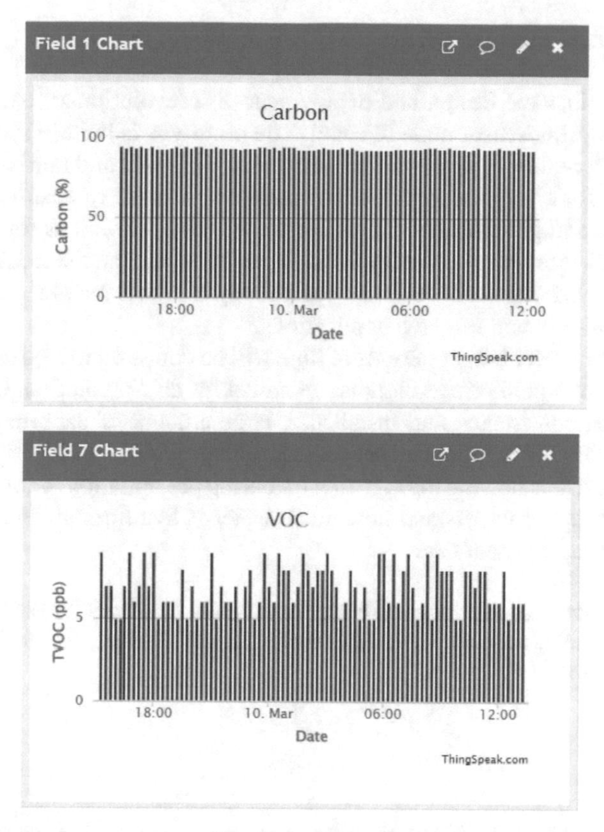

Fig. 6. (*continued*)

offered to Macquarie for future deployments. This is ideal as it is as close as possible to a real deployment area as possible whilst already having a network to make the data transmission and storage a much simpler process.

One of the developed test nodes was deployed alongside previously implemented sensors. This particular location also held a large rain gauge and it was deemed appropriate to deploy a node alongside it to compare and contrast collected data. Therefore, the tipping-bucket rain gauge was installed inside the gating, as seen in Fig. 5.

A first node was built, which used sensors to track ambient temperature and humidity, soil temperatures and moisture (%) at various depths, and rainfall. The temperature sensors were installed into the soil at various depths, coupled with environmental parameters as well as carbon and rainfall measurement testing. Figure 6 shows the results of the data collection, which included soil temperatures at three distinct depths over a singular day period.

During this period the node captured a rainfall event, in which saw changes in temperature and humidity, as well as soil moisture identifying that water was seeping through the soil. Soil temperature and carbon levels remained quite stable throughout the day.

5 Conclusion

This study describes the design and deployment of a revolutionary IoT-enabled environmental monitoring sensor node in detail. The prototype is built in such a way that it can collect ambient temperature and humidity, as well as soil and rainfall data, for long periods of time for a very cheap price. The node can be used to examine the interplay between surface and groundwater during recharge events, as well as for dynamic modelling of soil moisture for irrigation systems. It can be stated that the created sensor node is an energy-independent, low-cost, and compact system that may be quickly used in a variety of agricultural and research applications.

Data collection over longer periods of time will be collected in the future, as will the implementation of a predictive soil moisture model for data prediction. Once numerous nodes have been constructed and installed, a large amount of data may be collected, evaluated, and possibly interpolated between them to provide a much better picture of the amount of water that is flowing through the soil during precipitation events. This will also help researchers better understand how much water is lost through evapotranspiration under various weather situations.

Acknowledgements. The authors would like to thank NSSN, USYD, NSW DPIE, and Macquarie University for providing the research facilities needed to fabricate the sensor and carry out all the necessary experiments.

References

1. Masood, E., Tollefson, J.: COP26 climate pledges: what scientists think so far. Nature (2021)
2. Ward, M.: The principal greenhouse gases and their sources. Nat. Environ. Educat. Foundat (2015). https://www.neefusa.org/weather-and-climate/climate-change/principalgreenhouse-gases-and-their-sources
3. Fujisaki, K., Chevallier, T., Chapuis-Lardy, L., Albrecht, A., Razafimbelo, T., Masse, D., Ndour, Y.B., Chotte, J.L.: Soil carbon stock changes in tropical croplands are mainly driven by carbon inputs: a synthesis. Agricul. Ecosyst. Environ. **259**, 147–158 (2018)
4. Akhter, F., Alahi, M.E.E., Siddiquei, H.R., Gooneratne, C.P., Mukhopadhyay, S.C.: Graphene oxide (GO) coatedimpedimetric gas sensor for selective detection of carbon dioxide (CO_2) with temperature and humidity compensation. IEEE Sens. J. **21**(4), 4241–4249 (2020)
5. Akhter, F., Siddiquei, H.R., Alahi, M.E.E., Mukhopadhyay, S.C.: An IoT-enabled portable sensing system with MWCNTs/PDMS sensor for nitrate detection in water. Measurement **178**, 109424 (2021)
6. Aysha, F., Emma, J., Simon, F., Bruce, T., Justine, L., Andrew, T., Cara, S.: Foresighting Australian digital agricultural futures: applying responsible innovation thinking to anticipate research and development impact under different scenarios. Agricul. Syst. **190**, 103120 (2021). https://doi.org/10.1016/j.agsy.2021.103120
7. Yudhana, A., Rahmayanti, J., Akbar, S.A., Mukhopadhyay, S., Karas, I.R.: Modification of manual raindrops type observatory ombrometer with ultrasonic sensor HCSR04. Int. J. Adv. Comput. Sci. Appl. **10**(12), 277–281 (2019). https://doi.org/10.14569/IJACSA.2019.0101238
8. Yu, L., Gao, W., Redmond, S., Sha, T., Yanzhao, R., Zhang, Y., Su, G.: Review of research progress on soil moisture sensor technology. Int. J. Agricul. Biol. Eng. **14**, 32–42 (2021). https://doi.org/10.25165/j.ijabe.20211404.6404

9. Halloran, L., Rau, G., Andersen, M.: Heat as a tracer to quantify processes and properties in the vadose zone: a review. Earth Sci. Rev. **159**, 358–373 (2016). https://doi.org/10.1016/j.earscirev.2016.06.009
10. Quichimbo, E.A., Michael, S., Katerina, M., Daniel, H., Rafael, R., Cuthbert, M.: DRYP 1.0: a parsimonious hydrological model of DRYl and Partitioning of the water balance (2021). https://doi.org/10.5194/gmd-2021-137
11. Rau, G.C., Andersen, M.S., McCallum, A.M., Roshan, H., Acworth, R.I.: Heat as a tracer to quantify water flow in near-surface sediments. Earth Sci. Rev. **129**, 4058 (2014). https://doi.org/10.1016/j.earscirev.2013.10.015
12. Todorovic, M.: Crop evapotranspiration. In: Lehr, J.H., Keeley, J. (eds.) Water Encyclopedia: Surface and Agricultural Water, AW-57, pp. 571–578. Wiley, USA (2005)
13. Kanchana, S.: IoT in Agriculture: smart farming. Int. J. Sci. Res. Comput. Sci. Eng. Inform. Technol. 181–184 (2018). https://doi.org/10.32628/CSEIT183856
14. Sensirion_Humidity_Sensors_SHT3xDatasheet_digital 971521.pdf
15. Arduino: Arduino Uno WiFi (2017). https://store.arduino.cc/usa/arduino-uno-wifi
16. https://www.sparkfun.com/products/14348?ga=2.64212395.50694745.1646889615-255784082.1646787730
17. AD5933: Impedance Analyzer (2021). http://www.analog.com/media/en/technicaldocumentation/data-sheets/AD5933.pdf
18. https://cdn.sparkfun.com/datasheets/Sensors/Proximity/HCSR04.pdf
19. https://www.seeedstudio.com/Solar-Charger-Shield-v2-2.html [21] Dragino: LG01-S LoRa Gateway (2017). http://www.dragino.com/products/lora/item/119lg01-s.html
20. Shearan, B., Akhter, F., Mukhopadhyay, S.C.: Development of an IoT-Enabled Aqueous Sulphur Sensor with a rGO/AgNp Composite. In: Mandal, J.K., Roy, J.K. (eds.) Proceedings of International Conference on Computational Intelligence and Computing. AIS, pp. 233–239. Springer, Singapore (2022). https://doi.org/10.1007/978-981-16-3368-3_22
21. Fowzia, A., Hasin, S., Md, A., Krishanthi, J., Mukhopadhyay, S.C.: An IoT-enabled portable water quality monitoring system with MWCNT/PDMS multifunctional sensor for agricultural applications. IEEE Internet Things J. 1(2021). https://doi.org/10.1109/JIOT.2021.3069894
22. Akhter, F., Alahi, M., Mukhopadhyay, S.C.: Design and development of an IoT-enabled portable phosphate detection system in water for smart agriculture. Sens. Actuat. A **330** (2021). https://doi.org/10.1016/j.sna.2021.112861
23. Akhter, F., Khadivizand, S., Siddiquei, H.R., Alahi, M.E.E., Mukhopadhyay, S.: IoT enabled intelligent sensor node for smart city: pedestrian counting and ambient monitoring. Sensors **19**(15), 3374 (2019)
24. Akhter, F., Khadivizand, S., Lodyga, J., Siddiquei, H.R., Alahi, M.E.E., Mukhopadhyay, S.C.: Design and development of an IoT enabled pedestrian counting and environmental monitoring system for a smart city. In: 2019 13th International Conference on Sensing Technology (ICST), p. 16. IEEE (2019)
25. Akhter, F., Siddiquei, H.R., Alahi, M.E.E., Mukhopadhyay, S.C.: Recent advancement of the sensors for monitoring the water quality parameters in smart fisheries farming. Computers **10**(3), 26 (2021)
26. Shearan, B., Fowzia Akhter, S.C., Mukhopadhyay,: Development of an IoT-Enabled Aqueous Sulphur Sensor with a rGO/AgNp Composite. In: Mandal, Jyotsna Kumar, Roy, Joyanta Kumar (eds.) Proceedings of International Conference on Computational Intelligence and Computing. AIS, pp. 233–239. Springer, Singapore (2022). https://doi.org/10.1007/978-981-16-3368-3_22

Initial Studies for a Novel Electromagnetic Sensor for Detection of Carbon Content in Soil

Fowzia Akhter[1(✉)], K. P. Jayasundera[1], Anil Kumar A. S.[1,3], Brady Shearan[1], Waqas Ahmed Khan Afridi[1], Ignacio Vitoria[2], Boby George[3], and S. C. Mukhopadhyay[1]

[1] Department of Engineering, Macquarie University, Sydney, NSW 2109, Australia
fowzia.2k3@gmail.com

[2] Deptartment of Electrical, Electronic and Communication Engineering, Public University of Navarra, Arrosadía Campus, 31006 Navarra, Spain

[3] Department of Electrical Engineering, Indian Institute of Technology Madras, Chennai 600036, India

Abstract. Soils for agricultural activities are becoming a greenhouse gas source because of increasing atmospheric CO_2 concentration caused by soil organic carbon (SOC) stock depletion. Regular SOC monitoring at different parts of farmland is a useful technique to keep track of the SOC stock, but this can only be possible if a cost-effective, reliable SOC sensor is readily available. This research proposes a novel, low-cost planar electromagnetic sensor for the detection of SOC. Electrochemical Impedance Spectroscopy (EIS) has been used to characterize the sensor's performance for a range of SOC concentrations from 0.65% to 6.88% . The sensor is able to detect and distinguish between various SOC concentrations based on their impedance values. The outcomes are highly positive and bode well for creating a low-cost smart SOC monitoring device for agricultural uses.

Keywords: Global Warming · SOC · Electromagnetic Sensor · Agriculture · Carbon Credit

1 Introduction

Global warming is the long-term heating of Earth's climate system observed since the pre-industrial period (between 1850 and 1900). It is one of the vital environmental issues worldwide, which results mainly from the emission of greenhouse gases, particularly Carbon dioxide (CO_2), which contributes two-thirds to it [1]. We need to conduct agricultural activities as we need food for our survival. Soil organic carbon (SOC) loss due to agricultural activities is Australia's second-highest source of CO_2 generation [2]. Hence at the 26th UN Climate Change Conference (COP26), Nations promised to reduce the world temperature by taking necessary measures in all the contributing sectors of CO_2 generation [3]. In addition to generating clean energy, Australia's primary focus is enhancing SOC storage in the farmlands by adapting intelligent agricultural systems. It is well understood that for our survival, we need to produce crops, raise animals and

N. K. Suryadevara et al. (Eds.): ICST 2022, LNEE 1035, pp. 394–406, 2023.
https://doi.org/10.1007/978-3-031-29871-4_40

continue agricultural activities in which soil carbons are not disturbed and produce CO_2 [4]. The depletion of Carbon (C) stock in the soil significantly increases the concentration of CO_2 in the atmosphere, and Australian soils have become a greenhouse gas source [5]. One solution to this problem is regular monitoring of the soil carbon using distributed intelligent Internet of Things (IoT) enabled smart carbon sensing devices [6]. Hence, there is a need to develop a low-cost novel sensing system that will be modular and provide a change of carbon stock with time for farmers irrespective of changes in environmental parameters. Along with other sensors such as soil moisture, soil temperature, and other nutrients, the carbon sensor will make it an integrated monitoring system and provide insightful information on soil health throughout the year. Carbon stock sensing will not only address climate change but also make Australian farmland more fertile by increasing SOC storage. SOC is a nutrient source that helps aggregate soil particles, protects soil from erosion, enhances water storage capability, improves microbial activity, and provides resilience to physical degradation [7].

Numerous kinds of research have been attempted to estimate soil carbon in agricultural farms. The traditional method is bulk density estimate, in which soil-carbon stocks are calculated in tonnes of carbon per hectare. The dry weight of a known volume of soil is its bulk density. It can be obtained by driving a core, exhaust tube, or pipe to a specific depth into the earth [8]. The major drawback is that samples need to be sent to a laboratory, or an expert need to be on the farm to conduct the measurements, which are very time-consuming. Currently, it costs about $30/ha to measure soil carbon by applying the traditional method [9]. Some researchers have developed soil-carbon sensors using near-infrared (NIR) and electrochemical techniques. NIR sensors measure the soil carbon level using a specific wavelength of light. The operating mechanism is based on carbon's infrared signal adsorption characteristics [10]. Although these sensors detect a wide range of carbon concentrations, they have limited accuracy, the production cost is very high, and they consume much power and need frequent calibration [11].

Electrochemical sensors detect carbon based on the chemical reaction between the sensing surface and target molecules. The reported electrochemical sensors provide selective detection of Carbon under controlled laboratory conditions but suffer from inconsistent measurement and have limited lifetimes; they are also unsuitable for field application [12, 13]. Some researchers have developed prediction models to determine soil carbon by applying machine learning algorithms such as regression analysis. But these methods lack reliability because of unreliable detection methods such as soil color [14, 15]. Some researchers have used LiDAR (light detection and ranging) to determine biomass change in the soil remotely. However, there are no reported systems for remotely sensing soil carbon using LiDAR technology [16, 17]. Hence, it is essential to develop a cost-effective sensor to detect soil carbon correctly and intelligent sensing systems for monitoring soil carbon and other environmental parameters from any remote location to be meaningful to farmers.

Therefore, this research aims to transform the way the measurements are currently done by fabricating a novel carbon sensor for soil charcoal carbon measurements with the future potential of SOC determination.

2 Proposed Sensor

Figure 1 shows the design of the proposed planar electromagnetic sensor. A few designs have been tried before finalizing this design. The main advantage of this design is soil samples exposed to the sensor will interact with both the electric and magnetic fields. Hence, the sensor will be able to detect the amount of carbon content correctly. The proposed sensor has been designed using Altium Designer 6 and fabricated using simple printed circuit board (PCB) fabrication technologies with a thickness of 0.25 mm. After that, a coating layer is applied on the sensor with Wattyl In-cralac Killrust spray to prevent direct contact with the water sample. The top layer of the planar electromagnetic sensor is depicted in Fig. 1. The meander and interdigital coils are connected in series. When an AC voltage is applied across the terminals, the meander coil produces a magnetic field, and the interdigital coil produces an electric field. An electromagnetic field that interacts with the test material is created by combining both meander and interdigital coils [18–20].

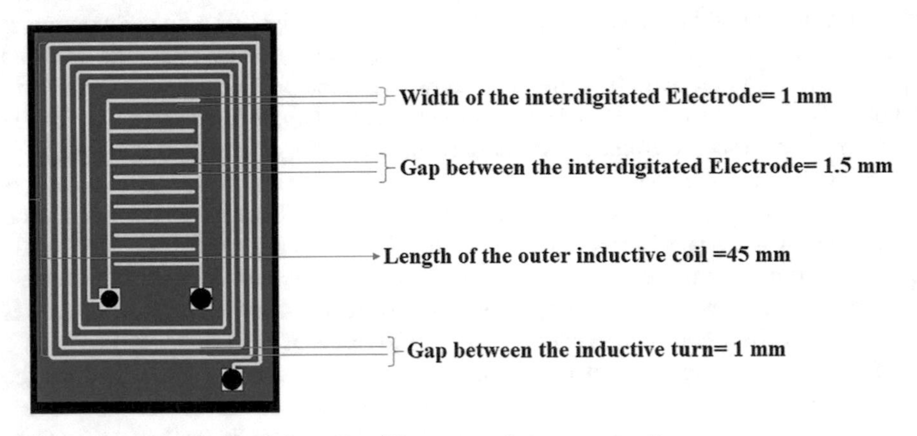

Fig. 1. Schematic of the proposed electromagnetic sensor.

In order to validate the proposed sensor, the finite element analysis (FEA) of the proposed sensor system has been carried out using the AC/DC module of the COMSOL Multiphysics software. A 3D model of the sensor has been designed with a length of the outer coil of 45 mm and a width of is 35 mm. As real soil organic carbon cannot be created in a simulation environment, four carbon-based materials have been chosen to analyze the sensor's response toward carbon. A sinusoidal excitation of 5 V, 1 kHz has been injected into the sensor, and the change in impedance of the sensor and the test materials have been placed over the sensor with an air gap of 0.5 mm. The effect of change in the volume of the soil samples in the output has been analyzed in this study. The gap between the interdigitated electrodes is 1.5 mm and the maximum range of detection is 5 mm due to the limited electric field distribution as per the simulation. Hence, the sensor's performance has been analyzed for two different soil thicknesses: 2 mm and 4 mm, (within the detection limit) without altering the surface area of the soil. In this study, four carbon-based materials have been considered: (1) graphite (2) carbon nanotube (CNT) (3) graphene oxide (GO), and (4) reduced graphene oxide (RGO). In the simulation, the graphite material has been taken from the COMSOL Multiphysics library, and its electrical characteristics (conductivity, permittivity, and permeability) have been changed for the other materials according to the information from [21–23]. For each material, the sensor has been simulated, and the impedance (Z) of the sensor has been obtained and plotted, as shown in Fig. 2. It is clear from Fig. 2, that the sensor can differentiate different carbon materials and distinguish the various amount of carbon contents. However, the difference in impedance obtained, for soil thickness of 2 mm and 4 mm, for CNT, GO, and RGO has been found to be small compared to that of Graphite.

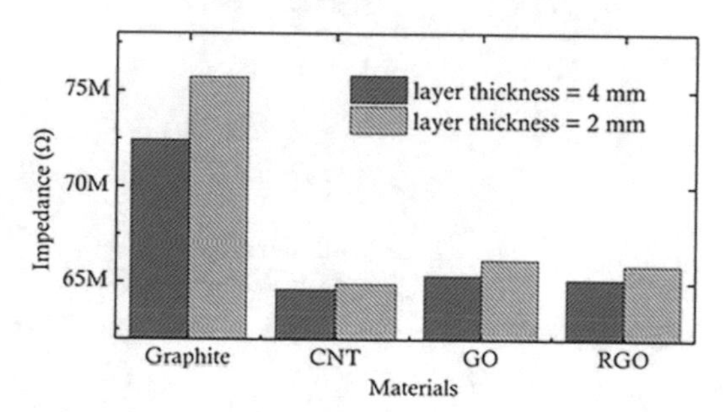

Fig. 2. The change in impedance of the sensor for different carbon-based materials, recorded from the finite element analysis.

3 Experimental Procedure

The experiments involving the sensor patch and soil samples have been conducted in a laboratory setting with constant relative humidity (40%) and temperature (25 °C) levels. Soil samples have been collected from various areas of New South Wales (NSW), Australia. Those have been then dried in the oven at 100 °C and finely ground. After that, the soil samples have been passed through a 2 mm sieve to filter out large organic debris. The samples have been then analyzed using the Walkley-Black (WB) chromic acid wet oxidation method and loss of ignition (LOI) method to know the amount of SOC in those samples. In the WB method, Oxidizable matter in the soil is oxidized by 1 N Potassium dichromate ($K_2Cr_2O_7$) solution. The reaction has been assisted by the heat generated when two volumes of Sulfuric Acid (H_2SO_4) have been mixed with one volume of the dichromate. The remaining dichromate has been titrated with Ferrous Sulphate ($FeSO_4$) in the presence of an indicator. Figure 3 shows the step-by-step procedure to conduct the titration experiment. The amount of soil organic carbon (SOC) has been calculated using the following formula:

$$\mathrm{SOC\%} = \frac{0.003\,\mathrm{g} \times N \times 10\,\mathrm{mL}\left(1 - \frac{T}{S}\right) \times 100}{OWD} = \frac{3 \times \left(1 - \frac{T}{S}\right)}{OWD} \tag{1}$$

Here, N = Normality of $K_2Cr_2O_7$ solution
T = Volume of $FeSO_4$ used in sample titration (mL)

S = Volume of $FeSO_4$ used in blank titration (mL)
OWD = Oven-dry sample weight

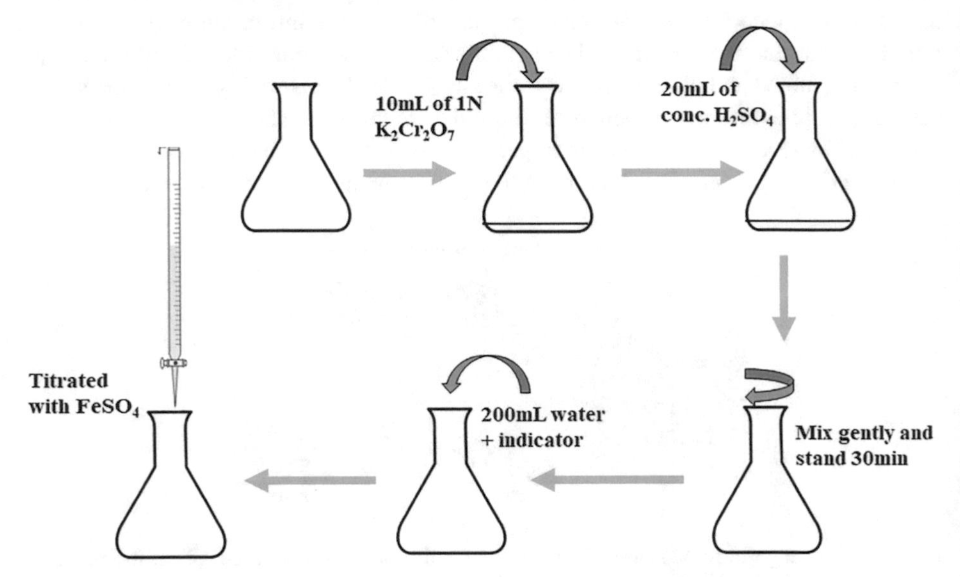

Fig. 3. Procedure followed as per Walkley-Black (WB) chromic acid wet oxidation method to determine the amount of soil organic carbon (SOC).

The titrant is inversely related to the amount of organic carbon present in the soil sample [24].

In addition to the WB method, the Loss of Ignition (LOI) dry combustion method has been applied to determine the SOC concentrations in soil samples as the WB method produces chemical waste. The amount of soil organic matter has been measured directly as the weight loss upon combustion at 350ºC. Initially, a 5 g scoop of soil samples has been placed into a 20 ml crucible and dried at 105 °C for 24 h. The weight of the sample has been recorded. After that, the furnace temperature has been increased to 350 °C for two hours. The new weight of the soil has been measured once it cools down to <150 °C.

$$\text{LOI}(\%) = \frac{((\text{wt.at}150\,^\circ\text{C}) - (\text{wt.at}350\,^\circ\text{C}))}{(\text{wt.at}150\,^\circ\text{C})} \times 100 \tag{2}$$

LOI method provides the amount of soil organic matter (SOM). However, the amount of SOC can be used calculated using the conventional factor of 1.724 [25].

$$\text{SOC}(\%) = \frac{\text{SOM}(\%)}{1.724} \tag{3}$$

In this way, the amount of SOC in the collected samples has been measured. The reason for analyzing the soil samples with two methods is to set the benchmark for the sensor. The sensor will be reliable if the results obtained from the sensor match with one of the method's results or with the average of the two methods.

After applying the traditional methods, it is found that the various soil samples have various concentrations. Among those, six soil samples having different concentrations (0.65%, 1.84%, 2.7%, 3.84%, 4.98%, and 6.88%) have been considered for analysis using the proposed sensor.

The experimental setup for carrying out tests with the sensor patch is shown in Fig. 4. An impedance analyzer (IM3536) has been used to characterize the sensor using the EIS method. It is a powerful technique for both qualitative and quantitative analysis. This has been used in this research to analyze electrochemical events occurring at the electrode surface to characterize the soil properties. In this method, the frequency-dependent real and imaginary impedance plays a vital role in determining various quantities that cannot be measured using geometric structures and material interfaces [26, 27].

Fig. 4. Experimental setup for the characterization of the sensor.

The experiment has been correctly carried out with the necessary precautions, such as cleaning and drying the container and sensor after each measurement and conducting calibration tests of the impedance analyzer to prevent instrumentation and measurement errors.

The impedance analyzer is connected to a computer to store the data in excel files via an automated data acquisition algorithm. The sensor is initially placed in the middle of the beaker, and then the soil is slowly poured inside the container to fill up the sensing area of the sensor. The beaker is filled with the samples in such a way that the sensor's whole sensing area is covered with the soil. The sensors patch and beaker are properly cleaned using MilliQ water and ethanol after each experiment cycle, then dried for a short period in preparation for the next measurement.

4 Results and Discussion

The proposed sensor's responses toward various soil samples containing between 0.65% and 6.88% SOC concentrations have been analyzed using Electrochemical Impedance Spectroscopy (EIS).

The samples after LOI experiments have also been tested using the proposed sensor. At 25 °C and 40% humidity, the electromagnetic sensor's response is measured over a frequency range of 10 Hz to 1 MHz. The experimental results for several soil samples are displayed in Figs. 5, 6, 7, and 8. Even though data have been gathered for a wide range of frequencies, only the data for a small range of frequencies that produce the most distinct results, has been shown in the paper. It is clear from Figs. 5, 6, 7, and 8 that the sensor can differentiate various SOC concentrations based on impedance, resistance, reactance, and phase angle.

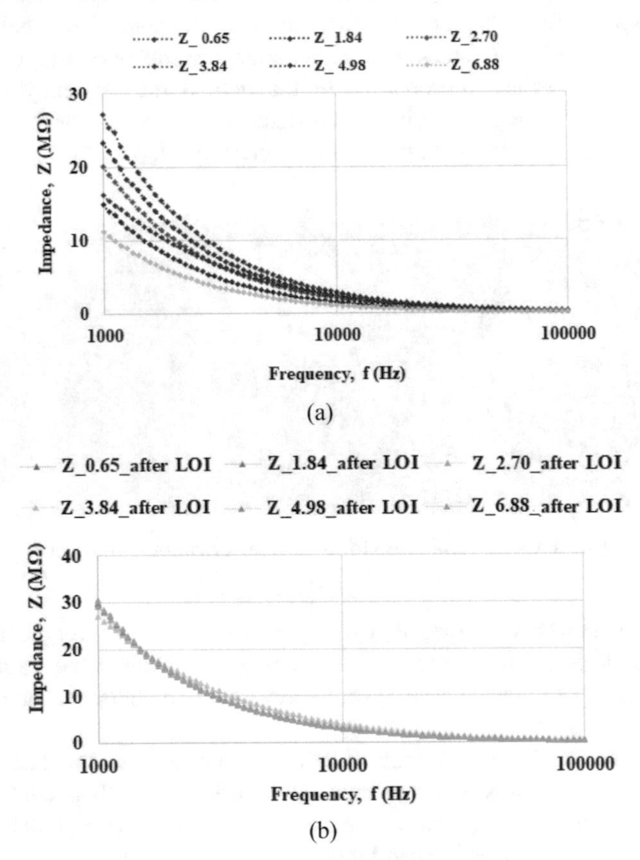

Fig. 5. Impedance (Z) of various soil samples, (a) before LOI and (b) after LOI experiments.

The soil samples after LOI experiments have also been analyzed using the WB method and it is found that the samples do not contain any SOC. Hence, impedance parameters after LOI experiments have been considered for 0% SOC and their average value has been taken as a reference value for further consideration to develop the calibration equation.

The calibration curve to quantify the SOC concentration from unknown soil samples is obtained based on the difference in impedance values from the reference value as shown in Fig. 9. The coefficient of determination (R^2) is 0.9971. The value of R^2 is nearer to 1 signifies the possibility of obtaining more accurate results using the calibration equation [28, 29].

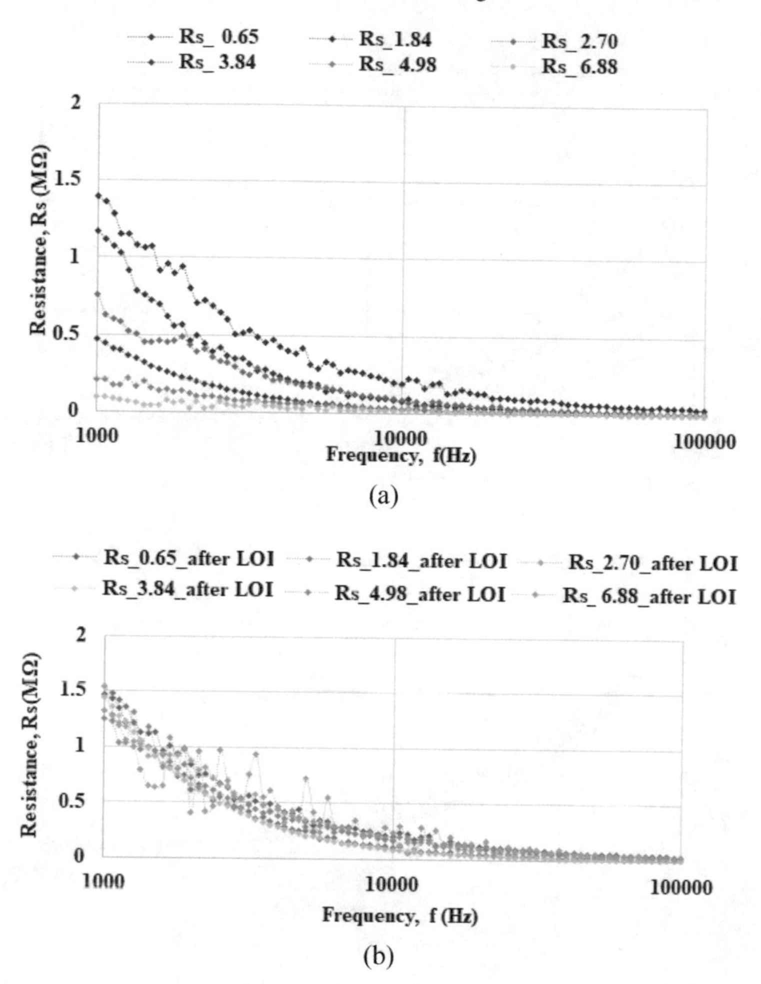

Fig. 6. Resistance (R_S) of various soil samples, (a) before LOI and (b) after LOI experiments.

According to Fig. 9, there is a strong correlation between the sensor's impedance and the SOC concentration. Therefore, soil charcoal carbon content can be determined using the following equation.

$$\mathrm{SOC}(\%) = \frac{\Delta Z - 3 \times 10^6}{3 \times 10^6} \tag{4}$$

where SOC (%) is the soil organic carbon percentage in the soil sample calculated based on deviation of impedance value from the reference value. The SOC coefficient, $\alpha_S = 3 \times 10^6\%\mathrm{SOC}$.

After developing the calibration standard, some samples were measured using the proposed sensor and also using the WB method. Figure 10 shows the experimental outcomes. It is clear from the figure that the actual value and predicted values are almost

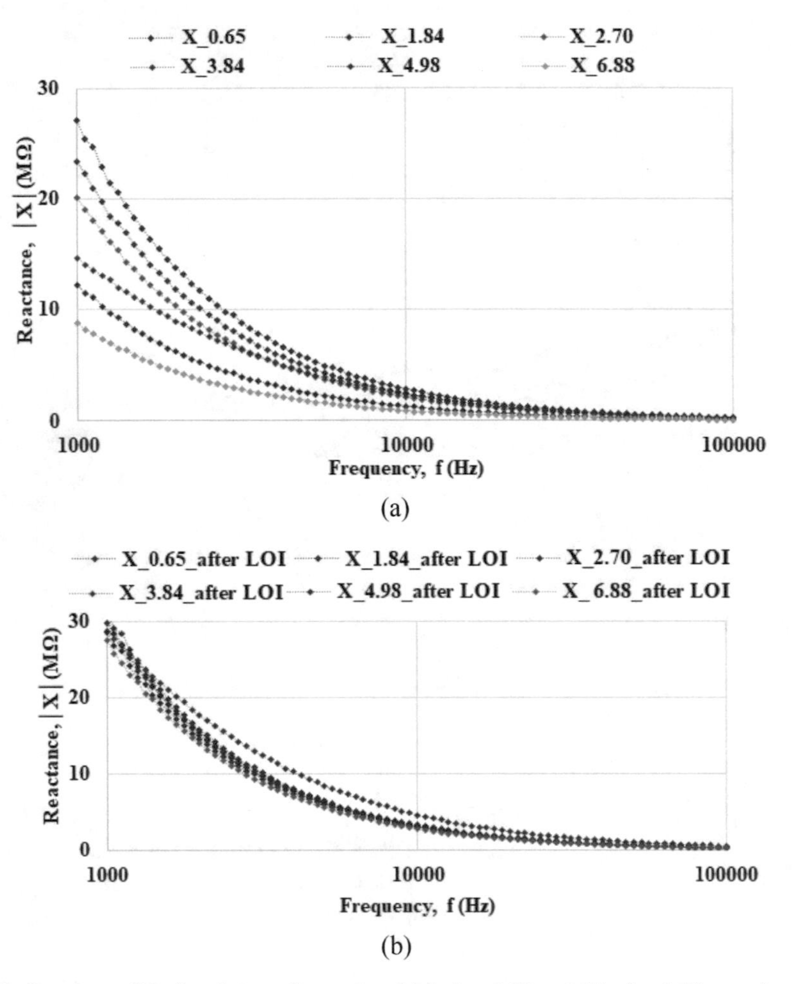

Fig. 7. Reactance |X| of various soil samples, (a) before LOI and (b) after LOI experiments.

the same. The value of the maximum error in SOC measurement using the calibration equation is, $\varepsilon r = 13.63\%$.

To understand the repeatability, the sensor responses to a soil sample have been determined five times. Figure 11 displays the experimental findings. The sensor's response to a soil sample has not shown any appreciable changes. The relative standard deviation (RSD) value has been calculated to investigate the consistency of the sensor's measurements. The maximum value of RSD is found as 2.30%. The lower the value of RSD, the more precise the measured data are [30, 31]. Hence, it can be said that the sensor is highly reliable.

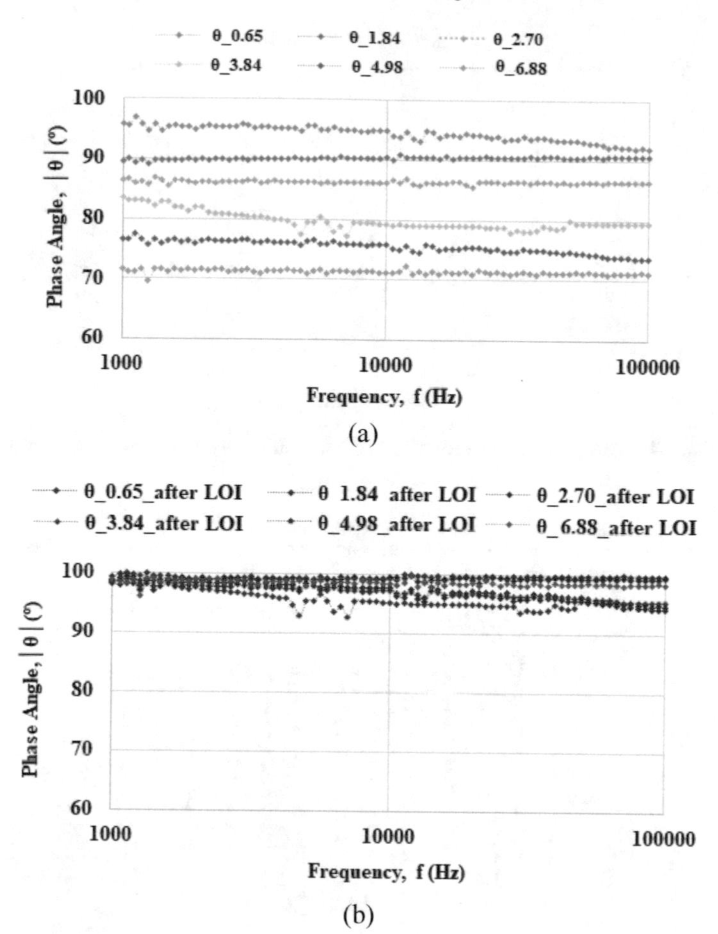

Fig. 8. Phase angle (θ) of various soil samples, (a) before LOI and (b) after LOI experiments.

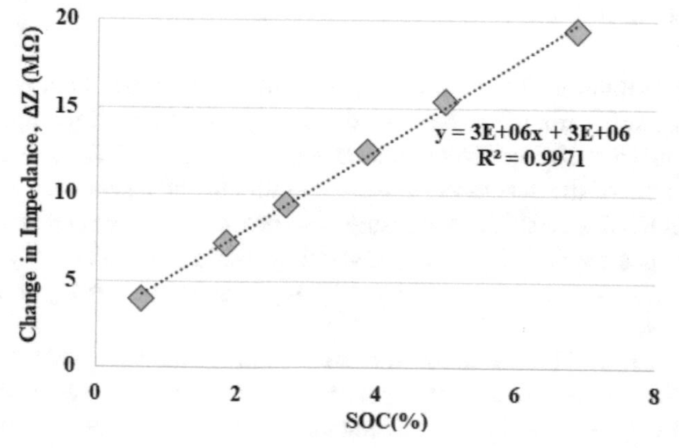

Fig. 9. Determination of calibration equation for SOC measurement based on the impedance difference from the reference value.

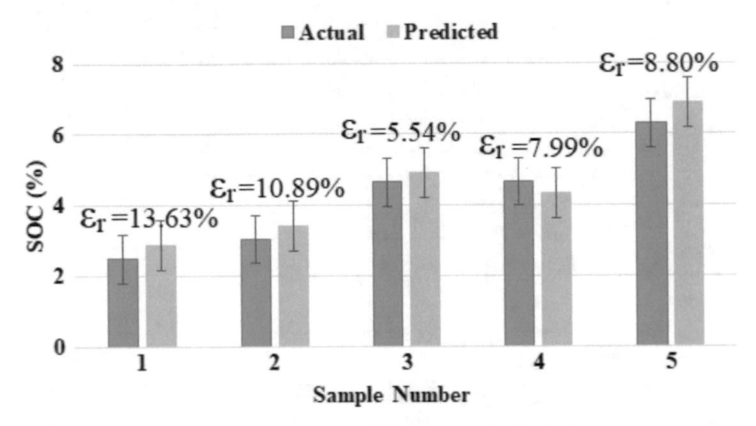

Fig. 10. Comparison between the WB method and sensor's measurements.

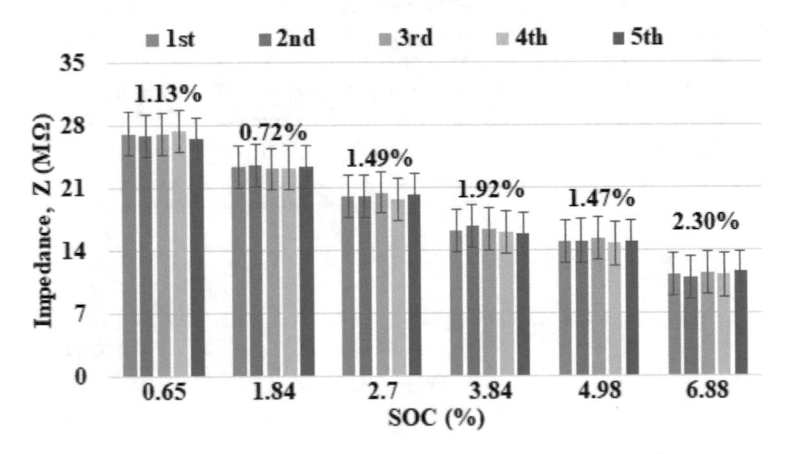

Fig. 11. Repeatability of the results for various soil samples.

5 Conclusion

A successful presentation of the fabrication and characterization of a novel electromagnetic sensor for soil carbon has been made. The proposed electromagnetic sensor is capable of identifying SOC concentrations at a specific concentration from various samples. A key feature of the developed sensor is its simple and easy operating mechanism for detecting SOC. Repeatability test results also prove that the sensor is highly reliable and produces repeatable results. Sensor responses within a second clarify that it will be very useful to monitor soil quality to identify any changes occurring on an acceptable time scale.

Our future work will be developing machine learning models for selectively detecting SOC in real-life samples. Moreover, a compact intelligent IoT-based smart SOC monitoring system will be developed to understand the change in SOC stock in Australian farmlands.

References

1. Akhter, F., Alahi, M.E.E., Siddiquei, H.R., Gooneratne, C.P., Mukhopadhyay, S.C.: Graphene oxide (GO) coated impedimetric gas sensor for selective detection of carbon dioxide (CO_2) with temperature and humidity compensation. IEEE Sens. J. **21**(4), 4241–4249 (2020)
2. Abbas, F., Hammad, H.M., Ishaq, W., Farooque, A.A., Bakhat, H.F., Zia, Z., Fahad, S., Farhad, W., Cerdà, A.: A review of soil carbon dynamics resulting from agricultural practices. J. Environ. Manage. **268**,110319 (2020)
3. Masood, E., Tollefson, J.: COP26 climate pledges: what scientists think so far. Nature (2021)
4. Ma, J., Rabin, S.S., Anthoni, P., Bayer, A.D., Nyawira, S.S., Olin, S., Xia, L., Arneth, A.: Assessing the impacts of agricultural managements on soil carbon stocks, nitrogen loss, and crop production–a modelling study in eastern Africa. Biogeosciences **19**(8), 2145–2169 (2022)
5. Sun, W., Canadell, J.G., Yu, L., Yu, L., Zhang, W., Smith, P., Fischer, T., Huang, Y.: Climate drives global soil carbon sequestration and crop yield changes under conservation agriculture. Global Change Biol. **26**(6), 3325–3335 (2020)
6. Akhter, F., Khadivizand, S., Lodyga, J., Siddiquei, H.R., Alahi, M.E.E., Mukhopadhyay, S.C.: December. design and development of an IoT enabled pedestrian counting and environmental monitoring system for a smart city. In: 2019 13th International Conference on Sensing Technology (ICST), pp. 1–6. IEEE (2019)
7. Oechaiyaphum, K., Ullah, H., Shrestha, R.P., Datta, A.: Impact of long-term agricultural management practices on soil organic carbon and soil fertility of paddy fields in Northeastern Thailand. Geoderma Reg. **22**, e00307 (2020)
8. Al-Shammary, A.A.G., Kouzani, A.Z., Kaynak, A., Khoo, S.Y., Norton, M., Gates, W.: Soil bulk density estimation methods: a review. Pedosphere **28**(4), 581–596 (2018)
9. Li, Y., Li, Z., Cui, S., Zhang, Q.: Trade-off between soil pH, bulk density and other soil physical properties under global no-tillage agriculture. Geoderma **361**, 114099 (2020)
10. Lequeue, G., Draye, X., Baeten, V.: Determination by near infrared microscopy of the nitrogen and carbon content of tomato (Solanum lycopersicum L.) leaf powder. Sci. Rep. **6**(1), 1–9 (2016)
11. Joe, H.E., Zhou, F., Yun, S.T., Jun, M.B.: Detection and quantification of underground CO_2 leakage into the soil using a fiber-optic sensor. Opt. Fiber Technol. **60**, 102375 (2020)
12. Cioffi, A., Mancini, M., Gioia, V., Cinti, S.: Office paper-based electrochemical strips for organophosphorus pesticide monitoring in agricultural soil. Environ. Sci. Technol. **55**(13), 8859–8865 (2021)
13. Ali, M.A., Dong, L., Dhau, J., Khosla, A., Kaushik, A.: Perspective—electrochemical sensors for soil quality assessment. J. Electrochem. Soc. **167**(3), 037550 (2020)
14. Schmidt, S.A., Ahn, C.: Predicting forested wetland soil carbon using quantitative color sensor measurements in the region of northern Virginia, USA. J. Environ. Manage. **300**, 113823 (2021)
15. Stiglitz, R.Y., Mikhailova, E.A., Sharp, J.L., Post, C.J., Schlautman, M.A., Gerard, P.D., Cope, M.P.: Predicting soil organic carbon and total nitrogen at the farm scale using quantitative color sensor measurements. Agronomy **8**(10), 212 (2018)
16. Reddy, A.D., et al.: Quantifying soil carbon loss and uncertainty from a peatland wildfire using multi-temporal LiDAR. Remote Sens. Environ. **170**, 306–316 (2015)
17. Tóth, S.F., Oken, K.L., Stawitz, C.C., Andersen, H.E.: Optimal survey design for forest carbon monitoring in remote regions using multi-objective mathematical programming. Forests **13**(7), 972 (2022)
18. Yunus, M.A.M., Mukhopadhyay, S.C.: Novel planar electromagnetic sensors for detection of nitrates and contamination in natural water sources. IEEE Sens. J. **11**(6), 1440–1447 (2010)

19. Mukhopadhyay, S.C.: A novel planar mesh-type microelectromagnetic sensor. Part I. Model formulation. IEEE Sens. J **4**(3), 301–307 (2004)
20. Mukhopadhyay, S.C.: A novel planar mesh-type microelectromagnetic sensor. Part II. Estimation of system properties. IEEE Sens. J. **4**(3), 308–312 (2004)
21. Zhang, W., Xiong, H., Wang, S., Li, M., Gu, Y.: Electromagnetic characteristics of carbon nanotube film materials. Chin. J. Aeronaut. **28**(4), 1245–1254 (2015)
22. Alfonso, M., Yuan, J., Tardani, F., Neri, W., Colin, A., Poulin, P.: Absence of giant dielectric permittivity in graphene oxide materials. J. Phys. Mater. **2**(4), 045002 (2019)
23. Hong, X., Yu, W., Chung, D.D.L.: Electric permittivity of reduced graphite oxide. Carbon **111**, 182–190 (2017)
24. Enang, R.K., Yerima, B.P.K., Kome, G.K., Van Ranst, E.: Assessing the effectiveness of the Walkley-Black method for soil organic carbon determination in tephra soils of cameroon. Commun. Soil Sci. Plant Anal. **49**(19), 2379–2386 (2018)
25. Schulte, E.E., Hopkins, B.G.: Estimation of soil organic matter by weight loss-on-ignition. Soil Organ. Matter: Anal. Interpret. **46**, 21–31 (1996)
26. Clemens, C., Radschun, M., Jobst, A., Himmel, J., Kanoun, O.: Detection of density changes in soils with impedance spectroscopy. Appl. Sci. **11**(4), 1568 (2021)
27. Sun, F., Chen, Z., Bai, X., Wang, Y., Liu, X., He, B., Han, P.: Theoretical and experimental bases for the equivalent circuit model for interpretation of silty soil at different temperatures. Heliyon e12652 (2022)
28. Akhter, F., Nag, A., Alahi, M.E.E., Liu, H., Mukhopadhyay, S.C.: Electrochemical detection of calcium and magnesium in water bodies. Sens. Actuat. A **305**, 111949 (2020)
29. Akhter, F., Siddiquei, H.R., Mukhopadhyay, S.C.: A graphene-based composite for selective detection of ethylene at ambient environment. In: 2021 IEEE Sensors, pp. 1–4. IEEE (2021)
30. Akhter, F., Siddiquei, H.R., Alahi, M.E.E., Mukhopadhyay, S.C.: Design and development of an IoT-enabled portable phosphate detection system in water for smart agriculture. Sens. Actuat. A **330**, 112861 (2021)
31. Akhter, F., Siddiquei, H.R., Alahi, M.E.E., Mukhopadhyay, S.C.: An IoT-enabled portable sensing system with MWCNTs/PDMS sensor for nitrate detection in water. Measurement **178**, 109424 (2021)

Performance Evaluation of Interdigitated Electrodes for Electrochemical Detection of Nitrates in Water

Kartikay Lal(✉), Tinu Thomas, and Khalid Arif

Department of Mechanical and Electrical Engineering, SF&AT, Massey University, Auckland 0632, New Zealand
{k.lal,k.arif}@massey.ac.nz

Abstract. Water is an indispensable natural resource necessary for the survival of life on Earth which is why it is crucial to maintain the quality of our most precious natural resource. Different physical, chemical, and biological parameters of water are continuously monitored to assess the quality of water. Nitrate is one of the toxic contaminants in water that is harmful to aquatic life as well as human health. There are many ways of detecting nitrates in water, but electrochemical analytical technique of detection has been the most common combined with the use of interdigitated electrodes due to ease of use, cost, and portability. This research assesses the performance of two types of electrodes such as copper on fiberglass substrate, and silver electrodes on acrylic sheet with different concentrations of Nitrates in deionized water. The measurements were taken using a LCR meter.

Keywords: Electrodes · interdigitated · water quality · nitrates · LCR meter

1 Introduction

Natural water resources are getting polluted every day by different activities, both natural and manmade. According to the survey of National Rivers Water Quality Network (NRWQN), the water quality of the New Zealand rivers is declining [1]. It is found that there is a significant rise in the level of nitrogen (nitrate or nitrite), and overall phosphorous levels in the rivers [2]. The main reason for the degradation of water quality is the expelling of insecticides and pesticides from agricultural lands that are rich in phosphorous and nitrogen with other reasons involving animal waste, discharge of harmful chemicals from industries and urban wastewater [3]. Excessive levels of nitrate in water would cause imbalance to the aquatic ecosystem as well as adversely affect human health.

Research shows that 70% of the New Zealand Rivers had exceeded the safe nitrate levels according to ANZG 2018 [4]. Urbanization, draining of wetlands, modifications of rivers alter normal functioning of river ecosystem. In recent years, the nitrate levels in river water have increased by 29%. It was also reported that nitrate concentrations are a lot higher in urban areas compared to rural [5]. WHO suggests that the maximum

N. K. Suryadevara et al. (Eds.): ICST 2022, LNEE 1035, pp. 407–413, 2023.
https://doi.org/10.1007/978-3-031-29871-4_41

contamination level of nitrate in water is 50 ng/L [6] whereas, the Environmental Protection Agency (EPA) of America suggested 45 mg/L [7]. In New Zealand, the level of nitrate in our water sources is getting worse [8] which is a growing concern. Therefore, continuous monitoring of nitrate levels in our rivers is crucial. Figure 1 shows nitrate levels in New Zealand rivers.

This paper evaluates the performance of three different patterns of electrodes named interdigitated, square spiral, and serpentine and three different materials of electrodes. Copper and silver electrodes were fabricated in house while the gold electrodes were procured. Copper electrodes were prepared on Massey University's LPKF PCB fabricator on fiberglass substrate and laser etched ridges filled with conductive silver paste. The three patterns that were prepared are shown below.

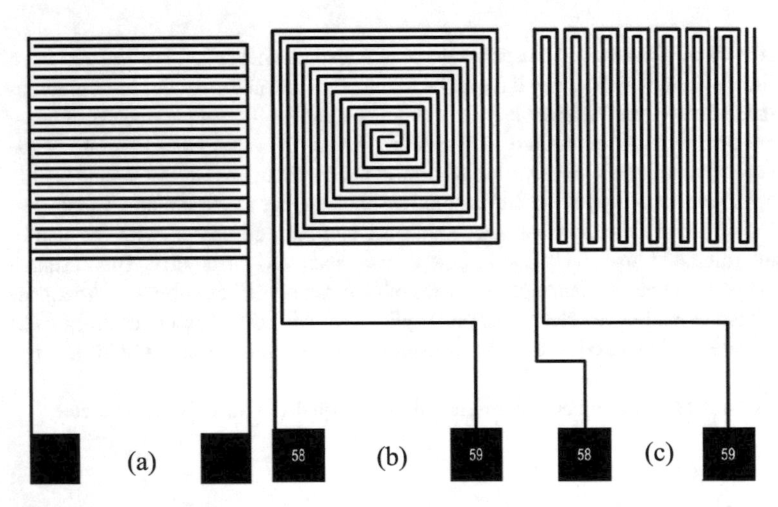

Fig. 1. Three different types of electrode patterns. Interdigitated (a), Square spiral (b), Serpentine (c)

1.1 Operating Principle of Interdigitated Sensor

Interdigitated sensors comprise of parallel electrodes constructed from electrically conductive material on a solid or flexible substrate [9]. Interdigitated planar electrodes comprise of two primary terminals that act as a parallel plate capacitor. These electrodes are usually placed close to each other, with pitch typically ranging from a few µm to nm. When a time-dependent voltage signal is applied to two parallel electrodes, an electric field is generated which protrudes from the electrodes [10]. This electric field can be used to determine the capacitance between the two electrodes using a LCR meter for instance. When an external material comes in close proximity to the electrodes, the electric field between the electrodes gets affected by the physical and chemical properties of the materials which alters the electric field [11]. This change in the electric field can be used to compute the capacitance which can be used to analyze the properties of the material.

Copper, silver, and gold electrodes are the three materials used for nitrate sensing in this paper. Since, copper is easily oxidized, a thin layer of Polydimethylsiloxane (PDMS) is coated on the electrodes to keep the metal from oxidation due to moisture and faradic currents.

2 Design of Proposed System

The sensors were designed and tested to analyze the nitrate concentration in 100 mL of tap water. The sensing area of the sensor is dipped in water and the impedance measurements were taken using GWINSTEK LCR-6000 precision LCR meter when the electrodes were submerged in water for 120 s. The electrodes were fabricated using copper and silver which are both excellent electrically conductive materials. Copper electrodes were manufactured on fiberglass substrate, also known as FR4 through the milling process. The process is graphically illustrated in Fig. 3. Silver electrodes were prepared using HPS-021LV screen printing silver ink from Novacentrix Metalon as illustrated in Fig. 2. Gold electrodes with an interdigitated pattern were also used which were purchased as shown in Fig. 4 (Fig. 5).

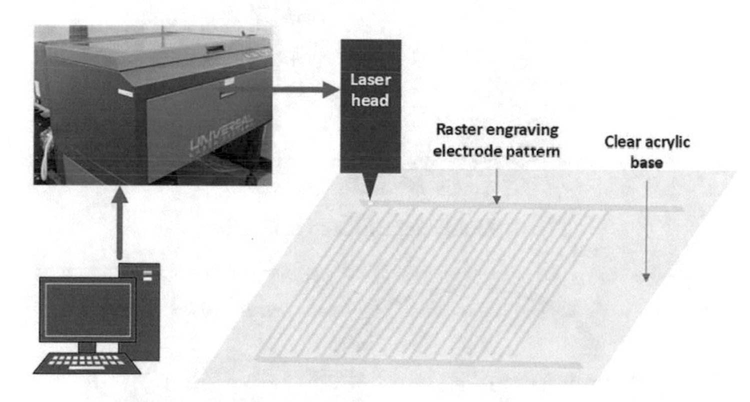

Fig. 2. Graphical representation of laser engraving on clear acrylic base.

The sample solutions were prepared using a mixture of potassium nitrate (KNO_3) and water. KNO_3 was procured from Sigma Aldrich has a concentration of 0.01 M of nitrate ions (NO_3^-). Using a pipette, 0.05 mL of nitrate solution was added to a test tube and added 125 mL of tap water to dilule the nitrate solution. The readings from the LCR meter were recorded after the sensor had been submerged in the solution for 120 s. For the other three tests, additional water added, while the volume of nitrate was kept the same.

As mentioned earlier, PDMS mixture was coated onto the electrodes which was a combination of silicone elastomer base and curing agent in the 10:1 ratio. A spin coater was used to uniformly spread PDMS mixture over the electrodes with the spin speed of 5000 RPM for 300 s, followed by placing the electrodes in an oven with hot bake function at 100 °C for 70 min.

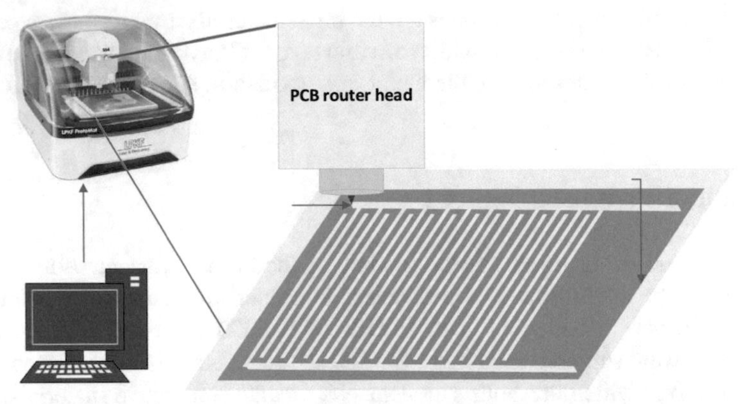

Fig. 3. Graphical design of a CNC PCB router, showing milling of thin interdigitated electrodes. The PCB router head holds the cutting tool which has a tip diameter of 0.2 mm.

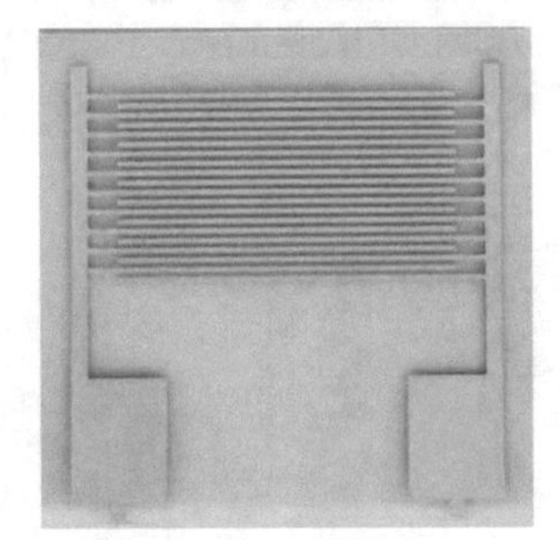

Fig. 4. Gold interdigitated electrodes

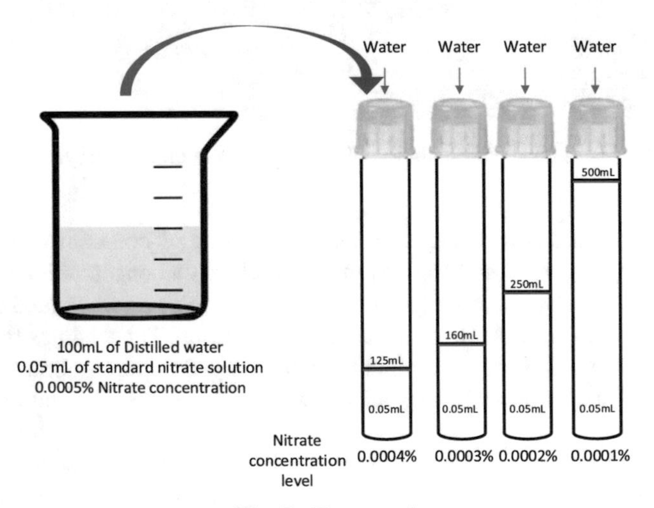

Fig. 5. Test samples

3 Results

The sensor was designed and tested initially to analyze the nitrate concentration in 100 mL of tap water. Initially, sensors were tested with and without PDMS. The sensing part of the sensor is dipped in water for 120 s and then the impedance was measured using LCR meter (Table 1).

Table 1. Impedance measurement in tap water

	Impedance (Z)
Bare electrode	1436.78
PDMS coated electrode	2873.56

3.1 Bare Electrode Testing

The bare electrode was dipped in two different samples with varying concentrations of nitrate solution and impedance values were recorded (Table 2).

Table 2. Impedance measurement in tap water

Water (mL)	Nitrate conc. (%)	Sample 1 imp. (Ω)	Sample 2 imp. (Ω)
100	0.0005	1811.59	1834.24
125	0.0004	1785.71	1824.82
160	0.0003	1760.56	1798.56
250	0.0002	1748.25	1785.71
350	0.0001	1724.14	1773.05

3.2 PDMS Coated Electrode Testing

A similar process to bare electrode testing was carried out for PDMS coated electrodes. Two samples of water with varying concentrations of nitrate were tested (Table 3).

As seen from Fig. 6 and Fig. 7, there is a significant difference between the impedance measurements between bare electrodes and PDMS coated electrodes. This is because the thin coating of PDMS prevents direct physical interaction between the nitrate ions and the surface of gold electrodes. It is also noticeable that impedance varies linearly with nitrate concentration.

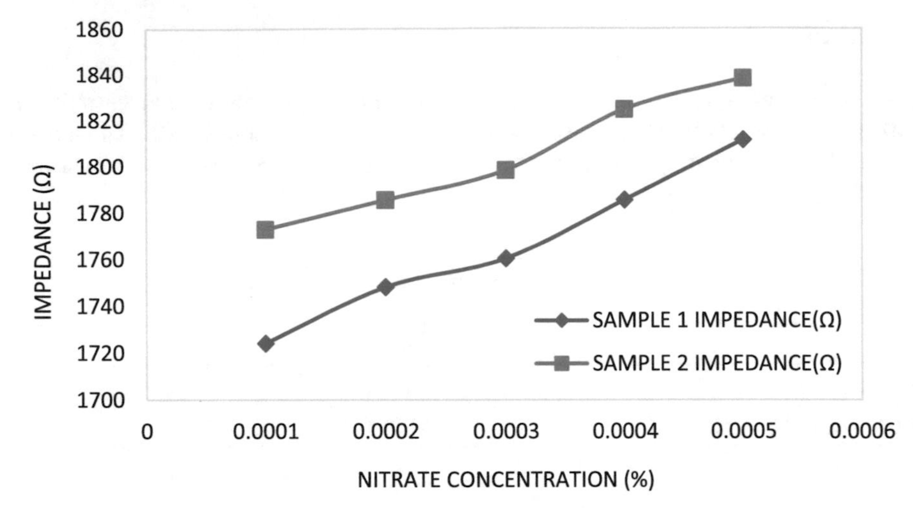

Fig. 6. Comparison between two samples with bare gold electrodes.

Table 3. Impedance measurement in tap water

Water (mL)	Nitrate conc. (%)	Sample 1 imp. (Ω)	Sample 2 imp. (Ω)
100	0.0005	4541.14	4385.96
125	0.0004	4098.36	4166.67
160	0.0003	4032.25	4098.36
250	0.0002	3968.25	4032.25
350	0.0001	3787.71	3968.25

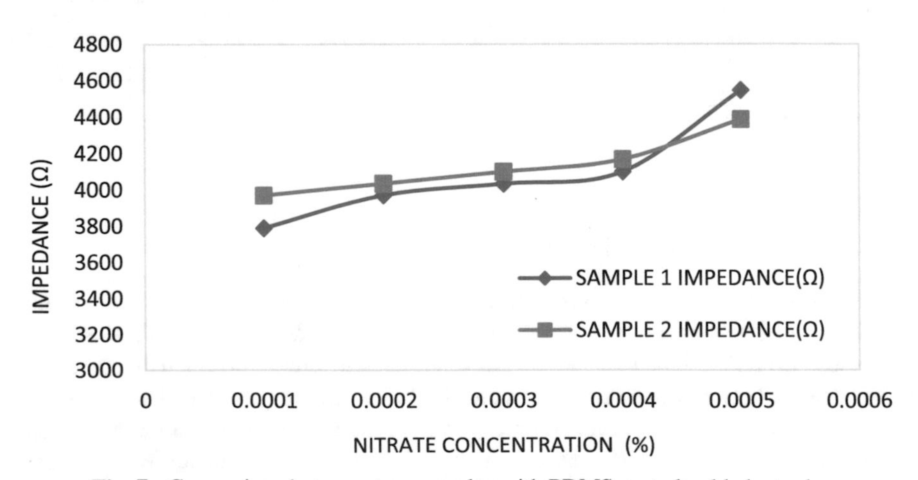

Fig. 7. Comparison between two samples with PDMS coated gold electrodes.

4 Conclusion

This research focuses on developing a low cost, portable nitrate sensor for water. The interdigitated gold electrodes are used as the sensor. As seen from the results, electrochemical sensor is efficient to measure nitrate levels in water. PDMS coated electrodes are found to be reusable, stable, and more precise than bare electrodes due to their ability of producing similar results at different times. The electrochemical sensor using the interdigitated gold electrodes are also found to be efficient to detect small variations in nitrate content in water, which can also produce stable results.

Acknowledgments. This work was supported by the Massey University Strategic Research Excellence Fund 2020 (RM22934) and the New Zealand Product Accelerator (RM23459).

References

1. Ballantine, D.J., Davies-Colley, R.J.: Water quality trends in New Zealand rivers: 1989–2009. Environ. Monit. Assess. **186**(3), 1939–1950 (2013). https://doi.org/10.1007/s10661-013-3508-5
2. Dymond, J., Ausseil, A.-G., Parfitt, R., Herzig, A., McDowell, R.: Nitrate and phosphorus leaching in New Zealand: a national perspective. N. Z. J. Agric. Res. **56**(1), 49–59 (2013)
3. https://www.doc.govt.nz/about-us/statutory-and-advisory-bodies/nz-conservation-authority/publications/protecting-new-zealands-rivers/02-state-of-our-rivers/water-quality/#17
4. https://www.stats.govt.nz/indicators/river-water-quality-nitrogen
5. Gadd, J., Snelder, T., Fraser, C., Whitehead, A.: Current state of water quality indicators in urban streams in New Zealand. NZ J. Mar. Freshwat. Res. **54**(3), 354–371 (2020)
6. Who, G.: Guidelines for drinking-water quality. World Health Organ. **216**, 303–304 (2011)
7. Kolpin, D.W., Burkart, M.R., Thurman, E.M.: Herbicides and nitrate in near-surface aquifers in the midcontinental United States, 1991, US Government Printing Office (1994)
8. Thomson, B., Nokes, C., Cressey, P.: Intake and risk assessment of nitrate and nitrite from New Zealand foods and drinking water. Food Addit. Contam. **24**(2), 113–121 (2007)
9. Igreja, R., Dias, C.: Analytical evaluation of the interdigital electrodes capacitance for a multi-layered structure. Sens. Actuat. A **112**(2–3), 291–301 (2004)
10. Olthuis, W., Streekstra, W., Bergveld, P.: Theoretical and experimental determination of cell constants of planar-interdigitated electrolyte conductivity sensors. Sens. Actuat. B Chem. **24**(1–3), 252–256 (1995)
11. Yunus, M.A.M., Mukhopadhyay, S.C.: Novel planar electromagnetic sensors for detection of nitrates and contamination in natural water sources. IEEE Sens. J. **11**(6), 1440–1447 (2010)

Coupled Monopole Antenna Array for IoT Based Smart Home Devices and Sensors

Ilyas Saleem[1](✉), Syed Muzahir Abbas[1], Subhas Mukhopadhyay[1], and M. A. B. Abbasi[2]

[1] School of Engineering, Faculty of Science and Engineering, Macquarie University Sydney, Sydney, NSW 2109, Australia
ilyas.saleem@hdr.mq.edu.au, {syed.abbas, subhas.mukhopadhyay}@mq.edu.au

[2] Centre for Wireless Innovation, Institute of Electronics, Communications and IT, Queen's University Belfast, Belfast, UK
m.abbasi@qub.ac.uk

Abstract. A compact, dual-band 1 × 2 coupled monopole printed antenna array design is proposed in this paper for indoor Internet of Things (IoT) based smart home devices and sensors. With an overall dimension of 30 × 70 × 1.5 mm^3, the presented design is operational at two frequency bands ranging from 3.6 to 4.3 GHz and from 5.6 to 6.6 GHz with stable impedance matching. The coupled planar monopole configuration ensures a consistent radiation performance over wide bandwidth. To verify its practical implementation on an IoT platform's device, the effects of dielectric height and ground plane variation on impedance matching over a wide frequency of operation are also discussed.

Keywords: Coupled Monopole · Printed Antenna · Mutual Coupling · Ground Plane Configuration · IoT · Smart Home

1 Introduction

When multiple sensors, objects, nodes or devices connects wirelessly to each other and to an internet-enabled network, their online accessibility for a user becomes possible. Such devices are then known as the Internet of Things (IoT) [1, 2]. The multi-layer service which allows users to manage, monitor, and automate IoT is called IoT platform [3, 4]. Through IoT platforms, multiple devices collaborate and interact with each other [5] to collect and analyze data; to optimize and predict patterns [6]. Within the next 2 years, there will be more than 80 billion IoT connections [7] resulting in a better-connected world. Millions of these connections will originate from smart home devices and with many IoT based wireless hardware systems; integrated sensors and antennas will be an integral element for optimum performance.

Printed antennas have been extensively used in wireless communication systems. They are simple in structure, small, easy to fabricate, economical to produce, exhibit reasonable gain & have reliable radiation performance, generally for a wide bandwidth

N. K. Suryadevara et al. (Eds.): ICST 2022, LNEE 1035, pp. 414–420, 2023.
https://doi.org/10.1007/978-3-031-29871-4_42

[8–10]. Implementation of multiband printed antennas in smart home devices is always preferred in order to support 5G, LTE MIMO, Wi-Fi, Bluetooth, and Zigbee technologies simultaneously [11, 12]. On the other hand, due to their compact size; smart home devices has limited space for antenna installation. This proximity of the antenna with electronic circuitry results in cross-coupling and degradation of performance. It is desired to reduce this coupling effect and subsequently reduce the co-channel interferences [13]. Multiple antenna configurations have been proposed to fulfill the demands of higher data rates over multiple wireless technologies [14–16], however, the stability of antenna elements in the presence of other electronic components and relatively high mutual coupling between the radiating surfaces is still one of the major issue revolving around IoT services and applications [17, 18]. Also, co-channel interferences has direct impact on data rate performance because multiple radiating elements operates over adjacent frequencies [19].

In this paper, a dual-band 1×2 printed antenna is presented. The proposed antenna is relatively small and has sufficient bandwidth and gain over the entire operating band which is achieved using coupled monopole technique. Two planar radiating monopoles facing each other are simultaneously excited via a transmission line, making antenna structure. The coupling is prominent along the vertical axis. When two copies of the same antennas are replicated in the horizontal axis, reduction in cross-coupling between the antennas is realized. Implementation of the proposed array structure in an IoT device is also verified by detailed full-wave electromagnetic analysis.

2 Antenna Design

Selection of dielectric substrate and resonant frequency are key design elements as they effect the antenna's performance and dimensions.

The milimeterWave spectrum of 5G offers remarkable data transfer speed with least delays but only at shorter distances and if there are no path obstructions. Whereas Sub-6GHz networks provides farther coverage but with a limited speed. To exploit the best features of both frequency bands while overcoming their shortcomings, C band (4–8 GHz) is preferred by the domestic device manufactures as it has a practical sweep of mid-range frequencies which allows 5G experience at longer distances. Therefore, resonant frequency (f_r) of C band (4 GHz) was selected for this design. Similarly, the planar monopole and transmission line is devised on FR4 substrate with dielectric constant (ε_r) of 4.8 and height (H) of 1.5 mm. FR4 substrate exhibits extraordinary mechanical strength and electrical insulation characteristics over high radio frequencies in humid & dry environments, which is why it is widely used in smart home hardware.

f_r, ε_r and H were used in the formulas listed in [20] to calculate the practical width and length of radiating element. Configuration of proposed design and its dimensions are shown in Fig. 1.

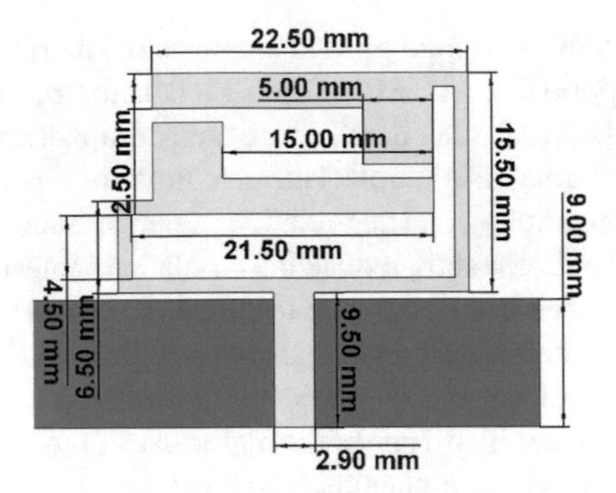

Fig. 1. Antenna Geometry

A quasi-rectangular antenna structure was simply fed by a 50 Ω microstrip feed line (outlined in yellow color) on the front of substrate. To achieve the desired results for intended IOT applications, several design iterations were performed comprised of appending different current paths for enhanced surface current density. During first iteration, two slots were introduced, one at the bottom and the other at the center of radiating element for band notching and gain enhancement. Then, two vertical slots were added to control the center frequency. Later on, current's path was truncated by two horizontal slits which led to the dual-band functionality. Also partial ground plane (outlined in green color) was incorporated on the backside of substrate for bandwidth enhancement, while impedance of the planar monopole is controlled for simultaneous matching in two frequency bands.

All these design iterations were performed in CST Studio Suite, where each parameter was fine-tuned through simulations. In the next section, a contemporary parametric and array configuration study is given to support the arguments made in this paper.

3 Analysis and Results

3.1 Effect of Substrate Dimensions

Height of substrate (H = 1.5) is added ~5 times with the length and width of the patch to have the length and width of substrate respectively [21]. We have modified this assumption for our analysis as given below:

$$Substrate\ Length = Patch\ Length + 3.3 \times H \tag{1}$$

$$Substrate\ Width = Patch\ Width + 6.6 \times H \tag{2}$$

Figure 2 shows the effect of substrate dimensions on impedance bandwidth when H is varied with a step size of 1 mm. It was observed that resonating bands in lower frequencies were negatively affected when H increased from 1.5, whereas the higher frequency resonant bands started shifting towards lower frequencies with an overall reduced operating bandwidth but with improved return loss.

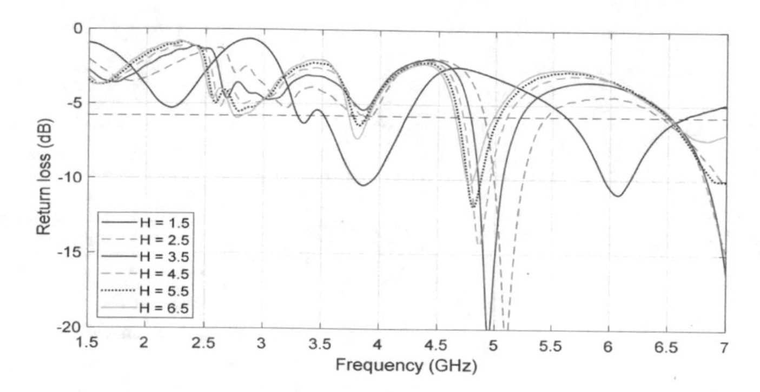

Fig. 2. Frequency responses for different H values

3.2 2 × 1 Array Configuration

The proposed antenna shows reasonable characteristics to be employed as an array antenna. The major issue in multi-antenna configurations is the mutual coupling. With high coupling values, communication between IoT hardware and the remote access point becomes weaker since radiating elements would have more interference.

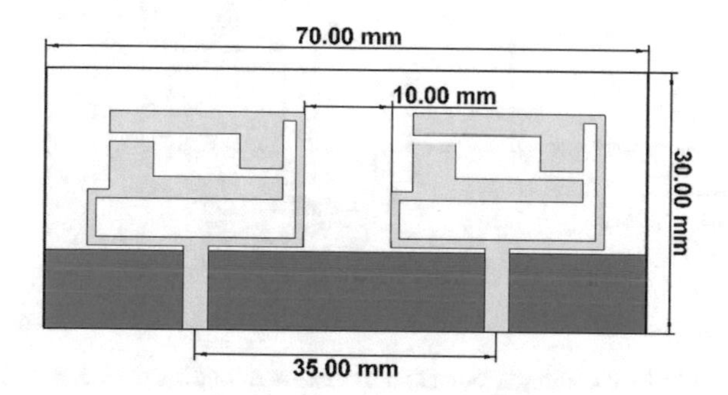

Fig. 3. Antenna Array Geometry

The distance at which radiating elements are placed is critical. Increasing the distance between them rectifies the coupling problem but introduces issues related to device footprint, since portable devices always have limited space to fit antenna into. To cater this dilemma, the distance (in terms of λ, i.e., wavelength at the highest frequency of operation) between the transmission lines is kept constant to 35 mm as shown in Fig. 3. Three ground plane configurations namely, Connected ground, Extended ground and Separated ground shown in Fig. 4 are applied to investigate low-complexity mutual-coupling reduction options selectable depending upon the IoT platform. From Fig. 4, it can be deduced that in connected and extended ground configurations, array behavior in terms of resonating frequency was similar to single antenna element but return loss for second band improved in the connected ground structure. Whereas, in separated ground configuration, return loss improved for both frequency bands.

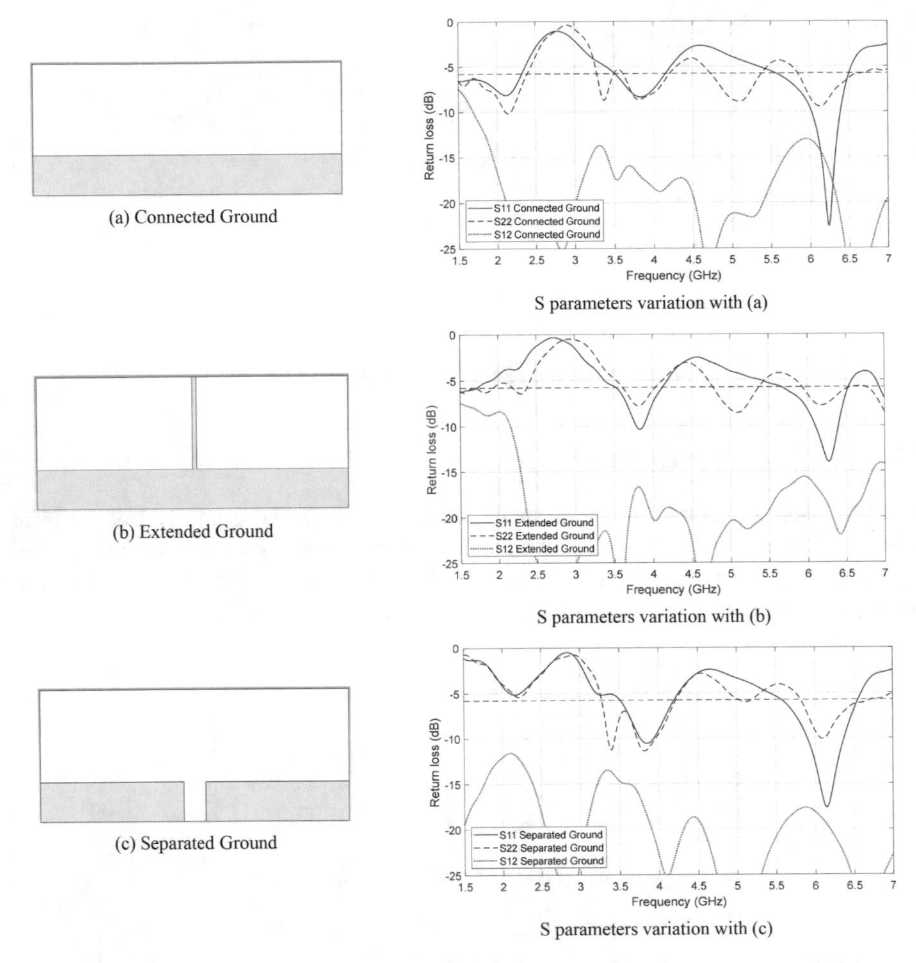

Fig. 4. Ground Plane Configurations and their corresponding S parameters variations

Note that MIMO configuration is also implementable if separate ground structure (Fig. 4c) is used with which low coupling of the order of <-12 dB can be expected.

Radiation Efficiency, 3 dB Beamwidth, Gain and Total Isotropic Sensitivity (TIS) of the antennas in array configuration are illustrated in Fig. 5. First, each radiating element was excited separately while the second element was terminated to 50 Ω load, and then both were excited simultaneously to confirm that the proposed design is radiating with reasonable radiation efficiency. 3 dB Beamwidth and Gain are also adequate over the entire operational frequency range. TIS is receiving antenna's average sensitivity when measured across an entire 360° spherical coverage. The simulated TIS in both azimuth and elevation is favorable enough for the IoT devices to operate in any orientation.

Finally, the operational resilience of the antenna can be affirmed by observing the involvement of the coupled planar monopoles in maintaining the high radiation efficiency of antenna. The *E*-field was analyzed to validate the surface current density as shown

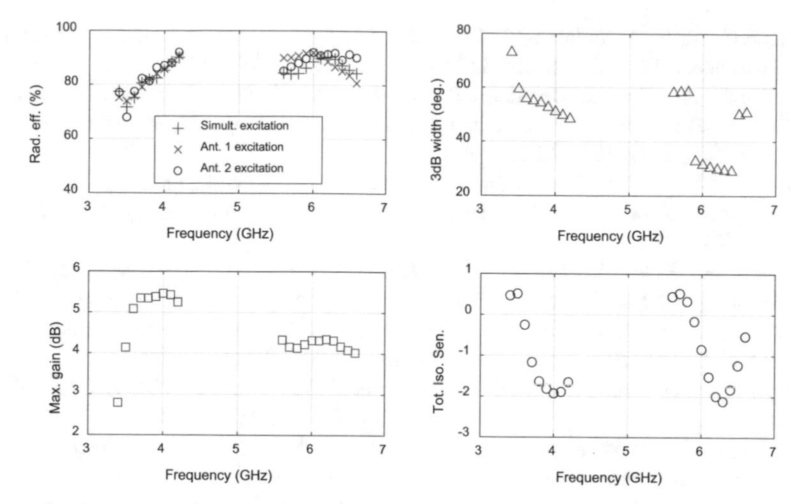

Fig. 5. Radiation Efficiency, 3 dB Width, Gain, Total Isotropic Sensitivity

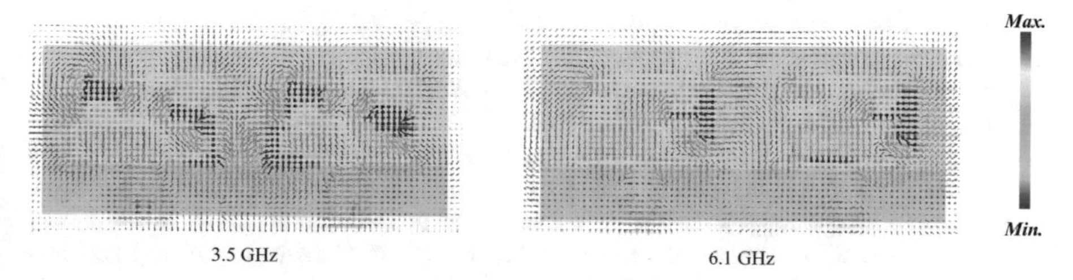

Fig. 6. E-field at 2 mm above the ground floor. Color scale mapped from 0–5000 V/m.

in Fig. 6. Here, it can be observed that for the lower frequency band, entire coupling monopole structure is assisting in the radiation, while smaller footprint of coupling monopoles is radiating at higher frequencies.

4 Conclusion

In this work, we proposed a microstrip line fed, coupled monopole printed antenna array operating over two bands which are suitable for almost all applications functional in C band; making this design a good solution to be integrated in an IoT based smart home devices and sensors. The mutual coupling was also analyzed, while three operations to reduce coupling are provided based on the IoT platform that the antenna is expected to be mounted on.

References

1. Lim, T.W.: The Internet of Things (IoT). In: Industrial Revolution 4.0, Tech Giants, and Digitized Societies, pp. 33–49. Springer, Singapore (2019). https://doi.org/10.1007/978-981-13-7470-8_3

2. Mahmood, A., et al.: Industrial IoT in 5G-and-beyond networks: vision, architecture, and design trends. IEEE Trans. Indust. Inform. **18**(6), 4122–4137 (2022)
3. Mijuskovic, A., Ullah, I., Bemthuis, R., Meratnia, N., Havinga, P.: Comparing apples and oranges in IoT context: a deep dive into methods for comparing IoT platforms. IEEE Internet Things J. **8**(3), 1797–1816 (2021)
4. Barros, T.G.F., Neto, E.F., Neto, J.A.D.S., De Souza, A.G.M., Aquino, V.B., Teixeira, E.S.: The anatomy of iot platforms - a systematic multivocal mapping study. IEEE Access **10**, 72758–72772 (2022)
5. Quadri, N.S., Yadav, K.: Efficient data classification for IoT devices using AWS kinesis platform. In: 21st Saudi Computer Society National Computer Conference, pp. 1–5 (2018)
6. Coutinho, R.W.L., Boukerche, A.: Transfer learning for disruptive 5G-enabled industrial Internet of Things. IEEE Trans. Indust. Inform. **18**(6), 4000–4007 (2022)
7. Juniper Research Whitepaper: IoT ~ The Internet of Transformation (2020). https://www.juniperresearch.com/whitepapers/iot-the-internet-of-transformation-2020
8. Chen, Y., et al.: Landstorfer printed log-periodic dipole array antenna with enhanced stable high gain for 5G communication. IEEE Trans. Anten. Propag. **69**(12), 8407–8414 (2021)
9. Nikkhah, A., Oraizi, H.: Efficient 2×2 printed circuit band-gap cavity-backed slot antenna. IEEE Trans. Anten. Propag. **68**(4), 2550–2556 (2020)
10. Li, Y., et al.: 3-D printed high-gain wideband waveguide fed horn antenna arrays for millimeter-wave applications. IEEE Trans. Anten. Propag. **67**(5), 2868–2877 (2019)
11. Cheng, K., Liang, Z., Hu, B.: Design of miniaturized microstrip antenna for smart home wireless sensor. In: IEEE 4th International Conference on Electronic Information and Communication Technology, pp. 234–236 (2021)
12. Göçen, C., Akdağ, İ., Palandöken, M., Kaya, A.: 2.4/5 GHz WLAN 4×4 MIMO dual band antenna box design for smart white good applications. In: 4th International Symposium on Multidisciplinary Studies & Innovative Technologies, pp. 1–5 (2020)
13. Kim, S., Nam, S.: A compact and wideband linear array antenna with low mutual coupling. IEEE Trans. Anten. Propag. **67**(8), 5695–5699 (2019)
14. Vishvaksenan, K.S., Mithra, K., Kalaiarasan, R., Raj, K.S.: Mutual coupling reduction in microstrip patch antenna arrays using parallel coupled-line resonators. IEEE Anten. Wireless Propag. Lett. **16**, 2146–2149 (2017)
15. Sharma, S.B., et al.: A novel u-slot aperture coupled annular-ring microstrip patch antenna for multiband GNSS applications. In: 14th European Conference on Antennas and Propagation, pp. 1–3 (2020)
16. Mukti, P., Farahani, H., Paulitsch, H., Bösch, W.: Mutual coupling reduction of aperture-coupled antenna array using UC-EBG superstrate. In: 13th European Conference on Antennas and Propagation, pp. 1–5 (2019)
17. Vamseekrishna, A., Madhav, B.T.P., Anilkumar, T., Reddy, L.S.S.: An IoT controlled octahedron frequency reconfigurable multiband antenna for microwave sensing applications. IEEE Sens. Lett. **3**(10), 1–4 (2019)
18. Abdulkawi, M., Sheta, A., Elshafiey, I., Alkanhal, A.: Design of low-profile single and dual-band antennas for IoT applications. In: MDPI RF/Microwave Circuits for 5G and Beyond: Electronics, vol. 22 (2021)
19. Qi, H., Liu, L., Yin, X., Zhao, H., Kulesza, W.J.: Mutual coupling suppression between two closely spaced microstrip antennas with an asymmetrical coplanar strip wall. IEEE Anten. Wireless Propag. Lett. **15**, 191–194 (2016)
20. Balanis, C.A.: Antenna Theory: Analysis and Design. Wiley (2016)
21. Wong, K.L.: Compact & Broadband Microstrip Antennas. Wiley (2004)

Flexible Tactile Sensors Based on 3D Printed Moulds

Aniket Chakraborthy[1,2], Suresh Nuthalapati[1,3], Rico Escher[1,3], Anindya Nag[1,2(✉)], and Memet Ercan Altinsoy[1,2]

[1] Faculty of Electrical and Computer Engineering, Technische Universität Dresden, 01062 Dresden, Germany
anindya1991@gmail.com
[2] Centre for Tactile Internet With Human-in-the-Loop (CeTI), Technische Universität Dresden, 01069 Dresden, Germany
[3] 6G-Life Research Hub, Technische Universität Dresden, 01062 Dresden, Germany

Abstract. The paper highlights the work done on the implementation of flexible sensors formed using 3D printing moulds. 3D printing process, being one of the state-of-the-art techniques, has been used to generate moulds. These moulds were subsequently used to develop the sensors with zinc oxide (ZnO) and polydimethylsiloxane (PDMS) as the processed materials. These two materials were considered due to their low cost, easy processing and enhanced electromechanical characteristics. The developed prototypes were characterized to determine their electrical performance with respect to mechanical deformations. The results for these sensors are promising and form a podium to form full-integrated tactile sensing systems.

Keywords: 3D printing · mould · PDMS · ZnO · Tactile

1 Introduction

The inclusion of sensors with the objects associated with daily life has revolutionized human life. The amount of time and energy required for the completion of a single task has been reduced to a great amount. Sensors differing in structure, dimension, working principle and applications have been developed and tested. Initially, after the popularization of semiconducting sensors [1, 2], silicon substrates-based prototypes have been fabricated and utilized for applications related to the complementary metal oxide semiconductor (CMOS) integrated circuits [3, 4]. Single crystal silicon elements have been mostly used to develop the substrates of the sensors. These prototypes were designed and developed using the microelectromechanical systems (MEMS) technique [5, 6]. Although these types of sensors were used for various industrial [7, 8] and environmental [9, 10] applications, certain limits deterred their further use. Some of the limitations of these silicon sensors included the high fabrication cost, degradation of sensitivity over time, the influence of ambient temperature and humidity on the response of the sensors and short life cycles [11]. This lets the researchers opt for other types of sensors. Flexible

N. K. Suryadevara et al. (Eds.): ICST 2022, LNEE 1035, pp. 421–430, 2023.
https://doi.org/10.1007/978-3-031-29871-4_43

sensors have been an alternative to perform similar applications with high efficiency [12, 13]. Some of the advantages of these sensors include low fabrication cost, high electrical conductivity, high mechanical flexibility, faster response to stimuli and easy customization of the sensory attributes [14, 15]. These sensors have been capable of performing a wide application spectrum compared to silicon sensors.

While the materials used to develop the silicon electrodes mostly included certain metallic nanoparticles like gold, platinum and chromium, the materials involved in devising the flexible sensors are much wider. The elements used to formulate the electrodes of the flexible sensors can be roughly divided into two categories, including carbon allotropes [16] and metallic nanomaterials [17]. While the former part includes elements like Carbon Nanotubes (CNTs) [18], graphene [19] and graphite [20], the latter has metallic nanowires [21] and nanoparticles [22]. Apart from these two categories, a third one consists of metallic oxides, which include certain metals like titanium, zinc, tin, copper, iron and others [23]. Among them, researchers have been largely preferring zinc oxide (ZnO) [24] over its counterparts due to their higher sensing response, long-term stability, lower power consumption, non-toxic nature and higher chemical stability [25]. Similarly, a lot of polymers have also been considered to form the substrates of the flexible sensors. These polymers can be categorized under natural, synthetic and semi-synthetic polymers. Some of the common ones include polydimethylsiloxane (PDMS) [26], polyethene terephthalate (PET) [27], polyimide (PI) [28], polyurethane (PU) [29] and poly(3,4-ethylenedioxythiophene) polystyrene sulfonate (PEDOT: PSS) [30]. Each of these polymers has been considered due to its mechanical integrity and flexibility, and ability to form strong interfacial bonds with the nanomaterials. Out of these polymers, PDMS has been the most popular one due to their hydrophobic nature, biocompatibility, low cost and easy customization [31]. This paper highlights using ZnO and PDMS as processed materials to form flexible sensors.

The processing of the nanomaterials and polymers to form flexible sensors has been done using a varied range of printing techniques. Some of the common printing techniques include screen printing [32], inkjet printing [33], 3D printing [34] and laser ablation [35]. Even though all of these types are well appreciated for developing flexible sensors, 3D printing has been one process that has transformed the outlook of sensor fabrication. 3D printing not only reduces the prototyping time but also minimizes electronic waste and makes the sensing system cost-effective. The use of 3D-printed moulds for forming flexible sensors is an extension of the 3D printing process [36]. This technique is also advantageous as the degree of customization and ability to integrate the raw materials is very high. The fabrication of the ZnO/PDMS-based sensors shown in this paper is done using the 3D printed moulds as the base templates. In terms of application of the sensors, the sensors are broadly used in three categories, including electrochemical [37], strain [38] and electrical [39] sensing applications. The novelty of this paper lies in the fabrication of novel resistive sensors using 3D-printed moulds. Also, this is the first instance where the chosen printing filament was able to generate sensors with high robustness and mechanical flexibility. Due to the strain-induced nature of the sensors, they were used for tactile sensing applications. The paper has been organized as follows. After the introduction to sensors, processed materials and fabrication technique, sections two and three elucidates the fabrication process and its responses to mechanical characterization and tactile-sensing applications. The conclusion is drawn in the final section of the paper.

2 Fabrication of the Sensing Prototypes

The fabrication process of the 3D printed moulds was carried out in the laboratory with constant temperature (25 °C) and humidity (RH 50%) conditions. The process started with pre-heating the printing device (Prusa i3 MK3S +) for around ten minutes. Although the default diameter of the nozzle was 0.4 mm, customization was done on the diameter for higher precision and better finish. The value of the altered diameter of the nozzle was 0.25 mm. Among the different kinds of the available printing filament, some of the common ones are polylactic acid (PLA) [40], acrylonitrile butadiene styrene (ABS) [41] and polyethylene terephthalate glycol (PETG) [42]. Among these, PETG was chosen for forming the moulds due to its advantages of high chemical resistance, excellent durability and easy customization for manufacturing purposes. Figure 1 showcases the schematic diagram of the individual steps to develop the sensors. Initially, the moulds were formed with chosen design and had trenches with a depth of 300 microns.

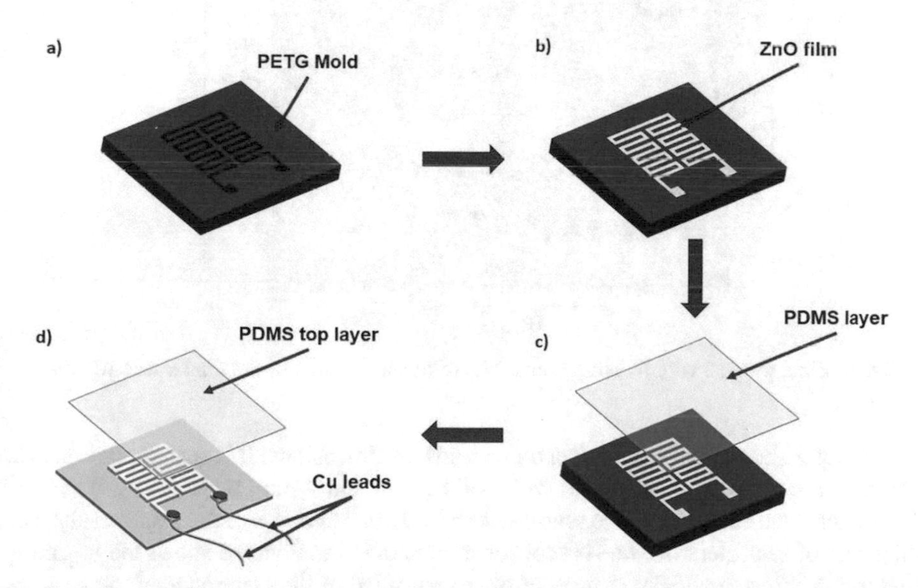

Fig. 1. Schematic diagram of the fabrication steps of the sensors. a) After the design of the sensors was printed on the PETG mould, b) ZnO was cast on the mould and copper leads were attached to them. c) Finally, a PDMS layer was cast on top of the mould and the samples were cured to form the prototypes.

Then, ZnO was cast on them and the additional ZnO nanopowder was removed from the moulds using a knife. Then, a layer of PDMS was cast on top of the mould. The PDMS ((SYLGARD ® 184, Silicon Elastomer Base) was formed by mixing the base polymer (pre-polymer) and curing agent (cross-linker) at a ratio of 10:1. The two parts were mixed uniformly and were cast on the mould. The height of this PDMS layer was adjusted to 800 microns with a casting knife (80 mm, EQ-Se-KTQ-80F). The samples were then desiccated for 30 min to remove the trapped air bubbles. Then, the samples were cured at 80 °C for 2 h to solidify the substrate. After the liquid PDMS was cast, it seeped through ZnO nanopowder on the trenches and formed nanocomposites. This assisted in solidifying the electrodes and attaching them to the PDMS substrates. The samples were then peeled off the PETG moulds. This was followed by forming the electrical contacts of the sensors by attaching copper leads (7440-50-8, 0.1 mm) to the bonding pads of the ZnO/PDMS sensors with the help of silver paste (735825-25G). Finally, the sensors were encapsulated with an additional PDMS layer to increase their overall robustness.

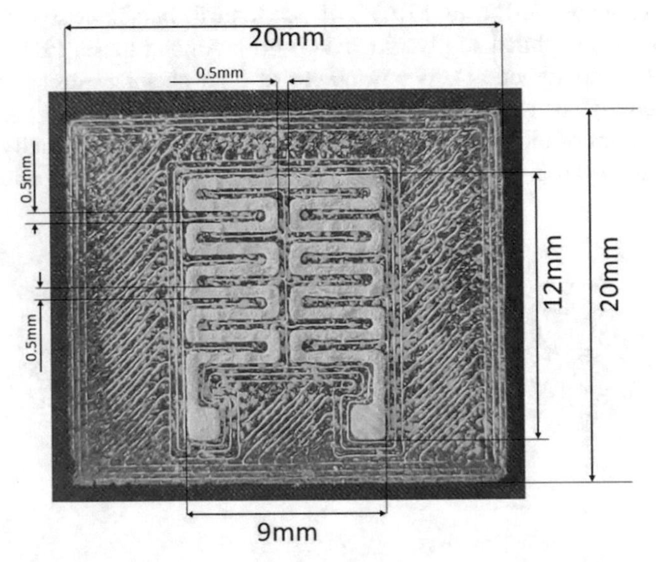

Fig. 2. Final product with its dimensions. The total surface area of the sensor was 108 mm^2.

Figure 2 shows the final product along with its dimensions. It is seen that the sensor is small in size, with a length and width of 12 mm and 9 mm, respectively. The total and effective surface area of the sensors were 400 mm^2 and 108 mm^2, respectively. The thickness of each electrode line is kept constant at 0.5 mm. Figure 3 shows the Scanning Electron Microscopy (SEM) image of the top-surface of the electrodes of the sensors. The images were taken with an input voltage of 10 kV. The thickness in the middle of the film was around 79.6 μm, with a roughness accuracy of 2 μm. It is seen that the PDMS matrix is present within the ZnO nanopowder, thus solidifying the electrodes. Figure 4 [43] shows the working principle of the sensors. The operation of these prototypes was based on the piezoresistive principle. In normal conditions, the ZnO nanopowder is located within the PDMS polymer matrix in a certain orientation. When pressure is exerted on the sensing area of the prototypes, the shape of the sensor changes. This simultaneously alters the orientation of the nanoparticles within the polymer matrix, leading to the change in response of the sensors. The sensor comes back to its original position after the pressure is lifted from the sensor. Some of the advantages of these sensors are their low fabrication cost, high durability and high dynamic range.

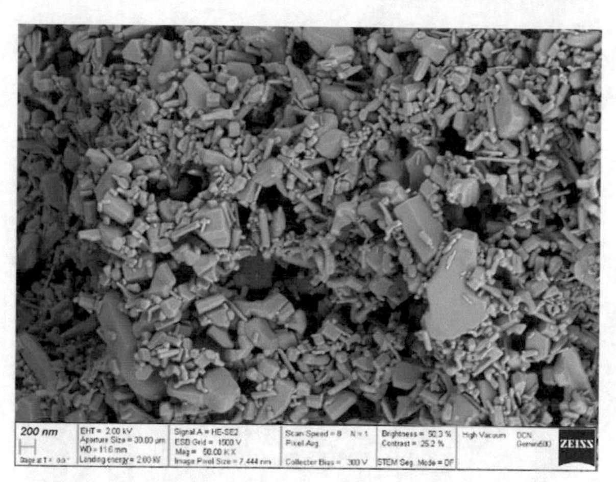

Fig. 3. Scanning electron microscopy (SEM) image of the top-surface of the electrodes of the developed sensors.

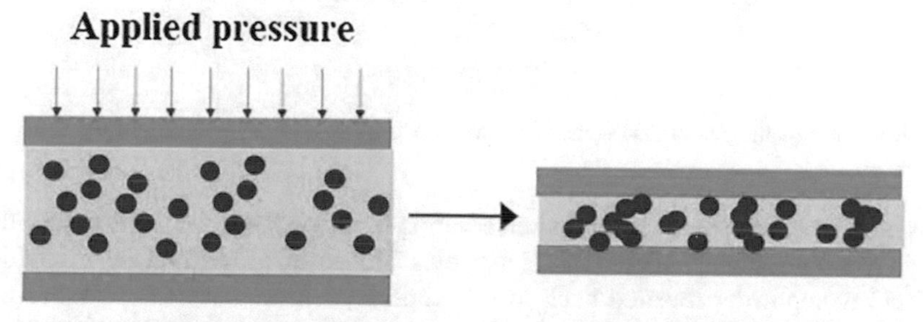

Applied pressure leads to the change in the orientation of the nanoparticles within the polymer matrix

Fig. 4. The working mechanism of these ZnO/PDMS-based sensors is based on the piezoresistive sensing principle [43].

3 Experimental Results

The experiments with these sensors were carried out in laboratory conditions. The measurements were taken with a HIOKI IM 3536 LCR meter with an input alternating voltage and frequency of 1 Vrms and 1 kHz, respectively. One end of the analyzer was connected to the sensor via Kelvin probes and the other end was connected to the laptop using a USB-USB cable. The data was collected in Microsoft Office Suite using an automatic data acquisition algorithm. The changes in the response of the sensors were determined in terms of relative impedance. The sensors were tested to detect the stability and reproducibility of their responses. Since the prototypes were flexible in nature, manual oscillatory bending of the sensors was carried out. Figure 5 shows the response of the sensors with the two positions given in its inset. The relative impedance values were reduced when the sensors were bent. This happens because as the sensors are bent, there is an increase in the effective connection between the nanopowder in the electrodes.

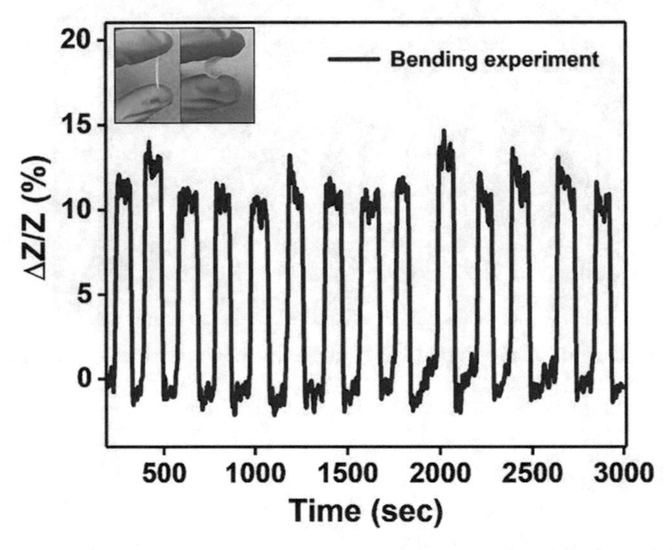

Fig. 5. Response of the ZnO/PDMS-based sensors towards oscillatory bending of the sensors.

This increases the overall current, thus reducing the impedance values. It is seen that the output was stable enough for the fourteen cycles. This result proves that the orientation of the ZnO nanopowder reverted back to its original position when the pressure was released every time. The sensors were then tested for tactile sensing applications, where pressure was exerted on the sensing area of the prototypes. Figure 6 shows the response of the sensors with respect to time. The pressure and contact area were kept almost constant during the entire process. Gloves were worn during the tactile process to negate the dielectric effect on the skin. Similar to the bendability test, the relative impedance values were reduced when the pressure was applied to the sensor. The glitches present in the response were caused due to a couple of possible reasons. The wires of the impedance analyzer could have been loose. It could have also been due to the fact that the copper wires were very thin, and maybe their connection with the Kelvin probes was not stable. Further work would be done on the electromechanical characteristics of the sensors and their application for tactile sensing applications.

A comparative study highlighting the materials used to develop the printed sensors, their advantages and applications have been provided in Table 1. It is seen that the proposed sensors are advantageous as given a combination of some of the materials and applications of other sensors. These prototypes can be considered for their simple fabrication, working principle and non-invasive nature.

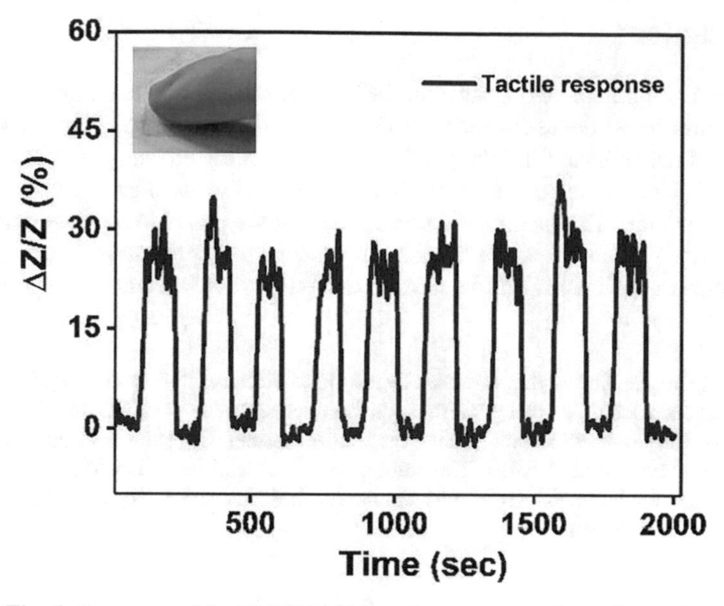

Fig. 6. Response of the ZnO/PDMS-based sensors towards tactile sensing.

Table 1. Comparison between the printed sensors in terms of processed materials, advantages and applications.

Materials	Advantages	Applications	Ref
Graphene, PDMS	High sensitivity, short response time, high cyclic stability	Tactile and pressure sensing	[44]
Graphene, Polyimide	Fast response and recovery, low detection limit and good cyclic stability	Force sensing	[45]
Graphene	High sensitivity and good reproducibility	Tactile sensing	[46]
TPU*	High flexibility	Tactile sensing	[47]
Carbon Nanotubes, PLA**	High directional sensitivity	Tactile sensing	[48]
CNTs, Ag NPs***, mulberry paper	High sensitivity, wide operating range		[49]
ZnO, PDMS	High cyclic stability and reproducibility, fast response	Tactile sensing	Proposed work

* TPU: Thermoplastic polyurethane, ** PLA: Polylactic acid, *** Ag NPs: silver nanoparticles

4 Conclusion

The paper highlights the work done on the design, development and utilization of 3D printed mould-based flexible sensors. ZnO nanopowder and PDMS polymer were used as processed materials to form the prototypes. Some of the advantages of these sensors include their fabrication cost, high mechanical flexibility and high reproducibility and stability of results. The tactile response of the prototypes also looks promising and validates the combination of the fabrication process and application. Further work on strain-induced applications will be carried out with these sensors and reported in the near future.

Acknowledgement. The work was funded by the German Research Foundation (DFG, Deutsche Forschungsgemeinschaft) as part of Germany's Excellence Strategy – EXC 2050/1 – Project ID 390696704 – Cluster of Excellence "Centre for Tactile Internet with Human-in-the-Loop" (CeTI) of Technische Universität Dresden. The authors are also thankful to Dr. Markus Löffler from Dresden Center for Nanoanalysis, Center for Advancing Electronics Dresden, TU Dreden, for providing the SEM images.

References

1. Frasco, M.F., Chaniotakis, N.: Semiconductor quantum dots in chemical sensors and biosensors. Sensors. **9**(9), 7266–7286 (2009)
2. Sze, S.M.: Semiconductor Sensors. John Wiley & Sons, Hoboken (1994)
3. Chung, S., Lee, T.: Towards flexible CMOS circuits. Nat. Nanotechnol. **15**(1), 11–12 (2020)
4. Pandey, A., Yadav, D., Singh, R., Nath, V.: Design of ultra low power CMOS temperature sensor for space applications. Int. J. Adv. Res. Electr. Electron. Instr. Energy **2**(8), 4117–4125 (1970)
5. Nag, A., Zia, A.I., Mukhopadhyay, S., Kosel, J.: Performance enhancement of electronic sensor through mask-less lithography. In: 2015 9th International Conference on Sensing Technology (ICST). IEEE (2015)
6. Tilli, M., Paulasto-Krockel, M., Petzold, M., Theuss, H., Motooka, T., Lindroos, V.: Handbook of Silicon Based MEMS Materials and Technologies. Elsevier, Amsterdam (2020)
7. Nag, A., Zia, A.I., Li, X., Mukhopadhyay, S.C., Kosel, J.: Novel sensing approach for LPG leakage detection: part I—operating mechanism and preliminary results. IEEE Sens. J. **16**(4), 996–1003 (2015)
8. Nag, A., Zia, A.I., Li, X., Mukhopadhyay, S.C., Kosel, J.: Novel sensing approach for LPG leakage detection—part II: effects of particle size, composition, and coating layer thickness. IEEE Sens. J. **16**(4), 1088–1094 (2015)
9. Alahi, M.E.E., Xie, L., Mukhopadhyay, S., Burkitt, L.: A temperature compensated smart nitrate-sensor for agricultural industry. IEEE Trans. Ind. Electron. **64**(9), 7333–7341 (2017)
10. Alahi, M.E.E., Mukhopadhyay, S.C., Burkitt, L.: Imprinted polymer coated impedimetric nitrate sensor for real-time water quality monitoring. Sens. Actu. B Chem. **259**, 753–761 (2018)
11. Han, T., Nag, A., Afsarimanesh, N., Mukhopadhyay, S.C., Kundu, S., Xu, Y.: Laser-assisted printed flexible sensors: a review. Sensors. **19**(6), 1462 (2019)
12. Afsarimanesh, N., Nag, A., Sarkar, S., Sabet, G.S., Han, T., Mukhopadhyay, S.C.: A review on fabrication, characterization and implementation of wearable strain sensors. Sens. Actu. A **315**, 112355 (2020)

13. Nag, A., Mukhopadhyay, S.C., Kosel, J.: Wearable flexible sensors: a review. IEEE Sens. J. **17**(13), 3949–3960 (2017)
14. Nag, A., Nuthalapati, S., Mukhopadhyay, S.C.: Carbon fiber/polymer-based composites for wearable sensors: a review. IEEE Sens. J. **22**, 10235–10245 (2022)
15. Nag, A., Simorangkir, R.B., Sapra, S., Buckley, J.L., O'Flynn, B., Liu, Z., et al.: Reduced graphene oxide for the development of wearable mechanical energy-harvesters: a review. IEEE Sens. J. **21**, 26415–26425 (2021)
16. Gao, J., He, S., Nag, A., Wong, J.W.C.: A review of the use of carbon nanotubes and graphene-based sensors for the detection of aflatoxin M1 compounds in milk. Sensors. **21**(11), 3602 (2021)
17. Sanderson, P., Delgado-Saborit, J.M., Harrison, R.M.: A review of chemical and physical characterisation of atmospheric metallic nanoparticles. Atmos. Environ. **94**, 353–365 (2014)
18. Nag, A., Alahi, M., Eshrat, E., Mukhopadhyay, S.C., Liu, Z.: Multi-walled carbon nanotubes-based sensors for strain sensing applications. Sensors. **21**(4), 1261 (2021)
19. He, S., Zhang, Y., Gao, J., Nag, A., Rahaman, A.: Integration of different graphene nanostructures with PDMS to form wearable sensors. Nanomaterials **12**(6), 950 (2022)
20. Nag, A., Alahi, M.E.E., Feng, S., Mukhopadhyay, S.C.: IoT-based sensing system for phosphate detection using Graphite/PDMS sensors. Sens. Actu. A **286**, 43–50 (2019)
21. Serra, A., Valles, E.: Advanced electrochemical synthesis of multicomponent metallic nanorods and nanowires: fundamentals and applications. Appl. Mater. Today **12**, 207–234 (2018)
22. Zhang, Y., Li, N., Xiang, Y., Wang, D., Zhang, P., Wang, Y., et al.: A flexible non-enzymatic glucose sensor based on copper nanoparticles anchored on laser-induced graphene. Carbon **156**, 506–513 (2020)
23. Moseley, P.T.: Progress in the development of semiconducting metal oxide gas sensors: a review. Meas. Sci. Technol. **28**(8), 082001 (2017)
24. Zhang, W., Zhu, R., Nguyen, V., Yang, R.: Highly sensitive and flexible strain sensors based on vertical zinc oxide nanowire arrays. Sens. Actu. A **205**, 164–169 (2014)
25. Pineda-Reyes, A.M., Herrera-Rivera, M.R., Rojas-Chávez, H., Cruz-Martínez, H., Medina, D.I.: Recent advances in ZnO-based carbon monoxide sensors: role of doping. Sensors. **21**(13), 4425 (2021)
26. Nag, A., Afasrimanesh, N., Feng, S., Mukhopadhyay, S.C.: Strain induced graphite/PDMS sensors for biomedical applications. Sens. Actu. A **271**, 257–269 (2018)
27. Nag, A., Mukhopadhyay, S.C., Kosel, J.: Tactile sensing from laser-ablated metallized PET films. IEEE Sens. J. **17**(1), 7–13 (2016)
28. Nag, A., Mukhopadhyay, S.C., Kosel, J.: Sensing system for salinity testing using laser-induced graphene sensors. Sens. Actu. A **264**, 107–116 (2017)
29. Stan, F., Stanciu, N.-V., Constantinescu, A.-M., Fetecau, C.: 3D printing of flexible and stretchable parts using multiwall carbon nanotube/polyester-based thermoplastic polyurethane. J. Manuf. Sci. Eng. **143**, 1–33 (2020)
30. Thana, K., Petchsang, N., Jaisutti, R.: Electrical and mechanical properties of PEDOT: PSS strain sensor based microwave plasma modified pre-vulcanized rubber surface. MS&E. **773**(1), 012049 (2020)
31. Nag, A., Afsarimanesh, N., Nuthalapati, S., Altinsoy, M.E.: Novel surfactant-induced MWCNTs/PDMS-based nanocomposites for tactile sensing applications. Materials. **15**(13), 4504 (2022)
32. Xiao, Y., Jiang, S., Li, Y., Zhang, W.: Screen-printed flexible negative temperature coefficient temperature sensor based on polyvinyl chloride/carbon black composites. Smart Mater. Struct. **30**(2), 025035 (2021)
33. Nayak, L., Mohanty, S., Nayak, S.K., Ramadoss, A.: A review on inkjet printing of nanoparticle inks for flexible electronics. J. Mater. Chem. C. **7**(29), 8771–8795 (2019)

34. Kalkal, A., Kumar, S., Kumar, P., Pradhan, R., Willander, M., Packirisamy, G., et al.: Recent advances in 3D printing technologies for wearable (bio) sensors. Addit. Manuf. **46**, 102088 (2021)
35. Nag, A., Mukhopadhyay, S.C.: Fabrication and implementation of printed sensors for taste sensing applications. Sens. Actu. A **269**, 53–61 (2018)
36. Han, T., Kundu, S., Nag, A., Xu, Y.: 3D printed sensors for biomedical applications: a review. Sensors. **19**(7), 1706 (2019)
37. Shirhatti, V., Nuthalapati, S., Kedambaimoole, V., Kumar, S., Nayak, M.M., Rajanna, K.: Multifunctional graphene sensor ensemble as a smart biomonitoring fashion accessory. ACS Sens. **6**(12), 4325–4337 (2021)
38. Nuthalapati, S., Kedambaimoole, V., Shirhatti, V., Kumar, S., Takao, H., Nayak, M., et al.: Flexible strain sensor with high sensitivity, fast response, and good sensing range for wearable applications. Nanotechnology **32**(50), 505506 (2021)
39. Kedambaimoole, V., Kumar, N., Shirhatti, V., Nuthalapati, S., Kumar, S., Nayak, M.M., et al.: Reduced graphene oxide tattoo as wearable proximity sensor. Adv. Electron. Mater. **7**(4), 2001214 (2021)
40. Liu, Z., Wang, Y., Wu, B., Cui, C., Guo, Y., Yan, C.: A critical review of fused deposition modeling 3D printing technology in manufacturing polylactic acid parts. Int. J. Adv. Manuf. Technol. **102**(9–12), 2877–2889 (2019). https://doi.org/10.1007/s00170-019-03332-x
41. Kuo, C.-C., Liu, L.-C., Teng, W.-F., Chang, H.-Y., Chien, F.-M., Liao, S.-J., et al.: Preparation of starch/acrylonitrile-butadiene-styrene copolymers (ABS) biomass alloys and their feasible evaluation for 3D printing applications. Compos. Part B Eng. **86**, 36–39 (2016)
42. Kasmi, S., Ginoux, G., Allaoui, S., Alix, S.: Investigation of 3D printing strategy on the mechanical performance of coextruded continuous carbon fiber reinforced PETG. J. Appl. Polym. Sci. **138**(37), 50955 (2021)
43. Chou, H.-H., Lee, W.-Y.: Tactile Sensor Based on Capacitive Structure: Functional Tactile Sensors, pp. 31–52. Elsevier, Amsterdam (2021)
44. Wang, H., Cen, Y., Zeng, X.: Highly sensitive flexible tactile sensor mimicking the microstructure perception behavior of human skin. ACS Appl. Mater. Interfaces. **13**(24), 28538–28545 (2021)
45. Wu, D., Cheng, X., Chen, Z., Xu, Z., Zhu, M., Zhao, Y., et al.: A flexible tactile sensor using polyimide/graphene oxide nanofiber as dielectric membranes for vertical and lateral force detections. Nanotechnology (2022)
46. Zhu, L., Wang, Y., Mei, D., Wu, X.: Highly sensitive and flexible tactile sensor based on porous graphene sponges for distributed tactile sensing in monitoring human motions. J. Microelectromech. Syst. **28**(1), 154–163 (2018)
47. Eijking, B., Sanders, R., Krijnen, G.: Development of whisker inspired 3D multi-material printed flexible tactile sensors. In: 2017 IEEE Sensors. IEEE (2017)
48. Mousavi, S., Howard, D., Zhang, F., Leng, J., Wang, C.H.: Direct 3D printing of highly anisotropic, flexible, constriction-resistive sensors for multidirectional proprioception in soft robots. ACS Appl. Mater. Interfaces. **12**(13), 15631–15643 (2020)
49. Lee, T., Kang, Y., Kim, K., Sim, S., Bae, K., Kwak, Y., et al.: All paper-based, multilayered, inkjet-printed tactile sensor in wide pressure detection range with high sensitivity. Adv. Mater. Technol. **7**(2), 2100428 (2022)

Fundamentals of Bio-Signal Sensor Design and Development in Medical Applications

Amir Shahbazi[1(✉)], Nasrin Afsarimanesh[1], Tele Tan[1,2], Ghobad Shafiei Sabet[3], and Gabriel Yin Foo Lee[4,5]

[1] School of Civil and Mechanical Engineering, Curtin University, WA Perth, Australia
{amir.amir,nasrin.afsarimanesh}@curtin.edu.au
[2] School of Electrical Engineering, Computing and Mathematical Sciences, Curtin University, WA Perth, Australia
[3] DGUT-CNAM Institute, Dongguan University of Technology, Dongguan, China
[4] St John of God Subiaco Hospital, WA Subiaco, Australia
[5] School of Surgery, University of Western Australia, WA Perth, Australia

Abstract. This paper reports the design and development of a durable bio-signal processing circuit which utilizes active dry electrodes. The idea behind dry electrodes is to manufacture a bio cap/helmet with numerous electrodes which cover the subject's scalp. The weak bio-signals acquired from the subject will be processed to minimize the common-mode noises and the results would be de-noised bio-signals. Signal processing methods applied in software, such as Wavelet Transform, enable us to remove the bio artifacts such as muscle movement or eye blinking, ultimately producing artifact-free bio-signals.

Keywords: Bio-signal · EEG · Active dry electrode · Digital signal processing

1 Introduction

In different research on brain-computer interfaces, EEG is a kind of psychophysiological test used to inspect the relations between mental and bodily procedures. Generally, EEG is created by inspecting the electrical potential of bio-signal sensors which are placed at allocated points on the scalp. EEG is commonly used to investigate medical, including epileptic seizures [1, 2], sleep disorders [3, 4], and attention-deficit/hyperactivity disorder (ADHD) [5, 6]. Recently, EEG has been employed in other applications, such as computer interface projects, computer games, and neuromarketing investigations [7–11].

The objective of this work is to create a durable low-component bio-signal processing circuit which utilises active dry electrodes [12, 13]. The goal behind employing dry electrodes is to manufacture a bio cap/helmet with numerous amount of electrodes that encapsulates the subject's scalp. The weak bio-signals acquired from the subject will be processed in a way to minimize the common-mode noises and the results would be de-noised bio-signals. Signal processing methods applied in software, such as Wavelet Transform, enable us to remove the bio artifacts such as skin artifacts, muscle movement or eye blinking and the results would be in close proximity to the bio-signals [14–17].

N. K. Suryadevara et al. (Eds.): ICST 2022, LNEE 1035, pp. 431–439, 2023.
https://doi.org/10.1007/978-3-031-29871-4_44

Bio-signals obtained from passive wet electrodes such as Ag/AgCl or gold-plated electrodes are extremely vulnerable to environmental noises, mostly radio frequency and powerline noises [18, 19]. Moreover, they require the application of conductive gel or paste to keep the skin-electrode impedance at a low level. The gel or paste might get dry during the bio-data acquisition process. Skin-electrode impedances at 10 Hz using Ag/AgCl electrodes, with the skin properly prepared, are typically about 5 k Ω [20]. In many cases, this value cannot be kept during the whole time of recording the bio-signals which results in poor signal quality. Another common problem is polarizing the electrode. In this situation, the poor skin-electrode contact can polarize the electrode which will affect the quality of the captured bio-signal [21].

One option which can be considered to resolve these problems and enhance the signal quality is designing an appropriate active dry electrode in a way to provide some levels of immunity against the common-mode noises and at the same time increasing the signal-to-noise ratio (SNR). This removes the DC offset due to poor connection between the skin and the electrode. Since the output of the active dry electrode has a very low impedance, common-mode noises hardly can be presented as differential signal at the pre-amplification stages. In the latter, the front-end instrumentation amplifier can use most of its capability to reject the common mode noise (usually above 100 dB) [22].

2 Materials and Methods

2.1 Active Dry Electrode Design Considerations

Designing a successful bio active sensor requires paying attention to the level of the bio-signals and the level of the environmental noises. A very good practice of designing an active electrode is utilising JFET input amplifiers as they provide an immensely high input impedance (in terms of Tera ohms or ×1012 Ω). JFET input amplifiers usually have a very small bias current in terms of 1–20pA to make them a suitable choice to deal with weak bio-signals.

Four different active electrode schemes have been tried in this design with the most suitable JFET op-amps. These four circuits are shown in Fig. 1.

AD8244, a 4-channel JFET buffer and AD8626/7, a JFET input op-amp are chosen for this purpose. Both amplifiers have a very high input impedance, very low output impedance in terms of tens of ohms and the lowest output voltage offset as well as micro-amp current consumption.

Design ACTIVE_ELEC1 is the simplest design and uses only one buffer channel, the output is a buffered input signal with no further signal processing. Design ACTIVE_ELEC2 still uses a single channel of AD8244 but there is a second-order Sallen-Key high pass filter with cut-off frequency $Fc = 0.5$ Hz at the input of the buffer to suppress the DC offset to the lowest level, enabling the front-end instrumentation amplifier to achieve high gains (G> 200). Designs ACTIVE_ELEC3 and ACTIVE_ELEC4 are similar to previous designs, the only difference is the buffer op-amp AD8244 is replaced with AD8626/7.

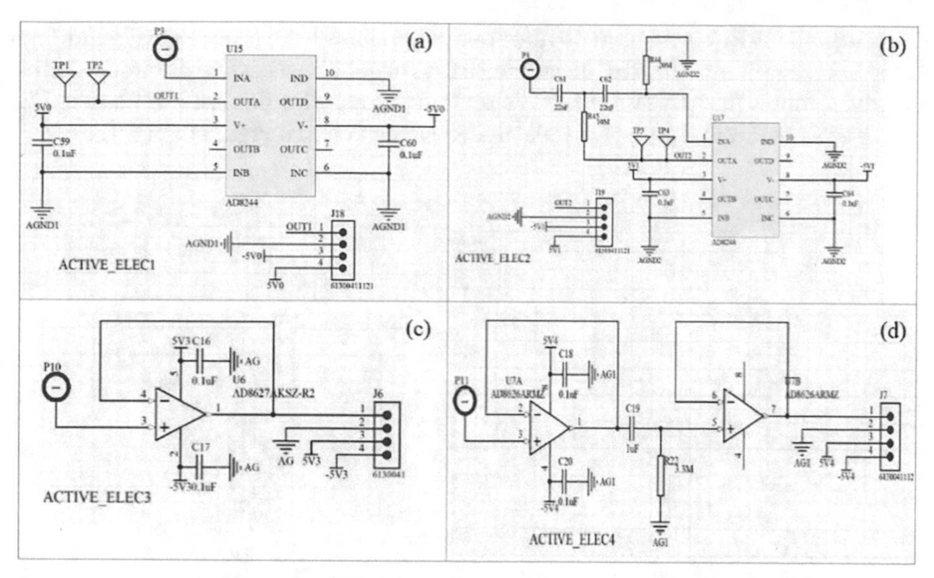

Fig. 1. (a) Single channel JFET buffer, (b) Single channel JFET buffer combined with a second-order high pass filter at the input to remove DC offset, (c) Single channel JFET buffer using AD8627, and (d) Single channel JFET buffer combined with a first-order high pass filter at the input to remove DC offset.

2.2 Bio-Signal Processing Circuits

Complete design circuitry for a single channel bio acquisition is shown in Fig. 2. The circuit operates with a 9V battery supply and uses a ± 5V dual supply as the rail-to-rail system supply. The circuit is designed in two types, both with the same circuitry apart from the instrumentation amplifier. Type 1 utilizes INA128 as the front-end pre-amplifier and type 2 utilizes AD8420 as the front-end pre-amplifier. A sample EMG signal acquiring circuit is prototyped on the breadboard by using INA126, a medical instrumentation amplifier and TL072, a general-purpose JFET op-amp.

Figure 2 demonstrates the signal processing stages of the experimental circuit. No 50 Hz notch filter is used in this circuit.

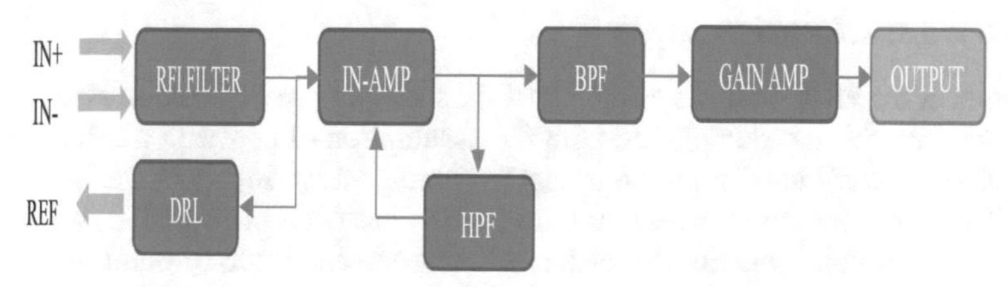

Fig. 2. Signal processing stages of the experimental circuit.

Both passive wet and active dry electrodes have been tested with the given circuit. Both types of electrodes are connected to the biceps to provide inputs to INA128 and the third electrode (REF) is connected to the bone at the elbow to provide a signal return path to the DRL amplifier. The REF electrode always should be a passive electrode. Active electrodes follow the ACTIVE_ELEC3 circuit in this report which is a JFET buffer. The results achieved from both types of electrodes are depicted in Figs. 6 to Fig. 13.

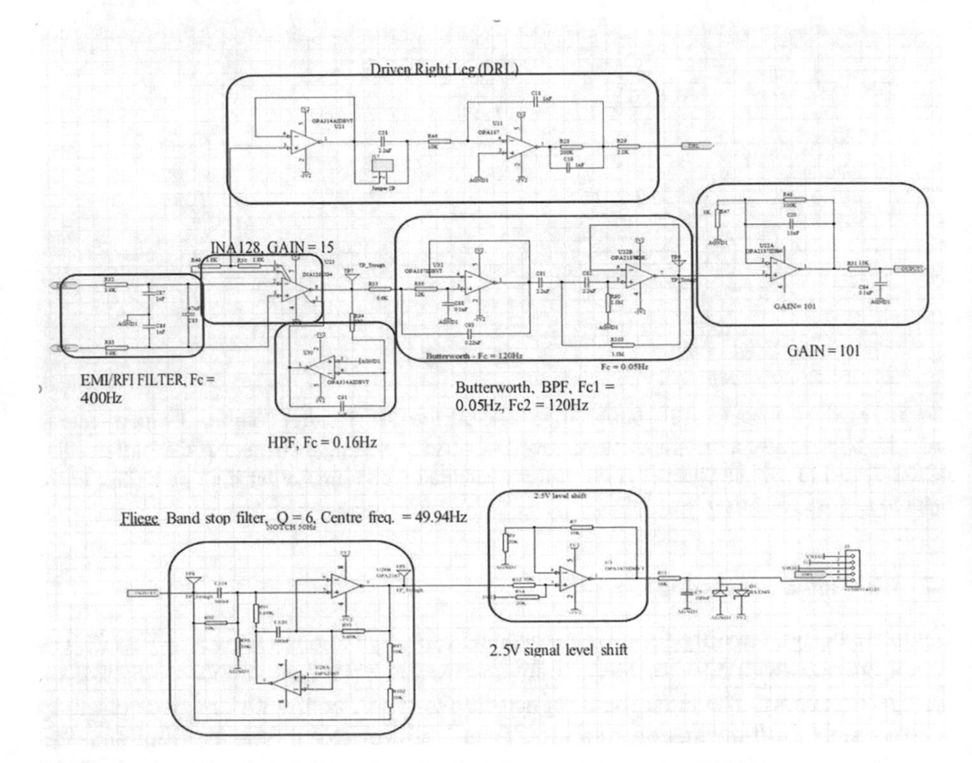

Fig. 3. Bio/EMG signal processing type 1 using INA128

3 Results and Discussion

3.1 DRL Circuit SPICE Analysis

The results of a SPICE analysis using TINA-TI software are depicted in Figs. 4 and 5. The analysis has been accomplished with the assumption of applying a common mode noise (50 Hz powerline noise) to both negative and positive inputs of the in-amp. The output (VF1) common-mode rejection ratio (in dB) has been measured in two different scenarios, in the first scenario, the system's reference electrode (ground electrode) is directly connected to the circuit's ground, and, in the second scenario, the reference electrode is connected to the output of a DRL circuit.

As it can be understood from Figs. 3 and 4, the results clearly show a considerable improvement in the common-mode rejection ratio when the reference electrode is connected to a DRL circuit.

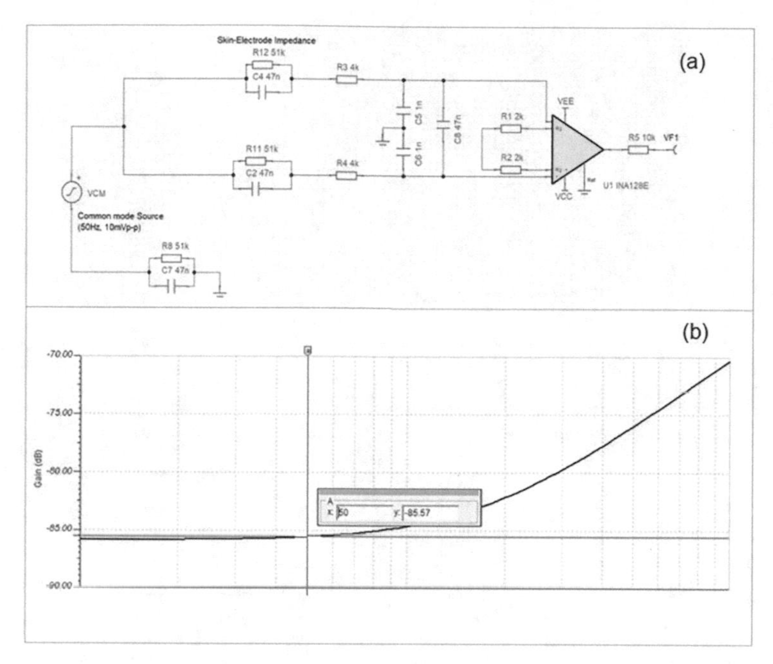

Fig. 4. (a) Applying common-mode source to the in-amp in absence of the DRL circuit. (b) CMRR at 50 Hz.

CMRR at 50 Hz is measured at −85.57dB when the reference electrode is directly connected to the circuit's ground. By utilising a tuned DRL circuit the CMRR can be improved to −135.5dB as shown in Fig. 4.

Cautions must be taken when utilising the DRL circuit in bio-potential applications. It is mandatory to connect the inputs of non-operational in-amps to the ground of the circuit when there are parallel channels measuring signals from different points, otherwise floating in-amp inputs can cause the DRL op-amp to be saturated and the output of the DRL line reaches to the negative supply rail. If this happens, the output of all channels will reach one of the supply rails making it impossible to retrieve the bio-potential signals.

3.2 Digital Signal Processing (DSP) and Signal De-Noising

Although analogue signal processing assists us to achieve the bio-potential signals, however, the acquired signal still consists of a considerable amount of noise. A successful bio-signal processing system must utilize at least an 8th-order low pass filtering stage to improve the SNR to desirable levels. Since our circuit it is aimed to use the minimum number of components in order to save space, achieving a 6th order low pass and a 2nd order notch filter is accomplished by employing mathematical methods and DSP. The technique used includes converting the low pass and notch filter generic transfer functions (in S-domain) into a discrete domain (Z-domain) and calculating corresponded Z-domain equations using MATLAB software. At the next stage, having the Z-domain

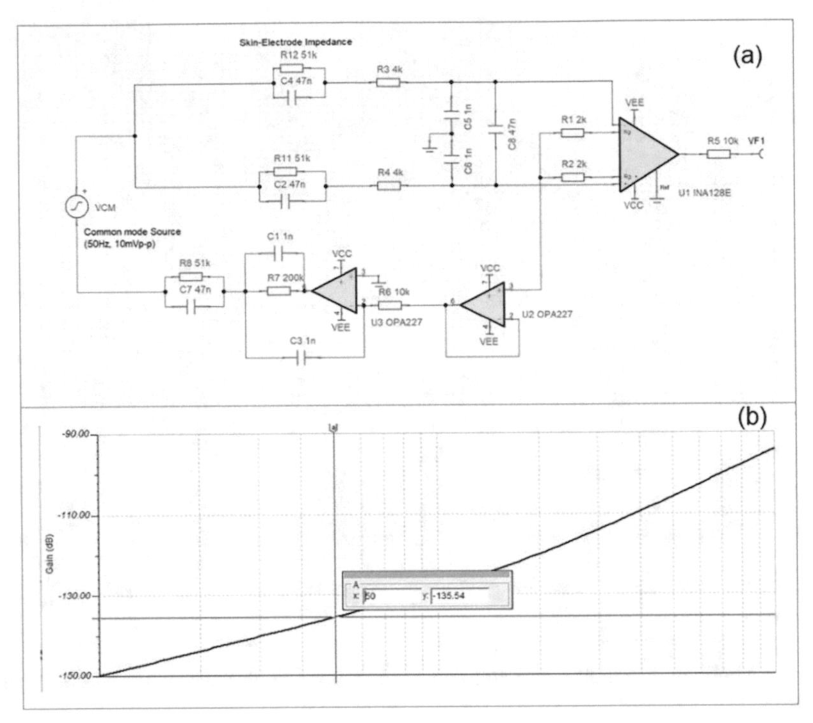

Fig. 5. (a) Applying common-mode source to the in-amp in presence of a DRL circuit. (b) CMRR at 50 Hz.

equations in hand, an appropriate microcontroller DSP library has been created in the CPP programming language.

3.3 Experimental Bio-Potential Signal Processing Circuit

In this experimental circuit, the high pass and the low pass filter cut-off frequencies are set to $f_c = 0.15$ Hz and Fc = 200 Hz, respectively. The in-amp gain was set to Av = 21 and the second stage non-inverting gain was set to Av = 1 – 201 utilizing a potentiometer. The illustrated signals in the pictures below were obtained from the total gain of Av = 21 × 115 = 2415. Since the DRL was connected, there was no effect of powerline noise on the signal, and the signal shape in muscle relax mode (no muscle contraction) was achieved with minimum noise and deviations from the oscilloscope's voltage axis. With the gain, Av = 2415, and with two stages of high pass filtering, the output DC offset was measured at 560 mV. Adding an extra high pass filter at the back end will remove this offset. From the DRL output measurement, it was understood the rejected common-mode noise mainly occurred at 50Hz as we have expected (Fig. 7, Fig. 8. Fig. 9, Fig. 10, Fig. 11 and Fig. 12)

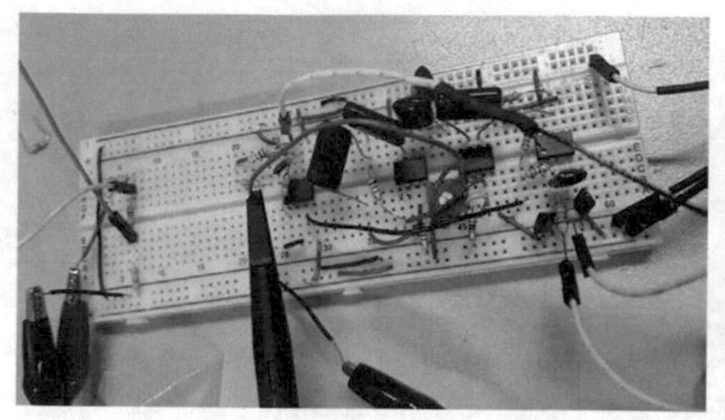

Fig. 6. Experimental bio-potential signal processing circuit using INA126 and TL072

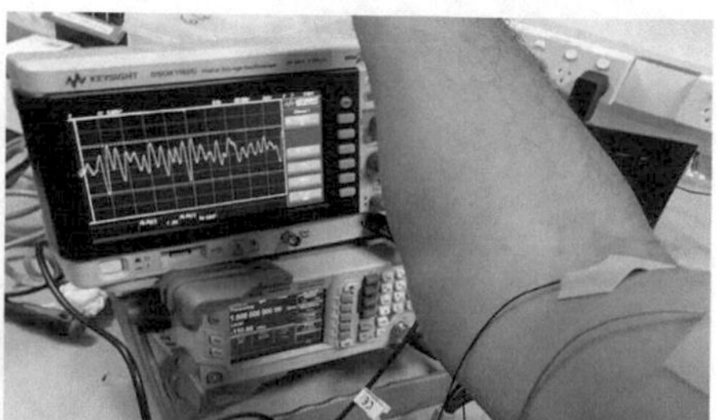

Fig. 7. Passive electrodes are connected to the biceps, the biceps is contracted and the corresponded signal can be observed on the oscilloscope, signal is pure EMG with no additional 50 Hz noise

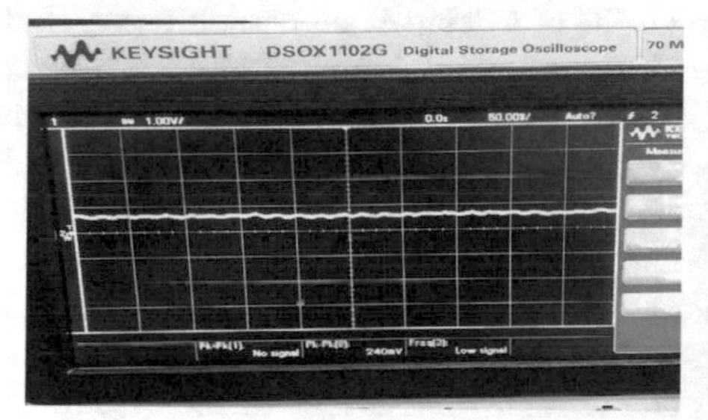

Fig. 8. Output signal of muscle relaxed mode. Since DRL is connected, 50 Hz noise doesn't have a significant effect on the acquired signal

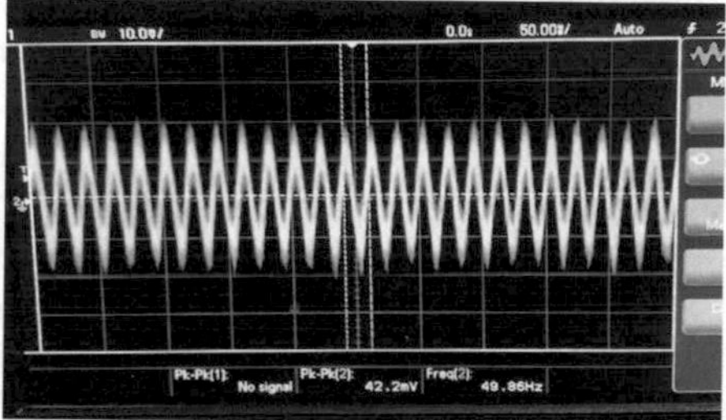

Fig. 9. Output of DRL circuit demonstrates the major contributor in common mode noise is the powerline noise

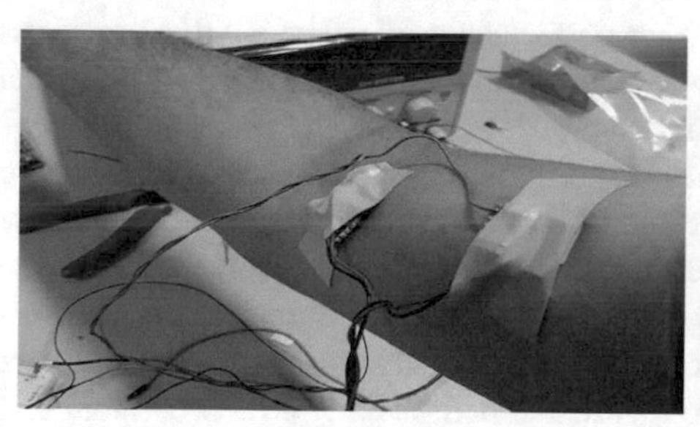

Fig. 10. Active dry electrodes connected to the biceps

Fig. 11. Active dry electrodes made using TL072 and male pin headers as signal collectors

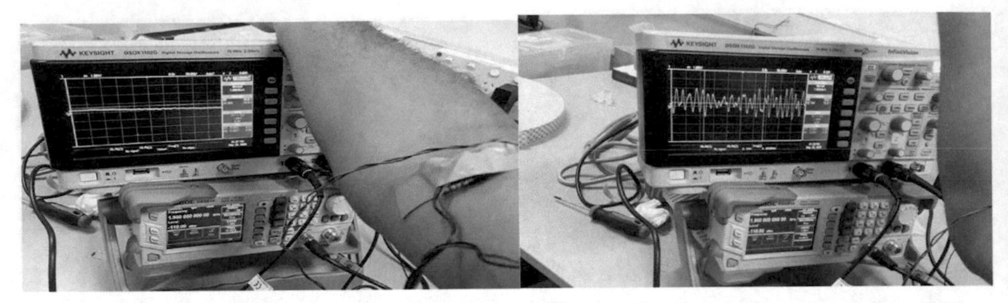

Fig. 12. EMG measurement using active dry electrodes, muscle relaxed.

Fig. 13. EMG measurement using active dry electrodes, muscle contracted

4 Conclusions and Future Work

The circuits presented in this report are the results of a detailed study on bio-potential signal processing and every single component in the proposed circuits plays a vital role in the appropriate functioning of the system. Different methods are considered to confront distinct obstacles including common-mode noises, DC offset, electrode polarizing, and so on. The successful outcome of these designs is also observed from a detailed analysis in SPICE software and making a sample circuit on a breadboard. Described circuits in this report together comprise a single channel to measure and analyse the bio-potential signals. Multi-channel signal processing circuits require further attention to the total common-mode noise present in the circuit and to improve the SNR of the system. A complete bio-signal processing system also includes digital processing circuits such as ADC, Microcontroller, etc. At the current stage, an 8-channel ADC with 24-bit resolution and the SPI communication speed of 30 KSPS is chosen to be integrated with the analogue system. The microcontroller which is responsible for processing the incoming data is chosen to be the STM32F4 series a high-performance 32-bit M4-ARM processor with 168 MHz CPU speed and a wide variety of communication/peripheral options. The needful software to demonstrate the measured data is developed in the visual studio environment and at the moment can process and demonstrate fast speed charts with the rate of 8KSPS. It also executes the FFT on incoming data and demonstrates the frequency range of bio-signals. However, the processing speed of data in the software must be increased to 16 KSPS, minimum, which requires extra attention to the software development. Finally, every section of the system including digital, analogue and software must be integrated properly to enable us to record and demonstrate pure bio-signals.

References

1. Boubchir, L., Daachi, B., Pangracious, V.: A review of feature extraction for EEG epileptic seizure detection and classification. In: 2017 40th International Conference on Telecommunications and Signal Processing (TSP), pp. 456–460. IEEE (2017)
2. Tzallas, A.T., Tsipouras, M.G., Fotiadis, D.I.: Epileptic seizure detection in EEGs using time–frequency analysis. IEEE Trans. Inf Technol. Biomed. **13**(5), 703–710 (2009)
3. Peter-Derex, L., et al.: Automatic analysis of single-channel sleep EEG in a large spectrum of sleep disorders. J. Clin. Sleep Med. **17**(3), 393–402 (2021)

4. Siddiqui, M.M., Rahman, S., Saeed, S.H., Banodia, A.: EEG signals play major role to diagnose sleep disorder. Int. J. Electron. Comput. Sci. Eng. (IJECSE) **2**(2), 503–505 (2013)
5. Lenartowicz, A., Loo, S.K.: Use of EEG to diagnose ADHD Current psychiatry reports, vol. 16(11), pp. 1–11 (2014)
6. Snyder, S.M., Rugino, T.A., Hornig, M., Stein, M.A.: Integration of an EEG biomarker with alinician's ADHD evaluation. Brain Behav. **5**(4), e00330 (2015)
7. Alakus, T.B., Gonen, M., Turkoglu, I.: Database for an emotion recognition system based on EEG signals and various computer games–GAMEEMO. Biomed. Signal Process. Control **60**, 101951 (2020)
8. Kerous, B., Skola, F., Liarokapis, F.: EEG-based BCI and video games: a progress report. Virtual Reality **22**(2), 119–135 (2017). https://doi.org/10.1007/s10055-017-0328-x
9. Abiri, R., Borhani, S., Sellers, E.W., Jiang, Y., Zhao, X.: A comprehensive review of EEG-based brain–computer interface paradigms. J. Neural Eng. **16**(1), 011001 (2019)
10. Yadava, M., Kumar, P., Saini, R., Roy, P.P., Prosad Dogra, D.: Analysis of EEG signals and its application to neuromarketing. Multimedia Tools Appli. **76**(18), 19087–19111 (2017). https://doi.org/10.1007/s11042-017-4580-6
11. Khurana, V., et al.: A survey on neuromarketing using EEG signals. IEEE Trans. Cognitive Developm. Syst. **13**(4), 732–749 (2021)
12. Mancuso, M., et al.: Transcranial evoked potentials can be reliably recorded with active electrodes. Brain Sci. 11(2), 145 (2021). https://www.mdpi.com/2076-3425/11/2/145
13. Narayanan, A.M., Bertrand, A.: Analysis of miniaturization effects and channel selection strategies for EEG sensor networks with application to auditory attention detection. IEEE Trans. Biomed. Eng. **67**(1), 234–244 (2019)
14. Khateb, F., Prommee, P., Kulej, T.: MIOTA-based filters for noise and motion artifact reductions in biosignal acquisition. IEEE Access **10**, 14325–14338 (2022)
15. Bennett, S.L., Goubran, R., Knoefel, F.: Examining the effect of noise on biosignal estimates extracted through spatio-temporal video processing. In: 2019 41st Annual International Conference of the IEEE Engineering in Medicine and Biology Society (EMBC), pp. 4504–4508. IEEE (2019)
16. Islam, M., Rastegarnia, A., Sanei, S.: Signal artifacts and techniques for artifacts and noise removal. In: Signal Processing Techniques for Computational Health Informatics, pp. 23–79, Springer (2021)
17. Cheng, J.Y., Goh, H., Dogrusoz, K., Tuzel, O., Azemi, E.: Subject-aware contrastive learning for biosignals, arXiv preprint arXiv:2007.04871, (2020)
18. Niu, X., Gao, X., Liu, Y., Liu, H.: Surface bioelectric dry electrodes: a review. Measurement **183**, 109774 (2021)
19. Naim, A.M., Wickramasinghe, K., De Silva, A., Perera, M.V., Lalitharatne, T.D., Kappel, S.L.: Low-cost active dry-contact surface emg sensor for bionic arms. In: 2020 IEEE International Conference on Systems, Man, and Cybernetics (SMC), pp. 3327–3332. IEEE (2020)
20. Li, G., Wang, S., Duan, Y.Y.: Towards gel-free electrodes: A systematic study of electrode-skin impedance. Sens. Actuators, B Chem. **241**, 1244–1255 (2017)
21. Huang, Y.-J., Wu, C.-Y., Wong, A.M.-K., Lin, B.-S.: Novel active comb-shaped dry electrode for EEG measurement in hairy site. IEEE Trans. Biomed. Eng. **62**(1), 256–263 (2014)
22. Kim, I., Lai, P.-H., Lobo, R., Gluckman, B.J.: Challenges in wearable personal health monitoring systems. In: 2014 36th Annual International Conference of the IEEE Engineering in Medicine and Biology Society, pp. 5264–5267. IEEE (2014)

Improved Interfacing Circuit for Planar Coil-Based Thin Angle Sensors

A. S. Anil Kumar[1,2(✉)], Boby George[1], and Subhas C. Mukhopadhyay[2]

[1] Indian Institute of Technology Madras, Chennai 600036, India
anil.appukuttannairsyamalaamma@hdr.mq.edu.au
[2] School of Engineering, Macquarie University, Sydney, Australia

Abstract. This paper presents improved interfacing schemes for a recently reported variable reluctance-based thin planar angle sensor. The design and analysis of two inductance-based measurement schemes appropriate for the planar angle sensor are presented. The first one is based on differential measurement, and the second method realizes a ratiometric measurement. The suggested interfacing circuits measure the differential or ratiometric inductances of the differential sensing coils directly from the sensor rather than measuring the individual inductances of the sensing coils separately. As a result, the sensor performs better, and measurement speed is improved ten times. Studies were conducted with a variable reluctance-based angle sensor, and the designs are validated. The functionality of the proposed circuit was verified for different rotating speeds. The worst-case linearity error for the differential measurement circuit was found to be 2.3.%, for a rotating speed close to 500 rpm.

Keywords: Inductance Measurement · Differential Measurement · Ratiometric Measurement · Angle Sensing · Variable Reluctance

1 Introduction

In robotic and automotive applications, it is necessary to measure the absolute angular position of the shaft with good precision and high resolution at a low cost. Angle sensors used in these applications must be small in size, easy to install, resistant to dust and pollution, and sufficiently robust against electromagnetic interference [1–5]. A certain amount of stress and axial vibration will be experienced by the majority of angle sensors during regular use. These factors need to be considered at the sensor design stage. In addition, well-known sensors such as variable reluctance type angle sensors [6], rotary variable differential transformers [7], and resolvers [8] are comparatively large and have pronounced axial thicknesses.

1.1 A Variable Reluctance-Based Thin Angle Sensor

A thin, non-contact angle sensor based on the variable reluctance approach is presented in [9]. With relatively small space usage, this sensor offers exceptional performance. The

N. K. Suryadevara et al. (Eds.): ICST 2022, LNEE 1035, pp. 440–449, 2023.
https://doi.org/10.1007/978-3-031-29871-4_45

sensor's overall thickness is less than 5 mm, and the output is insensitive to the shaft's vertical misalignment. Figure 1(a) and (b) illustrate the structural details of this sensor. This sensor's stationary part is made up of four planar circular coils (L_A to L_D) that are arranged in a circle with a 90° offset between each subsequent coil. The eccentric disc-shaped rotor of the proposed sensor is built of a high permeability material. The angular position (θ) of the eccentric disc, which rotates with the shaft, modulates the self-inductances of the four planar coils. This sensor incorporates a dual coil configuration that helps to avoid the effects of axial shaft misalignment. To accomplish this, the rotor disc is positioned between two parallel-mounted circular PCBs that are similar in shape, as shown in Fig. 1(b). The coils (L_{TA} to L_{TD}) in the top PCB and the coils (L_{BA} to L_{BD}) from the bottom PCBs are connected serially, resulting in four inductances: L_A to L_D. The change in inductance L_A to L_D, against (θ), over the full circular range, is depicted in Fig. 1(c). After that, a quadrant identification algorithm was used to compute the value of θ using the obtained inductance values. With a low power consumption of 25 mW, this sensor exhibits high linearity (worst-case linearity error of 0.7%) and good resolution (0.06°). Moreover, this sensor requires a far smaller vertical thickness than other angle sensors on the market.

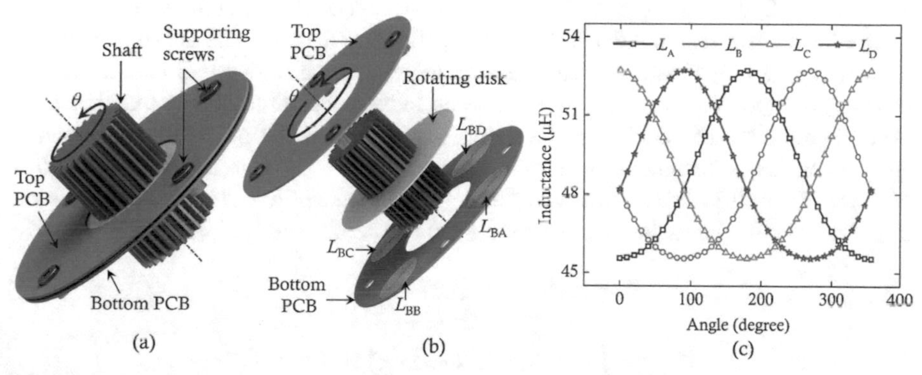

Fig. 1. Diagram illustrating the variable reluctance-based angle sensor's structural elements [9]. (a) The sensor's structure. (b) An exploded view of the sensor. The coils from the top PCB (LTA to LTD) are not visible in (b). (c) A graph that illustrates the change in inductance for each coil, from LA to LD

Even though this sensor performs exceptionally well at low speeds, its accuracy is observed to be declined at speeds of more than 50 rpm. This is mainly caused by the low update rate of the LDC1614 IC used in the measurement system. This method utilizes the linear portions of the inductance curves in Fig. 1(c). In that approach, it is necessary to use different coils in each quadrant to compute the angle. The quadrant identification algorithm [9] used in this sensor measures the inductance of each coil separately, and it then calculates the final angular position based on the magnitude of each inductance value. This calls for the sequential measurement of each of the four inductance values independently using a four-channel LDC (LDC 1614), which significantly slows down the sensor system's overall update rate. It is important to note that the sensor elements are not what restricts the sensing speed. By substituting another inductance measurement device with a faster update rate compared to LDC1614, the sensor can be used high-speed applications.

Such rapid inductance measurement methods, suitable for the sensing scheme described in [9], are presented in this study. The suggested approaches help to compute the angular displacement information at a faster rate, without sacrificing the accuracy or other performance parameters.

2 Differential Measurement Method

Differential measurement of the push-pull pair of waveforms is advantageous as it increases sensitivity and the resulting signal-to-noise ratio [4, 10]. In this sensor, two such push-pull pairs exist; L_A-L_C, and L_B-L_D. This is seen in Fig. 1(c). The differences in inductance in these coil pairs, $L_{CA} = L_C - L_A$ and $L_{DB} = L_D - L_B$, plotted against θ are shown in Fig. 2. Both differential inductance curves exhibit excellent sinusoidal properties, as shown by the curve fits, which have an R^2 value of 1. From the curve fitting, L_{CA} and L_{DB} can be expressed as $L_{CA}(\theta) = 0.00563 + 7.173\, sin(0.0174\theta + 0.158)$, and $L_{DB}(\theta) = -4.34 \times 10^{-6} + 7.169\, sin(0.0174\theta + 0.0069)$, respectively. As shown in Fig. 2, the curves result in a sine and cosine waveform, with respect to θ.

One differential pair of coils, let's say L_{CA}, allows us to determine the angle value for the entire circle range using the curve fitting technique that was previously discussed. However, two positions can provide the same inductance value due to the sinusoidal curve's symmetry. The curve L_{CA} is symmetrical around the point 180°, as seen in Fig. 2. Therefore, a technique is needed to determine if $0° \leq \theta < 180°$ or $180° \leq \theta < 360°$. The other differential inductance curve (L_{DB}) can be utilized to clear up this issue. As a result, the final angular displacement (θ r) can be obtained using (1) and (2):

$$\theta = A \sin^{-1}[\alpha L_{CA}(\theta) - \beta] - \gamma \tag{1}$$

$$\theta_r = \begin{cases} \theta & : \text{ if } \operatorname{sgn}(L_{DB}) = 1 \\ 360 - \theta : & \text{ if } \operatorname{sgn}(L_{DB}) = -1 \end{cases} \tag{2}$$

In (1), the values of A, α, β, γ are found to be 57.47, 0.14, 0.00084, and, 9.08, respectively. In (2), sgn (L_{DB}) will be + 1, when the angle of rotation will be less than 180°, and will be − 1 otherwise. In conclusion, the cosine curve can be used to derive angle information, and the sinusoidal curve's polarity can be used to resolve the ambiguity that results from the curve's symmetry. As a result, the output can be derived as a linear function of θ_r.

A measurement circuit is required to determine the difference in inductance coming from each coil pair with respect to θ. The details of the proposed differential inductance measurement circuit are provided next.

2.1 Measurement System

In Fig. 3, a circuit is depicted that provides the differential inductance of the coil pairs C-A and D-B. Here, L_A, L_B, L_C, and L_D represent the inductance, and R_A, R_B, R_C, and

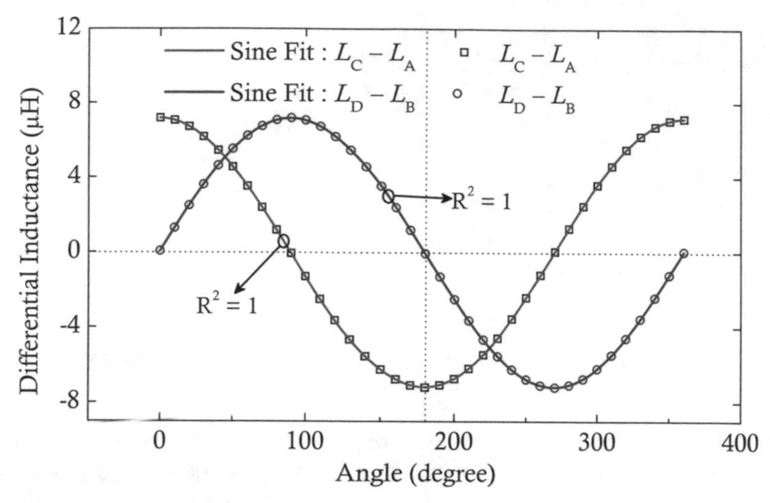

Fig. 2. Difference in inductances ($L_{CA} = L_C - L_A$ and $L_{DB} = L_D - L_B$) and the corresponding sine fit curve for the sensor presented in [9].

R_D represent the internal resistance of the coils A, B, C, and D respectively. In Fig. 3, in order to get the difference in inductance $L_{CA} = L_C - L_A$, L_C is connected to voltage source $V_s\angle 0°$ and L_A is connected to voltage source $V_s\angle 180°$. The sum of the currents flowing through coils C and A flows through the feedback resistor R_F of the op-amp OA_1. This current ($I_{F1} = I_C - I_A$) produces a voltage V_{O1} at the output of OA_1. The output V_{O1} can be expressed as:

$$V_{O1} = \frac{\sqrt{2}\, V_S\, R_F}{\omega}\left(\frac{L_C - L_A}{L_C L_A}\right)\sin(\omega\, t \pm 90°) \tag{3}$$

Similarly, L_{DB} ($= L_D - L_B$) can be obtained by connecting the coil L_D to voltage source $V_s\angle 0°$ and L_B to voltage source $V_s\angle 180°$. Here, the corresponding output voltage from OA_2 can be expressed as (4):

$$V_{O2} = \frac{\sqrt{2}\, V_S\, R_F}{\omega}\left(\frac{L_D - L_B}{L_D L_B}\right)\sin(\omega\, t \pm 90°) \tag{4}$$

In (3) and (4), $\sqrt{2}V_S$ and ω represent the peak value of the signal, and the frequency of the signal, respectively. It is clear from (3) and (4) that the output voltages are proportional to.

the difference in the associated inductance values. The nominal inductance value of all the coils (L_A, L_B, L_C, and L_D) is considered as L_0. The inductance values of these coils are equal in the absence of the rotor disc, where $L_A = L_B = L_C = L_D = L_0$. In that condition, the current flowing through both the coils, in a coil pair, will be equal in magnitude and opposite in direction, and the output voltages, V_{O1} and V_{O2} will be zero. Depending on the rotor's angular position, the inductances of coils A, B, C, and D will change to $L_0 + \Delta L_A$, $L_0 + \Delta L_B$, $L_0 + \Delta L_C$, and $L_0 + \Delta L_D$, respectively. Here, the changes in inductance due to the rotor's rotation, in L_A, L_B, L_C, and L_D, are

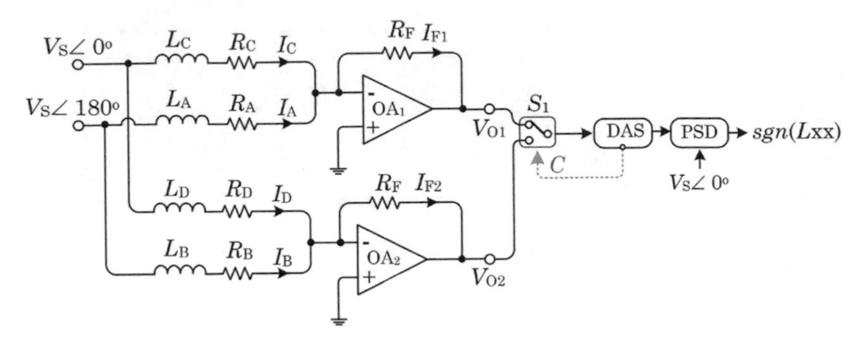

Fig. 3. Circuit to measure the differential inductance L_{CA} and L_{DB}

represented by ΔL_A, ΔL_B, ΔL_C, and ΔL_D. In general, if the change in inductance (ΔL) is relatively small in comparison to its nominal value (L_0), the term $1/(L_0 + \Delta L)$ can be approximated as $(1/L_0)(1 - (\Delta L/L_0))$. . Then, (3) and (4) can be simplified and the RMS values of V_{O1} and V_{O2} can be expressed as (5) and (6).

$$V_{O1R} = \frac{V_S R_F}{\omega}(\Delta L_C - \Delta L_A) \tag{5}$$

$$V_{O2R} = \frac{V_S R_F}{\omega}(\Delta L_D - \Delta L_B) \tag{6}$$

In (5) and (6), V_{O1R} and V_{O2R} represent the RMS values of voltages V_{O1} and V_{O2}, respectively. By dividing (5) by ($V_S.R_F/\omega$), L_{CA} is obtained. Similar to that, L_{DB} is derived from (6). The estimated values of L_{CA} and L_{DB} vary as cosine and sine functions respectively. For a range of 0° to 180°, $L_D < L_B$, and the phase of the signal L_{DB} will be + 90° and sgn(L_{DB}) = + 1. For 180° $< \theta \leq$ 360°, the phase value of the L_{DB} will be − 90°, and sgn(L_{DB}) = − 1. Similarly, sgn(L_{CA}) = − 1, if 90° $< \theta \leq$ 270°, and sgn(L_{CA}) = + 1, otherwise. These phase changes can be measured with the help of a phase-sensitive detector (PSD) [11]. The obtained value of V_{O1} together with the phase information of V_{O1} and V_{O2} can be used to determine the angular displacement (θ_r), using (1) and (2). The obtained angular displacement is shown, together with its error characteristics, in Fig. 4. It is important to note from Fig. 4 that, as we increase the value of offset inductance (L_0) of the coils, the nonlinearity of the sensor reduces significantly. The linearity of the sensor improves by a factor of 7.5 when L_0 = 400 μH compared to the linearity obtained with L_0 = 45 μH.

3 Ratiometric Measurement

Ratiometric methods are widely used for sensor signal processing. The advantage of ratiometric measurement is that it eliminates errors brought on by ambient temperature drift and enhances sensor linearity [4, 12]. Furthermore, this technique eliminates the excitation source dependence of the sensor output. The delta-over-sum approach [13], which calculates the sensor output as the ratio of the difference between two signals and the sum of two signals, is a typical technique for ratiometric measurement.

For the sensor reported in [9], the ratiometric calculation of two coil pairs, C-A and D-B, can be expressed as $R_{CA} = (L_C - L_A)/(L_C + L_A)$ and $R_{DB} = (L_D - L_B)/(L_D + L_B)$ respectively. These ratio metric measurements provide an excellent R^2 value of 0.9999 when fitted with a sinusoidal curve. R_{CA} and R_{DB} can be expressed as $R_{CA}(\theta) = 1.26 \times 10^{-4} + 0.0732\, sin(0.0174\theta - 1.56)$, and $R_{DB}(\theta) = -6.84 \times 10^{-7} + 0.0732\, sin(0.0174\theta - 3.13)$, respectively, based on the curve fitting. Since the ratio metric values for coil pairs, C-A and D-B, are seen to be sine and cosine functions, respectively, the θ_r value can be obtained from the curve fitting by applying a similar procedure used in (1) and (2).

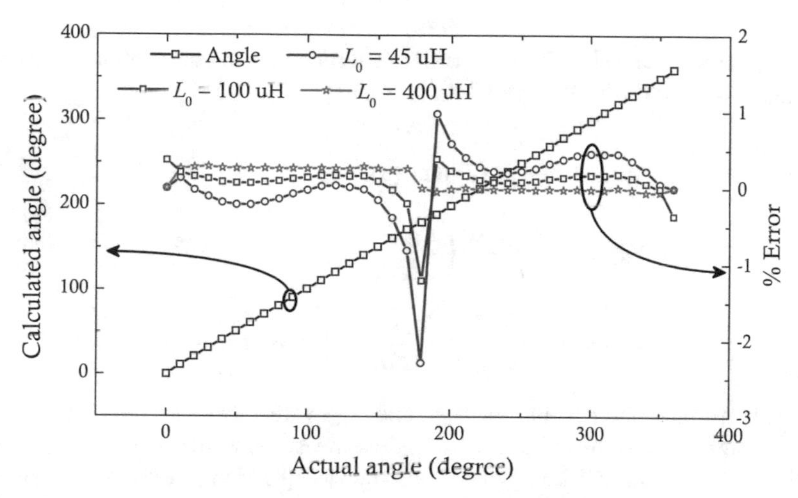

Fig. 4. Calculated output and error characteristics. The percentage nonlinearity of the output is plotted in the secondary y-axis, for different values of offset inductance (L_0).

3.1 Measurement Circuit

The ratiometric inductance measurement is carried out via a circuit presented in Fig. 5. Both coil pairs, C-A and D-B, are connected to their corresponding current-to-voltage (*i*-to-*v*) converters. The coils L_C and L_D are constantly connected to the sinusoidal voltage source $V_S\angle 0^o$ while the coils L_A and L_B are connected either to $V_S\angle 0^o$ or $V_S\angle 180^o$ using single pole double throw (SPDT) switches, S_1 and S_2 respectively. Switch S_1 is controlled by C_1 from the microcontroller (µC), while S_2 is controlled by C_2.

The proposed measurement consists of two measurement cycles. During the first measurement cycle, the microcontroller sets switches S_1 and S_2 to position 1. Thus the coils L_A and L_B are excited from $V_S\angle 180°$. As a result, the signals V_{R1} and V_{R2} follow (3) and (4) respectively. The values of V_{R1} and V_{R2} will be stored in the microcontroller memory. During the second measurement step, the microcontroller sets switches S_1 and S_2 to position 2. In this condition, all the coils are excited using $V_S\angle 0°$. Now, it is

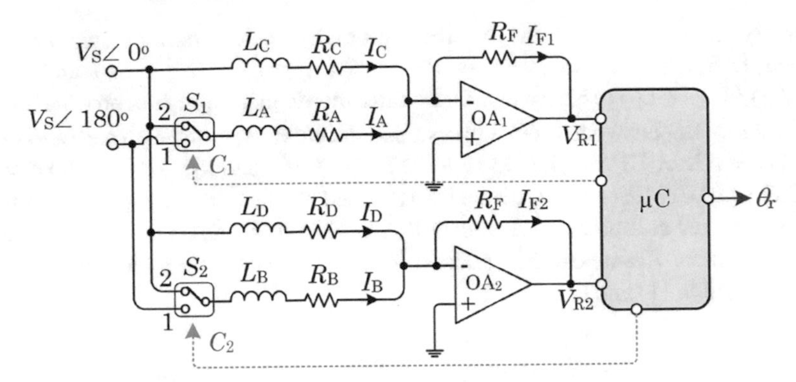

Fig. 5. Measurement circuit for the ratiometric calculation of inductance for the coil pairs L_{CA} and L_{DB}. μC represents the microcontroller.

possible to express V_{R1} and V_{R2} as (7) and (8).

$$V_{R1} = \frac{\sqrt{2}\,V_S\,R_F}{\omega}\left(\frac{L_C + L_A}{L_C L_A}\right)\sin(\omega\,t + 90°) \tag{7}$$

$$V_{R2} = \frac{\sqrt{2}\,V_S\,R_F}{\omega}\left(\frac{L_D + L_B}{L_D L_B}\right)\sin(\omega\,t + 90°) \tag{8}$$

Now, if we take into account the V_{R1} and V_{R2} values from both the measurement cycles and perform the ratio of the corresponding measured values of the first cycle to the second cycle, the results will be equal to (3)/(7) and (4)/(8). Here, we can see that the RMS values of (3)/(7) and (4)/(8) are equal to R_{CA} and R_{DB}, respectively, and independent of V_S, ω, and R_F. Here, the phase information of R_{CA} and R_{DB} can be measured using the same micro controller. After determining the magnitude and phase values of V_{R1} and V_{R2} from the two subsequent measurement steps, the values of R_{CA} and R_{DB} are calculated, and the final angular displacement is obtained. From the simulated results, the output is determined as a linear function of displacement (θ_r) with the worst-case linearity error of 0.42%.

4 Experimental Results

In order to test the practicality of the differential measurement schemes, a prototype of the circuit was built and tested in the laboratory. We have fabricated a prototype of the first method (differential measurement scheme) to examine the sensor performance because the second technique (ratiometric approach) also derives from it. According to Fig. 3, the coil pairs C-A and D-B are linked to the *i*-to-*v* converter. The *i*-to-*v* converters, OA_1 and OA_2, were implemented in the prototype using the LF347 IC. The Tektronix AFG 3022B signal generator was used to create the sinusoidal excitation signals, $V_S\angle 0°$ and $V_S\angle\, 180°$. The excitation frequency, 15 kHz, is significantly lower than the self-resonant frequency (SRF) of the coils. An experimental setup was implemented, as depicted in Fig. 6, to assess the sensor performance at various rotaing speeds. The sensor's stator part was fixed to a 3D-printed support structure. The rotor of a dc servo motor, whose speed is controlled by a motor drive, was used to mount the prototype sensor. In order to provide the necessary speed and position control, the motor shaft was mechanically connected to a 500-line optical quad encoder with an accuracy of 0.18°. This encoder acted as a reference measurement and gave position feedback to the motor drive.

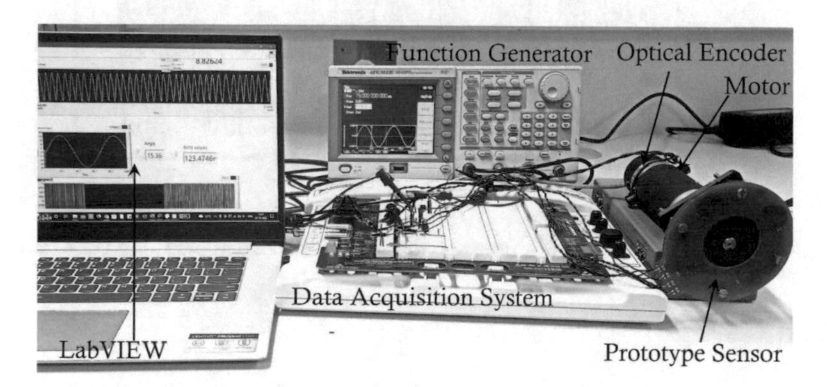

Fig. 6. Photograph of the experimental setup and measurement system

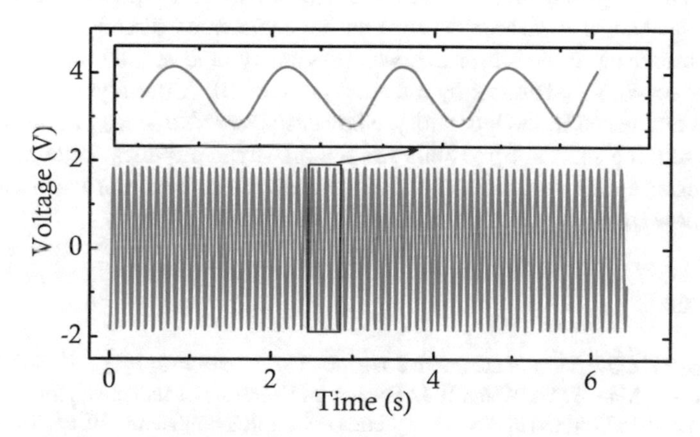

Fig. 7. Output voltage, V_{O1}, obtained from the prototype for the rotational speed of 475 rpm

The tests were performed for different rotating speeds. Then the voltages V_{O1} and V_{O2} were acquired and then processed using a single pole double throw (SPDT) switch and a data acquisition system (DAS)-NI ELVIS II$^+$. The SPDT switch was realized using IC DG419. θ is determined by the magnitude of V_{O1} and the sign of V_{O2}. Figure 7 shows the change in V_{O1}, against θ, for the rotational speed of 475 rpm. As expected, the RMS value of V_{O1} shows sinusoidal characteristcs with an R^2 value of 0.979, as illustrated in Fig. 7. Figure 8 shows the final output from the prototype. The worst-case error in the output is 2.3%. The prototype gave accurate results for speeds up to 500 rpm.

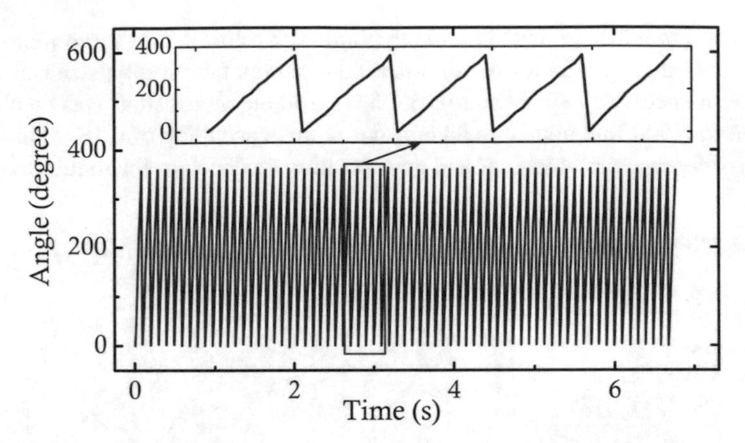

Fig. 8. Measured final output θ r, from the prototype, for the rotational speed of 475 rpm.

5 Conclusions

In this paper, two inductance measuring circuits have been developed and presented that are suitable for angle sensors that have differential sensing coils and sinusoidal responses. The proposed sensing circuits do not measure the individual inductances separately; instead, it measures the differential inductance directly from the pair of coils. In comparison to the sensor that was previously reported, this approach helps to improve the sensor's update rate by a factor of about 10. A prototype circuit has been constructed and tested in the laboratory. The sensor's performance has been evaluated for various rotating speeds. Up to 500 rpm, the prototype provides accurate results.

Adopting an appropriate data acquisition system can help to improve this even more. The worst-case linearity error of the sensor is found to be 2.3%.

References

1. Fleming, W.J.: Overview of automotive sensors. IEEE Sensors J. **1**(4), 296–308 (2001)
2. Horig, C.-F., Mao. S.-H., Wang, P. J.: Design and analysis of inductive reluctance position sensor. In: 2018 Progress in Electromagnetics Research Symposium (PIERS-Toyama), 2018, pp. 621–626 (2018)
3. GmbH, R.B.: Bosch automotive electrics and automotive electronics, pp. 208–231. Springer, Cham (2014). https://doi.org/10.1007/978-3-658-01784-2
4. Kumar, A.S.A., George. B., Mukhopadhyay, S.C.: Technologies and applications of angle sensors: a review. IEEE Sensors J. 21(6), 7195–7206 (2021). https://doi.org/10.1109/JSEN.2020.3045461
5. Anil Kumar, A.S., George, B.: A noncontact angle sensor based on Eddy current technique. IEEE Trans. Instrument. Measurem. 69(4), 1275–1283 (2020). https://doi.org/10.1109/TIM.2019.2908508
6. Noboru, T., Nobuo, S.: Variable reluctance type angle sensor, U.S. Patent 8947075B2, (3 Feb 2015)
7. Rotary Variable Differential Transformer (RVDT), document AS-827- 001, Moog Components Group (Dec 2016)

8. Sun, L., Taylor, J., Callegaro, A.D., Emadi, A.: Stator-PM-based variable reluctance resolver with advantage of motional back-EMF. IEEE Trans. Ind. Electron **67**(11), 9790–9801 (2020)
9. Kumar, A.S.A., George, B., Mukhopadhyay, S.C.: Design and development of a variable reluctance based thin planar angle sensor. IEEE Trans. Ind. Electron. (Accepted)
10. Philip, N., George, B.: Design and analysis of a dual-slope inductance-to-digital converter for differential reluctance sensors. IEEE Trans. Instrum. Meas. **63**(5), 1364–1371 (2014)
11. Pallas-Areny, R., Webster, J.G.: Sensors signal conditioning, Hoboken, NJ. Wiley, USA (2000)
12. Anandan, N., Varma Muppala, A., George, B.: A flexible, planar-coil-based sensor for through-shaft angle sensing. IEEE Sensors J. 18(24), 10217–10224 201
13. Vemuri, A.T., Matthew, S.: Ratiometric measurements in the context of LVDT-sensor signal conditioning Texas Instruments Incorporated, https://www.ti.com/lit/an/slyt680/slyt680.pdf?ts=1665204702537&ref_url=https%253A%252F%252Fwww.google.com%252F, (Accessed 08 Oct 2022)

Multi-sensing Platform Design with a Grating-Based Nanostructure on a Coverslip Substrate

J. J. Imas[1,2], Ignacio Del Villar[1,2], Carlos R. Zamarreño[1,2], Subhas C. Mukhopadhyay[3], and Ignacio R. Matías[1,2(✉)]

[1] Department of Electrical, Electronics and Communications Engineering, Public University of Navarra, 31006 Pamplona, Navarra, Spain
natxo@unavarra.es

[2] Institute of Smart Cities, Public University of Navarra, 31006 Pamplona, Navarra, Spain

[3] School of Engineering, Macquarie University, Sydney, NSW 2109, Australia

Abstract. Two different thin film designs with a grating pattern are simulated on a soda lime coverslip, which acts as optical waveguide, with the purpose of generating both a lossy mode resonance (LMR) in transmission and reflection bands. This way both phenomena can be made sensitive to different parameters, leading to a multi-sensing device. The first design consists of a grating patterned in a SnO_2 thin film deposited on the coverslip. The performance of the device in both transmission and reflection is numerically studied in air for different values of the grating pitch. Small grating pitches (in the order of the μm) are more suitable for generating the reflection bands while larger values (500 μm or more) are required to produce the LMR, when the reflection bands are no longer visible. Due to the inability to obtain both phenomena with this design, a second design is assessed, where the grating is combined with a section of constant thickness. In this case the desired response is obtained, which opens the path to use this device for multi-sensing applications, measuring several parameters at the same time.

Keywords: Sensor · lossy mode resonance (LMR) · coverslip · gratings · Waveguides

1 Introduction

Lossy mode resonances (LMRs) occur when the real part of the permittivity of the thin film material is positive and higher in absolute value than its imaginary part and both the real part of the permittivity of the optical waveguide and the surrounding medium. Metal oxides and polymers are the materials that fulfil these conditions [1]. LMR based sensors have multiple applications including refractive index sensing, gas detection, humidity sensing, pH sensing or biosensing [1–3], where the waveguide is usually an optical fiber.

However, in recent years, LMRs have also been obtained employing planar waveguides (glass slides, coverslips). Their main advantage over optical fibers is that they are more robust and cost-effective. In addition, only a polarizer is needed in the case of

N. K. Suryadevara et al. (Eds.): ICST 2022, LNEE 1035, pp. 450–459, 2023.
https://doi.org/10.1007/978-3-031-29871-4_46

planar waveguides, as opposed to optical fibers, where both an inline polarizer and a polarization controller are required to properly tune the polarization state of light [4].

Different sensors have been developed based on planar waveguides, most of the times measuring two parameters at the same time as in [5] (refractive index and temperature), [6] (temperature and humidity), or [7] (refractive index in two different liquids). This is achieved by depositing two thin films with different thicknesses on the same side [5, 7] or on both sides [6] of the planar waveguide. The thicknesses have to be carefully chosen so that both resonances are in the wavelength range under study but far enough so they do not overlap.

Following this tendency of developing multi-sensing platforms based on planar waveguides and LMRs, the purpose of this work is to explore, by means of theoretical simulations, the possibility of introducing a grating pattern on a thin film deposited on a coverslip. This way, the thin film will produce the LMR in the transmission spectrum while the grating pattern will generate bands in the reflection spectrum. Each of these phenomena could be employed to measure a different parameter. In the first place, a design consisting of a pure grating pattern is assessed, and then, the combination of a grating pattern and a section of constant thickness is analyzed.

2 Theory and Simulations

The structure under study is a soda-lime coverslip on which a SnO_2 thin film is deposited and different thin film patterns are considered. The material employed for the films is SnO2, which fulfils the conditions for generating LMRs [1, 8]. Two different designs have been studied, whose results are shown in the following section: a grating pattern with a 50% duty cycle (only half of each pitch is covered by the thin film) and the combination of a grating pattern with a section of constant thickness. Two possibilities have been considered for a subsequent practical implementation: photolithography and a focused ion beam scanning electron microscope (FIB-SEM).

A schematic representation of the two structures that are simulated, as well as a close-up of the section and the top-view of the grating pattern are shown in Fig. 1. Regarding the section, the thin film has a thickness of 160 nm (this choice will be explained afterwards) while the coverslip and the poly(methyl methacrylate) (PMMA) substrate have a thickness of 150 μm and 5 mm, respectively. The coverslip acts as the optical waveguide, and the role of the PMMA substrate is to support the structure. A total length of 20 mm is considered, which is a standard value for a coverslip.

The gratings patterned on the thin films are known as Bragg gratings [9, 10] (other names are also used such as Bragg stacks [11], multilayer stacks [12, 13], or micro-gratings [14]) and they produce bands in the reflection spectrum whose central wavelength is given by the Bragg-Snell equation for normal incidence:

$$m\lambda = 2\left(n_{effH} w_H + n_{effL} w_L\right)(1) \tag{1}$$

where m is the diffraction order and n_{effH}, n_{effL}, w_H and w_L are the effective refractive indices and widths of the sections with the high- (H) and low- (L) refractive index materials respectively. As it has be decided to employ a 50% duty, $w_H = w_L$.

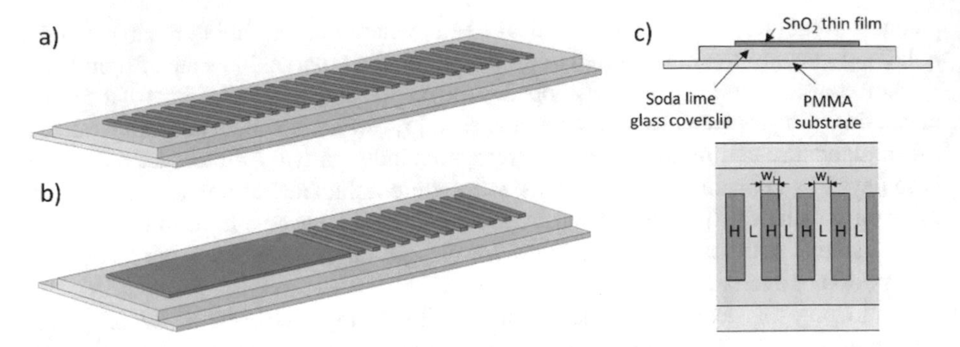

Fig. 1. 3D schematic diagram of a coverslip with a grating pattern on a thin film. b) Same as a) for a coverslip with a thin film with a section of constant thickness and a section with a grating pattern. c) Section of the coverslip and top view of the gratings.

In the structures under study, n_{effH} and n_{effL} correspond to the effective refractive index of the sections with and without thin film, respectively. Nevertheless, the SnO_2 thin film does not appreciably affect the effective refractive index of the corresponding section. Therefore, it can be considered that $n_{effH} = n_{effL} = n_{eff}$, where n_{eff} is defined as the effective refractive index of the structure. If it is considered that the addition of w_H and w_L is equal to the pitch of the grating (Λ), then Eq. (1) can be simplified in our case to:

$$m\lambda = 2n_{effH}\Lambda(2) \tag{2}$$

The structures studied in this work were analyzed with the commercial software FIMMWAVE. The propagation was calculated with FIMMPROP, a module integrated with FIMMWAVE. For the section, a non-uniform mesh of 500 elements was used in the Y direction. In the case of the PMMA substrate, only a thickness of 30 μm is considered in the simulation (as stated above, it only supports the structure, not affecting the propagation of light, so it is not necessary to consider the total thickness of this layer). The non-uniform mesh is used in the Y direction because the SnO2 thin film is much thinner than the rest of the layers. With respect to the X direction, only one element is required in the mesh as the program establishes that there are not going to be changes in power along the X axis.

Regarding the materials, for the SnO_2 thin film, a refractive index of 1.9 + 0.01i was used based on ellipsometric measurements in [15]. The coverslip is made of soda lime glass, whose refractive index model was taken from [16], while the model for PMMA is obtained from [17]. The surrounding medium refractive index (SRI) is equal to 1 (air).

A schematic diagram of the setup that will be used for interrogating the structures under study is depicted in Fig. 2, in which it will be required to measure both the reflection and the transmission spectra. A halogen light source (for example, TAKHI Halogen source, from Pyroistech S.L., was employed in [5]) will be connected to input 1 of an optical circulator (for instance, WMC3L1S from Thorlabs). One end of the device under study will be connected to output 2 of the optical circulator and the other end will be connected to a spectrometer to measure the transmitted power (such as a NIRQuest

NIR spectrometer, from OceanOptics). A linear polarizer (LPVIS050 from Thorlabs was used in [5]) will be placed between the output of the optical fiber that couples the light into the coverslip and the coverslip itself [6] to control the polarization state of light. Output 3 of the optical circulator will be connected to another spectrometer in order to measure the reflected power. All the connections are made with multimode optical fibers.

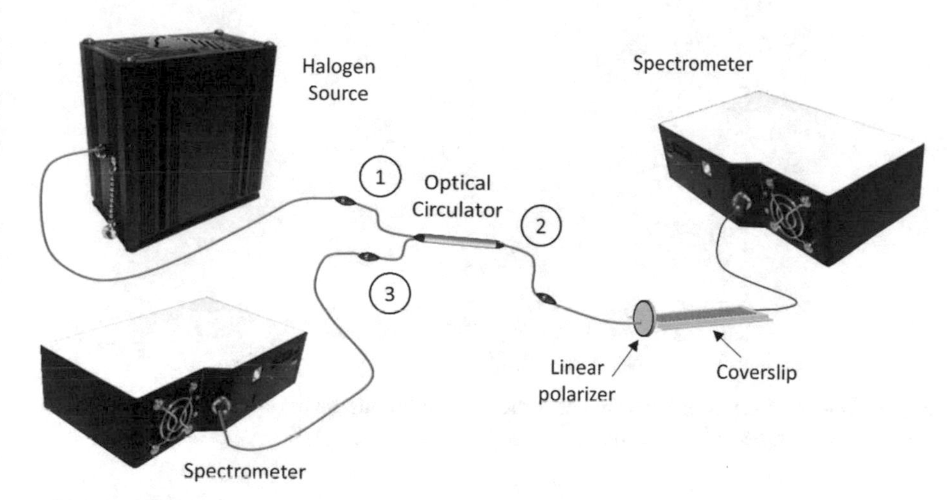

Fig. 2. Schematic of the setup that would be required for the interrogation of the device.

3 Results

3.1 Grating Pattern

In the first place, the structure consisting of a coverslip with a SnO_2 grating pattern on top of it was studied. Different grating pitch values, from 4 μm to 20,000 μm, were assessed, keeping in all cases a total length of 20 mm. Therefore, the number of periods varies from 500 (4 μm grating period) down to 1 (20,000 μm). The thickness of the SnO_2 thin film is 160 nm so the LMR is centered in the studied wavelength range (1200–1800 nm). Transverse electric (TE) polarized light is employed. It has been decided to use TE polarization as it gives better results than transverse magnetic (TM) polarization in the case of the D-shaped optical fiber for the same thin film design [18].

The transmitted power for each simulated value of the grating pitch is shown in Fig. 3. In all the cases, the transmission spectrum has been obtained after subtracting the transmitted power corresponding to the blank device (coverslip with a length of 20 mm and no thin film). It can be appreciated that, with the lowest pitch values (4 μm, 8 μm, and 10 μm), no resonance appears. If the grating pitch is increased by one order of magnitude (50 μm, 80 μm, and 100 μm cases), it seems that a resonance starts to appear, but it still is not very clear. It is required to increase the pitch another order of magnitude (500 μm, 800 μm) to clearly see the generation of an LMR. If the grating pitch is further increased (20,000 μm; which corresponds to only one period and can be hardly considered a grating), the LMR gains in definition, and it can be observed that it has shifted to the red in comparison with the previous cases.

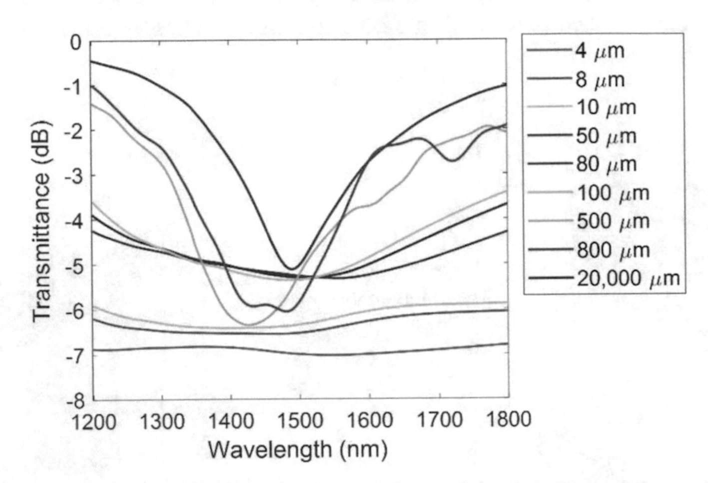

Fig. 3. Transmitted power for different values of the grating pitch (from 4 μm to 20,000 μm) over a total grating length of 20 mm.

Regarding the reflection bands, Table 1 includes their expected position using Eq. (2) and the ones obtained with the simulations (represented in Fig. 4a) for grating pitch values of 4, 8 and 10 μm, and it can be checked that they are very similar. The theoretical positions of the reflection bands have been calculated using $n_{eff} = 1.508$ in Eq. (2), which is the average refractive index value of soda lime glass in the studied range (from 1200 nm to 1800 nm) [16], as the effective index of the structure will be close to this value. For the sake of simplicity, only the values that belong to the studied wavelength range are shown in the table.

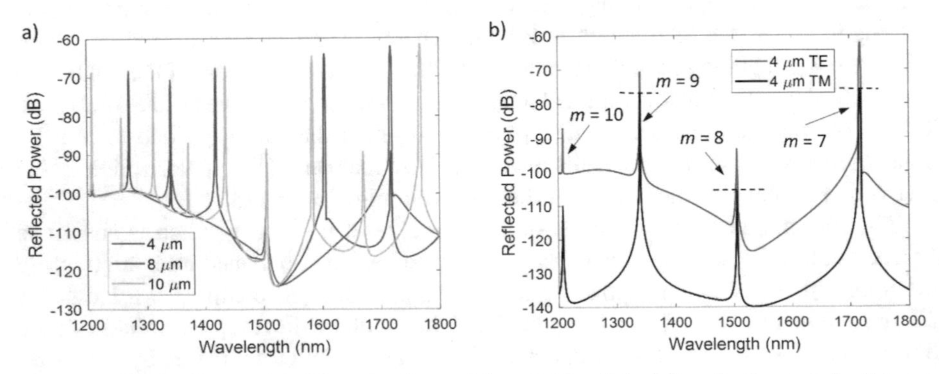

Fig. 4. Reflection bands for different values of the grating pitch a) Results for a pitch of 4 μm, 8 μm, and 10 μm (TE polarization in all cases) b) Results for a pitch of 4 μm for both TE and TM polarization.

Table 1. Theoretical and simulated wavelengths for the reflection bands for different values of the grating pitch.

m	Theoretical wavelengths from formula (nm)			Wavelengths obtained in simulation (nm)		
	Pitch			Pitch		
	4 μm	8 μm	10 μm	4 μm	8 μm	10 μm
7	1723			1718		
8	1508			1506		
9	1340			1341		
10	1206			1208		
11–13						
14		1723			1717	
15		1609			1605	
16		1508			1506	
17		1419	1774		1419	1768
18		1340	1676		1341	1671
19		1270	1587		1271	1584
20		1206	1508		1208	1506
21			1436			1435
22			1371			1371
23			1311			1312
24			1257			1258
25			1206			1208

It can also be checked in Fig. 4b, where the results for a grating pitch of 4 μm are represented for both TE and TM polarization, that the reflection bands reach a higher intensity in the TE polarized case, as it was expected (the dashed lines in Fig. 4b indicate the position of the maximum in the TM bands). All the other cases that are shown in this work correspond to TE polarization.

On the other hand, it can be observed both in Fig. 4a, and Fig. 4b that the reflected power of the bands with odd m is higher than that of the bands with even m (their values are indicated for the 4 μm grating pitch in Fig. 4b and the values of m for the 8 μm, 10 μm cases in Fig. 4a are included in Table 1). For instance, in the case of the 4 μm grating pitch, the reflection bands at 1718 nm ($m = 7$) and 1341 nm ($m = 9$) have a higher intensity than the ones at 1506 nm ($m = 8$) and 1208 nm ($m = 10$). These last two bands can be seen in Fig. 4b, but not in Fig. 4a, where they are hidden by more intense reflection bands corresponding to other values of the grating pitch.

However, considering this distinction between bands with odd and even m, there are no relevant differences among the bands corresponding to different values of the grating pitch in Fig. 4a (peak values of around −70 dB for odd m values and −90 dB for even m values).

If the grating pitch is increased up to 50 μm (see Fig. 5), more reflection bands are obtained in the same wavelength range, simply due to Eq. (2). Obviously, these reflection bands are closer to each other in comparison with the previously shown cases for lower grating pitch values. It can still be appreciated the difference in power between the bands with odd and even m (although this difference is less important at shorter wavelengths). Similar results (not shown for the sake of clarity) are obtained for 80 μm, and 100 μm (cases studied in in Fig. 3).

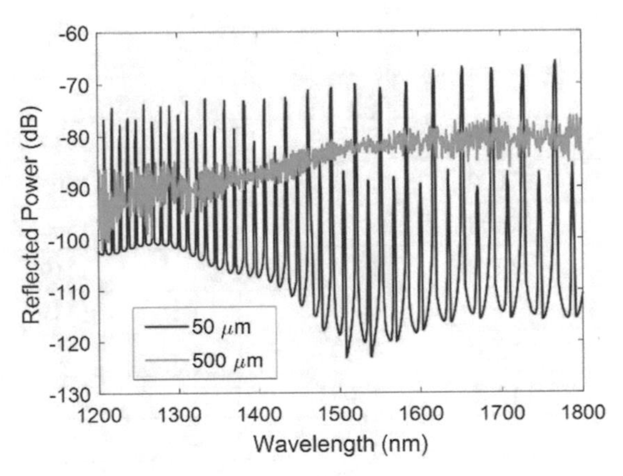

Fig. 5. Reflection bands for different values of the grating pitch (50 μm, 500 μm).

If the grating pitch is further increased to 500 μm (see again Fig. 5), the reflection bands are so close to each other that they are no longer distinguishable with the employed resolution. It must be considered that it was necessary to use this value for the grating pitch in order to generate an LMR (see Fig. 3). The same result (no reflection bands) is obtained for the remaining grating pitch values (800 μm, 20,000 μm) that have been studied in transmission.

The main conclusion for this device is that a value of the grating pitch that is suitable for generating the LMR (500 μm or higher), is not adequate for generating the reflection bands (they get too close to each other). On the other hand, when the reflection bands are clearly distinguishable (for instance, with a grating pitch in the order of the few μm), there is no LMR in transmission.

3.2 Grating Pattern with a Constant Thickness Section

The easiest option to solve the problems from the previous design seems to substitute a part of the grating by a section of constant thickness (see Fig. 1b). This way, the grating section (with a grating pitch in the order of the few μm) can produce the reflection bands, while the section of constant thickness generates the LMR. As in the previous case, TE polarization is used, the SnO_2 thin film thickness is 160 nm and the surrounding medium is air. The length of both sections (grating and section of constant thickness) is 10 mm, and three cases have studied: grating pitch of 4 μm (2500 periods), 8 μm (1250 periods) and 10 μm (1000 periods). The transmission spectrum and the reflection bands are shown in Fig. 6a, and Fig. 6b respectively.

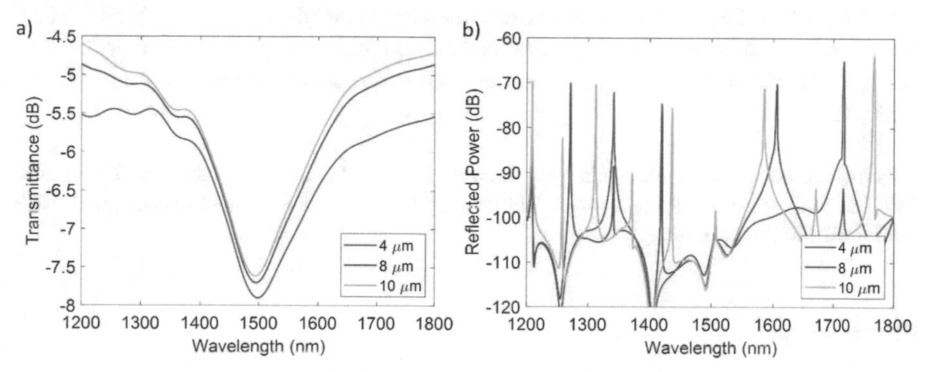

Fig. 6. Coverslip with a thin film with a section of constant thickness and a section with a grating pattern. a) Transmission spectrum for pitch values of 4 μm, 8 μm, and 10 μm. b) Reflection bands for pitch values of 4 μm, 8 μm, and 10 μm.

It can be observed that the desired response is obtained, that is, bands appear in the reflection spectrum due to the gratings and an LMR is generated in transmission thanks to the section of constant thickness. The position of the reflection bands is the same as in the previous device (see again Table 1 and Fig. 4a), which means we can still use Eq. (2). Comparing Fig. 6b with Fig. 4a, it can also be appreciated that the reflection bands in Fig. 6b are less intense when approaching 1500 nm due to the presence of the LMR at that wavelength. The reflection bands located at 1500 nm lose around 10 dB in Fig. 6b with respect to Fig. 4a, but the ones at around 1400 nm and 1600 nm also lose around 5 dB.

Nevertheless, even if the initial purpose has been achieved, it has to be considered that the LMR resonances are not very deep and the reflection bands are not very strong. The LMR resonance depth is in the range from 2.5 to 3 dB depending on the value of the pitch and the highest reflection bands achieve values around -70 dB, which are not very suitable for a practical implementation of this device. Different possibilities, including modifying the dimensions of the coverslip or employing other materials, could be explored in order to improve the response of the device.

4 Conclusions

Two different structures, both based on employing a coverslip as optical waveguide, have been theoretically studied in this work. The first one consisted on a grating pattern inscribed on a SnO_2 thin film deposited on the coverslip. It has been demonstrated that it is not possible to find a suitable value of the grating pitch for this design to produce both an LMR in transmission and reflection bands. This problem has been solved with the second design, which combines the grating with a section of constant thickness, generating both the LMR and the reflection bands.

The next step will be to manufacture this structure by means of a FIB-SEM or a photolithography process, although it is interesting to try to optimize the performance of the device in the first place, managing to obtain a deeper LMR and more intense reflection bands. Then, once the device has been fabricated the response of the LMR and the reflection bands should be characterized for two different parameters (for instance, the SRI and the temperature), opening the door to the development of a multi-sensing platform.

Acknowledgments. This research was funded by the Spanish Ministry of Universities (FPU18/03087 grant), and the Spanish Ministry of Economy and Competitiveness (PID2019-106231RB-I00 research fund).

References

1. Del Villar, I., et al.: Optical sensors based on Lossy-Mode resonances. Sens. Actuators B Chem. **240**, 174–185 (2017). https://doi.org/10.1016/J.SNB.2016.08.126
2. Chiavaioli, F., Janner, D.: Fiber optic sensing with Lossy Mode resonances: applications and perspectives. J. Lightwave Technol. **39**, 3855–3870 (2021). https://doi.org/10.1109/JLT.2021.3052137
3. Wang, Q., Zhao, W.M.: A comprehensive review of Lossy Mode resonance-based fiber optic sensors. Opt Lasers Eng **100**, 47–60 (2018)
4. Fuentes, O., Del Villar, I., Corres, J.M., Matias, I.R.: Lossy mode resonance sensors based on lateral light incidence in nanocoated planar waveguides. Scientific Reports **9**, 1–10, (2019):https://doi.org/10.1038/s41598-019-45285-x
5. Fuentes, O., Corres, J.M., Dominguez, I., Del Villar, I., Matias, I.R.: Simultaneous measurement of refractive index and temperature using LMR on planar waveguide. Proceedings of IEEE Sensors (2020), :https://doi.org/10.1109/SENSORS47125.2020.9278727
6. Dominguez, I., Del Villar, I., Fuentes, O., Corres, J.M., Matias, I.R.: Dually nanocoated planar waveguides towards multi-parameter sensing. Scientific Reports **11**(1), 11–8 (2021). doi:https://doi.org/10.1038/s41598-021-83324-8
7. Dominguez, I., Corres, J.M., Fuentes, O., Del Villar, I., Matias, I.R.: Multichannel refractometer based on Lossy Mode resonances. IEEE Sens. J. **22**, 3181–3187 (2022). https://doi.org/10.1109/JSEN.2022.3142050
8. Sanchez, P., Zamarreño, C.R., Hernaez, M., Matias, I.R., Arregui, F.J.: Optical fiber refractometers based on Lossy Mode resonances by means of SnO2 sputtered coatings. Sens. Actuators B Chem. **202**, 154–159 (2014). https://doi.org/10.1016/j.snb.2014.05.065
9. Ascorbe, J., Corres, J.M., Del Villar, I., Arregui, F.J., Matias, I.R.: Fabrication of bragg gratings on the end facet of standard optical fibers by sputtering the same material. J. Lightwave Technol. **35**, 212–219 (2017). https://doi.org/10.1109/JLT.2016.2640021
10. Gallego, E.E., Ascorbe, J., Del Villar, I., Corres, J.M., Matias, I.R.: Nanofabrication of phase-shifted bragg gratings on the end facet of multimode fiber towards development of optical filters and sensors. Opt Laser Technol **101**, 49–56 (2018). https://doi.org/10.1016/J.OPTLASTEC.2017.11.001
11. Pavlichenko, I., Exner, A.T., Guehl, M., Lugli, P., Scarpa, G., Lotsch, B. v.: Humidity-Enhanced Thermally Tunable TiO 2/SiO 2 Bragg Stacks. J. Phys. Chem. C **116**, 298–305 (2012). doi:https://doi.org/10.1021/jp208733t

12. Wilkens, V., Koch, C., Molkenstruck, W.: Frequency response of a fiber-optic dielectric multilayer hydrophone. In: Proceedings of the IEEE Ultrasonics Symposium, vol. 2, pp. 1113–1116 (2000). https://doi.org/10.1109/ULTSYM.2000.921520
13. Poxson, D.J., Schubert, E.F., Mont, F.W., Kim, J.K., Schubert, M.F., Chhajed, S.: Design of multilayer antireflection coatings made from co-sputtered and low-refractive-index materials by genetic algorithm. Optics Express **16**(8), 5290–5298 (2008). doi:https://doi.org/10.1364/OE.16.005290
14. Arregui, F.J., Matias, I.R., Cooper, K.L., Claus, R.O.: Fabrication of microgratings on the ends of standard optical fibers by the electrostatic self-assembly monolayer process. Optics Lett. 26(3), 131–133 (2001). doi:https://doi.org/10.1364/OL.26.000131
15. Chiavaioli, F., et al.: Femtomolar detection by nanocoated fiber label-free biosensors. ACS Sens. **3**, 936–943 (2018). https://doi.org/10.1021/acssensors.7b00918
16. Rubin, M.: Optical properties of soda lime silica glasses. Solar Energy Materials **12**, 275–288 (1985). https://doi.org/10.1016/0165-1633(85)90052-8
17. Sultanova, N., Kasarova, S., Nikolov, I.: Dispersion properties of optical polymers. Acta Phys. Pol A **116**, 585–587 (2009). https://doi.org/10.12693/APHYSPOLA.116.585
18. Imas, J.J.. Zamarreño, C.R., Del Villar, I., Matías, I.R.: Optimization of fiber bragg gratings inscribed in thin films deposited on D-shaped optical fibers. Sensors 21, 4056–4056 (2021). doi:https://doi.org/10.3390/S21124056

A Technology Review and Field Testing of a Soil Water Quality Monitoring System

Waqas A. K. Afridi[1](✉), Fowzia Akhter[1], Ignacio Vitoria[2], and S. C. Mukhopadhyay[1]

[1] Department of Engineering, Macquarie University, Sydney, NSW 2109, Australia
waqas.afridi@mq.edu.au

[2] Department of Electrical, Electronic and Communication Engineering, Public University of Navarra, Arrosadía Campus, 31006 Pamplona, Spain

Abstract. Soil water quality is one of the most influential factors in ensuring the productivity of agricultural farms. Soil water quality and soil quality are hugely dependent on each other. Hence, it is essential to have a clear understanding of the essential soil quality parameters and the existing technologies to detect those parameters. This paper briefly discusses the vital soil quality parameters for their significance towards fostering sustainable agriculture. Moreover, a technology review of recent studies has been critically analyzed, and their strengths and weaknesses have been addressed. Moreover, an Internet of Things (IoT)- enabled low-cost, low-power soil monitoring system has been proposed to overcome the drawbacks of the existing technologies. The initially developed system has been deployed in a residential garden for preliminary testing and results. However, the findings of the proposed system satisfy the expected outcome as the testing soil parameters, such as soil moisture content and temperature, vary accordingly with the increase in depth underneath the surface. Also, environmental parameters such as ambient temperature, carbon dioxide and humidity vary expectedly over day and night. Data obtained from this system will be beneficial to derive realistic water-balance estimations and sustainable agriculture decision-making.

Keywords: Groundwater · soil quality parameters · IoT · smart sensing systems

1 Introduction

Groundwater is an essential water source to meet the community's needs in many countries [1]. It helps supply water in urban and rural areas and reduces surface water scarcity [2]. The leading causes of groundwater pollution, however, are thought to be anthropogenic activities, including farming, industrial effluents, and improper waste disposal on the land surface [3, 4]. Global crop production has recently seen a progressive increase in the heavy use of agrochemicals (fertilizers and pesticides) in agricultural fields [5]. Agrochemicals are used by both large- and small-scale farmers to grow crops and boost agricultural yields. They have thus increased the rate at which they apply fertilizers and pesticides, which may affect the groundwater quality. However, other variables affect groundwater quality, including geological formation, soil type and permeability, depth

N. K. Suryadevara et al. (Eds.): ICST 2022, LNEE 1035, pp. 460–475, 2023.
https://doi.org/10.1007/978-3-031-29871-4_47

to the water table, precipitation levels, the aquifer's hydraulic conductivity, and the solubility of the rock components [6]. Groundwater is susceptible to contamination from various sources, including industrial and agricultural operations, changes in land use, and other activities. Poor groundwater management can result in multiple severe issues with water quality, including water that is unfit for eating by humans or other animals [7]. The primary groundwater quality parameters are salinity, acidity, nutrients, and contaminants such as heavy metals, industrial chemicals, and pesticides. Poor groundwater quality can have significant economic effects by lowering agricultural and horticultural productivity [8]. When polluted groundwater enters waterways and wetlands, it may harm the environment and affect ecosystems that depend on groundwater. Moreover, poor groundwater quality can seriously end people's health [9].

Groundwater quality can be managed by employing Internet of Things (IoT)-enabled sensing systems to track soil quality indicators such as moisture, temperature, pH, nitrate, phosphate, potassium, salinity, and organic carbon [10]. Storing the data into the cloud server and sharing those with specialists will enable farmers to receive professional advice from anywhere in the world. A typical schematic of significant groundwater components is shown in Fig. 1 [11]. For this reason, agricultural nations like Australia, New Zealand, Japan, and the USA are now interested in fusing technology with agriculture. It is essential to have a clear understanding of the crucial aspects of soil quality and their effects to develop and implement intelligent agricultural systems.

This review article addresses the factors that affect groundwater quality, the ideal value for each parameter to ensure sustainable agriculture, and the technologies currently used to measure those parameters. It also proposes a cost-effective, energy-efficient, innovative sensing system for intelligent farming. Adapting this system will help farmers identify issues affecting farm conditions and make decisions to boost output effectively based on real-time data.

2 Essential Soil Water Quality Parameters

More than seventy per cent of the earth's surface is comprised of water, which is fundamental to life on this planet. Typically, the soil inside the earth's surface is composed of a three-phase system that is solid (soil particles), liquid (water and solutes), and gas (air). When the soil is void of water, all the pores are filled with air; however, if the air is replaced with water, the soil is said to be saturated. When soil is water-saturated and the atmosphere is excluded for long periods, many soil organisms suffer from a lack of oxygen [12].

The relationship between soil and water is essential to soil organisms and plant life. Not only does soil water contain various chemicals that influence soil's physical, chemical, and biological properties, but its flow and retention inside the soil also play a vital role in the soil formation process over time [13]. This section will highlight some essential soil water quality parameters useful in water balance investigation and precision agriculture.

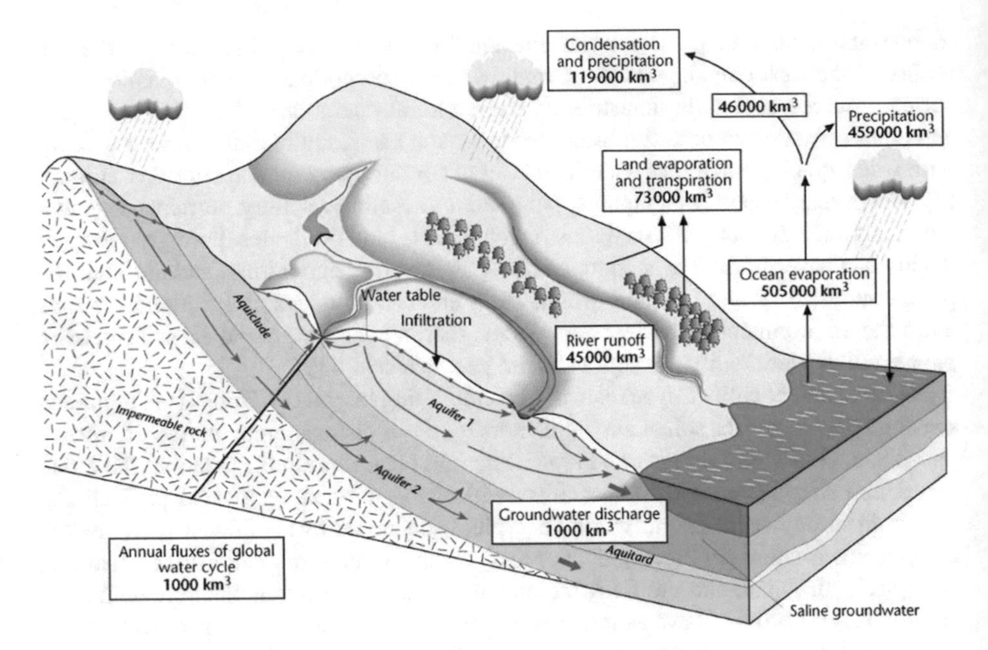

Fig. 1. Groundwater components in a typical hydrological cycle [11].

2.1 Soil Moisture Content

The soil moisture content is a prime environmental variable directly related to the process of evapotranspiration. Its measurement helps determine potential soil conditions for scheduling irrigation and crop yielding activities, predicting flood conditions, forecasting precipitation patterns, and eliminating wasteful water use. There is a range of scientific methods of measuring soil moisture content. Many devices have been developed over time for indirect soil moisture measurement, such as time domain reflectometry, the dielectric constant of soil, neutron scattering, thermalization of H atoms in soil, low-cost moisture resistance cells, electrical conductivity electrodes, watermark sensors, and more [14]. In our prototype system, we have used analogue capacitive soil moisture sensors at different depths to detect the moisture content in the soil.

2.2 Soil Temperature

Soil temperature is a crucial factor in all physical, chemical, and microbiological processes in soil. Especially in the evaporation process, it directly affects the water movement patterns and distribution in soil, aeration in pores such as the conversion of liquid to gaseous water in the form of water vapors. It also has a good influence on agriculture crop breeding and optimum vegetation capacity. Optimum temperatures are best suited for healthy soil organisms as it prominently increases the nitrification rate [15]. Electrical-resistance digital thermometers are widely used in soil monitoring applications to measure temperatures within the soil, depending on the degree of precision required. In our prototype system, we have used one-wire digital temperature sensors to find out inside soil temperatures.

2.3 Soil pH

The pH value is a defining quality parameter that controls the availability of essential chemical nutrients in the soil, which is helpful for optimum crop productivity and land treatment. It is a measure of hydrogen ions (H+) concentration that determines the acidic or alkaline nature of the soil. Though it usually ranges from 1–14 for different crop types, the ideal range is between 5.5 (slightly acidic) and 7.5 (slightly alkaline) for most crops' maximum productivity. Moreover, soils that face heavy rainfall are usually acidic since rainwater being somewhat acidic, reacts with carbon dioxide in the atmosphere to form carbonic acid and percolates through pores inside soil as bicarbonates which are critical for determining growth [16, 17]. The most common method for soil pH measurement is the colorimetric test using the test kit.

2.4 Soil Nitrate

A type of inorganic nitrogen (N) found in soils naturally is called nitrate (NO_3^-). Some earth sources are exudates from growing plants, chemical fertilizers, animal manure/compost, rainfall, and lightning [18]. The building blocks of life, nucleotides, amino acids, and proteins, including nitrogen, are among the required nutrients for plants. Only a few plant species can form symbiotic relationships with particular bacteria, which allows them to use atmospheric nitrogen. The resources for the other species are found in the soil, which contains nitrogen in various forms [19]. For instance, the soil solution might include several types of organic N, including soluble proteins or amino acids produced by proteolytic processes. In temperate climates, inorganic N forms predominate, and fertilizers are frequently delivered as nitrate, ammonium, or urea. Nitrate levels between 20–30 mg/Kg are acceptable for fertile soil [20].

2.5 Soil Phosphate

Phosphorus is one of the major plant nutrients in the soil. It is a constituent of plant cells, essential for cell division and the development of the plant's growing tip [21]. For this reason, it is vital for seedlings and young plants. Without phosphorus, plant growth is retarded. Plants have stunted roots and are checked and spindly. Deficiency symptoms include flat greyish-green leaves, the red pigment in leaf bases, and dying leaves [22]. Phosphorus deficiency is difficult to diagnose, and it may be too late to do anything when it is recognized. If plants are starved of phosphorus as seedlings, they may not recover when phosphorus is applied later—healthy levels of P in soil range from 25 to 50 ppm [23].

2.6 Soil Potassium

One of the essential soil minerals for plants is phosphorus. It contributes to the development of the plant's growing tip and cell division depending on this plant cell component [24]. It is essential for young plants and seedlings because of this. In the absence of phosphorus, plant growth is slowed. Plants are checked and wiry, with stunted roots [25]. Flat, greyish-green leaves, the red pigment in leaf bases, and decaying leaves are all signs of deficiency. When phosphorus deficiency is discovered, it is too late to act healthy P levels in soil range from 25 to 50 ppm. Therefore, if plants are starved of phosphorus as seedlings, they might not recover when phosphorus is added later [26].

2.7 Soil Salinity

The amount of salts in the groundwater influence osmosis as water enters plant roots. Moisture can return to the soil from plant roots if the salt content of the soil water is too high [27]. The plant becomes dehydrated as a result, which lowers yields or perhaps kills it. Even though the consequences of salinity may not be apparent, crop production losses might nonetheless happen. The capacity of a particular crop to draw water from salinized soils determines how well it can tolerate salt. Because salinity interferes with nitrogen uptake, stunts growth, and prevents plant reproduction, it impacts the production of crops, pastures, and trees [28]. Some ions, most notably chloride, are poisonous to plants, and when their concentration rises, the plant becomes poisoned and perishes. There are about 1600 mS/m of salt per meter where plants can survive [29].

2.8 Soil Carbon

Organic carbon is the term used to describe the carbon found in soil organic matter. An essential element of productive agriculture is soil organic carbon [30]. Numerous soil properties, including stability, enhanced water infiltration, aeration, and nutrient and water holding capacity, are influenced by organic carbon. Microbial activity is crucial for enhancing soil structure because it provides food for soil microbes and is an essential metabolite produced by bacteria. Soil microflora creates macroaggregates by using their secretions to bind soil particles together. These macroaggregates function as the foundation for bettering soil structure [31]. The capacity of the soil to retain water is increased through improved soil structure. For sustainable agriculture, the more carbon stored in the earth, the better. Although SOC can range from 0.3% in desert soils to 14% in intensive dairy soils, dryland agricultural soils typically have an organic carbon content of 0.7% to 4% [32].

3 Recent Advancements on Soil Water Quality Monitoring Systems

Soil water quality monitoring has been the prime component in recent years to ensure a sustainable natural environment while protecting land fertility and water wastage. Numerous research studies have been carried out in the second decade of the twenty-first century towards developing a robust and cost-bearing soil sensing system that can deter accurate water quality parameters inside the soil surface [33, 34]. However, conventional soil water monitoring techniques are somewhat laborious and time-consuming, require consistent laboratory instrumentation and are also not feasible when the soil sampling site is far from the testing laboratory. These limitations have been suppressed to a certain extent with the introduction of portable testing techniques such as microwave spectroscopy, remote sensing, and GIS methods without compromising measurement accuracy and instrument sensitivity. This section of the paper intends to review the existing studies conducted towards developing real-time soil monitoring systems with variable cost expenditure and appropriate system accuracy. The reviewed studies are then compared in a tabular form for their testing capabilities and technological drawbacks against our prototype soil sensing system, which is under ongoing design improvement and to be pre-tested for real field deployment.

C. Cojocaru et al. (2020) have used a Teralytic-made commercial three-layered soil probe to gather real-time ground information using the LoRaWAN communication network. The probe measures various parameters at different depths, such as; microclimate parameters like temperature and humidity at surface level, soil parameters including moisture, salinity, temperature, pH, and NPK nutrient monitoring, as well as gas parameters CO2 and O2. The system also features an automated dashboard warning application that ensures real-time alerts for farm users. The major drawback of the study is that it leased already established commercialized soil probes yearly for monitoring soil quality and maximizing crop yields requiring ongoing funding. Real insights are limited to the third-party company, which diminishes its significance in the research domain [35].

Y. Xu et al. (2022) have proposed an experimental investigation on the application of Software Defined Radio (SDR) based wireless soil sensing system by measuring the magnitude and phase responses at discrete frequency levels and applying a Fourier transform to visualize the time-domain reactions to track soil nutrient information. For this, the system utilizes a Surface Acoustic Wave (SAW) device with interdigitated transducers to convert the obtained electrical signal into an acoustic signal and reverse propagation at the output signal to excite the polymer sensor. LimeSDR-mini has been used as a low-frequency carrier to measure in-phase and quadrature (IQ) modulation signals to extract the output. Researchers have also simulated resistance variations of the polymer sensor using the surface mount device resistors on a designed circuit board. Later, the study used RMSE and R-Square analytical techniques to evaluate SDR experimental results with the standard Vector Network Analyzer (VNA) reference values to validate system results and performance. However, the overall experimental setup's signal strength is too weak to analyze, mainly due to smaller gain range settings resulting in clipping and unstable output power [36].

K. Y. Raneesh et al. (2021) have evaluated soil macronutrient detecting sensors and proposed a 3-in-1 prototype sensor gadget to aid farmers in maximizing crop yields by independently measuring on-farm soil NPK, moisture, and temperature. Different soil types (sandy, loam, silty clay, and sandy clay) have been tested for classification and estimating the optimum nutrients and irrigation required for each soil type based on equipped sensor readings in the study. Thereby, soil moisture content and temperature values are determined by measured resistance values in soil. Moreover, the readings from the three-legged instrument are communicated using Wi-fi and can be observed on a mobile application. However, the developed system is a small-scale testing tool that only can measure up to a 1-m distance in soil and is not a typical representation of the farm field. Moreover, the study hasn't used data analytics to validate instrument performance and results [37].

B. Kempegowda (2016) have integrated various available soil testing sensors with an ATmega328 microcontroller to develop a real-time soil monitoring system that can contribute to optimal crop production. The multi-sensor prototype system can measure a small area's soil moisture, pH, temperature, humidity, light intensity and carbon dioxide level. In the study, parameter values were captured using LCD for six days and compared against the standardized data to analyze different crop field degradation patterns. Although the study has not integrated any wireless communication protocol for remote monitoring and field implementation, the work done is a scientific contribution to improving agricultural practices [38].

M. Khaydukova et al. (2021) have carried-out traditional physicochemical quantification methods in the laboratory on a range of twenty soil samples extracted from different locations for the quick evaluation of soil macronutrients such as Nitrogen, Phosphorus, Potassium and for estimating quality parameters like pH, conductivity, and organic carbon contents. For this, a compact multisensory system comprising 26 potentiometric sensors was designed for immediate fertility testing in soil-water extracts. A multichannel digital voltmeter with high input impedance is used for sample data collection to determine soil properties. While multivariate regression methods have been implied on the acquired parameter dataset using the Partial Least Square modelling tool for interpreting parameter values and reliability assessment of NPK. The multisensory system was proposed for one-shot simultaneous quantification and estimation of the main soil nutrient parameters that are essential for soil fertility. However, the proposed testing method is performed in a controlled laboratory environment which is time-consuming and laborious, where the multi-sensor system is powered through a laptop and is not representative of actual circumstances. Hence various influencing factors may be contemplated when incorporating actual field deployment [39].

S. Bhaskar et al. (2021) have developed a multi-functional flower harvesting movable robot named as AGROBOT to assist farmers reduce their workload and risks in the field. The system incorporates existing advanced technologies like Image Processing, AI, ML, and IoT and an integrated electronic circuitry mainly comprised on microcontrollers, sensors, and drive motors to perform algorithm-based successive field applications in a farm such as detecting flowers, cutting and placing them into basket, detecting soil moisture, pH, and fertility, detecting pests and spraying the pesticides on plants, detecting the trespassers, and sending real-time alert messages to farm owners. Although, the prototype system has demonstrated ambitious range of farm applications for field farmers but the system has clear drawbacks such as sensor detection readings may not be accurate in many instances creating false alerts as well as incorrect mechanical operations. Moreover, there is a high risk of equipment malfunctioning while performing electro-mechanical operations in uncontrolled farming environment [40].

W. Zhao et al. (2022) carried-out an experimental study by creating different water-stress levels in winter wheat gradient fields to illustrate the importance of having an IoT based intelligent irrigation control system that can determine precise irrigation strategies and regulated treatments based on the physiological indicators and water-stress conditions. The IICS system was developed to monitor real-time soil moisture at different profile layers in the plotted fields and to perform automated irrigation application using PLC controller. The system incorporates Hydra Probe-II as multi-depth soil moisture sensors, MC302L as a low-cost low-power data collector integrated with solar charging controller, po-li battery, gprs/gps, true color touch screen, and other instruments. Furthermore, researchers have performed statistical modelling analysis using ANNOVA and LSD test to evaluate the response of tested indicators. The study concluded that the biomass, yield, and water use efficiency of winter wheat are not much affected in the mild water-stressed fields and are more suitable for irrigation applications than moderate or severe water-stressed fields. However, the limitation of the system is that the findings are limited to the tested crop type and it is not suitable for all soil types, and there is a wastage of fresh water due to the entire area being irrigated multiple times, also the li-po battery is minimal for system application [41].

C. Rusu et al. (2019) have developed a miniaturized real-time soil monitoring sensor to detect soil water content and electrical conductivity. The developed sensor was designed in a lab to use with two set frequency electrodes and a ground to measure electrical impedance of the soil described as resistance and capacitance that are translated into soil conductivity and soil water content, respectively. The soil sensor was tested and calibrated in a potted Jiffy soil media with various known volumetric soil water contents. The microprocessor is also used to digitally communicate the received responses from sensor electrodes via UART interface, while sensor data is read on a pc connected through usb-port also powers the soil sensor. Further, simulation analysis using COMSOL were also performed by the testing researchers to verify the obtained results from the sensor. However, the clear system limitation is that the sensor calibration is only specific to the small area of the tested soil pot with a conductivity range of 0–200 mS m − 1, also the researcher has not implemented any network protocol for real-time remote monitoring [42].

S. Millán et al. (2019) have proposed a water-balance algorithm using an automated irrigation control system mainly comprised of soil moisture sensors to automate irrigation scheduling under plum crop field conditions. The device comprises 15 capacitive soil moisture sensors placed at different depths and distances for continuous soil water content measurements. A cloud-hosted interactive web platform IRRIX was used to capture daily parameter data, data processing and analysis, and irrigation applications based on the feedback control algorithm, combining crop water-balance estimations and sensor readings adjustments. Other field equipment includes an air-temperature sensor, datalogger, solar panel with a voltage regulator, lead battery, digital water meter, and a relay controller for solenoid valve application. In the study, researchers have used three irrigation strategies to cover the crop water needs throughout the crop cycle. However, the weaknesses in the system are that it had not considered integrating soil moisture sensors data to determine automated irrigation applications. Also, the monitoring method adopted is not a real-time information system [43].

P. Placidi et al. (2021) have demonstrated the application of self-built IoT-enabled low-cost soil sensing nodes using LoRaWAN network protocol comprised of off-the-shelf components, including sensors, mainly capacitive soil moisture sensors v1.2 and soil temperature sensors. In the study, researchers have compared their obtained experimental results of two types of soils with an already established reference sensor bought from Sentek company to verify the reliability and performance of their sensor node. Measurements of the soil temperature showed good agreement with the referenced sensory system. In contrast, the capacitive soil moisture sensor has shown inconsistency in detecting accurate volumetric water content in the soil. The study has adopted a water infiltration and redistribution modelling approach and has performed statistical analysis to determine the possible correlation between the data values obtained at different depths in the experimented soil types. At the same time, a cloud-based virtual machine server is employed for database management. However, the testing sensors are calibrated in a controlled lab environment for a small bucket scenario and are not conceived for power optimization for long-term, large-scale applications [44]. Table 1 shows the comparative study of the existing groundwater quality monitoring systems.

Table 1. Comparison of recent soil water quality monitoring systems worldwide.

Testing place	Sensing tool	Measured parameters	Comm. technology	Type of analytics	Weakness	Ref.
Farm field, Romania	Meter long Teralytic soil probe	Moisture, salinity, temperature, pH, and NPK	LoRaWAN	Summarized dashboard analytics	Leased probe, constant funding, limited data access	[35]
Lab environment, USA	SAW device as RF detector	Soil nutrient	LimeSDR-mini as GPR reader	RMSE and R-Square analysis	Weak signal strength, unstable output power	[36]
Multiple soil samples, India	3-legged prototype sensor gadget	NPK, moisture, and temperature	Wi-Fi	Not performed	Small scale measurements, no system validation	[37]
Red soil sample, India	Multi-sensor prototype system	Soil moisture, pH, temperature, humidity, light intensity and CO2	Wired LCD	Data readings for decision making	No standard calibration, no wireless monitoring	[38]
Lab environment with twenty soil samples, Russia	Potentiometric multisensor system	NPK, pH, conductivity, carbon	Digital mV-meter	PLS regression modelling, reliability test	Powered by laptop	[39]
Farm field, India	Moisture hygrometer, pH sensor, electro chemical sensor	Soil moisture, pH, and fertility	Wi-Fi	Image Processing, AI, and ML	Inaccurate readings, false alerts, malfunctioning	[40]
Wheat gradient field, China	Hydra Probe-II	Soil moisture	GPRS	Statistical ANNOVA and LSD test	Crop specific, insufficient battery	[41]
Potted Jiffy soil media,	PCB elect. Impedance sensor	Soil moisture, and EC	Bluetooth	COMSOL simulated FE analysis	Limited calibration range, no remote access	[42]
Plum crop field, Spain	10HS capacitance probe	Soil moisture	Wi-Fi	Water-balance algorithm	No real-time sensor integration	[43]
Small bucket, Italy	Off-the-shelf sensors	Soil moisture, and soil temperature	LoRaWAN (TTN)	Statistical comparative analysis	Inconsistent moisture data, No power optimization	[44]

(continued)

Table 1. (*continued*)

Testing place	Sensing tool	Measured parameters	Comm. technology	Type of analytics	Weakness	Ref.
Residential garden, Australia	Low-cost, low power electrical sensors	Soil moisture, soil temperature, CO_2. tVOC, Amb. Temp, Humidity, Bar.pressure	LoRaWAN	Statistical water balance modeling (to be performed)	Battery optimization and performance validation	Proposed system

4 Proposed System

Our prototype system is a 3D-printed three-layered 600 mm long and 50 mm broad multi-functional soil sensing node instrument designed to be buried inside the soil in actual farming conditions for measuring soil water content and temperature. Each node layer comprised an anti-corrosive capacitive soil moisture sensor and a digital soil temperature sensor. The multi-functional sensor node also entails CCS811/BME280 environmental-combination sensor that can measure CO_2 (ppm), tVOC (ppb), humidity (g/Kg), temperature (°C), and barometric pressure (kPa). A high-strength digital tipping bucket rain sensor is the centerpiece of the node instrument to determine accurate rainfall patterns.

The real-time wireless sensing system utilizes a low-power, low-cost LoRa transceiver with a center-fed external dipole antenna as a WAN communication protocol for continuous data transmission over considerable distances. In our designed type, the LoRa communication shield is compatible with working on a 915 MHz frequency band with a data rate of <50 Kbps at a line-of-site of 10–15 km. The ThingSpeak cloud server is used as an IoT analytics platform to visualize and analyze the live data streams of the developed sensor node. Other main components of the prototype system include; the Arduino Mega 2560 microcontroller, which is the heart of the system that has 54 digital i/o's, 16 analogue inputs, a Stackable 6 V 6 W solar charger shield with environmentally friendly 3.7 V 6000 mAh rechargeable polymer lithium-ion battery used for adaptive power consumption and energy harvesting in the field environment, Qwiic shield to enable I2C bus on Arduino MCU in series to connect the environmental combination sensor. Figure 2 shows the connection diagram of the proposed system.

The electronic shields are glued and enclosed inside an electrical junction box to be used as a protective mounting device. The box enclosure is screwed properly and sealed with silicon epoxy coating to make the hardware design more water resistant, UV protective, and environmentally safe. For outdoor field installation, the system hardware is being designed to be deployed on a pointed metal fence post using pole mounting brackets to ensure a firm and robust hardware installation that can withstand harsh weather conditions.

At this study stage, the prototype system is under ongoing hardware design improvements, code development, and potential functional enhancements to accomplish long-term, large-scale application capability. The initially developed system was deployed in a residential garden for twenty days to perform preliminary testing and statistical analysis and to identify technical challenges encountered during remote installation.

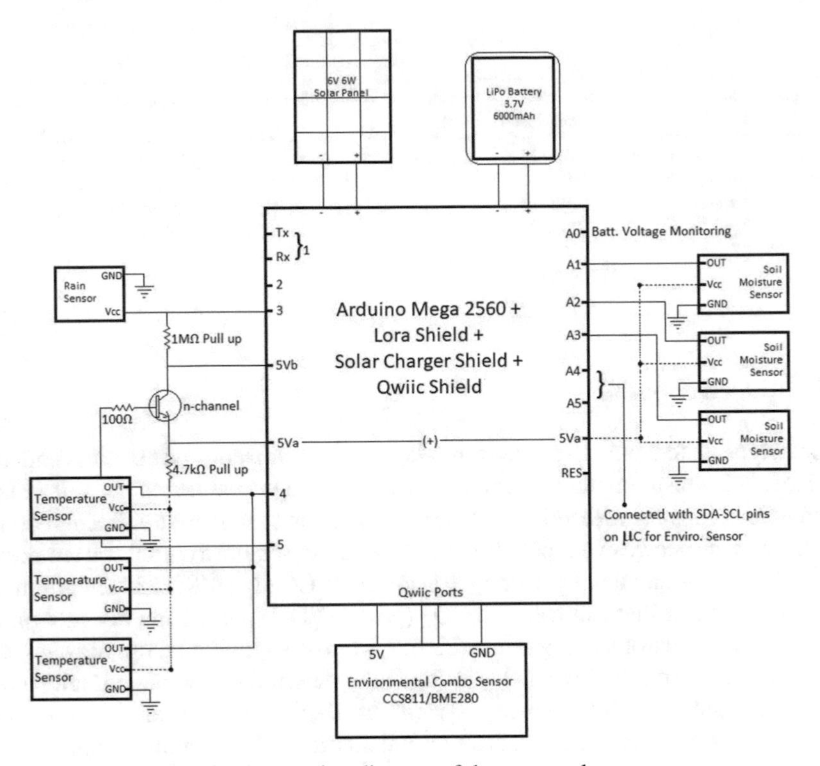

Fig. 2. Connection diagram of the proposed system.

4.1 Field Installation and Preliminary Results

The initially developed two soil sensing nodes were first experimented in the lab for electronic circuit testing, calibration, and reference value measurements before being deployed in a real field for obtaining experimental data in outside uncontrolled environment. Below Fig. 3 shows the installation of first two sensor probes at a residential garden. The installation time was more than two hours for two instruments. The main tool used for digging a 600mm deep and 50mm wide hole in the soil was a heavy-duty steel Giantz Power Augur which is compatible with most of power drills along with other necessary equipment available onsite.

Figure 4 shows the readings of the volumetric water content data obtained from the capacitive soil moisture sensors at three different depths of the soil that is top (20 cm), middle (40 cm), and bottom (60 cm). As can be seen the top layer of the surface shows more variations in moisture readings than the second and third layer of the probe due to mostly exposed to direct rainfall, sunlight, and other climatic conditions. Since moisture data is inversely proportional to the capacitive sensor readings. Thereby, on seventh and last day of the testing we had a good rain precipitation onsite which is translated into higher water content data on those days at the top layer of the surface from 48% to 72% and 40% to 68% of moisture readings. However, the middle and bottom layer sensors being less exposed to direct rainwater and sunlight show unfluctuated steadily declining readings maintained throughout the testing period which can be translated as

Fig. 3. Installation of two sensor nodes at a residential garden for pre-testing

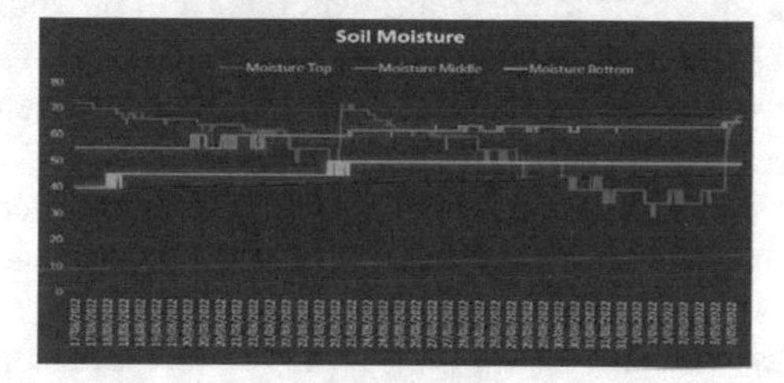

Fig. 4. Three-layered Soil moisture readings.

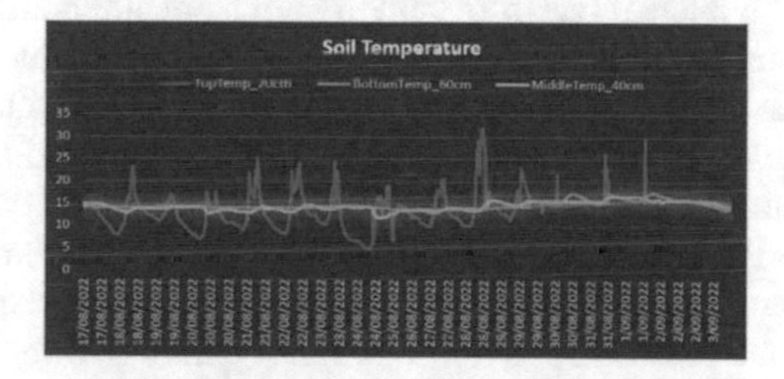

Fig. 5. Three-layered Soil temperature readings.

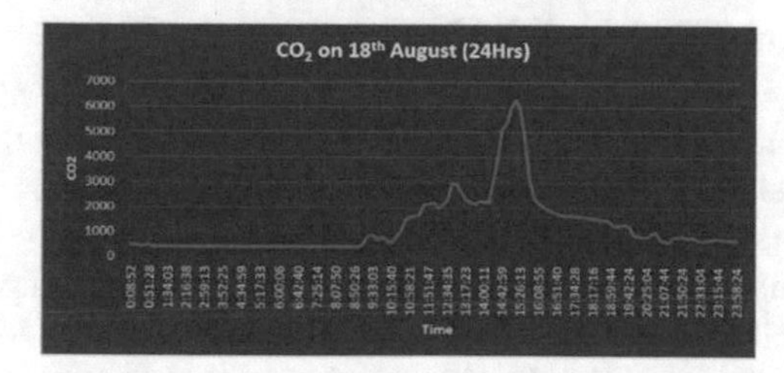

Fig. 6. Carbon dioxide emission during 24 h.

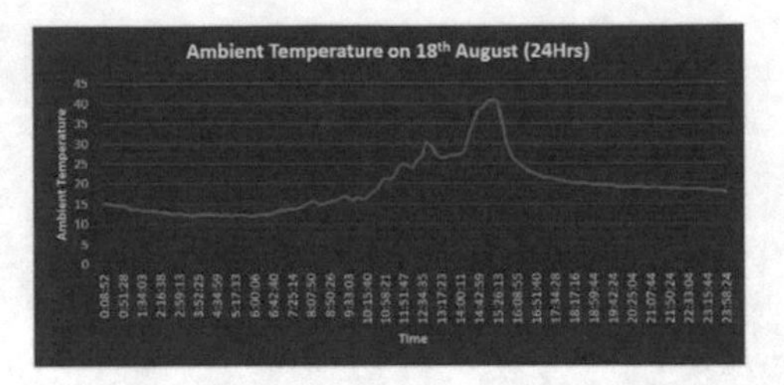

Fig. 7. Ambient temperature during 24 h.

middle layer being moderately moistured from 55% to 66% while the bottom layer is interpreted as close to a less moistured surface from 40% to 50% moisture readings.

Alongside, Fig. 5 shows the temperature readings from the digital soil temperature sensor along with the moisture sensor at three subsequent layers of study surface. In parallel with the top soil moisture sensor, the soil temperature sensor at the top layer also shows periodic variations in temperature readings from 5 °C to 30 °C due to first point of contact to rainwater and to daylight heat. However, the middle and bottom soil temperature sensors exhibit constant readings 15 °C maintained at 40cm and 60cm depth inside soil throughout this period.

Figure 6 shows the carbon dioxide CO_2 (ppm) concentration present in air over a full-day period acquired from the environmental sensor CCS881 used in the system. As can be seen the carbon dioxide in air remained unchanged during the night till 8:50am and has increased with fluctuations from 9am to 2pm from 400 ppm to 2200 ppm. After that, it has a further rapid increase from 2200 ppm to 6200 ppm during 2–3 pm while a quick drop in volume to 2200 ppm again in the next hour perhaps due to the increase in gas emissions by vehicles passing through the site. In the last 8 h till midnight, the CO_2 concentration continued to decrease gradually from 2200 ppm to 400 ppm. On the other hand, Fig. 7 presents the ambient temperature readings acquired from the environmental sensor BME280 over a full-day period. The ambient temperature seems to have higher values during day time reaches up to 40 °C while it exhibits lower values 15 °C amid night.

5 Conclusion and Future Works

All the critical parameters of groundwater and their impact on soil quality have been successfully demonstrated. Moreover, the optimum level of each factor to ensure healthy soil for sustainable agriculture has also been discussed. Additionally, technologies existing for soil quality monitoring systems have been discussed and compared. Furthermore, a self-contained intelligent soil sensor node has been proposed to monitor the condition of the soil and groundwater. The main benefits of using the proposed system compared to existing systems are incorporating inexpensive, low-power sensors and optimisation of wake-up time to increase the lifetime of the sensor node. The proposed system's implementation fee of USD 300 includes the cost of buying a few electronic components. The overall cost can significantly decrease if the product is produced in larger quantities. Having said that, the current IoT based project has significant future plans to implement in the proposed system. At the moment, the current soil sensor node is undergoing significant design and functional improvements for optimizing system longevity and results accuracy. Also, the fabrication of a soil quality detection sensor and its integration in a microcontroller-based IoT system is under investigation to be included in the system in near future. Moreover, with the help of real-time data analytics and evapotranspiration modeling, the system will be more capable of predicting accurate climatic conditions and profitable agricultural decision making. Owing to the fact, the expected outcomes of the project can be successfully achieved with distinctive quality and in minimum study period if it can attract substantial industry collaboration and commercialization. On the whole, the systems suggested in the study fulfil agriculture 4.0 goals, not only applying pesticides but also managing the farms using cutting-edge technology, such as sensors, machinery, gadgets, and IoT based communication. Any agricultural farm using these systems will be able to use modern technologies efficiently to make more profitable and productive decisions in the future.

References

1. Libra, R.D., Hallberg, G.R., Hoyer, B.E.: Impacts of agricultural chemicals on ground water quality in Iowa. In: Ground Water Quality and Agricultural Practices, pp. 185–215, October 2020. https://doi.org/10.1201/9781003069782-14
2. Riaz, U., et al.: Evaluation of ground water quality for irrigation purposes and effect on crop yields: a GIS based study of Bahawalpur. Pakistan J. Agric. Res. **31**(1) (2018). https://doi.org/10.17582/JOURNAL.PJAR/2018/31.1.29.36
3. Lenin Sundar, M., et al.: Simulation of ground water quality for noyyal river basin of Coimbatore city, Tamilnadu using MODFLOW. Chemosphere **306** (2022). https://doi.org/10.1016/J.CHEMOSPHERE.2022.135649
4. Rana, R., Ganguly, R., Gupta, A.K.: Indexing method for assessment of pollution potential of leachate from non-engineered landfill sites and its effect on ground water quality. Environ. Monit. Assess. **190**(1), 1–23 (2017). https://doi.org/10.1007/s10661-017-6417-1
5. Maurya, J., Pradhan, S.N., Seema, Ghosh, A.K.: Evaluation of ground water quality and health risk assessment due to nitrate and fluoride in the Middle Indo-Gangetic plains of India. Hum. Ecol. Risk Assess. An Int. J. **27**(5), 1349–1365 (2020). https://doi.org/10.1080/10807039.2020.1844559
6. Talebiniya, M., Khosravi, H., Zohrabi, S.: Assessing the ground water quality for pressurized irrigation systems in Kerman Province, Iran using GIS. Sustain. Water Resour. Manag. **5**(3), 1335–1344 (2019). https://doi.org/10.1007/s40899-019-00318-1
7. Hasan, M.N., Rahman, K., Tajmunnaher, Bhuia, M.R.: Assessment of ground water quality in the vicinity of Sylhet City, Bangladesh: a multivariate analysis. Sustain. Water Resour. Manag. **6**(5), 1–16 (2020). https://doi.org/10.1007/S40899-020-00448-X
8. Vijay, S., Kamaraj, K.: Ground water quality prediction using machine learning algorithms in R. Int. J. Res. Anal. Rev. **6**(1), 743–749 (2019). Accessed 02 Oct 2022. http://ijrar.com/
9. Sandhu, R.S., Sehgal, S.K., Amrit, K., Singh, N., Singh, D.: Environmental monitoring to assess ground water quality and its impact on soils in southwestern India. Acta Geophys. **70**(1), 349–360 (2022). https://doi.org/10.1007/S11600-021-00702-6/TABLES/4
10. Akhter, F., Siddiquei, H.R., Alahi, M.E.E., Mukhopadhyay, S.C.: Recent advancement of the sensors for monitoring the water quality parameters in smart fisheries farming. Computers **10**(3), 1–20 (2021). https://doi.org/10.3390/computers10030026
11. UNEP: Hydrogeological Environments (2003)

12. Lowery, B., Hickey, W.J., Arshad, M.A.C., Lal, R.: Soil water parameters and soil quality. Methods Assess. Soil Qual., 143–155 (1996). https://doi.org/10.2136/sssaspecpub49.c8
13. Goldhaber, M., Banwart, S.A.: Soil formation, January 2014. https://doi.org/10.1079/9781780645322.0082
14. Evett, S.R.: Some aspects of time domain reflectometry (TDR), neutron scattering, and capacitance methods of soil water content measurement. In: Comparison of Soil Water Measurement Usingthe Neutron Scattering, Time Domain Reflectometry Andcapacitance Methods, pp. 5–49, IAEA-TECDOC-1137 (2000). http://www.cprl.ars.usda.gov/programs
15. Tan, X., Shao, D., Gu, W.: Effects of temperature and soil moisture on gross nitrification and denitrification rates of a Chinese lowland paddy field soil. Paddy Water Environ. **16**(4), 687–698 (2018). https://doi.org/10.1007/s10333-018-0660-0
16. Oh, N.H., Richter, D.D.: Soil acidification induced by elevated atmospheric CO2. Glob. Chang. Biol. **10**(11), 1936–1946 (2004). https://doi.org/10.1111/j.1365-2486.2004.00864.x
17. Odutola Oshunsanya, S.: Introductory chapter: relevance of soil pH to agriculture. Soil pH Nutr. Availab. Crop Perform., 3–6 (2019). https://doi.org/10.5772/intechopen.82551
18. Akhter, F., Siddiquei, H.R., Alahi, M.E.E., Mukhopadhyay, S.C.: An IoT-enabled portable sensing system with MWCNTs/PDMS sensor for nitrate detection in water. Meas. J. Int. Meas. Confed. **178**, 1 (2021). https://doi.org/10.1016/J.MEASUREMENT.2021.109424
19. Akhter, F., Siddiquei, H.R., Alahi, M.E.E., Jayasundera, K.P., Mukhopadhyay, S.C.: An IoT-enabled portable water quality monitoring system with MWCNT/PDMS multifunctional sensor for agricultural applications. IEEE Internet Things J. **9**(16), 14307–14316 (2021). https://doi.org/10.1109/JIOT.2021.3069894
20. Alahi, M.E.E., Pereira-Ishak, N., Mukhopadhyay, S.C., Burkitt, L.: An internet-of-things enabled smart sensing system for nitrate monitoring. IEEE Internet Things J. **5**(6), 4409–4417 (2018). https://doi.org/10.1109/JIOT.2018.2809669
21. Akhter, F., Siddiquei, H.R., Alahi, M.E.E., Mukhopadhyay, S.C.: Design and development of an IoT-enabled portable phosphate detection system in water for smart agriculture. Sens. Actuators A Phys. **330**, 1–11 (2021). https://doi.org/10.1016/J.SNA.2021.112861
22. Nag, A., Alahi, M.E.E., Feng, S., Mukhopadhyay, S.C.: IoT-based sensing system for phosphate detection using Graphite/PDMS sensors. Sens. Actuators A Phys. **286**, 43–50 (2019). https://doi.org/10.1016/J.SNA.2018.12.020
23. Akhter, F., Siddiquei, H.R., Alahi, M.E.E., Mukhopadhyay, S.C.: Design, fabrication, and implementation of a novel MWCNTs/PDMS phosphate sensor for agricultural applications. In: Mandal, J.K., Roy, J.K. (eds.) Proceedings of International Conference on Computational Intelligence and Computing. AIS, pp. 303–308. Springer, Singapore (2022). https://doi.org/10.1007/978-981-16-3368-3_29
24. Kassim, A.M., Nawar, S., Mouazen, A.M.: Potential of on-the-go gamma-ray spectrometry for estimation and management of soil potassium site specifically. Sustain. **13**(2), 1–17 (2021). https://doi.org/10.3390/su13020661
25. Bhandari, S., Singh, U., Kumbhat, S.: Nafion-modified carbon based sensor for soil potassium detection. Electroanalysis **31**(5), 813–819 (2019). https://doi.org/10.1002/ELAN.201800583
26. Nawar, S., Richard, F., Kassim, A.M., Tekin, Y., Mouazen, A.M.: Fusion of Gamma-rays and portable X-ray fluorescence spectral data to measure extractable potassium in soils. Soil Tillage Res. **223**, 105472 (2022). https://doi.org/10.1016/J.STILL.2022.105472
27. Corwin, D.L.: Climate change impacts on soil salinity in agricultural areas. Eur. J. Soil Sci. **72**(2), 842–862 (2021). https://doi.org/10.1111/EJSS.13010
28. Zhao, C., Zhang, H., Song, C., Zhu, J.K., Shabala, S.: Mechanisms of plant responses and adaptation to soil salinity. Innov. **1**(1), 100017 (2020). https://doi.org/10.1016/J.XINN.2020.100017

29. Peng, J., et al.: Estimating soil salinity from remote sensing and terrain data in southern Xinjiang Province, China. Geoderma **337**, 1309–1319 (2019). https://doi.org/10.1016/J.GEODERMA.2018.08.006
30. Schmidt, S.A., Ahn, C.: Predicting forested wetland soil carbon using quantitative color sensor measurements in the region of northern Virginia, USA. J. Environ. Manage. **300**, 113823 (2021). https://doi.org/10.1016/J.JENVMAN.2021.113823
31. Mukhopadhyay, S., Chakraborty, S.: Use of diffuse reflectance spectroscopy and Nix pro color sensor in combination for rapid prediction of soil organic carbon. Comput. Electron. Agric. **176**, 105630 (2020). https://doi.org/10.1016/J.COMPAG.2020.105630
32. Shetti, N.P., et al.: Electro-sensing base for herbicide aclonifen at graphitic carbon nitride modified carbon electrode – water and soil sample analysis. Microchem. J. **149**, 103976 (2019). https://doi.org/10.1016/J.MICROC.2019.103976
33. Poojary, A.A., Bhat, M.S., Aishwarya, C.S., Mahesh, B.M.N., Shruthi, K.R.: Soil, water and air quality monitoring system using IoT. IRJET **7**(02), 416–418 (2020)
34. Loganathan, G.B., Mohan, E., Kumar, R.S.: IoT based water and soil quality monitoring system. Int. J. Mech. Eng. Technol. **10**(2), 537–541 (2019)
35. Cojocaru, C., Ene, A., Gojgar, A.F.: Farm's soil quality using wireless Npk sensor, pp. 3–7, November 2020
36. Xu, Y., Amineh, R.K., Dong, Z., Li, F., Kirton, K., Kohler, M.: Software defined radio-based wireless sensing system. Sensors **22**(17), 6455 (2022). https://doi.org/10.3390/s22176455
37. Raneesh, K.Y., Niveditha, J.V., Pramodh, S.P., Zainudheen, S., Sivan, V., Hemalatha, S.: Evaluation of a wireless insitu soil monitoring data collection system evaluation of a wireless insitu soil monitoring data collection system, November 2021
38. Kempegowda, B.: Real-time soil monitoring system for the application of agriculture a precision farming for water optimization view project, January 2016. https://doi.org/10.4010/2016.1304
39. Khaydukova, M., et al.: One shot evaluation of NPK in soils by 'electronic tongue. Comput. Electron. Agric. **186**, 106208 (2021). https://doi.org/10.1016/j.compag.2021.106208
40. Bhaskar, S., Pradeep Kumar, M., Avinash, M.N., Harshini, S.B.: Real time farmer assistive flower harvesting agricultural robot. In: International Conference for Convergence in Technology, I2CT 2021, pp. 1–8 (2021). https://doi.org/10.1109/I2CT51068.2021.9417817
41. Zhao, W., et al.: An IoT-based intelligent irrigation control system for precision irrigation : a case study in the Fangshan comprehensive experimental station. IEEE Sens. J. (2022)
42. Rusu, C., et al.: Miniaturized wireless water content and conductivity soil sensor system. Comput. Electron. Agric. **167**, 105076 (2019). https://doi.org/10.1016/j.compag.2019.105076
43. Millán, S., Casadesús, J., Campillo, C., Moñino, M.J., Prieto, M.H.: Using soil moisture sensors for automated irrigation scheduling in a plum crop. Water (Switzerland) **11**(10) (2019). https://doi.org/10.3390/w11102061
44. Placidi, P., Morbidelli, R., Fortunati, D., Papini, N., Gobbi, F., Scorzoni, A.: Monitoring soil and ambient parameters in the iot precision agriculture scenario: an original modeling approach dedicated to low-cost soil water content sensors. Sensors **21**(15) (2021). https://doi.org/10.3390/s21155110

Vertical Trajectory Analysis Using QR Code Detection for Drone Delivery Application

Avishkar Seth[1](✉), Alice James[1], Endrowednes Kuantama[2], Subhas Mukhopadhyay[1], and Richard Han[2]

[1] School of Engineering, Macquarie University, Sydney, Australia
avishkar.seth@mq.edu.au
[2] School of Computing, Macquarie University, Sydney, Australia

Abstract. The advent of the internet and fast-processing computers have enabled drones to fly autonomously for a variety of applications. Most of the research focuses on horizontal trajectory planning and mapping for autonomous navigation. In this study, we propose a method to address the urban last-mile drone delivery problem. The paper suggests an autonomous vertical trajectory scanning method that could be used to analyse the appropriate level and unit in an apartment building. The QR code embedded with the level and unit number information is used as a marker that can be detected by the drone's visual recognition framework. The suggested method aims to conduct real-time detection of the apartment at every level using consistent trajectory tracking. The experiments are tested indoors for 3 levels and 10 unique QR codes, comparing with 4 different trajectory planning patterns to analyse the most efficient trajectory. The parallel path is observed to be the most optimum for maximum area coverage and the quickest arrival to the desired destination.

Keywords: Drone sensing · Trajectory scan · Path planning · Navigation · QR code · Drone delivery

1 Introduction

In the recent years, Unmanned Aerial Vehicles (UAVs) have become one of the most available and acceptable vehicles for research, delivery, photography, security, monitoring and a variety of other use [1]. The increase in research interest and efforts aimed at enhancing this technology, autonomy for drones has become an inevitable reality. Autonomous systems aid in perception and navigation allowing the UAVs to traverse through complex environments with little to no involvement of a human pilot [2, 3]. Even though a lot of work has gone into resolving the shortcomings, a more effective and efficient approach still needs to be designed. The development and use of UAVs therefore greatly depend on the high-performance autonomous navigation capacity. Drone delivery is a very promising solution to the increasing logistics and delivery issue. This idea has been deeply explored in the private sector as well as university research. Recent literature in the related field discusses the current stress on the e-commerce businesses with

N. K. Suryadevara et al. (Eds.): ICST 2022, LNEE 1035, pp. 476–483, 2023.
https://doi.org/10.1007/978-3-031-29871-4_48

package delivery and customer satisfaction [4–6]. With the rising demand for a quick delivery and minimal cost solution, the competition to deliver a package to customers is rising. Parcels in the same locality are usually grouped together to minimize costs and save time. The transportation of a single parcel at the last mile is a challenge faced by these logistics providers. There are currently very limited solutions which address this issue. Most commonly, the road vehicle and drone approach are used for optimisation [7 8]. In this approach, a road vehicle (truck or van) carries the packages to a set location with multiple deliveries. Upon reaching the location, the last mile delivery is completed by the drones. Each drone carries a single package and transports it to the nearest parcel collection area.

Recently, there has also been research in urban last mile problem further addressing the delivery problem related to energy consumption [9] and delivery of advertising material [10]. A very popular method is adding visual markers strategically placed on an area such as balcony or window for the drone to find using visual sensors [11, 12]. The authors in [11] test out their custom drone in real world and show how it can search a drop off point on a balcony. An IoT based drone system for last mile delivery focussing on collision avoidance near building units has been tested[12]. Most of the literature methods used solely GPS to deliver the parcels without any visual navigation or recognition [13, 14], and isolated areas [15]. Thus, a need for more vision-based scanning is prevalent in the literature.

A very crucial part of drone delivery in an apartment type space is scanning for the correct unit number and floor number. A similar method of scanning is used in asset and inventory management in logistics businesses [16–19]. The drone-aided routing problems mostly occur in context of drone delivery application [20]. Their review discusses the positive aspects of drone delivery on the environment.

There has been immense research on horizontal path planning algorithms relying on different sensors. In this paper, we describe a method to deliver parcels directly to a multi storey apartment building with an open balcony. The paper focusses on the development of a vertical trajectory for the drone with a barcode type scanning algorithm. The described technique is proven to provide an effective sensing mechanism for drone deliveries. Drones could drastically reduce the cost and time associated with performing last-mile deliveries and reacting to crises. The use of drones in crisis scenarios has the potential to save lives since they may convey urgently required supplies of food, water, and medical equipment over dangerous terrain while also deploying wireless sensors to offer real-time reports on the situation.

2 Proposed Screening Pattern

Deliveries in high-rise apartments cannot be localised only by GPS, as it does not provide information of the apartment levels and position. GPS is often used by drones to pinpoint their location. It requires further scanning on a vertical trajectory with unique identifiers for every apartment.

The proposed software framework is described in this section. The proposed screening pattern is shown in Fig. 1. The process commences with the customer order. The delivery order is specific to apartments with balcony, once confirmed, the system then checks for the location details, i.e. the address and apartment level and number. The drone then checks for the set waypoints and then flies to the given destination. Once arrived at the Point of Interest (POI), the drone then begins the proposed flight trajectory to check the different levels. During the level scan, the QR code is scanned for every apartment in approximately 1 s. When the QR code is recognized and matched with delivery address, the drone will deliver the package at the given balcony. On completion, the drone will then return to the launch site.

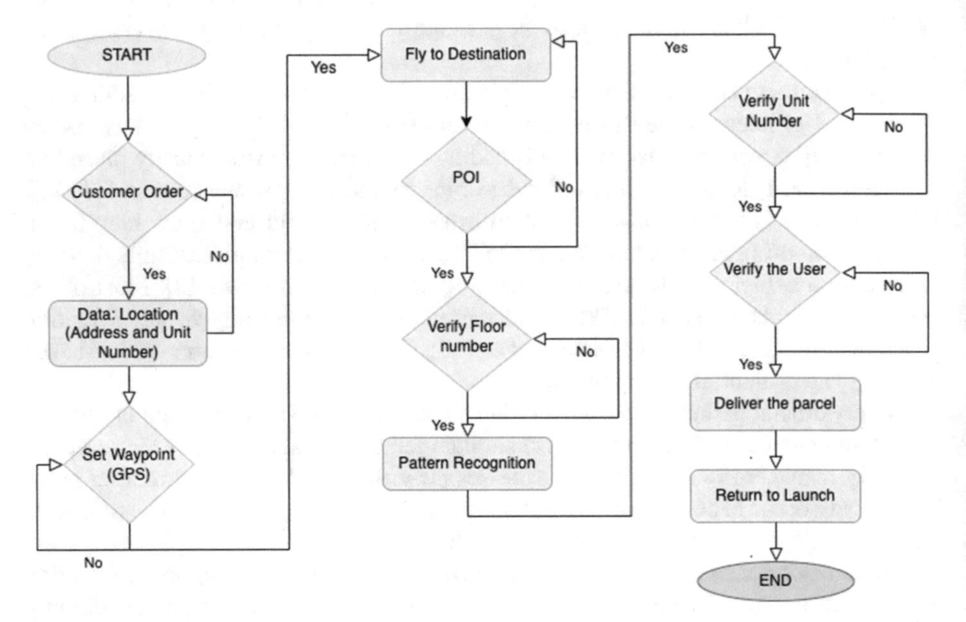

Fig. 1. Flowchart of the proposed drone delivery for apartments with balcony

The QR code identification process starts with the drone receiving the input image from its onboard camera connected to the microprocessor. The system consists of two processors, the on-board computer, and the flight controller. The processor is programmed to recognise the QR code from a predefined library of unique QR codes generated for this experiment. This message of the level and apartment number is matched with the stored value to the onboard computer. The microprocessor can run ROS software and uses the MAVROS and MAVLINK to create a message bus between flight controllers. The message sent to the microprocessor is decoded into IMU, GPS sensor outputs which autonomously move the drone into the desired direction. These values are then sent to the ESC (Electronic Speed Controller) which adjust the motor speed accordingly. At the ground station level, the drone is connected to the software via telemetry and RC Control for flight control and setup.

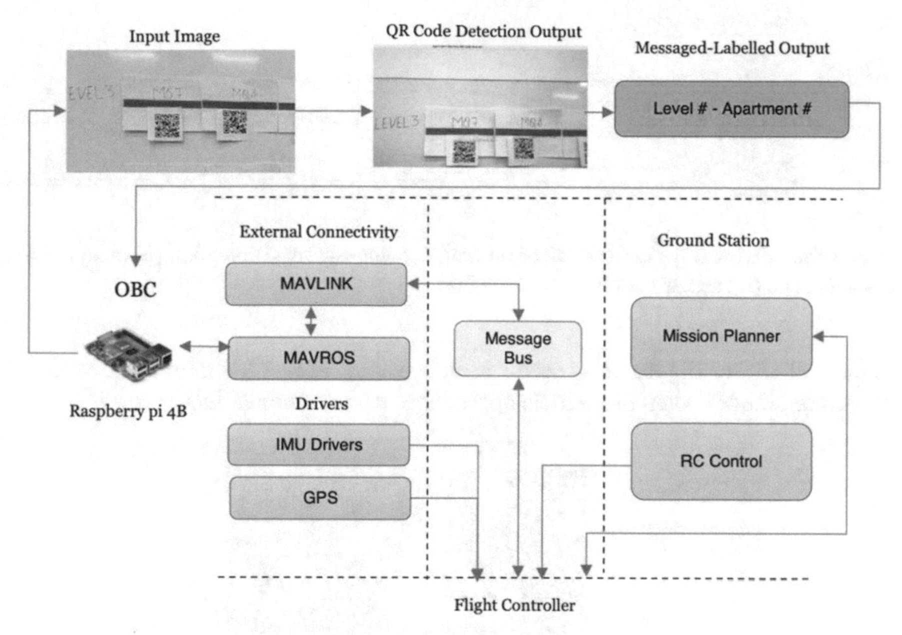

Fig. 2. The block diagram to implement QR code recognition for drone delivery system

3 Results and Tests

3.1 Drone Flight Test

Decomposition is often not necessary for coverage missions conducted with a single drone over regularly shaped and non-complex regions of interest. To investigate such locations, simple geometric designs are adequate. The spiral (SP) and back-and-forth (BF) patterns are the most prevalent types [20]. The Mission Planner, the most widely used flight-control programme, uses the back-and-forth pattern.

The parallel search pattern or the creeping line search pattern are both preferred when the information is not known with a high degree of accuracy and the search region is vast. The parallel pattern is preferable when the delivery apartment is equally likely to be in any section of the search area, whereas the creeping line is the ideal option when the apartment is more likely to be in one end of the search area than the other. The choice of which of these patterns to employ can also be influenced by the size of the search region. The delivery drone is anticipated to reach the delivery location using GPS, as every level of the apartment will be in the same location, the trajectory to determine the level and apartment is required.

As shown in Fig. 3 (i), (ii), the back-and-forth pattern in this study is divided into parallel and creeping lines, with the former being preferred as the search region is broad and there is no knowledge of the anticipated target meeting site for an apartment delivery. Straight lines and right-side 90-degree turn make up the square flight pattern. As seen in Fig. 2 (iii), the pattern is typically used when consistent area coverage is required. It begins at the centre and moves outwards towards the borders in a pattern resembling

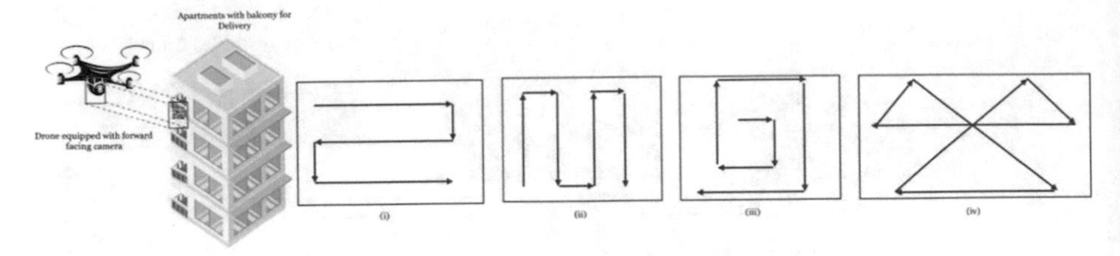

Fig. 3. Drone Flying trajectories in areas without decomposition: (i) parallel, (ii) creeping line, (iii) square, and (iv) sector search

an elliptical shape. Figure 2(iv) sector search pattern shows a straight line with 120 right-turning moves when the vehicle approaches the area's edge [20].

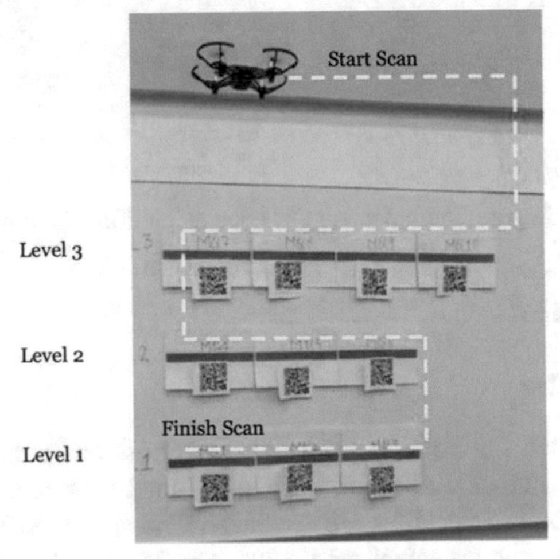

Fig. 4. An illustration of the drone and the parallel pattern programmed to perform the QR code detection

The proposed method aims to minimise the distance while an algorithm chooses the apartment for delivery and estimates the motions at maximum altitude in accordance with position of the marker. According to the, flying at an ideal speed can reduce the quantity of energy used. The Fig. 4 demonstrates the indoor test of the drone with an inclined forward-facing camera that is flying in the parallel pattern to determine the QR Code for delivery to the apartment. The pattern allows for sufficient time to scan the QR Code and determine the apartment for delivery.

3.2 Classification of Barcode at Different Levels

QR (Quick Response) codes have become very popular in many areas of applications due to its large storage capacity in a small area and high-speed scanning and reading process. Although QR codes can be found in many colours, the default black and white colour contrast provides an easier to scan method since these will be placed outdoors in different lighting conditions.

Scanning and recognition of a QR code is an elaborate process which is explained briefly here. Each QR code generated has the apartment information coded inside. The pattern co-ordinates embedded are recognised by the camera on the drone and sent to the on-board computer. The algorithm for reading QR codes consists of three steps: Pre-processing, binarization, and a decoding method are the three steps. It's important to remember that pre-processing happens after binarization. Figure 5 presents the key steps. A QR code input for a reading application is a grayscale image with a rotatable QR code. The position pattern localisation is used initially. Then, we use a re-sampling procedure to precisely determine the size and orientation of the QR code. The identification of white and black modules should then be finished before using the common decoding technique. The extracted message is what is produced.

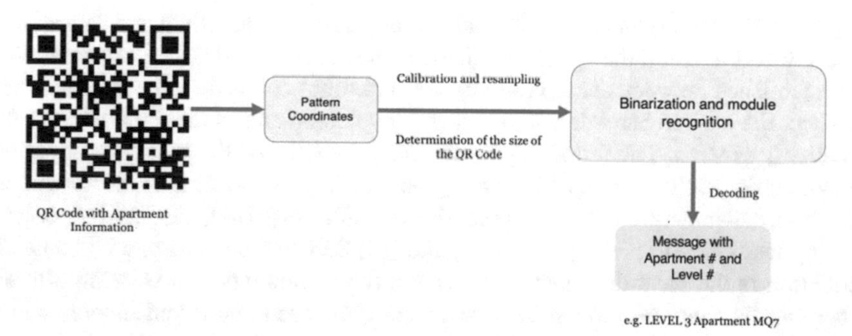

Fig. 5. The key steps of QR code recognition for the parallel scanning algorithm

As discussed in Sect. 3.1 earlier, the drone carries out a parallel scanning algorithm to detect the accurate level and unit number in the building. Figure 6 shows a photograph of the drone taken in real time detecting the correct level and apartment from the QR code placed. The enlarged image is the screenshot taken from the ground station computer which receives the real-time feed from the drone's onboard computer connected via the WiFi network.

Fig. 6. Illustration of the autonomous drone flight near apartments to detect the correct delivery location using barcode scanning for 3rd level, MQ8 Apartment.

4 Conclusion

The proposed method can efficiently read barcode during drone flight and determine the delivery location using the optimised flight pattern. An effortless communication has been established between the microprocessor and flight controller. The microprocessor can detect the desired barcode within 1 s and give command to the flight controller for the delivery process. The following are the main benefits of the suggested system: [i] efficient vertical deliveries capable of being able to scan and determine the apartment in large areas, and underground (if drone services are allowed); [ii] Quick delivery services, requiring less effort to pick up the items, and [iii] Satisfied customers with contactless distribution is the most desirable method, and the technology that is being offered is contactless. The current process and experiment has been conducted indoors and this will be further tested and optimised for outdoor settings. The drone delivery path will also be tested against various environmental factors such as wind, excessive light, heat, etc. The project aims to further include a security verification for parcel delivery between the drone and the human recipient.

References

1. Ahmed, F., et al.: Recent Advances in Unmanned Aerial Vehicles: A Review. Arab. J. Sci. Eng., 1–22 (2022)
2. Quan, L., et al.: Survey of UAV motion planning. IET Cyber-Syst. Rob. **2**(1), 14–21 (2020)
3. Cabreira, T.M., Brisolara, L.B., Ferreira Jr Paulo, R.: Survey on coverage path planning with unmanned aerial vehicles. Drones **3**(1), 4 (2019)
4. Benarbia, T., Kyamakya, K.: A literature review of drone-based package delivery logistics systems and their implementation feasibility. Sustainability **14**(1), 360 (2021)
5. Xie, W., Chen, C., Sithipolvanichgul, J.: Understanding e-commerce customer behaviors to use drone delivery services: a privacy calculus view. Cogent Bus. Manag. **9**(1), 2102791 (2022)

6. Perreault, M., Behdinan, K.: Delivery drone driving cycle. IEEE Trans. Veh. Technol. **70**(2), 1146–1156 (2021)
7. Wang, C., et al.: On optimizing a multi-mode last-mile parcel delivery system with vans, truck and drone. Electronics **10**(20), 2510 (2021)
8. Choi, Y., Schonfeld, P.M.: A comparison of optimized deliveries by drone and truck. Transp. Plan. Technol. **44**(3), 319–336 (2021)
9. Rodrigues, T.A., et al.: In-flight positional and energy use data set of a DJI Matrice 100 quadcopter for small package delivery. Sci. Data **8**(1), 1–8 (2021)
10. Ullah, F., et al.: Advertising through UAVs: optimized path system for delivering smart real-estate advertisement materials. Int. J. Intell. Syst. **36**(7), 3429–3463 (2021)
11. Brunner, G., et al.: The urban last mile problem: Autonomous drone delivery to your balcony. In: 2019 International Conference on Unmanned Aircraft Systems (ICUAS). IEEE (2019)
12. Chen, K.-W., et al.: DroneTalk: an internet-of-things-based drone system for last-mile drone delivery. IEEE Trans. Intell. Transp. Syst. (2022)
13. Deutsche Post DHL Group: DHL'S PARCELCOPTER: CHANGING SHIPPING FOREVER, 30 October 2022. https://www.dhl.com/discover/en-my/business/business-ethics/parcelcopter-drone-technology
14. Kim, S.-H., et al.: Design and flight tests of a drone for delivery service. J. Inst. Control Rob. Syst. **22**(3), 204–209 (2016)
15. Deaconu, A.M., Udroiu, R., Nanau, C.-Ş.: Algorithms for delivery of data by drones in an isolated area divided into squares. Sensors **21**(16), 5472 (2021)
16. Xu, L., Kamat, V.R., Menassa, C.C.: Automatic extraction of 1D bar-codes from video scans for drone-assisted inventory management in warehousing applications. Int. J. Log. Res. Appl. **21**(3), 243–258 (2018)
17. Mukkirwar, R., Nandini, R., Tembhurde, P., Bhagat, R.: Real-time barcode detection from streamlined video from a drone. In: Kumar, R., Dohare, R. K., Dubey, H., Singh, V. P. (eds.) Applications of Advanced Computing in Systems. AIS, pp. 241–246. Springer, Singapore (2021). https://doi.org/10.1007/978-981-33-4862-2_26
18. Gago, R.M., Pereira, M.Y.A., Pereira, G.A.S.: An aerial robot-ic system for inventory of stockpile warehouses. Eng. Rep. **3**(9), e12396 (2021)
19. Radácsi, L., et al.: A path planning model for stock inventory using a drone. Mathematics **10**(16), 2899 (2022)
20. Macrina, G., et al.: Drone-aided routing: a literature review. Transp. Res. Part C Emerging Technol. **120**, 102762 (2020)

Causality Assessment for IIoT Sensing System and Digital Twin

R. K. Sanayaima Singh and Nagender Kumar Suryadevara(✉)

University of Hyderabad, Hyderabad 500046, Telangana, India
{21mcpc05,nks}@uohyd.ac.in

Abstract. This paper presents details about the need to implement causality strategies for the prediction of sensor failures in an Industrial Internet of Things environment. The cause-effect relationships between the electro-mechanical systems, augmented with the sensing units have been realized with the help of different types of causality methods. The implementation details related to the causality effect and the propagation of defectiveness in a series of bearings through a shaft of the four bearings of the IIoT system were illustrated for the prediction of the near failure of the system.

Keywords: Causality · Digital Twin · IIoT · Structural Vector Autoregressive

1 Introduction

The majority of the applications related to the Industrial Internet of Things (IIoT) encompass the hardware components such as Digital Twin, and the software methods Causality [2]. The importance of causality is that it can help us to predict the abnormality conditions ahead of the fatal situation that may arise from the physical objects' usage. When the physical objects are augmented with the appropriate sensing systems, then the data collected from the usage of the physical object will help us to predict the vulnerability situation with the application of causality strategies so that appropriate action can be taken ahead of the situation.

1.1 Digital Twin (DT)

The Digital Twin refers to the ability to clone a physical object into a software counterpart. The concept of DT was first introduced by Michael Grieves in 2003 [1]. A digital twin (DT) is a comprehensive software representation of a physical object (PO). A DT is a realistic model simulating the object behaviours in the deployed environment [6]. A DT refers to a PO and captures all the necessary information and behaviours of that PO in the form of model and data. The DT can encompass the historical data of a PO and its lifecycle.

N. K. Suryadevara et al. (Eds.): ICST 2022, LNEE 1035, pp. 484–489, 2023.
https://doi.org/10.1007/978-3-031-29871-4_49

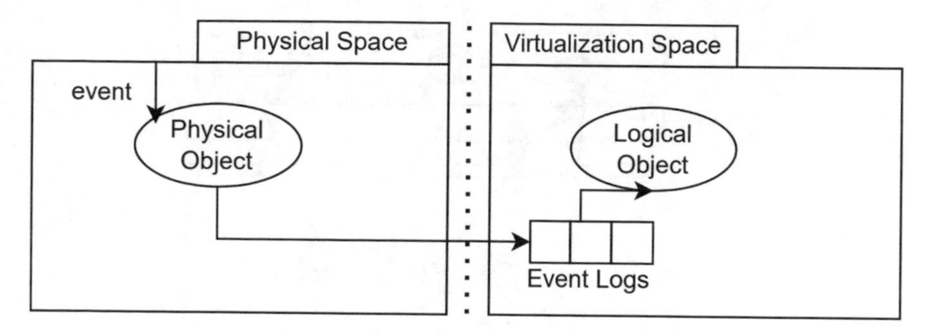

Fig. 1. Representation of Digital Twin [6].

1.2 Digital Twin Representation

As shown in Fig. 1, a physical object (PO) exists in the physical space. Events happening in the physical space are captured by the PO. In the virtual counterpart of the physical space, the PO is represented by a logical object (LO). The events captured by the PO is registered in the event logs. This log is consumed by the LO.

The LO generally models the identity, behaviour and cause-effect of the PO and the historical data about the changes in its lifecycle. The degree of resemblance and detail of nature and behaviours of the PO to be modeled as LO varies from application to application. It is desirable to capture every aspect of the PO in the LO so that the simulation of event processing and interaction among POs can be modeled precisely. This representation of a PO and its behaviour in the virtual space as a LO is termed as Digital Twin (DT).

1.3 Architecture

The architecture of a digital twin application consists of mainly three spaces; the physical space, the virtualization space and the application space [6]. The POs in the physical space are modeled as LOs in the virtualization space through the virtualization layer. A PO is described by its attributes, properties and behaviour. The LO should represent at least the attributes, properties and behaviour of the PO that are necessary and sufficient to be considered as its representation in the given context. The LOs may represent one or more POs or even a combination of POs and LOs as shown in Fig. 2.

The LOs are exposed to the application space as programmable objects through the digital twin API. The interface gives access to the objects in the virtualization space. Through this interface, the physical objects can be accesses and managed indirectly through logical objects. The interface gives the necessary mechanism to use and manipulate the LOs. The LOs represented in the virtualization space are consumed by applications and remote LOs.

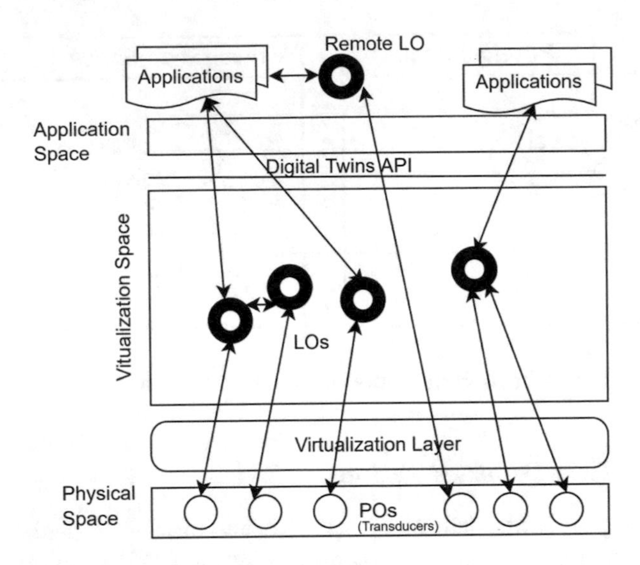

Fig. 2. Architecture of Digital Twin [6].

1.4 Application and Advantages

Digital twins are used in the design and production phase of complex products. This enables operational data collection and check the working of the product. Any variation or unexpected behaviour can be identified easily. In an IoT system, the transducers - sensors and actuators are the POs and they can be modeled as LOs in the corresponding DT system. Each transducers are given unique identification and their behaviours such as data collection, sampling rate, etc. are modeled in the LOs.

One of the main advantage of DT is that the entire system can be simulated in the virtual environment and prediction of behaviour, monitoring and orchestration can performed without being deployed physically. The LOs can be exposed as on demand usable variation of the POs. This makes it possible to serve the physical product as a service to the customers and new functionality and user interactions can be implemented easily.

1.5 Causal Effect

If X has a causal effect on Y, the distribution of Y after intervening on X and setting it to x differs from the distribution of Y after setting X to x' [5]. The causal effects have directional and symmetry. If x has causal effect on y it is written as $x \rightarrow y$, and if y has causal effect on x, it is written as $x \leftarrow y$. Sometimes there may be a confounding factor z which has causal effect on both x and y, i.e. $x \leftarrow z \rightarrow y$ [3].

The observational analysis of a system usually is done by studying the correlation of the factors involved. This gives information about events that happen together but little information about the reason behind the events happening together and their interrelations. Interventional study of the system gives a causal model.

Causal models are usually represented as directed graphical models. A bayesian network can be used to represent a causal model with edges denoting the notion of direct causal effect, causal graphical models. Without a causal model, the root cause of a problem is difficult to fix.

In Sect. 2, we discuss about the structural and Granger causality for IoT digital twins, followed by a section about the experimental setup of the system under observation. The paper is concluded by Sect. 4 where the results and analysis of the observed data is discussed.

2 Structural and Granger Causality for IoT Digital Twins

The multichannel IoT time series data is modelled as SVAR (Structural Vector Autoregressive). Estimation of structural and Granger causality factors from measured multichannel sensor data [4].

Mathematically,

$$y_t = S^0 y_t + \sum_{d=1}^{D} S^d y_{t-d} + e_t \tag{1}$$

where, S^0 is the structural causal factor matrix whose diagonal entries are zero. S^d are the lagged causal factors, y_t is the time series and e is the error term.

For two time series $[y_1(n), y_2(n)]^T$, the granger causality is defined as the ratio of prediction error variance of one time series over the reduced prediction error variance [5].

3 Experimental Setup

To study the propagation of defectiveness in series of bearings through a shaft, the authors in [7] designed a specific test rig as shown in Fig. 3. Several run-to-failure tests were performed under normal load conditions. Four test bearings are installed on the shaft. The shaft is driven by a motor. The vibration data of the bearings was sampled every 20 min. The test was performed for 35 days.

4 Results and Analysis

It is observed from the test data that the bearing 2 has positive Granger effect on the bearing 1. Bearing 2 also has structural causal effects on bearing 1 and 3. However, the strength of the effect varies.

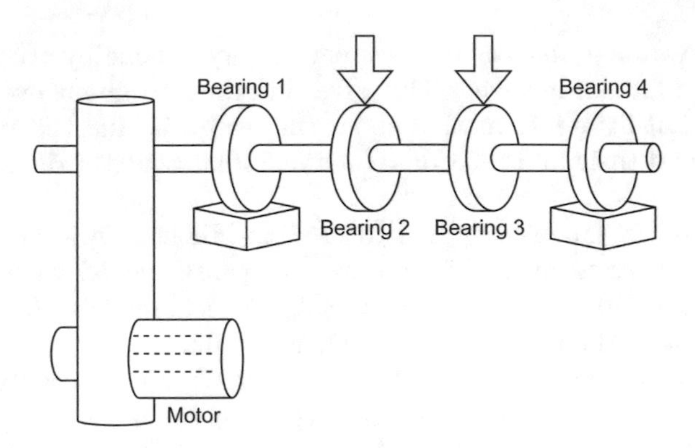

Fig. 3. NASA Bearing test data collection setup [5,7].

The result also shows that there is complex structural and granger causality pattern among the bearings. This is an indication of some coupling in the bearings which is not obvious. This coupling affects the lifetime of the bearings.

The couplings can not only happen in the instant time but also across be delayed (lagged). The Granger causality pattern becomes more complex as the lag time is increased. This information can be used to understand and predict the failure of bearing which can not be seen in structural causality.

5 Conclusion

The multichannel IoT time series data, also known as the vector time series data can be modelled as SVAR (Structural Vector Autoregressive). This parameters learned by this model is further used to infer the causal coefficients and causal relationship of the underlying factor. With such modeling designs, the IoT system is modelled as a Digital Twin. The application of causality in IoT allows for better diagnostics of the problem and prediction of failures.

References

1. Githens, G.: Product lifecycle management: driving the next generation of lean thinking by Michael Grieves (2007)
2. Grieves, M., Vickers, J.: Digital twin: mitigating unpredictable, undesirable emergent behavior in complex systems. In: Kahlen, F.-J., Flumerfelt, S., Alves, A. (eds.) Transdisciplinary Perspectives on Complex Systems, pp. 85–113. Springer, Cham (2017). https://doi.org/10.1007/978-3-319-38756-7_4
3. Kaddour, J., Lynch, A., Liu, Q., Kusner, M.J., Silva, R.: Causal machine learning: a survey and open problems (2022). https://doi.org/10.48550/ARXIV.2206.15475
4. Madhavan, P.: Evidence-based prescriptive analytics, causal digital twin and a learning estimation algorithm (2021). https://doi.org/10.48550/ARXIV.2104.05828
5. Madhavan, P.: Structural & granger causality for IoT digital twin (2022). https://doi.org/10.48550/ARXIV.2203.04876

6. Minerva, R., Lee, G.M., Crespi, N.: Digital twin in the IoT context: a survey on technical features, scenarios, and architectural models. Proc. IEEE **108**(10), 1785–1824 (2020)
7. Qiu, H., Lee, J., Lin, J., Yu, G.: Wavelet filter-based weak signature detection method and its application on rolling element bearing prognostics. J. Sound Vib. **289**(4–5), 1066–1090 (2006)